Graduate Texts in Physics

AF148230

More information about this series at
www.springer.com/series/8431

Graduate Texts in Physics

Graduate Texts in Physics publishes core learning/teaching material for graduate- and advanced-level undergraduate courses on topics of current and emerging fields within physics, both pure and applied. These textbooks serve students at the MS- or PhD-level and their instructors as comprehensive sources of principles, definitions, derivations, experiments and applications (as relevant) for their mastery and teaching, respectively. International in scope and relevance, the textbooks correspond to course syllabi sufficiently to serve as required reading. Their didactic style, comprehensiveness and coverage of fundamental material also make them suitable as introductions or references for scientists entering, or requiring timely knowledge of, a research field.

Series Editors

Professor Richard Needs

Cavendish Laboratory
JJ Thomson Avenue
Cambridge CB3 0HE, UK
rnll@cam.ac.uk

Professor William T. Rhodes

Department of Computer and Electrical Engineering and Computer Science
Imaging Science and Technology Center
Florida Atlantic University
777 Glades Road SE, Room 456
Boca Raton, FL 33431, USA
wrhodes@fau.edu

Professor Susan Scott

Department of Quantum Science
Australian National University
Science Road
Acton 0200, Australia
susan.scott@anu.edu.au

Professor H. Eugene Stanley

Center for Polymer Studies Department of Physics
Boston University
590 Commonwealth Avenue, Room 204B
Boston, MA 02215, USA
hes@bu.edu

Professor Martin Stutzmann

Walter Schottky Institut
TU München
85748 Garching, Germany
stutz@wsi.tu-muenchen.de

Michel Rieutord

Fluid Dynamics

An Introduction

 Springer

Michel Rieutord
Institut de Recherche en Astrophysique et Planétologie
Université Paul Sabatier
Toulouse
France

Revised and expanded translation from the French language edition of: Une introduction à la Dynamique des Fluides, © 1997 Masson, France.

ISSN 1868-4513   ISSN 1868-4521 (electronic)
Graduate Texts in Physics
ISBN 978-3-319-36071-3   ISBN 978-3-319-09351-2 (eBook)
DOI 10.1007/978-3-319-09351-2
Springer Cham Heidelberg New York Dordrecht London

Printed on acid-free paper

Springer is part of Springer Science+Business Media (www.springer.com)

Preface

The idea that guided the first French edition of the present book was to give to newcomers in Fluid Dynamics a presentation of the field that was anchored in Physics rather than in Applied Mathematics as it had been the case so often in the past. Presently, however, connections with Physics are getting stronger and this is fortunate. Indeed, Physics is, etymologically, the science of Nature and fluids occupy a major place in Nature. They are everywhere around us and their motion (their mechanics) influences our everyday life, at least through the weather. Any physicist can hardly escape being fascinated by the sight of some remarkable fluid flows like breaking waves or the gently travelling smoke ring.

The connection between Fluid Mechanics and Applied Mathematics is certainly understandable by the very small number of equations that control a fluid flow. This is fascinating for an applied mathematician, especially if keen on the theory of partial differential equations. Actually, a few decades ago, expertise in asymptotic expansions, singular perturbations, and other mathematical technics was a necessary condition to make progress in the theory of fluid flows. But the pressure of maths has certainly lessened in the recent times because of the strong (exponential) growth of numerical simulations. It is now easier to experiment numerically a fluid flow and get a detailed description of the solutions of Navier–Stokes equation. Interpretation of the results may challenge the intuition of the physicist rather than the skill of the mathematician. But even in the pioneering times, when theoretical investigations of fluid flows were at the strength of the pencil, famous physicists like Newton, Maxwell, Kelvin, Rayleigh, Heisenberg, Landau, Chandrasekhar, and others made essential contributions to the field of Fluid Dynamics. As noted by Heisenberg himself, the theory of turbulence awaits to be written, and this is still the case.

The present book is based on the lectures I delivered at Paul Sabatier University in Toulouse during the last two decades. It is intended to beginners in the field and aims at providing them with the necessary basis that will allow them to attack most of Fluid Dynamics questions. I have tried, as much as possible, to illustrate the concepts with examples taken in natural sciences, often in Astrophysics, which is my playground. Some exercises are offered at the end of each chapter. The reader may thus check his/her understanding of the text. Some of the exercises are also meant to extend the subject in a different way. In that respect, I also give some references for further reading. As far as maths are concerned, the last chapter proposes some brief reminders or introduction to the mathematical tools that are used in the text. With the solutions of the exercises, the book should be self-contained.

As far as teaching is concerned, the first four chapters constitute the bulk of a Fluid Mechanics introduction to third year students. The four following chapters were typically taught to fourth year students, while part of the last ones are currently taught to students about to start a Ph.D. As the reader will note, some sections are tagged with ♠. They can be skipped at first reading and present other illustrations of the subject of the chapter.

Ending this short preface, I would like to thank the many colleagues who have, by various means, contributed to the achievement that a book writing represents. I would like to specially thank Alain Vincent and Hervé Willaime who provided me with original data of turbulent flows. I have much benefitted from the remarks of Arnaud Antkowiak, Pierre-Louis Blelly, Boris Dintrans, Katia Ferrière and Thierry Roudier. They helped me very much at improving various parts of the work. I cannot forget that this adventure of writing started, thanks to the support and help of José-Philippe Pérez. I know that my wife Geneviève and my children Clément and Sylvain will forgive me for the many hours spent outside the real world. The realization of the present book owes much to the kind support of Dr. Ramon Khanna of Springer; I thank him very much for his faith in the project. Finally, I should thank the many students who attended the performance written below, their questions were always beneficial, their enthusiasm always stimulating and their fear challenging for the teacher.

Toulouse, France   Michel Rieutord
May 2014

Contents

Chapter 1
The Foundations of Fluid Mechanics

1.1 A Short Historical Perspective

The first step in Fluid Mechanics was certainly carried out by Archimedes (-287, -212) who was a mathematician and a physicist in Antiquity. He formulated a now well-known theorem which says that a body immersed in a fluid supports an upward push equal to the weight of the displaced fluid. This is the first result in the theory of fluid equilibria. Knowledge did not evolve much until the works of Evangelista Torricelli (1608–1647) and Blaise Pascal (1623–1662). Torricelli did a famous experiment when he put upside down a tube full of mercury. The liquid went down a little and left a column of mercury 76 cm high. Hence, it was demonstrated the existence of atmospheric pressure, the weight of air and the existence of vacuum, much discussed at the time. Pascal gave a full account of all these phenomena in his treatise *L'équilibre des liqueurs* published in 1663. The static of fluids was almost set up. Fluid dynamics started with the work of Leonhard Euler (1707–1783) who formulated for the first time the equation of motion of an inviscid fluid. Daniel Bernoulli (1700–1782) contributed to the study of such fluids with theorems on energy conservation that revealed fundamental properties of fluid flows.

The next important step has been the formulation of the effects of viscosity. This was done during the nineteenth century with the work of Henri Navier (1785–1836), Georges Stokes (1819–1904) and Jean-Louis-Marie Poiseuille (1799–1869). Let us note that Isaac Newton (1642–1727) already showed the existence of viscosity with experiments and left his name associated with a kind of fluids (the most common ones like air and water) now known as the Newtonian fluids. Navier–Stokes equation, which controls the dynamics of viscous fluids, was first formulated by Navier in the case of a fluid with constant viscosity.

Then, fluid mechanics took various ways. We shall mention only the main ones. Studies of flows stability started with the works of Helmholtz (1868) and Lord Kelvin (1880). Heat transport was studied by W. Prout (1834), Rumford (1870), A. Oberbeck (1879), H. Bénard (1900), J. Boussinesq (1903) and Lord Rayleigh

© Springer International Publishing Switzerland 2015
M. Rieutord, *Fluid Dynamics*, Graduate Texts in Physics,
DOI 10.1007/978-3-319-09351-2_1

(1916). Turbulence focused the interest of O. Reynolds (1883), L. Prandtl (1920), A. Kolmogorov (1940), etc.

Presently, fluid dynamics is even more diversified, but some old problems resist: turbulence remains one of the main unsolved problem of classical physics.

1.2 The Concept of a Fluid

1.2.1 Introduction

Fluids gather a very large number of forms of matter which, as far as the most common are concerned (liquids and gases), can be characterized by the ease with which one can deform them. This point of view expresses the difference between the "solid state" and the "fluid state" which are both "mechanical states". Common experience assumes that such a difference is obvious, however we shall see that a more detailed inspection somehow blurs the difference between solids and fluids. In fact, this difference is contained in a law, known as the rheological law, which states how the matter reacts to (is deformed by) a stress (a force per surface unit). Generally speaking, matter is considered as fluid if the internal (shear) stresses only depend on velocity gradients.

We shall come back on these concepts, but let us give a simple example to appreciate what is behind this definition. We consider a cork floating on water in a container. The cork is in A and we wish to move it to B. Because water is a fluid, the force we need to apply on the cork just depends on velocity gradients. It can be made as small as we wish; we just need to move the cork more slowly.

From the point of view of thermodynamics, we may say that the states A and B and cannot be distinguished: the energy and entropy variations needed to make the system passing from one state to the other can be made vanishingly small. On the contrary, if we did a similar experiment of a solid, some work would have been needed and the energy of the states A and B would differ.

1.2.2 Continuous Media

To describe the motion of a fluid, obviously we cannot (and do not wish to!) describe the motion of all its molecules or atoms individually. We are only interested in their mean motion. This means replacing the set of atoms or molecules which constitutes the fluid by a medium that behaves as this mean motion. Such an assumption is valid when the scale L, which we are interested in, is large compared to the mean free path ℓ of atoms or molecules. The ensuing approximation is measured by the *Knudsen number*

$$\mathrm{Kn} = \frac{\ell}{L} \tag{1.1}$$

which needs to be small compared to unity. The cases where Kn ≥ 1 is the subject of the dynamics of rarefied gases based on the kinetic theory of gases. In general, it is not included in fluid mechanics. An introduction to the relation between fluid dynamics and gas kinetics is given in Chap. 11.

Thermodynamics and the dynamics of continuous media share many similarities in their description of matter. Indeed, in these two approaches, the microscopic components of matter are ignored and only mean values are retained. The drawback of such a way of doing is that some quantities can only be obtained by experiments, like heat capacity in thermodynamics or viscosity in fluid mechanics, the equation of state in thermodynamics or the rheological law in fluid mechanics. Their theoretical determination needs a more detailed approach based on the statistical properties of the microscopic components of matter (see Chap. 11).

1.3 Fluid Kinematics

The first step needed to understand the motion of fluids is to find the right tools that will allow us to describe a fluid flow. This is the role of fluid kinematics that study fluid motion without worrying about its causes.

1.3.1 The Concept of Fluid Particle

Very often, we shall use the concept of *fluid particle* or *fluid element*. This is an idealized view of a piece of fluid: it is so small that fluid properties are uniform inside. However, it is big enough so as to contain a large number of atoms or molecules, allowing us to assume that the fluid is locally in thermodynamic equilibrium (hence the temperature is defined). Such a particle is not a point mass: it owns a surface which authorizes contact forces with other particles.

1.3.2 The Lagrangian View

A first way for describing the fluid motion is to give the trajectories of all the fluid particles. Such a description is the *Lagrangian* one. It may be summarized by the set of trajectories of all fluid particles that were in the domain $\mathcal{D}_{t_0}$ at $t = t_0$. Mathematically, it is the set

$$\mathcal{D}_t = \{\mathbf{x}(t, \mathbf{x}_A) \mid \mathbf{x}_A \in \mathcal{D}_{t_0}, t \geq t_0\}$$

Such a description is used in some specific studies where it provides simplifications. However, generally speaking, its use is not very popular because of some intrinsic difficulties in the expression of stresses. We refer the reader to the end of this chapter for a more detailed presentation.

1.3.3   The Eulerian View

The most natural way to represent the motion of a fluid is certainly the intuitive one used to describe the flow of water in a river. In such a case, we would say that the stream is fast in one place at a given time and hardly noticeable in some other place and some other time. Thus doing, one describes the velocity field as a function of space and time. Actually, if the vector function

$$\mathbf{v}(x, y, z, t)$$

is known, then everything is known about the fluid flow. The use of the velocity field as a function of position and time gives *the Eulerian description* of the flow. This is the most popular way of doing and we shall use it almost all the time from now on.

1.3.4   Material Derivatives

Quite often, we shall consider the time evolution of some quantity, like ϕ, which is attached to a fluid particle that one follows.

To express the variations of ϕ, we introduce the derivative

$$\frac{D\phi}{Dt}$$

also called *material derivative* or *Lagrangian derivative* (see Sect. 1.10). ϕ may be any quantity (scalar, vector, tensor) depending on space coordinates and time $\phi \equiv \phi(x, y, z, t)$. When we attach ϕ to the motion of a fluid particle, the coordinates x, y, z are functions of time

$$\phi \equiv \phi(x(t), y(t), z(t), t)$$

where $x(t), y(t), z(t)$ represent the trajectory of the fluid particle. Now, the total variation of ϕ is

$$D\phi = \phi(t + dt) - \phi(t) = \frac{\partial \phi}{\partial t} dt + \frac{\partial \phi}{\partial x} dx + \frac{\partial \phi}{\partial y} dy + \frac{\partial \phi}{\partial z} dz$$

$$= \left(\frac{\partial \phi}{\partial t} + v_x \frac{\partial \phi}{\partial x} + v_y \frac{\partial \phi}{\partial y} + v_z \frac{\partial \phi}{\partial z} \right) dt$$

where we observed $(dx/dt, dy/dt, dz/dt)$ are just the xyz-components of the velocity of the fluid particle that we are following. Thus,

$$\frac{D\phi}{Dt} = \frac{\partial\phi}{\partial t} + v_x\frac{\partial\phi}{\partial x} + v_y\frac{\partial\phi}{\partial y} + v_z\frac{\partial\phi}{\partial z} \tag{1.2}$$

or

$$\frac{D\phi}{Dt} = \frac{\partial\phi}{\partial t} + (\mathbf{v}\cdot\nabla)\phi \tag{1.3}$$

The operator D/Dt therefore symbolizes a differentiation along a curve, namely along the trajectory of a fluid element. We shall see that this quantity appears very often in the equations governing fluid flows. The term $(\mathbf{v}\cdot\nabla)\phi$ is called the *advection term* and represents the transport of the quantity ϕ by the velocity field $\mathbf{v}$. In a steady flow, it represents the variations of ϕ along a streamline (a curve which is everywhere tangential to the velocity field, see Sect. 1.3.7).

Another illustration of the role played by the advection term is given by the case where ϕ is conserved by each particle. Thus,

$$\frac{D\phi}{Dt} = 0$$

In such a case, an observer measuring ϕ at a given point as a function of time will see the variations of ϕ corresponding to the passing particles displaying their value of ϕ. Formally, this means

$$\frac{\partial\phi}{\partial t} = -(\mathbf{v}\cdot\nabla)\phi$$

where the second equation shows that the temporal variations of ϕ at a given point of space is only due to the advection term characterizing the transport. If $\mathbf{v}$ is uniform then any function $\phi(\mathbf{r} - \mathbf{v}t)$ is such that $D\phi/Dt = 0$.

1.3.5 Distortion of a Fluid Element

An important aspect of the motion of fluid particles is their proper motion. Indeed, we mentioned above that, although of vanishingly small size, fluid particles own a surface and a volume. Thus, they can be distorted by a non-uniform flow. To visualize this effect, it is convenient to consider a fluid particle of parallelepipedic

Fig. 1.1 Evolution of a fluid element

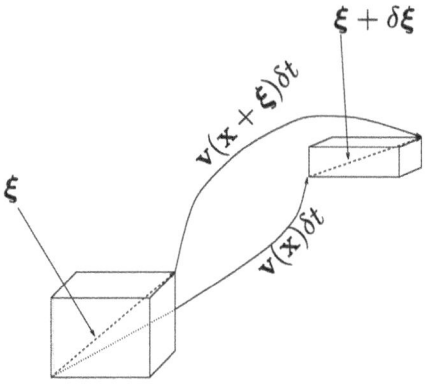

shape which we may characterize by a vector $\boldsymbol{\xi}$ (see Fig. 1.1). If we express the variation of $\boldsymbol{\xi}$ with time we find

$$\boldsymbol{\xi} + \delta\boldsymbol{\xi} = \boldsymbol{\xi} + \mathbf{v}(\mathbf{x} + \boldsymbol{\xi})\delta t - \mathbf{v}(\mathbf{x})\delta t$$

Assuming that $\boldsymbol{\xi}$ is very small compared to $\mathbf{x}$, we have

$$\mathbf{v}(\mathbf{x} + \boldsymbol{\xi}) = \mathbf{v}(\mathbf{x}) + (\boldsymbol{\xi} \cdot \nabla)\mathbf{v} + \mathcal{O}(\xi^2)$$

at first order and thus

$$\delta\boldsymbol{\xi} = (\boldsymbol{\xi} \cdot \nabla)\mathbf{v}\,\delta t \qquad (1.4)$$

Using indicial notations[1]

$$\delta\xi_i = \xi_j \partial_j v_i\, \delta t$$

In this expression we discover the tensor "velocity gradient" whose components are $\partial_i v_j$. As any second order tensor, it can be decomposed into its symmetric and antisymmetric parts:

$$\partial_i v_j = \frac{1}{2}(\partial_i v_j + \partial_j v_i) + \frac{1}{2}(\partial_i v_j - \partial_j v_i) = s_{ij} + a_{ij}$$

These two parts play very different roles. Let us first focus on the antisymmetric one. We note that it is represented by only three components (a_{12}, a_{23}, a_{31}). Actually,

---

[1] We shall often use these notations which are very handy. In Chap. 12 "Mathematical complements", we give a summary of what is needed to go ahead with these notations. Let us recall here that we always use the implicit summation on repeated indices. Thus $\mathbf{a} \cdot \mathbf{b} = \sum_{i=1}^{3} a_i b_i$ is just noted $a_i b_i$.

these components are related to the curl of the velocity field. Indeed,

$$a_{ij} = \frac{1}{2}(\partial_i v_j - \partial_j v_i) = \frac{1}{2}(\delta_{ik}\delta_{jl} - \delta_{il}\delta_{jk})\partial_k v_l = \frac{1}{2}\epsilon_{ijm}\epsilon_{mkl}\partial_k v_l$$

where we introduced the Kronecker symbol δ_{ij} and the completely antisymmetric tensor ϵ_{ijk} (its components are just ± 1 or 0, but see Chap. 12 for basic properties). Therefore,

$$a_{ij} = \frac{1}{2}\epsilon_{ijm}(\nabla \times \mathbf{v})_m \tag{1.5}$$

This expression shows that the three components (a_{12}, a_{23}, a_{31}) are just the components of the curl of $\mathbf{v}$ up to a factor $1/2$, since ϵ_{ijk} is either zero or unity. Expression (1.5) is useful to understand the physical meaning of $[a]$, and consequently that of the curl. We may uncover it by calculating the variation of $\boldsymbol{\xi}$ associated with $[a]$ only, namely

$$\delta\xi_i = \xi_j a_{ji}\delta t = \frac{1}{2}\epsilon_{jim}\xi_j(\nabla \times \mathbf{v})_m\delta t = \left(\frac{1}{2}(\nabla \times \mathbf{v}) \times \boldsymbol{\xi}\right)_i \delta t$$

This expression shows that the variation of a fluid element associated with $[a]$ is just a solid body rotation at the angular velocity

$$\boldsymbol{\Omega} = \frac{1}{2}\nabla \times \mathbf{v}$$

This result gives us the physical interpretation of the vector $\nabla \times \mathbf{v}$, also called the vorticity and enlights the wording *curl* for the operator $\nabla\times$.

Since a solid body rotation does not distort our parallelepipedic fluid element, its deformation must be contained in the remaining part of the velocity gradient tensor, namely $[s]$. This symmetric part of the velocity gradient tensor is called the *rate-of-strain tensor*. The distortion effect of $[s]$ on a fluid element can be explicited by considering the variations of the length of the vector $\boldsymbol{\xi}$. Indeed,

$$\delta(\boldsymbol{\xi}^2) = 2\xi_i\delta\xi_i = 2\xi_i\xi_j\partial_i v_j\delta t = 2\xi_i\xi_j s_{ji}\delta t$$

where we used the fact that $\xi_i\xi_j a_{ij} = 0$ since a_{ij} is antisymmetric and $\xi_i\xi_j$ is symmetric. Thus, only $[s]$ contributes to the variation of the length of $\boldsymbol{\xi}$ and thus to the distortion of the fluid particle. In order to appreciate more completely the effects of the deformation it is useful to express the variations of $\boldsymbol{\xi}$ in a basis where $[s]$ is diagonal. Such a basis always exists because $[s]$ is symmetric. In this new basis the variation of $\boldsymbol{\xi}$ associated with $[s]$ is

$$\delta\xi_i = \xi_j s_{ji}\delta t = \xi_i s_{ii}\delta t$$

where no summation is assumed on the indices in the last expression. This equation shows that the fluid element is stretched in the i-direction when $s_{ii} > 0$ or destretched when $s_{ii} < 0$. It is then easy to compute its volume variation at first order due to the velocity field:

$$\delta V = (\xi_1 + \delta\xi_1)(\xi_2 + \delta\xi_2)(\xi_3 + \delta\xi_3) - \xi_1\xi_2\xi_3 = \xi_1\xi_2\xi_3 \left(\frac{\delta\xi_1}{\xi_1} + \frac{\delta\xi_2}{\xi_2} + \frac{\delta\xi_3}{\xi_3} \right) + \mathcal{O}(\delta\xi^2)$$

$$= \xi_1\xi_2\xi_3(s_{11} + s_{22} + s_{33})\delta t + \mathcal{O}(\delta\xi^2)$$

But $s_{11} + s_{22} + s_{33}$ is just the trace of $[s]$ and we have

$$\mathrm{Tr}(s) = s_{ii} = \mathbf{\nabla} \cdot \mathbf{v} \, ,$$

so that the relative volume variation of the fluid element is

$$\frac{\delta V}{V} = (\mathbf{\nabla} \cdot \mathbf{v})\delta t \tag{1.6}$$

This result shows that the dilation or contraction of fluid elements is associated with the trace of the rate-of-strain tensor, which is also the divergence of the velocity field. Hence, expanding elements appear in flows with positive divergence while contracting elements are where divergence is negative. Note that this result is independent of the chosen basis as the trace of a tensor is invariant in basis changes.

1.3.6 Incompressible Fluids

An important model for the description of fluid flows is the so-called incompressible fluid. For such a fluid the density is assumed constant. This assumption is very popular as it much simplifies the equations of motion (and the physics of the fluid). In addition it is quite a good approximation for liquids which are only slightly compressible. Moreover, we shall see later that even gas flows can be modeled by such a fluid provided the velocity is small compared to sound velocity (see Sect. 3.2.5).

From (1.6) we see that if the volume of a fluid element does not (or cannot) vary, then

$$\mathbf{\nabla} \cdot \mathbf{v} = 0 \tag{1.7}$$

Fluid particles neither expand nor contract. This is the main constraint that must be met by a fluid whose density variations can be ignored. We shall find this relation again when studying the implications of mass conservation.

1.3.7   The Stream Function

In the foregoing section, we found out that the flow of an incompressible fluid must meet the constraint $\nabla \cdot \mathbf{v} = 0$. Let us now consider a further simplified case when the fluid flow is two-dimensional. If x and y are cartesian coordinates in this plane, $\nabla \cdot \mathbf{v} = 0$ reads

$$\frac{\partial v_x}{\partial x} + \frac{\partial v_y}{\partial y} = 0 \,.$$

It is easy to show (see exercise 3) that (1.7) implies the existence of a function $\psi(x, y, t)$ such that

$$v_x = \frac{\partial \psi}{\partial y}, \qquad v_y = -\frac{\partial \psi}{\partial x} \tag{1.8}$$

called the *stream function*. The expression of a two-dimensional flow is therefore contained in that of a unique scalar field, namely the stream function. This wording explicits the fact that isolines of the stream function are always tangential to the velocity field and therefore trace out the velocity field lines or the *streamlines*. Indeed, along a streamline

$$d\psi = 0 \quad \Longleftrightarrow \quad \frac{\partial \psi}{\partial x} dx + \frac{\partial \psi}{\partial y} dy = 0 \quad \Longleftrightarrow \quad -v_y dx + v_x dy = 0$$

$$\begin{vmatrix} v_x & dx \\ v_y & dy \end{vmatrix} = 0 \quad \Longleftrightarrow \quad \mathbf{v} \text{ parallel to } d\mathbf{l}$$

When the flow is three-dimensional this idea can be generalized but two stream functions are needed. This is easily understood if one remembers that in three dimensions the velocity field has three components that are constrained by one equation (1.7). Hence, one is left with two independent quantities. Without loss of generality, one can write

$$\mathbf{v} = \nabla \times (\chi \mathbf{a}) + \nabla \times \nabla \times (\psi \mathbf{a})$$

In this expression the choice of $\mathbf{a}$ is not imposed. In spherical geometry one usually choose the radial vector $\mathbf{r}$. The first term is then called the *toroidal* velocity field because the field lines are on a torus. The second term is the *poloidal field*; its field lines are generally not confined on a surface except in the axisymmetric case where they can be drawn in a meridian plane $\varphi = \text{Cst}$.

1.3.8 Evolution of an Integral Quantity Carried by the Fluid

In the following, we shall meet the time evolution of a quantity (mass, momentum, energy) associated with a fluid domain D (volume, surface, line). In the case where the points of the fluid domain have a velocity equal to that of the fluid at every instant, the domain is called *a material domain*. It is always made of the same fluid particles. In other words this domain is a kind of macroscopic particle.

We shall later need the expression of the material derivative of a quantity f integrated over a material domain $V(t)$. So, we first express

$$\frac{d}{dt} \int_{V(t)} f(\mathbf{r}, t) dV$$

as a function of the local derivatives of f. For this, we write

$$\frac{d}{dt} \int_{V(t)} f(\mathbf{r}, t) dV = \frac{\int_{V(t+dt)} f(\mathbf{r}+d\mathbf{r}, t+dt)dV' - \int_{V(t)} f(\mathbf{r}, t)dV}{dt} \qquad (1.9)$$

We develop the first integral to first order and since $d\mathbf{r} = \mathbf{v}dt$, we obtain

$$f(\mathbf{r}+d\mathbf{r}, t+dt) = f(\mathbf{r}, t) + \frac{Df}{Dt}dt$$

However, volume $V(t + dt)$ is slightly different from volume $V(t)$ become the velocity field distorts it. To take into account this distortion, we shall still integrate over $V(t)$ but with a distorted elementary volume dV'. From (1.6) we know that

$$dV' = dV + dV\boldsymbol{\nabla} \cdot \mathbf{v}\, dt = dV(1 + \boldsymbol{\nabla} \cdot \mathbf{v}\, dt) \ .$$

Using this expression in (1.9), we finally get

$$\frac{d}{dt} \int_{V(t)} f(\mathbf{r}, t) dV = \int_{V(t)} \left(\frac{Df}{Dt} + f\boldsymbol{\nabla} \cdot \mathbf{v} \right) dV = \int_{V(t)} \left(\frac{\partial f}{\partial t} + \boldsymbol{\nabla} \cdot (f\mathbf{v}) \right) dV \qquad (1.10)$$

The same exercise can be repeated with a contour entrained by the fluid. Let us evaluate

$$\frac{d}{dt} \oint_{C(t)} \mathbf{A}(\mathbf{r}, t) \cdot d\mathbf{l}$$

As above, we need considering the variations of the fluid element $d\mathbf{l}$, which is modified at time $t + dt$ just as $\boldsymbol{\xi}$ in (1.4). Hence,

$$d\mathbf{l}' = d\mathbf{l} + (d\mathbf{l} \cdot \nabla)\mathbf{v}dt$$

and thus

$$\frac{d}{dt} \oint_{C(t)} \mathbf{A}(\mathbf{r}, t) \cdot d\mathbf{l} = \oint_{C(t)} \left(\frac{D\mathbf{A}}{Dt} \cdot d\mathbf{l} + \mathbf{A} \cdot (d\mathbf{l} \cdot \nabla)\mathbf{v} \right) \tag{1.11}$$

which is more handy if we use indices

$$\frac{d}{dt} \oint_{C(t)} \mathbf{A}(\mathbf{r}, t) \cdot d\mathbf{l} = \oint_{C(t)} \left(\frac{DA_i}{Dt} + A_j \partial_i v_j \right) dl_i \tag{1.12}$$

This expression can be simplified by noting that

$$A_j \partial_i v_j = \partial_i (A_j v_j) - v_j \partial_i A_j \qquad \text{and} \qquad \partial_j A_i - \partial_i A_j = \epsilon_{jik} (\nabla \times \mathbf{A})_k$$

Hence, it turns out that

$$\frac{d}{dt} \oint_{C(t)} \mathbf{A}(\mathbf{r}, t) \cdot d\mathbf{l} = \oint_{C(t)} \left(\frac{\partial A_i}{\partial t} - (\mathbf{v} \times (\nabla \times \mathbf{A}))_i \right) dl_i \tag{1.13}$$

1.4 The Laws of Fluid Motion

In the foregoing section we presented the quantities that are used to describe a fluid flow. In this new section we shall express the laws of Physics that govern the evolution of these quantities. They are derived from the general principles expressing the conservation of mass, momentum and energy.

1.4.1 Mass Conservation

1.4.1.1 The Equation of Continuity

We consider a fixed volume of fluid V whose mass is

$$M = \int_{(V)} \rho \, dV .$$

Its variation with time is given by the mass flux density $\rho\mathbf{v}$ crossing the surface (S) bounding the volume (V). Let $d\mathbf{S}$ be the surface element oriented by the external normal $\mathbf{n}$ so that $d\mathbf{S} = \mathbf{n}dS$. Hence

$$\frac{dM}{dt} = -\int_{(S)} \rho\mathbf{v} \cdot d\mathbf{S} \quad \Longleftrightarrow \quad \int_{(V)} \frac{\partial\rho}{\partial t} dV = -\int_{(V)} \boldsymbol{\nabla} \cdot \rho\mathbf{v} \, dV \qquad (1.14)$$

$$\Longleftrightarrow \quad \int_{(V)} \left(\frac{\partial\rho}{\partial t} + \boldsymbol{\nabla} \cdot \rho\mathbf{v} \right) dV = 0 \qquad (1.15)$$

Note that the minus sign in (1.14) comes from the orientation of the surface (S): when $\mathbf{v}$ is parallel to $d\mathbf{S}$ the mass M decreases. Equation (1.15) being true for any volume the integrand must be vanishing and we have

$$\frac{\partial\rho}{\partial t} + \boldsymbol{\nabla} \cdot \rho\mathbf{v} = 0 \qquad (1.16)$$

which expresses locally the conservation of mass. This equation is often referred to as the *continuity equation*. One may also write it using the material derivative of ρ, namely

$$\frac{D\rho}{Dt} = -\rho\boldsymbol{\nabla} \cdot \mathbf{v} \qquad (1.17)$$

which shows that the density of a fluid element varies because of its volume variation expressed by $\boldsymbol{\nabla} \cdot \mathbf{v}$, since its mass is constant.

Equations (1.16) and (1.17) can also be derived directly from (1.10) using a volume attached to fluid particles and setting $f = \rho$. One may remark that the flux term $\int_{(S)} \rho\mathbf{v} \cdot d\mathbf{S}$ disappears then.

As expected, when $\rho =$ Cst, all these equations lead back to (1.7), namely the case of incompressible fluids.

1.4.1.2   Material Derivative with Mass Conservation

In most cases physical quantities like energy, momentum, are not attached to the volume of elements but to their mass $dm = \rho dV$. This implies that when writing the balance between losses and gains for a fixed volume as in (1.14), we always face integrals taking into account the flux of the quantity carried by the mass flux across (S). In general, if ϕ is such a physical quantity and S_ϕ its volumic sources, we have

Variations of ϕ in V = ϕ carried by $\mathbf{v}$ through S + Sources of ϕ

or, mathematically,

$$\frac{d}{dt} \int_{(V)} \rho\phi \, dV = -\int_{(S)} \rho\phi \, \mathbf{v} \cdot d\mathbf{S} + \int_{(V)} S_\phi \, dV$$

which can be rewritten as

$$\int_{(V)} \frac{\partial\rho\phi}{\partial t} \, dV = -\int_{(V)} \mathbf{\nabla} \cdot (\rho\phi\mathbf{v}) dV + \int_{(V)} S_\phi \, dV$$

$$\Longleftrightarrow \quad \int_{(V)} \left(\phi\frac{\partial\rho}{\partial t} + \rho\frac{\partial\phi}{\partial t} \right) dV = -\int_{(V)} (\phi\mathbf{\nabla} \cdot \rho\mathbf{v} + \rho\mathbf{v} \cdot \mathbf{\nabla}\phi) dV + \int_{(V)} S_\phi \, dV$$

While using the equation of continuity, we find that

$$\int_{(V)} \rho\frac{\partial\phi}{\partial t} \, dV + \int_{(V)} \rho\mathbf{v} \cdot \mathbf{\nabla}\phi \, dV = \int_{(V)} S_\phi \, dV \qquad (1.18)$$

or

$$\int_{(V)} \rho\frac{D\phi}{Dt} \, dV = \int_{(V)} S_\phi \, dV \qquad (1.19)$$

As (1.15) this relation is valid for any volume V and is therefore valid locally as

$$\rho\frac{D\phi}{Dt} = S_\phi \qquad (1.20)$$

We shall see that the equations of momentum, energy or entropy all share this structure. ϕ is then a velocity field (the momentum per unit mass), the internal energy or the entropy per unit mass.[2]

Let us now rederive (1.20) using a volume $V(t)$ attached to the fluid. From its definition, this volume contains the same fluid particles at any time. So,

$$\text{Variations of } \phi \text{ in V(t)} = \text{Sources of } \phi$$

or, mathematically, it reads

$$\frac{d}{dt} \int_{V(t)} \rho\phi \, dV = \int_{V(t)} S_\phi \, dV$$

---

[2]One may often find in literature the terminology "specific entropy" which also means entropy per unit mass.

From (1.10)

$$\frac{d}{dt} \int_{V(t)} \rho \phi \, dV = \int_{V(t)} \left(\frac{D(\rho\phi)}{Dt} + \rho\phi \nabla \cdot \mathbf{v} \right)$$

$$dV = \int_{V(t)} \left[\phi \left(\frac{D\rho}{Dt} + \rho \nabla \cdot \mathbf{v} \right) + \rho \frac{D\phi}{Dt} \right] dV$$

Using the equation of mass conservation (1.17),

$$\Longrightarrow \qquad \frac{d}{dt} \int_{V(t)} \rho \phi \, dV = \int_{V(t)} \rho \frac{D\phi}{Dt} \, dV \qquad\qquad (1.21)$$

We immediately find (1.20). This derivation is much faster than the preceding one and we shall prefer it in the following.

1.4.2   Momentum Conservation

1.4.2.1   The Stress Tensor $[\sigma]$

Before expressing momentum conservation, we need precising the way fluid elements interact. Contact forces are specific to the mechanics of continuous media. Their existence shows again the fact that fluid particles are not point masses but small volumes with a surface on which contact forces can be exerted. Let us consider an elementary surface $d\mathbf{S}$ on which an elementary force $d\mathbf{f}$ is applied. These two vectors are related in a very general way by

$$d\mathbf{f} = [\sigma] d\mathbf{S}$$

or with indices

$$df_i = \sigma_{ij} dS_j$$

We thus define the stress tensor $[\sigma]$ and at the same time *the stress* $\mathbf{T} = [\sigma]\mathbf{n}$ applied on a given point of the surface whose normal is $\mathbf{n}$. Thus defined, the stress is a force per unit surface. One then makes the hypothesis that $[\sigma]$ depends only on the local properties of the flow.[3]

---

[3]This implies in particular that the stress tensor is independent of the surface on which the stress is computed. It is independent of its orientation $\mathbf{n}$ and its curvature radii. That would not be the case if the given surface is the seat of surface tension at the interface between a gas and a liquid. Some additional terms must be taken into account (see (1.70)).

Let us show now that $[\sigma]$ is a symmetric tensor. If S is a surface covering some volume (V), we note that the resultant of stress forces on S can also be written as the resultant of volumic forces since (see Sect. 12.2):

$$\int_{(S)} \sigma_{ij} dS_j = \int_{(V)} \partial_j \sigma_{ij} dV$$

Hence, one can always associate a volumic force with a surface stress. If we consider a fluid element, the above identity just says that the resultant of contact forces on a fluid particle is equal to the divergence of the stress tensor. Let us now consider the torque exerted by the stress forces on the volume. We have

$$m_i = \int_{(S)} (\mathbf{r} \times d\mathbf{f})_i = \int_{(S)} (\mathbf{r} \times [\sigma] d\mathbf{S})_i$$

$$m_i = \int_{(S)} \epsilon_{ijk} x_j \sigma_{kl} dS_l = \int_{(V)} \partial_l (\epsilon_{ijk} x_j \sigma_{kl}) dV \qquad (1.22)$$

But this torque can also be expressed with the volumic force associated with the stress:

$$m_i = \int_{(V)} (\mathbf{r} \times Div[\sigma])_i \, dV = \int_{(V)} \epsilon_{ijk} x_j \partial_l \sigma_{kl} dV \qquad (1.23)$$

where we introduced the vectorial divergence Div of a symmetric tensor which is such that

$$(Div[\sigma])_i = \partial_j \sigma_{ij} = \partial_j \sigma_{ji}$$

The equality of expressions (1.22) and (1.23) implies that

$$\partial_l (\epsilon_{ijk} x_j \sigma_{kl}) = \epsilon_{ijk} x_j \partial_l \sigma_{kl} \iff \epsilon_{ijk} \delta_{lj} \sigma_{kl} + \epsilon_{ijk} x_j \partial_l \sigma_{kl} = \epsilon_{ijk} x_j \partial_l \sigma_{kl}$$

$$\iff \epsilon_{ijk} \sigma_{kj} = 0$$

which is equivalent to saying that the stress tensor is symmetric as shown in Chap. 12 (see (12.5)), so

$$\sigma_{ij} = \sigma_{ji} \, . \qquad (1.24)$$

The symmetry of $[\sigma]$ has been obtained after equating (1.22) and (1.23). This implicitly assumes that the fluid does not contain any torque density.

1.4.2.2   The Equation of Momentum

We are now in a position of using the conservation of momentum for some arbitrary fluid volume. The variation of momentum of a fluid volume thus reads

$$\frac{d}{dt}\int_{(V)}\rho\mathbf{v}dV = \int_{(V)}\mathbf{f}dV + \int_{(S)}[\sigma]d\mathbf{S}$$

or, with words,

> *Variation of total momentum = Resultant of volumic forces*
> *+ Resultant of stresses applied on (S)*

Using the theorem of divergence and (1.21), we find

$$\int_{(V)}\frac{D\mathbf{v}}{Dt}\rho dV = \int_{(V)}\mathbf{f}dV + \int_{(V)}\mathbf{Div}[\sigma]dV$$

or, locally,

$$\rho\frac{D\mathbf{v}}{Dt} = \mathbf{Div}[\sigma] + \mathbf{f} \qquad (1.25)$$

This equation is just Newton's second law applied to a fluid element of unit volume. We note in passing that the expression of acceleration

$$\mathbf{a} = \frac{D\mathbf{v}}{Dt} = \frac{\partial\mathbf{v}}{\partial t} + (\mathbf{v}\cdot\nabla)\mathbf{v}$$

is nothing but the material derivative of the velocity. This expression of the acceleration can be obtained in a more intuitive manner by considering a fluid particle whose trajectory is $(x(t), y(t), z(t))$. Its velocity at time t where the particle is at (x, y, z) is just the fluid velocity $\mathbf{v}(x, y, z, t)$. The velocity along the trajectory is thus $\mathbf{v}(x(t), y(t), z(t), t)$ while its acceleration, also along the trajectory, is

$$\frac{d\mathbf{v}}{dt} = \frac{\partial\mathbf{v}}{\partial x}\frac{dx}{dt} + \frac{\partial\mathbf{v}}{\partial y}\frac{dy}{dt} + \frac{\partial\mathbf{v}}{\partial z}\frac{dz}{dt} + \frac{\partial\mathbf{v}}{\partial t}$$

Since at the given point $(\frac{dx}{dt}, \frac{dy}{dt}, \frac{dz}{dt}) = \mathbf{v}$, we may write

$$\frac{d\mathbf{v}}{dt} = v_x\frac{\partial\mathbf{v}}{\partial x} + v_y\frac{\partial\mathbf{v}}{\partial y} + v_z\frac{\partial\mathbf{v}}{\partial z} + \frac{\partial\mathbf{v}}{\partial t} = \frac{\partial\mathbf{v}}{\partial t} + (\mathbf{v}\cdot\nabla)\mathbf{v} = \frac{D\mathbf{v}}{Dt}$$

which shows that the material derivative of the velocity is indeed the acceleration of the fluid particle.

The term $(\mathbf{v} \cdot \nabla)\mathbf{v}$ can also be written $(\mathbf{v} \cdot \nabla)\mathbf{v} = (\nabla \times \mathbf{v}) \times \mathbf{v} + \nabla \frac{1}{2}\mathbf{v}^2$ (see (12.42)) which gives another expression for the acceleration, namely

$$\frac{D\mathbf{v}}{Dt} = \frac{\partial \mathbf{v}}{\partial t} + (\nabla \times \mathbf{v}) \times \mathbf{v} + \nabla \frac{1}{2}\mathbf{v}^2 \qquad (1.26)$$

Now if we look back to (1.25), we see that the acceleration of a fluid element is controlled by two volumic forces, namely $\mathbf{f}$ and $Div[\sigma]$.

The volumic force $\mathbf{f}$ is specific to the problems at hands. Such a force may be the gravitational force (usually arising through buoyancy), the Laplace force if the fluid is electrically conducting (see Chap. 10) or an inertial force like the Coriolis one (see Chap. 8).

$Div[\sigma]$ is the volumic force due to stresses. Unlike $\mathbf{f}$ it is always there (except in very special cases like the one describe in Sect. 1.10.2) and represents contact forces between fluid elements. It depends on the nature of the fluid. Its expression needs a full discussion that will be presented in Sect. 1.5.

1.4.3   Energy Conservation

The equation translating the conservation of energy expresses the first principle of thermodynamics with a fluid element. The energy balance reads:

$$\frac{d}{dt} \int_{(V)} \rho(\frac{1}{2}v^2 + e)dV = \int_{(V)} \mathbf{f} \cdot \mathbf{v}dV + \int_{(S)} v_i\sigma_{ij}dS_j - \int_{(S)} \mathbf{F} \cdot d\mathbf{S} + \int_{(V)} QdV$$

where e denotes the specific internal energy,[4] $\mathbf{F}$ is the (surface density of) heat flux and Q the power of local heat sources. These sources may come from chemical reactions (burning), from nuclear reactions (in the central part of stars) or from a phase transition (latent heat release in water vapour condensation for instance).

Using words, the latter equation would read:

> *The variation of energy (kinetic and internal) per unit of time =*
> *the power of volumic forces + the power of stresses*
> *+ the heat flux through S + the power of local heat sources*

Transforming surface integrals into volume ones and using (1.21), we get

$$\int_{(V)} \rho\frac{D(\frac{1}{2}v^2 + e)}{Dt}dV = - \int_{(V)} \nabla \cdot \mathbf{F}dV + \int_{(V)} \mathbf{f} \cdot \mathbf{v}dV + \int_{(V)} \partial_j(v_i\sigma_{ij})dV + \int_{(V)} QdV$$

---

[4]The existence of internal energy for a fluid element assumes that the fluid is locally at thermodynamic equilibrium. We shall come back on this point thoroughly when we discuss the constitutive relations.

the volume being any volume, the equation is valid locally, hence:

$$\rho\frac{D(\frac{1}{2}v^2 + e)}{Dt} = -\nabla \cdot \mathbf{F} + \mathbf{f} \cdot \mathbf{v} + \partial_j(v_i\sigma_{ij}) + \mathcal{Q} \tag{1.27}$$

1.4.3.1 The Equation for Internal Energy

The foregoing equation of energy combines the internal and kinetic energies. However, the evolution of kinetic energy is governed by the sole equation of momentum. Indeed, taking the scalar product of (1.25) with $\mathbf{v}$, we find

$$\rho\frac{D(\frac{1}{2}v^2)}{Dt} = \mathbf{f} \cdot \mathbf{v} + v_i\partial_j\sigma_{ij} \tag{1.28}$$

which is just the expression of the kinetic energy theorem applied to a fluid element: the change of kinetic energy of a fluid element comes from the work of applied forces. Subtracting the evolution of kinetic energy from (1.27), we find the evolution of internal energy:

$$\rho\frac{De}{Dt} = -\nabla \cdot \mathbf{F} + \sigma_{ij}\partial_j v_i + \mathcal{Q} \tag{1.29}$$

which expresses locally the first principle of thermodynamics: the variation of internal energy of a fluid particle is equal to the heat received $(-\nabla \cdot \mathbf{F} + \mathcal{Q})$ plus the work of the stresses $\sigma_{ij}\partial_j v_i$. We should note that this work depends solely on the local rate-of-strain tensor $[s]$ since $[\sigma]$ is symmetrical, $\sigma_{ij}\partial_j v_i = \sigma_{ij}s_{ij}$.

1.4.3.2 The Equation for Entropy

Instead of using the internal energy to express the conservation of energy, it is often useful to choose the entropy. This is easily derived from the equalities relating the various thermodynamic quantities

$$de = Tds - PdV = Tds + Pd\rho/\rho^2$$

These equations link total derivatives and thus apply to all partial derivatives with respect to time and space. Hence they can be combined to yield a relation between material derivatives:

$$\frac{De}{Dt} = T\frac{Ds}{Dt} + \frac{P}{\rho^2}\frac{D\rho}{Dt} \tag{1.30}$$

Combined with mass conservation (1.16), it gives

$$T\rho\frac{Ds}{Dt} = -\nabla \cdot \mathbf{F} + \sigma_{ij}\partial_j v_i + P\nabla \cdot \mathbf{v} + \mathcal{Q} \qquad (1.31)$$

This equation expresses that the variation of entropy of fluid elements is the result of heat sources inside the elements plus the heat flux coming from the neighbouring elements, plus some heat generation due to (viscous) stresses (but see Sect. 1.6.2).

1.4.4 The Constitutive Relations

The foregoing (1.16), (1.25), (1.29) or (1.31) need now to be completed by the expression of the stress tensor $[\sigma]$ and the heat flux $\mathbf{F}$ as functions of the quantities used to describe the fluid (velocity, temperature, density, ...). Such relations are called *the constitutive relations* and are specific to the microscopic nature of the fluid. They will describe the thermodynamic, the mechanical and the thermal behaviour of the fluid. The constitutive relation(s) describing the mechanical behaviour is called the rheological law. It may also include solids. As we shall see below, the frontier between fluids and solids is not as neat as common sense would say.

1.5 The Rheological Laws

1.5.1 The Pressure Stress

In order to give an expression for $[\sigma]$, we shall first consider the case of a homogeneous and isotropic fluid in thermodynamic equilibrium. The isotropy of the fluid and the fact that the stress tensor depends only on the local properties of the fluid demands that the eigenvalues of $[\sigma]$ (which is always diagonalizable because of its symmetry) are identical in the three directions of space (Fig. 1.2). Hence, we can write:

$$\sigma_{ij} = -P\delta_{ij} \qquad (1.32)$$

where P is a scalar function that we shall identify to pressure. One may wonder whether such a definition gives the same function to which we are used to in Thermodynamics, namely the intensive variable associated with the volume. To check this point, we just need to consider the equation of internal energy (1.29)

Fig. 1.2 The pressure force

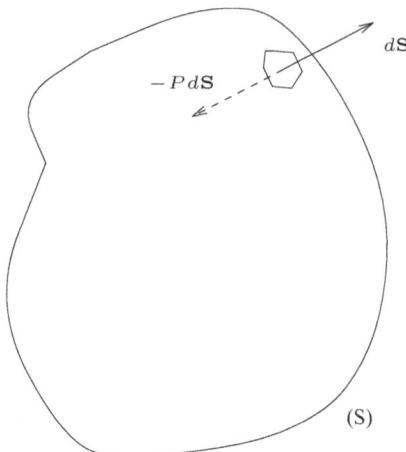

neglecting **F** and $\mathcal{Q}$. The variation of internal energy δe for a small volume ΔV during δt is

$$\Delta V \rho \delta e = \sigma_{ij} \partial_i v_j \Delta V \delta t = -P \mathbf{\nabla} \cdot \mathbf{v} \Delta V \delta t = -P \delta(\Delta V)$$

where we used (1.6). Since $\rho \Delta V$ is just the mass Δm of the small volume, $\delta(\Delta V)/\Delta m$ is the variation of the specific volume (the volume occupied by a unit mass). Hence, the foregoing relation, taken in the limit $\Delta V \to 0$, leads to the differential relation $-P = \frac{\partial e}{\partial v}$ where v is the specific volume. This expression is the well-known one of Thermodynamics.

From (1.32) we also find out the volumic force associated with the pressure. Taking the divergence of $[\sigma]$ yields:

$$f_i = \partial_j \sigma_{ij} = -\delta_{ij} \partial_j P = -\partial_i P = -(\mathbf{\nabla} P)_i$$

This expression shows that the volumic pressure force is the opposite of the pressure gradient: pressure forces push the fluid elements towards the low pressure regions as expected. The International System unit of pressure is the *Pascal* but many others are still in use (see box).

The units for pressure (or stress)

Pressure and more generally the stress is expressed in pascals, however other units are often used; we give here a short memo of this little zoo.

- The *pascal* (Pa) is the official unit of the International System; this is the stress exerted by a force of 1 Newton on 1 square metre. This is a rather small quantity since the atmospheric pressure is of the order of 10^5 Pa.
- The *bar* = 10^5 Pa is the appropriate multiple close to the atmospheric pressure.
- The *millibar* =100 Pa was the unit used in meteorology; it is now replaced by the *hectopascal*.
- The *barye* = 0.1 Pa is the pressure unit in the CGS system and represent the stress of one dyne per square centimeter.
- The *atmosphere* = 101,325 Pa is an old fashion unit which is the mean atmospheric pressure at sea level.
- The *kilogram force per square centimetre* is also an old fashion unit, which has been much used in engineering. This is the pressure exerted by the weight of a one kilogram mass on one square centimeter; thus 1 kgf/cm^2 $\simeq$ 98100 Pa which is approximately 1 bar.
- The *pound per square inch* = *psi* is the British equivalent of the kgf/cm^2. 1 psi = 6894.7 Pa
- Lastly, the *torr* ou *millimetre of mercury* : This is the pressure exerted by a 1 mm thick layer of mercury in the Earth gravity field. 1 torr $= \rho_{Hg} g 1 \, mm = 13595 * 9.8 * 0.001 \simeq$ 133.3 Pa. 760 torr $\sim$ 1 atm.

1.5.2   The Perfect Fluid

The foregoing discussion brings us to the case of an ideal fluid whose stress tensor would be composed solely of the pressure term. Such a fluid is called a *perfect fluid* or an *inviscid fluid*. We shall see below that it is actually a convenient idealization of real fluids.

If we write the momentum equation for such a fluid, it reads

$$\rho \frac{D\mathbf{v}}{Dt} = -\nabla P \tag{1.33}$$

which is known as *Euler equation*. Here, all the extra volumic forces, which are problem dependent, have been removed.

Perfect fluids are also endowed with the property that they do not allow any heat flux. Hence, $\mathbf{F}$ is vanishing. If no heat source is present ($Q = 0$), the equation of internal energy (1.29) now reads

$$\rho \frac{De}{Dt} = -P \nabla \cdot \mathbf{v} \tag{1.34}$$

and that of entropy

$$\frac{Ds}{Dt} = 0 \tag{1.35}$$

This latter relation shows us that the entropy of fluid particles in a perfect fluid remains constant. We find here the first conservation law specific to perfect fluids. This point will be developed in Chap. 3.

1.5.3 Newtonian Fluids

1.5.3.1 The Viscosities

Everybody has experienced the slow flow of honey compared to water. The two fluids being of similar density, the pull of gravity on fluid particles is similar and therefore the very different flowing behaviours they show must be associated with some intrinsic property. Such a property is manifestly related to the ability of fluid particles to slide on each other. Water particles slide much more easily than honey ones! This feature of the fluid is commonly called *viscosity*. Physically, we see that this property characterizes the interactions between fluid elements or, in other words, contact forces. As such, they are surface forces and thus should be included in the stress tensor, added to the pressure. Hence, the stress tensor components should read:

$$\sigma_{ij} = -P\delta_{ij} + \sigma_{ij}^{visc}$$

In order to find out the expression of this new term, we shall consider a fluid in equilibrium (both mechanical and thermodynamic). We perturb this equilibrium so that the fluid moves. We assume that the perturbation is small enough so that the fluid particles remain close to the thermodynamic equilibrium. We need now to find out which quantity is appropriate to measure the disequilibrium. At first glance, one may say that the velocity is the appropriate quantity. This is not the case (unfortunately!) because we note that if the velocity field is uniform, then a simple change of the reference frame will make it vanishing and thus fluid elements are still in equilibrium. Hence, non-uniformity of the velocity field is essential. We may thus think to the derivative of the velocity field, namely $\partial_i v_j$. This is still not the good quantity. Indeed, we have seen in Sect. 1.3.5 that the velocity gradient tensor is composed of two parts, the symmetric and antisymmetric ones, which describe a very different evolution of the fluid particles. The antisymmetric part describes the local rotation of fluid particles: it can be zeroed by the use of a rotating frame. We are thus left with the symmetric part of the velocity gradient tensor, the so-called rate-of-strain tensor, which cannot be nullified by any change of frame. The rate of strain thus appears as the most simple measure of the fluid mechanical disequilibrium at the level of a fluid particle. We therefore write

$$\sigma_{ij} \equiv f_{ij}(s_{kl})$$

The six functions f_{ij} are unknown and depends, a priori, on the local properties of the fluid. One way to simplify the problem is to use our assumption of the fluid being close to equilibrium. In such a case, the rate of strain is small and the functions can be expanded around zero:

$$\sigma_{ij} = f_{ij}(0) + L_{ijkl}s_{kl} + \cdots \tag{1.36}$$

with

$$L_{ijkl} = \left(\frac{\partial f_{ij}}{\partial s_{kl}} \right)_{([s]=0)} \tag{1.37}$$

$f_{ij}(0)$ is the value of the stress when the rate of strain is zero. It is obviously equal to $-P\delta_{ij}$, namely to the stress at equilibrium, which comes only from the pressure. The rank-4 tensor L characterizes also the equilibrium properties of the fluid. As before we assume that the fluid is isotropic. Since L is symmetric with respect to permutation of i and j and k and l, the only rank-4 tensor sharing these properties is of the form:

$$L_{ijkl} = a(\delta_{ik}\delta_{jl} + \delta_{jk}\delta_{il}) + b\delta_{ij}\delta_{kl} \tag{1.38}$$

where a and b are two scalar coefficients specific to the fluid. We thus find

$$\sigma_{ij} = -P\delta_{ij} + 2as_{ij} + bs_{kk}\delta_{ij}$$

or

$$\sigma_{ij} = -P\delta_{ij} + a(\partial_i v_j + \partial_j v_i) + b(\partial_k v_k)\delta_{ij}$$

In general, one gives the following equivalent form of $[\sigma]$:

$$\sigma_{ij} = -P\delta_{ij} + \mu \left(\partial_i v_j + \partial_j v_i - \frac{2}{3}(\partial_k v_k)\delta_{ij} \right) + \zeta(\partial_k v_k)\delta_{ij} \tag{1.39}$$

where we set $a = \mu$ and $b = \zeta - 2/3\mu$. μ is called the *dynamic shear viscosity* while ζ is the *dynamic volume viscosity* or *bulk viscosity* (sometimes also called second viscosity). These two coefficients are expressed in *pascal second* (Pa·s) also called the *poiseuille*. We may note that Pa s $\equiv$ kg m^{-1} s^{-1}. Now, we note that

$$\mathrm{Tr}\left(\partial_i v_j + \partial_j v_i - \frac{2}{3}(\partial_k v_k)\delta_{ij} \right) = 0 \quad .$$

so that it is natural to introduce

$$c_{ij} = \partial_i v_j + \partial_j v_i - \frac{2}{3}(\partial_k v_k)\delta_{ij} \ . \qquad (1.40)$$

These are the components of the *shear tensor* $[c]$ which corresponds to a deformation without change of volume.

All the fluids which verify the rheological law (1.39) are called the *Newtonian fluids*. Despite the severe simplifications that have been done, this constitutive relation is verified by a large number of gas or liquids. This comes from the fact that it depends solely on very general properties of the fluid at equilibrium, namely, isotropy and closeness to local thermodynamical equilibrium. In Sect. 1.9.1 we'll discuss the limits of these hypothesis and will introduce the non-Newtonian fluids.

We shall also very often use another viscosity coefficient, namely the *kinematic viscosity* defined as

$$\nu = \frac{\mu}{\rho}$$

which is expressed in $m^2 \cdot s^{-1}$.

1.5.3.2 The Microscopic Side

We introduced viscosity by considering the friction of fluid elements on each others. However, one may wonder what is going on at the microscopic level. Let us first remark that friction implies an exchange of momentum: we can slide indefinitely on the ground if we do not lose our momentum. Viscosity thus characterizes this exchange of momentum between fluid particles. Such an exchange is of course due to atoms or molecules which carry that quantity in their thermal random motion.

A simple dimensional argument gives us the order of magnitude of kinematic viscosity. Indeed, such a coefficient is dimensionally the product of a velocity by a length. At the microscopic level, typical velocity and length scales are the mean thermal velocity and the mean free path (see Chap. 11 for a more detailed introduction). These scaling imply that the kinematic viscosity of a gas increases with temperature and decreases with density. Conversely, the viscosity of liquids tends to decrease with temperature or increase with density, since in this case the exchange of momentum is rather due to attractive interactions than to collisions. Thus, for a given fluid, viscosity is minimum near the liquid–gas phase transition. The foregoing arguments show that the less viscous fluids will be found with low temperature gases.[5]

---

[5]The fluids with very low viscosities are extremely interesting experimentally as they allow us to reach very high Reynolds numbers in a small size experiment. This is the reason why many experiments have been realized with helium near its critical point (2.2 bars and 5.2 K). In these

Table 1.1 Transport coefficients and Prandtl numbers of some common fluids at a temperature of 20 °C (data are from various sources)

Fluid	μ (Pa s)	ν (m^2 s^{-1})	χ (W m^{-1}K^{-1})	$\mathcal{P}$
Air	$1.8\ 10^{-5}$	$1.5\ 10^{-5}$	0.026	0.71
H$_2$	$8.9\ 10^{-6}$	$1.0\ 10^{-4}$	0.19	0.69
Water	$1.0\ 10^{-3}$	$1.0\ 10^{-6}$	0.6	7.0
Ethanol	$1.2\ 10^{-3}$	$1.5\ 10^{-6}$	0.18	16
Glycerin	1.3	$1.0\ 10^{-3}$	0.28	10^4
Olive oil	0.1	$1.0\ 10^{-4}$	0.17	1,400
Mercury	$1.55\ 10^{-3}$	$1.14\ 10^{-7}$	8.7	0.025

In Table 1.1 we give values of the viscosities of some common fluids. We did not include values of the bulk viscosity. The reason is that this quantity has been scarcely measured; such a measurement is indeed difficult. One needs a flow without shear and with large values of $\nabla \cdot \mathbf{v}$. This is realized using sound waves. For nitrogen, Lighthill (1978) finds $\zeta \sim 0.8\mu$. On the theoretical side, using an approach based on Boltzmann equation, one can show that bulk viscosity is zero for monatomic gases (at least in the Boltzmann description).

In many cases, however, ζ is just neglected and speaking of viscosity refers to the shear viscosity μ. This approximation is usually known as *Stokes hypothesis*. We see that it is certainly well verified by monatomic gases and liquids.

1.5.3.3   The Momentum Equation and Navier–Stokes Equation

The expression of the rheological law for Newtonian fluids allows us to give an explicit form of the momentum equation for these fluids. It reads

$$\rho\frac{D\mathbf{v}}{Dt} = -\nabla P + \nabla(\mathbf{v}\cdot\nabla\mu) + \nabla\times(\mathbf{v}\times\nabla\mu) + \mu\Delta\mathbf{v} - \mathbf{v}\Delta\mu + \nabla((\mu/3 + \zeta)\nabla\cdot\mathbf{v}) + \mathbf{f}$$
(1.41)

---

conditions indeed, helium reaches its minimum viscosity. It is not a liquid, thus atoms interactions are weak and while still a gas, the velocity of atoms is minimized. The kinematic viscosity obtained in such conditions is $\nu \simeq 2\ 10^{-8}$ m^2/s.

We shall not discuss the case of superfluids which needs to be approached from the side of quantum mechanics and refer the reader to the book of Guyon et al. (2001) for an introduction.

This equation may also be written

$$\rho\frac{D\mathbf{v}}{Dt} = -\nabla P + \mu\left(\Delta\mathbf{v} + \frac{1}{3}\nabla\nabla\cdot\mathbf{v}\right) + 2(\nabla\mu\cdot\nabla)\mathbf{v}$$

$$+\nabla\mu\times\nabla\times\mathbf{v} + \left(\nabla\zeta - \frac{2}{3}\nabla\mu\right)\nabla\cdot\mathbf{v} + \mathbf{f} \qquad (1.42)$$

This is *Navier–Stokes equation*. This equation much simplifies for a fluid with constant density and constant viscosity. One then obtains the so-called Navier[6] equation:

$$\rho\frac{D\mathbf{v}}{Dt} = -\nabla P + \mu\Delta\mathbf{v} \qquad (1.43)$$

where we discarded external body forces $\mathbf{f}$. This equation is used either when density variations are negligible or, if they are not, as a first step before attacking the complications due to compressibility.

1.6 The Thermal Behaviour

1.6.1 The Heat Flux Surface Density

The next constitutive relation to be addressed is that prescribing the heat flux density as a function of the other variables. The heat flux essentially appears when the temperature field is non-uniform.[7] This means that the temperature gradient is the appropriate quantity to measure the distance to thermal equilibrium. We shall write:

$$\mathbf{F} \equiv \mathbf{F}(\nabla T)$$

As for the mechanical constitutive relation, we also assume that the fluid elements are not far from equilibrium and therefore that the temperature gradient is small; we can thus expand to first order the heat flux density, namely

$$F_i(\nabla T) = F_i(0) - \chi_{ij}\partial_j T$$

---

[6]Henri Navier (1785–1836) published this equation in 1822 in *Mémoire sur les lois du mouvement des fluides*, in Mém. de l'Acad. des Sciences.

[7]Other processes like gradients of chemical species may also generate a heat flux but these processes usually give a weak effect that will be neglected in this book.

where

$$\chi_{ij} = -\left(\frac{\partial F_i}{\partial(\partial_j T)}\right)_{\partial_j T=0}$$

is the tensor of thermal conductivities. If the fluid is isotropic, conductivity is the same in all directions and we can write $\chi_{ij} = \chi\delta_{ij}$. Noting that at equilibrium the flux is vanishing, we have $F_i(0) = 0$. We find the well-known *Fourier law*, namely

$$\mathbf{F} = -\chi\nabla T \ . \tag{1.44}$$

The minus sign has been introduced so that the thermal conductivity is positive. As shown in the next section, the sign of this quantity is controlled by the second law of Thermodynamics. As we introduced the kinematic viscosity for the momentum diffusion, it is also convenient, as far as heat transport is concerned, to introduce the heat diffusion coefficient or *thermal diffusivity*

$$\kappa = \frac{\chi}{\rho c_p} \tag{1.45}$$

which is also expressed in m^2/s. The ratio between the kinematic viscosity ν and this quantity is

$$\mathcal{P} = \frac{\nu}{\kappa} \tag{1.46}$$

and called the *Prandtl number* which is specific to each fluid. Note however that this number may vary with temperature and density since diffusion coefficients usually depend on the thermodynamic state of the fluid.

1.6.2   *The Equations of Internal Energy and Entropy*

With the foregoing Fourier and Newtonian rheological laws, we are in a position to write a complete equation for internal energy or entropy. If we consider (1.29) and replace $\mathbf{F}$ and $[\sigma]$ by their respective expression, we find that

$$\rho\frac{De}{Dt} = \nabla\cdot(\chi\nabla T) - P\nabla\cdot\mathbf{v} + \mathcal{D} + \mathcal{Q} \tag{1.47}$$

where

$$\mathcal{D} = \sigma_{ij}^{visq}\partial_j v_i = \partial_j v_i \left(\mu(\partial_i v_j + \partial_j v_i - \frac{2}{3}(\partial_k v_k)\delta_{ij}) + \zeta(\partial_k v_k)\delta_{ij}\right)$$

represents the *viscous dissipation*. This term may also be written as

$$\mathcal{D} = \left(s_{ij} + a_{ij}\right)\left(\mu c_{ij} + \zeta(\partial_k v_k)\delta_{ij}\right) = s_{ij}\left(\mu c_{ij} + \zeta(\partial_k v_k)\delta_{ij}\right)$$

because $a_{ij}c_{ij} = 0$ and $a_{ij}\delta_{ij} = \text{Tr}[a] = 0$. Using the definition (1.40) of c_{ij}, it turns out that

$$2s_{ij} = c_{ij} + \frac{2}{3}(\partial_k v_k)\delta_{ij} ,$$

which implies

$$\mathcal{D} = \frac{1}{2}\left(c_{ij} + \frac{2}{3}(\partial_k v_k)\delta_{ij}\right)\left(\mu c_{ij} + \zeta(\partial_k v_k)\delta_{ij}\right)$$

Developing this expression and using the fact that the trace of $[c]$ is zero ($\text{Tr}[c] = \delta_{ij}c_{ij} = 0$), we find

$$\mathcal{D} = \frac{\mu}{2}c_{ij}c_{ij} + \zeta(\partial_k v_k)^2 \tag{1.48}$$

or, explicitly

$$\mathcal{D} = \frac{\mu}{2}[c_{11}^2 + c_{22}^2 + c_{33}^2 + 2c_{12}^2 + 2c_{13}^2 + 2c_{23}^2] + \zeta(\nabla \cdot \mathbf{v})^2$$

Later, we shall use the more condensed expression

$$\mathcal{D} = \frac{\mu}{2}(\nabla : \mathbf{v})^2 + \zeta(\nabla \cdot \mathbf{v})^2 \tag{1.49}$$

Now, the entropy equation is deduced from (1.31), namely

$$\rho T\frac{Ds}{Dt} = \nabla \cdot (\chi\nabla T) + \mathcal{D} + \mathcal{Q} \tag{1.50}$$

This expression may be used to show that the Second Principle of Thermodynamics implies the positivity of transport coefficients like the viscosities and the thermal conductivity. For this purpose, we need considering a volume attached to the fluid particles. The Second Principle says that the entropy of this mass increases more than the entropy produced either by the internal heat sources or by the external heat flux. In mathematical terms this is expressed by

$$\frac{d}{dt}\int_{V(t)} \rho s dV \geq \int_{S(t)} \left(-\frac{\mathbf{F}}{T}\right) \cdot d\mathbf{S} + \int_{V(t)} \frac{\mathcal{Q}}{T} dV$$

Using (1.21) and (1.50), this inequality may be rewritten as

$$\int_{V(t)} \left[\chi(\nabla T)^2/T + \mathcal{D}\right] \frac{dV}{T} \geq 0$$

As it must be verified for any velocity or temperature field, it implies that

$$\chi \geq 0, \qquad \mu \geq 0, \qquad \zeta \geq 0$$

These inequalities show that the irreversibility of thermodynamic transformations is intimately associated with the diffusion phenomena that are represented by these coefficients.

To summarize, the equations of motion of a Newtonian fluid are

$$\frac{\partial \rho}{\partial t} + \nabla \cdot \rho \mathbf{v} = 0 \tag{1.16}$$

$$\rho \frac{D\mathbf{v}}{Dt} = -\nabla P + \nabla(\mathbf{v}\cdot\nabla\mu) + \nabla\times(\mathbf{v}\times\nabla\mu) + \mu\Delta\mathbf{v} - \mathbf{v}\Delta\mu + \nabla((\mu/3+\zeta)\nabla\cdot\mathbf{v}) + \mathbf{f} \tag{1.41}$$

$$\rho \frac{De}{Dt} = \nabla \cdot (\chi\nabla T) - P\nabla\cdot\mathbf{v} + \mathcal{D} + \mathcal{Q} \tag{1.47}$$

or

$$\rho T \frac{Ds}{Dt} = \nabla \cdot (\chi\nabla T) + \mathcal{D} + \mathcal{Q} \tag{1.50}$$

These equations are however still incomplete. We need to specify the thermodynamics relations, which relate the pressure, density and temperature, internal energy, etc. since the fluid is assumed to be locally at (or asymptotically close to) thermodynamic equilibrium. These are those characterizing the thermodynamics including the equations of state. We present them now.

1.7 Thermodynamics

In the foregoing sections we discussed the constitutive relations related to the mechanical and thermal behaviours. They told us the way the fluid behaves when it is slightly perturbed from equilibrium. We now complete them with the relations which specify the actual local thermodynamic equilibrium.

Let us recall that a medium in thermodynamic equilibrium is characterized by a relation like

$$e \equiv e(s, V, N, \ldots) \tag{1.51}$$

which expresses internal energy as a function of the various extensive quantities of the system (entropy, volume, number of particles, etc.). From this general relation, one derives the equations of state:

$$\begin{cases} T = \dfrac{\partial e}{\partial s} \\[2mm] P = -\dfrac{\partial e}{\partial v} \\[2mm] \pi_{ch} = \dfrac{\partial e}{\partial N} \end{cases} \tag{1.52}$$

which defines the intensive quantities of the system, the temperature T, the pressure P or the chemical potential π_{ch}.

1.7.1 The Ideal Gas

The expression of internal energy of an ideal gas as a function of extensive variables is

$$e = e_0 \left(\frac{\rho}{\rho_0}\right)^{\gamma-1} \exp\{(s - s_0)/c_v\} \tag{1.53}$$

Two classical relations come out of this expression:

$$PV = nRT \tag{1.54}$$

and

$$e = c_v T = \frac{1}{\gamma - 1}\frac{P}{\rho}, \quad \text{with} \quad \gamma = \frac{c_p}{c_v} \tag{1.55}$$

where $R = k_B \mathcal{N}$ is the ideal gas constant, or the macroscopic expression of Boltzmann constant k_B using Avogadro number $\mathcal{N}$. n is the mole number and c_v (resp. c_p) is the specific heat capacity at constant volume (resp. pressure).

Equation (1.54) may be written as

$$P = R_* \rho T \tag{1.56}$$

which is more convenient in fluid mechanics. $R_* = R/M$ where M is the mass of a mole of gas. Specific enthalpy reads

$$h = c_p T = \frac{\gamma}{\gamma - 1} \frac{P}{\rho}$$

while various expressions of entropy can be derived from

$$Tds = de - \frac{P}{\rho^2} d\rho = dh + \frac{dP}{\rho}$$

For instance:

$$s = c_v \ln(T/T_0) - R_* \ln(\rho/\rho_0) + s_0 \tag{1.57}$$

$$s = c_p \ln(T/T_0) - R_* \ln(P/P_0) + s_0 \tag{1.58}$$

$$s = c_v \ln(P/P_0) - c_p \ln(\rho/\rho_0) + s_0 \tag{1.59}$$

1.7.2 Liquids

If we focus on liquids, thermodynamics is simplified because liquids are little compressible. In most cases, the density variations mainly come from temperature variations. A simplified model consists in retaining only such a relationship like

$$\rho = \rho_0(1 - \alpha(T - T_0)) \tag{1.60}$$

which is completed by $e = cT$. α is the thermal dilation coefficient and $c = c_v \approx c_p$.

1.7.3 Barotropic Fluids

A symmetrical case to that of liquids appears when the density is solely a function of pressure

$$\rho \equiv \rho(P) \quad \text{or} \quad P \equiv P(\rho) \tag{1.61}$$

In most cases this is not an equation of state of the fluid, but an approximation well verified in certain circumstances.

Two examples are frequently met: the cases of an isothermal or of an isentropic ideal gas. For these very cases, pressure is just a function of density:

$$P = k\rho \quad \text{or} \quad P = K\rho^\gamma$$

Such a dependence arises when temperature or entropy variations can be neglected.

In the general case, such relations do not exist. The flows are called *baroclinic* because isobars and isotherms are inclined with respect to each other. Barotropicity and baroclinicity may have important consequences on the nature of the flows as it will be shown in Chaps. 2, 3 or 7.

1.8   Boundary Conditions

The laws of fluid motion that we have established are partial differential equations. Their solutions are completely defined when boundary conditions and initial conditions are given. These conditions describe the various interactions (mechanical or thermal) of the fluid with the outside, which can be a solid, another fluid, the vacuum or the fluid itself.

1.8.1   Boundary Conditions on the Velocity Field

Two types of boundary conditions are usually met by fluid flows. They describe respectively the interaction fluid–solid and fluid–fluid. They are called the *no-slip* and *free-surface* boundary conditions.

1.8.1.1   On a Solid Wall

The boundary conditions generally assumed at the frontier between a solid and a fluid is that the velocity of the fluid equals that of the solid.[8] If the solid is at rest, the fluid velocity must vanish on the boundary

$$\mathbf{v} = \mathbf{0} \quad \text{on the bounding surface} \quad (1.62)$$

This boundary condition is usually referred to as the *no-slip* boundary condition. This condition may be interpreted as if the fluid stick to the solid. This hypothesis is far not obvious. Actually, it has been much debated by the end of the nineteenth century. The question was largely solved by G.I. Taylor in 1923 when he studied the stability of a fluid flow between two rotating cylinders (the so-called Taylor–Couette flow). The agreement between theory (using these

---

[8]This assumption means, among other things, that the solid is impermeable which is not always the case. If the solid is a porous medium, some mass flux may occur through the boundary. Actually, flows through porous media are very much studied because of their numerous applications like oil or gas extraction.

boundary conditions) and experiment showed that the no-slip hypothesis was certainly quite relevant.[9]

We may note that if the fluid is perfect (no viscosity), no adherence is possible on the wall. Of the three conditions (1.62), only a single one remains, namely $\mathbf{v} \cdot \mathbf{n} = 0$ where $\mathbf{n}$ is the normal to the solid wall. The component of the velocity perpendicular to the wall vanishes, while the other components are unspecified.

1.8.1.2 On a Free Surface

The other type of boundary conditions on the velocity is the one called *the free surface* or free interface. This is the condition to be used when the fluid defines itself the surface, just like the sea surface is defined by that of the water. Let

$$S(\mathbf{r}, t) = \mathrm{Cst}$$

be the equation of this surface. At any point of this surface

$$dS = 0 = \frac{\partial S}{\partial t} dt + dx \frac{\partial S}{\partial x} + dy \frac{\partial S}{\partial y} + dz \frac{\partial S}{\partial z}$$

Similarly as (1.3), $(dx/dt, dy/dt, dz/dt)$ represents the velocity of the surface, which, by definition, is also the fluid velocity. Hence, a first boundary condition is

$$\frac{\partial S}{\partial t} + v_x \frac{\partial S}{\partial x} + v_y \frac{\partial S}{\partial y} + v_z \frac{\partial S}{\partial z} = 0 \qquad \text{on} \qquad S(\mathbf{r}, t) = \mathrm{Cst}$$

or

$$\frac{DS}{Dt} = 0 \qquad \text{on} \qquad S(\mathbf{r}, t) = \mathrm{Cst} \tag{1.63}$$

This last relation shows that the material derivative of the surface is zero at the surface. In other words, the surface is fixed for a fluid particle at the surface, or, a fluid particle initially at the surface remains attached to it.

We note that this boundary condition is purely geometrical. We did not use any physical law to write it down. In many situations, it is simplified because the surface is time-independent. In such a case it reads

$$\mathbf{v} \cdot \nabla S = 0$$

---

[9]A rather complete account of the history of the quest of the correct boundary conditions at a solid wall may be found in Goldstein (1938, 1965). The irony of the story is that scientists are presently looking for materials that let the fluid slipping on the walls. This is especially important when dealing with small pipes in microfluidic (see Tabeling, 2004).

But ∇S is a vector perpendicular to the surface. Setting $\mathbf{n}$ as the unit normal, the preceding relation is just

$$\mathbf{v} \cdot \mathbf{n} = 0 \tag{1.64}$$

expressing that the flow is tangential to the surface at the fluid boundary.

At this stage it is worth pointing out that this condition, much simpler than (1.63), is often used even if the surface is not strictly steady. This approximation is physically acceptable when the time scales or the length scales of the problem at hands are far larger than the ones arising from surface waves (capillarity or gravity waves).

As may be guessed, condition (1.63) is not sufficient to fully specify the solution of a problem. We need now expressing the continuity of the stress when crossing the surface. In other words, on each side of the surface the stress must be the same (up to the sign). For instance, if the surface separates the fluid from the vacuum, we write

$$[\sigma]\mathbf{n} = \mathbf{0} \qquad \text{on} \qquad S(\mathbf{r}, t) = \text{Cst}$$

Together with (1.63), this relation constitutes the free-surface boundary conditions. If we compare to (1.62), we may note that these boundary conditions are four. The additional equation is in fact the one that determines the surface $S(\mathbf{r}, t)$ which is also an unknown of the problem. We shall dwell on this problem more thoroughly when discussing the propagation of surface waves in Chap. 5.

1.8.1.3 The Stress-Free Boundary Conditions

In many situations the bounding surface is known and it is useful to assume that the fluid slips freely along the boundary, either because this boundary separates fluids of very different densities, or because in a first approach of a complex problem, one wishes to avoid boundary layers generated by a solid–fluid interface or waves allowed by a moving surface.

A fluid freely slipping on a surface does not exert any tangential stress. Mathematically, this is expressed by

$$\mathbf{n} \times ([\sigma]\mathbf{n}) = \mathbf{0} \qquad \text{on} \qquad S \tag{1.65}$$

This vectorial condition in fact amounts to two scalar conditions and needs to be completed by the kinematic one (1.64). Conditions (1.65) together with (1.64) now give three scalar conditions, just like (1.62). These conditions are known as *stress-free or free-slip conditions*.

1.8.2   Boundary Conditions on Temperature

The foregoing boundary conditions described the dynamics of the interaction of the fluid with its environment. They are related to the momentum equation and mass conservation. We should now ask for the conditions which are associated with the equation of energy. Such conditions express the way energy is exchanged through a bounding surface. Since we restrict our discussion to the case where the boundary does not allow for mass exchanges, fluxes of energy are only of microscopic origin, namely from thermal conduction. Generally speaking, these conditions require the continuity of temperature and energy flux, namely

$$T = T_{ext} \qquad \text{and} \qquad \mathbf{n} \cdot \mathbf{F} = \mathbf{n} \cdot \mathbf{F}_{ext}. \qquad (1.66)$$

For a fluid with constant conductivities, the second condition is also a condition on the temperature gradient.

When we study the equilibrium or the motion of fluids in presence of temperature gradient, we shall use the notion of *perfect conductor*. Such a medium is an idealization of a material that can accept any heat flux. Thus, when a fluid is in contact with a perfect conductor its temperature is fixed to that of the conductor.

The other extreme case is also useful: it is the *perfect insulator*. For this medium the heat flux is set to zero (or fixed to a given value), while the temperature can take any value. An example is given in Chap. 7.

1.8.3   Surface Tension

Free surface boundary conditions are often taken at the interface of two immiscible fluids. A complete description of free-surface boundary conditions thus calls for the introduction of surface tension. This phenomenon is the consequence of the fact that some energy must be spent to increase the surface of contact between two immiscible fluids. Only liquids own a surface tension at their boundaries because the liquid phase is characterized by an attractive interaction between the molecules (a van der Waals type force). The energy of the liquid is therefore minimized when each of its molecule is surrounded by other similar molecules. Those molecules on the boundary have a higher energy. Hence, a larger bounding surface demands more energy.

If we introduce γ, the ratio of the energy variation to the surface variation, namely

$$dE = \gamma dS, \qquad (1.67)$$

we note that γ is both an energy per unit surface and a force per unit length. Let us therefore consider a surface, delimited by a contour C, taken on the surface separating two immiscible fluids. If we decompose dS into $dldn$, dl being locally parallel to C and dn perpendicular to it, $\gamma dldn$ can be interpreted as the work done by a force $\gamma dl\mathbf{e}_n$ to extend the surface by Ldn (L is the length of C). Thus, the surface supports a resulting force

$$\mathbf{R} = \oint_{(C)} \gamma dl\mathbf{e}_n$$

where $\mathbf{e}_n$ is the outer normal unit vector of C. The use of the divergence theorem in two dimensions (see Sect. 12.2.3) allows us to transform this integral into

$$\mathbf{R} = \int_{(S)} \nabla\gamma \, dS \qquad (1.68)$$

which shows now that variations of surface tension are sources of a surface force, or, in other words, of a stress. This stress has the peculiarity of being purely tangential, which implies that if the surface separating two Newtonian fluids experiences variations of the surface tension, some flow will appear for no static constraint can compensate this stress. Such a phenomenon is at the origin of Marangoni–Bénard convection which is an instability coming from the dependance of γ with respect to temperature (see Sect. 6.3.5 for a detailed presentation).

The foregoing discussion focused on a first effect of surface tension. Indeed, we restricted the surface variation dS to the local tangent plane of C. This is just like the case where one pulls on a piece of rubber to increase its size. However, another simple way of extending the surface exists: this is by pushing it in a direction perpendicular to its actual surface. An easy way to make this idea quantitative is to consider a drop of liquid. If its radius varies of dR its surface varies of $dS = 8\pi RdR$, and the energy $dE = \gamma 8\pi RdR$ must be spent. As above, this energy may also be interpreted as the work of the surface tension $F = 8\pi R\gamma$, which has a surface density

$$\mathbf{f} = \frac{8\pi R}{4\pi R^2}\gamma\mathbf{e}_r = \frac{2\gamma}{R}\mathbf{e}_r$$

It works like a normal stress. Hence, inside a liquid drop at equilibrium, the pressure is slightly higher than outside the drop since

$$-P_{ext} = -P_{int} + \frac{2\gamma}{R}$$

$$\Longleftrightarrow \qquad P_{int} = P_{ext} + \frac{2\gamma}{R}$$

as demanded by the continuity of normal stress.

The foregoing formula is however specific to the sphere. With more general surfaces, two radii of curvature (R_1 and R_2) are necessary to describe the surface variations associated with a normal motion. This leads to the famous Laplace formula

$$P_{int} = P_{ext} + \gamma \left(\frac{1}{R_1} + \frac{1}{R_2} \right) \tag{1.69}$$

which is demonstrated in Landau and Lifchitz (1971) for instance.

Finally, the two effects of surface tension that we just described can be gathered in a single formula which states the dynamic boundary condition at a liquid–gas interface

$$[\sigma_{liq}]\mathbf{n} + \gamma \left(\frac{1}{R_1} + \frac{1}{R_2} \right) \mathbf{n} + \nabla \gamma = [\sigma_{gas}]\mathbf{n} \tag{1.70}$$

Here, $\mathbf{n}$ is the normal of the surface that is *oriented from the liquid to the gas*. Curvature radii are positive if the centre of curvature is inside the liquid.

We shall come back to surface tension in a few occasions: first, for some aspects of fluids equilibria, and then when considering the propagation of surface waves.

1.8.4   Initial Conditions

Finally, we should say a few words about the boundary conditions on time, in other words the initial conditions. The equations of motions are all of first order in time. This means that the initial state of the fluid completely determines its future evolution. This is true only in principle. The example of meteorology just shows that the behaviour of the fluid is unpredictable beyond a few days, essentially because the initial state is always imperfectly known and imperfections are amplified by the nonlinearities of the equations of motion.

1.9   More About Rheological Laws: Non-Newtonian Fluids ♠

1.9.1   The Limits of Newtonian Rheology

We have seen that the (mechanical) perturbations with respect to the thermodynamic equilibrium could be characterized by the rate-of-strain tensor $[s]$. We then admitted that such perturbations were small enough so as to justify a Taylor expansion of the stress tensor with respect to the rate of strain. We need now to precise what we mean by "small".

While we introduced [s], and more generally $\partial \mathbf{v}$, we in fact introduced a macroscopic time scale T_m. Indeed, if L and V are respectively a typical length scale and a typical velocity scale of the flow, the rate-of-strain tensor introduces

$$T_m = \frac{L}{V}$$

as a new time scale since $|s_{ij}| \equiv V/L$. This time scale expresses the rate at which a deformation is imposed to the fluid. But a material which is moved away its equilibrium state tends to come back to it on a relaxation time scale T_{relax}, through processes of microscopic origin. This time scale is of course specific to the fluid. The Newtonian behaviour is therefore the asymptotic limit when the relaxation time scale is vanishing. We now see that a new non-dimensional number has arose with the ratio of the macroscopic and microscopic time scales. This is known as the *Deborah number*:

$$De = \frac{T_{relax}}{T_m}$$

The Newtonian limit is thus $De = 0$ while the opposite limit $De = +\infty$ would rather describe a solid. Between these two extremes a huge variety of rheological laws exist, which we shall briefly present now.

1.9.2 The Non-Newtonian Rheological Laws

In the foregoing presentation of the Newtonian rheological law, we show from rather general arguments that $\sigma_{ij} = f_{ij}(s_{kl})$. Thus doing, we did not take the most general expression (so as to keep the argument as simple as possible). However, now that we realized that the small parameter was the relaxation time of the fluid, we may anticipate that the stress sustained by a fluid element is not only a function of the actual rate of strain but may also be a function of the strain itself (or the past rate of strain!); hence one would rather write:

$$\sigma_{ij} = f_{ij}\left(\ldots, \int_{-\infty}^{t} s_{kl} dt', s_{kl}, \frac{ds_{kl}}{dt}, \ldots \right) \tag{1.71}$$

where the dots designate either integral or derivatives of higher orders. Equation (1.71) is however not fully satisfactory yet. Indeed, the important strain is the one which the fluid element experiences during its trajectory. We thus see that in this perspective, the Lagrangian formulation is interesting for the description of non-Newtonian fluids. Finally, $[\sigma]$ is usually not an explicit function of $[s]$ and therefore one should solve something like

$$\mathcal{G}\left(\ldots,\oint_{-\infty}^{t}\sigma_{kl}dt',\sigma_{kl},\frac{D\sigma_{kl}}{Dt},\ldots\right)=\mathcal{F}\left(\ldots,\oint_{-\infty}^{t}s_{kl}dt',s_{kl},\frac{Ds_{kl}}{Dt},\ldots\right)$$
(1.72)

In this horrendous expression, $\oint_{-\infty}^{t}s_{kl}dt'$ means an integral along the path followed by the particle. In fact, this equation underlines the theoretical difficulties to be faced when modelling the motion of non-Newtonian fluids. This is one of the reasons why experiment is an important tool for the investigation of rheological laws.

In order to get a broad idea of these laws, we shall now review the main ones which have emerge throughout the exploration of non-Newtonian fluids.

1.9.3   Linear Viscoelasticity

Let us assume that (1.72) is linear with respect to each function and that the coefficients are constants. We thus write

$$a_0\sigma_{ij}+a_1\frac{D\sigma_{ij}}{Dt}+\cdots+a_n\frac{D^n\sigma_{ij}}{Dt^n}=b_0s_{ij}+\cdots+b_m\frac{D^m s_{ij}}{Dt^m}$$

which is the general law of linear viscoelasticity. Let us further simplify this relation by retaining only a_0,b_0,b_1 so that

$$\sigma_{ij}=\mu\left(s_{ij}+\tau_r\frac{Ds_{ij}}{Dt}\right)$$
(1.73)

When the fluid element faces a constant stress, its deformation is

$$s_{ij}=\frac{\sigma_{ij}}{\mu}\left(1-e^{-t/\tau_r}\right)$$

This expression shows that the rate of strain follows the stress with a delay of order τ_r. This is Kelvin's model.[10]

---

[10]In fact such a model rather applies to solids. The rate of strain is then replaced by the strain itself. Kelvin's solid does not react instantaneously to a stress and reaches its equilibrium after a relaxation time τ_r.

Another model is Maxwell's one. In some sense it is the symmetric of Kelvin's. The roles of stress and strain are exchanged. One sets $b_1 = 0$ and $a_1 \neq 0$ or

$$\sigma_{ij} + \tau_r \frac{D\sigma_{ij}}{Dt} = \mu s_{ij} \tag{1.74}$$

For a given rate of strain, the stress is delayed of τ_r. As an example, let us imagine a situation where the fluid is smoothly flowing. Suddenly, the shear is suppressed. The stress disappears only progressively according to

$$\sigma_{ij} = \mu s_{ij} e^{-t/\tau_r} \tag{1.75}$$

Typically, a fluid element "remembers" its past deformation and imposes a stress to its neighbourhood. Such fluids have some "memory". They are very common. Honey and jam are typical examples of our everyday life. Everybody has seen the droplet of honey rising up after the flow being cut. The stress does not vanish immediately after the flow being stopped and is able to move the fluid up. Such a behaviour is understood as the results of the intrication of macromolecules constituting the fluid.

The foregoing model, devised by Maxwell is certainly much simplified and needs to be complemented by nonlinear effects that we now discuss.

1.9.4   The Nonlinear Effects

Nonlinear effects play a major part in the dynamics of non-Newtonian fluids. To appreciate their influence, it is useful to consider very simple flows. Let us consider the basic shear flow

$$v_x = y/T$$

where T is the time scale imposed by the shear. For a Newtonian fluid the stress-tensor components would read

$$\sigma_{xx} = \sigma_{yy} = \sigma_{zz} = -p$$

$$\sigma_{xy} = \mu/T, \quad \sigma_{yz} = \sigma_{xz} = 0$$

The normal stress on a surface is the same in every direction (and equal to the pressure). Non-Newtonian fluids can generate an anisotropy controlled by the direction of the shear. This anisotropy is defined by two new quantities:

$$\sigma_{xx} - \sigma_{yy} = N_1, \quad \sigma_{xx} - \sigma_{zz} = N_2 \tag{1.76}$$

As this anisotropy is not depending on the sign of the velocity and therefore on the sign of s_{xy}, N_1 and N_2 are even functions of s_{xy}; hence

$$N_1 = \alpha_1 (s_{xy})^2 + \mathcal{O}((s_{xy})^4)$$

$$N_2 = \alpha_2 (s_{xy})^2 + \mathcal{O}((s_{xy})^4)$$

for low values of the shear. This is a nonlinear effect from the beginning. Experimentally, it turns out that $N_1 \gg |N_2|$ and $N_2 \leq 0$. The raise of such anisotropies due to shear is also called the *Weissenberg effect* and may have some spectacular effects (see the box).

Now, nonlinearities may come simply from the relation

$$\sigma_{xy} = f(s_{xy})$$

In general this relation is written $\sigma_{xy} = \mu(s_{xy})s_{xy}$ so as to emphasize the dependence of shear viscosity on shear. Fluids for which μ increases with s_{xy} are called *shear-thickening* fluids while fluids with the opposite behaviour are *shear-thinning* fluids. The nonlinearity of the relation may be interpreted as the demonstration of a change in the fluid structure.

In this category, one finds essentially diphasic fluids as for instance fluids containing solid particles or polymeric solutions. The actual behaviour of the fluid depends on the volume occupied by each phase. Clearly, such fluids have a relaxation time which is not macroscopically small.[11]

1.9.5   Extensional Viscosities

The foregoing discussion may seem a little restrictive. Many flows are not mere shear flows. Moreover, Newtonian fluid flows may also generate normal stresses. Let us consider the following two-dimensional flow:

$$v_x = x/T, \quad v_y = -y/T, \quad v_z = 0 \tag{1.77}$$

also shown in Fig. 1.4. It is associated with the following stress tensor components:

$$\sigma_{xx} = 2\mu/T, \quad \sigma_{yy} = -2\mu/T, \quad \sigma_{xy} = 0 \tag{1.78}$$

---

[11]Note that in the case of a fluid containing solid particles, the relaxation time is the characteristic time needed by a solid particle to reach the local fluid velocity when their initial velocities differ.

Fig. 1.3 The open-syphon
effect due to extension
viscosities in a 0.75 %
aqueous solution of
polyethylene oxide (credit
Barnes et al. 1989)

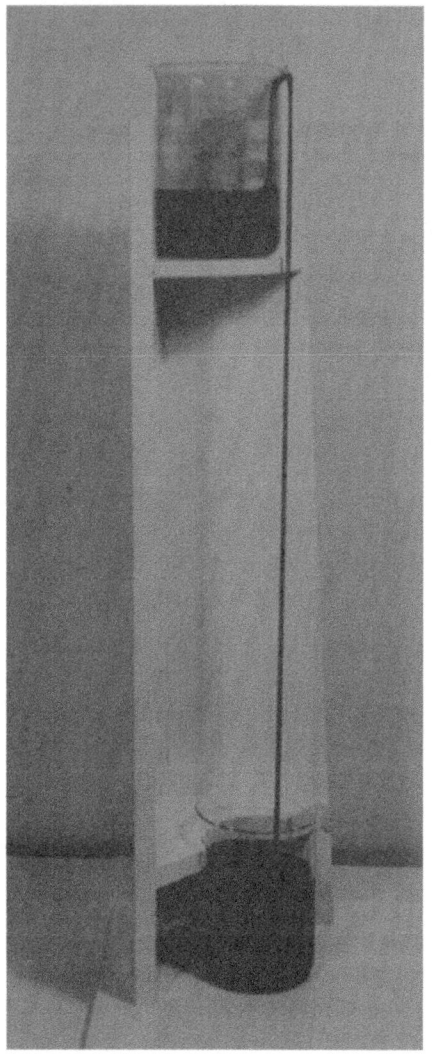

Within rheology, new viscosities are associated with these types of flows. These are the extensional viscosities which may have unusual effects (see Fig. 1.3). They are defined using three types of motions:

- The planar extension is shown in Fig. 1.4. One sets

$$\sigma_{xx} - \sigma_{yy} = \mu_P(T)/T$$

where μ_P is the *planar extensional viscosity*. For small values of the rate of strain $(T \to \infty)$, we should recover the Newtonian fluid; therefore, from (1.78)

Fig. 1.4 Flow corresponding
to a plane extension

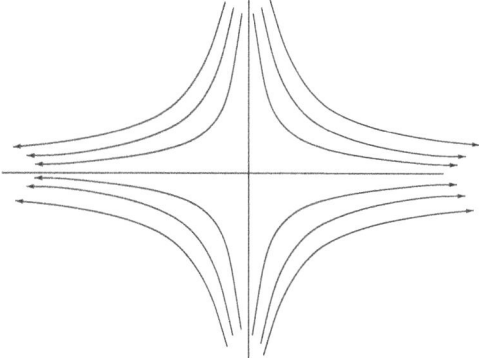

Fig. 1.5 Uniaxial extension

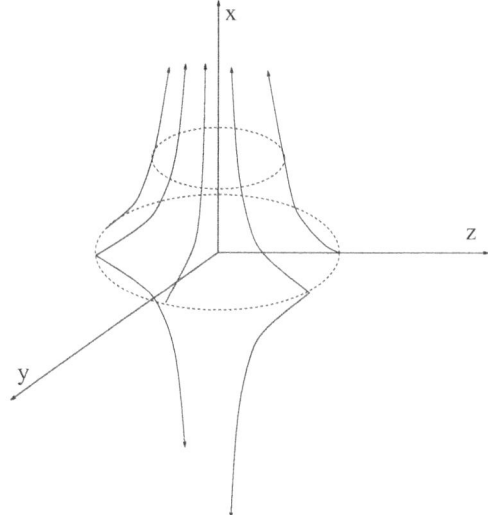

$$\lim_{T \to \infty} \mu_P(T) = 4\mu$$

- The uniaxial extension (see Fig. 1.5) flow has the following form

$$v_x = x/T, \quad v_y = -y/2T, \quad v_z = -z/2T$$

One then sets

$$\sigma_{xx} - \sigma_{yy} = \sigma_{xx} - \sigma_{zz} = \mu_E(T)/T \qquad (1.79)$$

In the Newtonian limit, it turns out that $\mu_E = 3\mu$.
- The biaxial extension flow (see Fig. 1.6) is canonically

$$v_x = x/T, \quad v_y = y/T, \quad v_z = -2z/T$$

Fig. 1.6 Biaxial extension

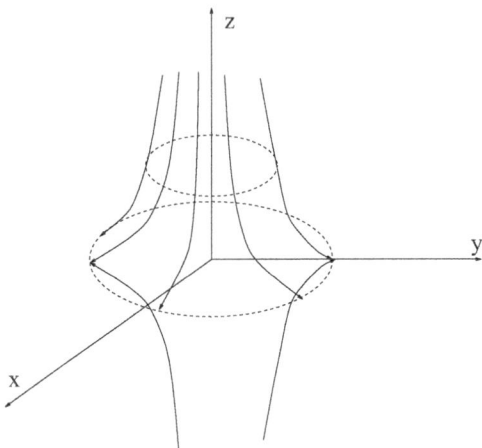

and one introduces μ_{EB} such that

$$\sigma_{xx} - \sigma_{zz} = \sigma_{yy} - \sigma_{zz} = \mu_{EB}(T)/T$$

One may notice that this latter motion can be obtained from the uniaxial extension flow by changing the sign of the rate of strain. This means that the two viscosities are related by

$$\mu_{EB}(T) = 2\mu_E(-T/2) \tag{1.80}$$

Non-Newtonian fluids in your kitchen!

We already mention honey and jam as typical common examples for the viscoelastic behaviour. But there are other examples which are worth mentioning: for instance consider the egg white. This is certainly a strange fluid. Visco-elasticity may easily be noticed but with slightly unusual tools one may put into evidence the Weissenberg effect, namely the rise of normal stresses after imposed shear strain. Just take an electric drill equipped with a rod and let the rod rotating in the egg white. You will note that instead of being expelled from the rod, like water, this fluid climbs along the rod. The shear imposed by the differential rotation generates normal stresses strong enough to overcome gravity.

Corn flour (Maizena^{TM}) mixed with a small amount of water gives another yet very non-Newtonian fluid. We suggest the following experiment. Let us mix 15 g of this flour with 2 cm^3 of water in a plate. When the mixture is smooth enough, let your finger slowly moving through this fluid. You will notice that the liquid just flows around it like any other viscous fluid. However, if you now increase the speed of your finger, you will immediately notice that the fluid thickens very strongly (be careful of not throwing away the plate!). This fluid may even be rolled between hands as a solid ball, but it will flow immediately after you cease rolling it.

1.9.6   The Solid–Fluid Transition

When a solid is stressed by increasing constraints, there is a first threshold beyond which the deformation is no longer reversible. This threshold marks the limit of elasticity. It is called *the yield stress*. With a still increasing stress the plastic behaviour of the solid leaves the place to a fluid behaviour when a new threshold (that of plasticity) is overcome.

This schematic behaviour shows up at very high values of the stress for most of the solids, however there exist some materials where these critical stresses are very low. For instance, a paint should behave as a fluid so as to be easily spread but as a solid for very small stresses so as not to drain when spread over a wall. All such fluids (or solids!) are called Bingham[12] fluids (or Bingham plastics). Their relaxation time is zero or infinite according to the value of the stress with respect to a critical stress. This is an ideal view of course!

1.10   An Introduction to the Lagrangian Formalism ♠

We briefly sketched out in Sect. 1.3.2 the idea of the Lagrangian description of fluid motion. We showed that it consists in describing the set of trajectories of fluid particles as we would do with a set of point masses. One thus no longer describes a velocity field but a field of displacements ξ indexed by the initial position of each particle. If $\mathbf{x}$ is the current position of the particle, we have:

$$\mathbf{x}(\mathbf{q}, t) = \mathbf{q} + \xi(\mathbf{q}, t) \tag{1.81}$$

The displacement $\xi(\mathbf{q}, t)$ is a function of the initial position $\mathbf{q}$ and time. The relation (1.81) may be interpreted as *a mapping*: it makes the correspondence between the Eulerian coordinate $\mathbf{x}$ and the Lagrangian one $\mathbf{q}$. Such a relation makes sense only if it is one-to-one. Particles cannot collide! This constraint is expressed by the fact that the Jacobian of the transformation (1.81) cannot vanish. This quantity is

$$J = \text{Det}[M] \tag{1.82}$$

where $[M]$ is the matrix

$$M_{ij} = \frac{\partial x_i}{\partial q_j} = \delta_{ij} + \frac{\partial \xi_i}{\partial q_j} \tag{1.83}$$

---

[12]Named after Eugen C. Bingham (1878–1945) who proposed the first mathematical description of these fluids.

1.10.1   The Equations of Motion

1.10.1.1   The Eulerian and Lagrangian Variations

We first need to introduce the notion of Eulerian and Lagrangian variations of some quantity ϕ (a scalar, vector, tensor).

The Eulerian variation of a quantity is the one obtained at a fixed Eulerian coordinate, so at a fixed $\mathbf{x}$. Therefore, this variation is simply

$$\delta\phi = \phi(\mathbf{x}, t + \delta t) - \phi(\mathbf{x}, t) = \frac{\partial\phi}{\partial t}\delta t$$

Following the same idea, the Lagrangian variation requires a variation at a fixed $\mathbf{q}$. We shall denote it $\Delta\phi$. It turns out that

$$\Delta\phi = \phi(\mathbf{q}, t + \delta t) - \phi(\mathbf{q}, t) = \phi(\mathbf{q}, t) + \boldsymbol{\xi} \cdot \nabla\phi + \frac{\partial\phi}{\partial t}\delta t - \phi(\mathbf{q}, t)$$

where we introduced $\boldsymbol{\xi}$ as the displacement of the particle during δt. The two variations are related of course:

$$\Delta\phi = \delta\phi + \boldsymbol{\xi} \cdot \nabla\phi \qquad (1.84)$$

Now if we observe that

$$\frac{\Delta\phi}{\delta t} = \frac{D\phi}{Dt}$$

we understand why the operator $\frac{D}{Dt}$ has also been called the Lagrangian derivative.

Relation (1.84) gives also the variation of a vectorial or tensorial quantity, however only if the projection basis is constant. In such a case, however, the formulation is not fully Lagrangian as the basis is not local. A more consistent formulation includes a frame dragging by the fluid. In this context, the Lagrangian variation of a vector reads

$$\Delta v^i = \delta v^i + \xi^j \partial_j v^i - v^j \partial_j \xi^i$$

One subtracts to the variation of each components the variation due to the changing frame. The quantity $\xi^j \partial_j v^i - v^j \partial_j \xi^i$ is called the Lie derivative of the velocity. This formulation allows an expression of the equation of motion in the most complex situations such as General Relativity or non-Newtonian fluids.

This is of course very specialized matter and we refer the reader to a reference like Friedman and Schutz (1978) for a discussion of this formalism.

1.10.1.2   Density Evolution

In the Lagrangian formulation the evolution of density is not controlled by a partial differential equation. One just needs to compute the Jacobian of the transformation relating the initial and final positions. Indeed, the conservation of the mass of each particle implies that

$$dm = \rho d^3\mathbf{q} = \rho'd^3\mathbf{x} = \rho'Jd^3\mathbf{q}$$

$$\Longleftrightarrow \qquad \rho' = \rho/J \tag{1.85}$$

This formula also shows that if the fluid is incompressible then ρ is a constant and $J = 1$: $[M]$ is a unitary matrix.

1.10.1.3   Momentum Evolution

Unlike density, the evolution of momentum is not easily obtained from the Lagrangian formulation. The main difficulty comes from the fact that the force field which applies to a fluid particle is generally a function of the instantaneous position of the particle, namely its Eulerian coordinates. A change of coordinates is thus necessary to express all the terms with the Lagrangian coordinates. For instance, a perfect fluid in a gravitational potential obeys

$$\rho' \frac{\partial^2\boldsymbol{\xi}}{\partial t^2} = -\nabla_{\mathbf{x}}P - \rho' \nabla_{\mathbf{x}}\phi_g \tag{1.86}$$

where ϕ_g is the gravitational potential and $\nabla_{\mathbf{x}}$ indicates that derivatives of the gradient are taken with respect to the Eulerian coordinates $\mathbf{x}$. If we now express every term with the Lagrangian coordinates $\mathbf{q}$, we need to use the matrix $[M]$ to make the coordinate change. Since $\nabla_{\mathbf{q}} = [M]\nabla_{\mathbf{x}}$ the Lagrangian form of the Euler equation is

$$\rho' \frac{\partial^2\boldsymbol{\xi}}{\partial t^2} = -[M]^{-1}\nabla_{\mathbf{q}}P - \rho' [M]^{-1}\nabla_{\mathbf{q}}\phi_g \tag{1.87}$$

This formulation is uneasy to use unless for some specific problems where simplifications occur.

1.10.2   An Example of the Use of the Lagrangian Formulation

The following example taken from cosmology shows that it is sometimes a wise idea to use the Lagrangian coordinates. The primordial gas which led to the formation of galaxies is usually modeled as a fluid without viscosity and pressure, solely subject to gravitational forces. Neglecting the expansion of the Universe, the evolution of the fluid is given by

$$\frac{\partial^2 \xi}{\partial t^2} = -[M]^{-1} \nabla_q \phi_g \tag{1.88}$$

If we restrict this equation to one dimension, its solution can be derived immediately. Indeed, in this case particles cannot cross and experience a constant gravitational pull related to the mass staying on left and right of the particle. Thus, gravitational acceleration is now a Lagrangian invariant and (1.88) leads to the solution

$$\mathbf{x} = \mathbf{q} + \mathbf{u}_0(\mathbf{q})t + \frac{1}{2}\mathbf{g}(\mathbf{q})t^2 \tag{1.89}$$

1.11 Exercises

1. Express the vorticity of the velocity field $\Omega \mathbf{e}_z \times \mathbf{r}$; what is its peculiarity?
2. What are the components of the rate-of-strain tensor in cartesian coordinates for the velocity field $\Omega \mathbf{e}_z \times \mathbf{r}$? What can be concluded? Same question for $\mathbf{v} = \lambda \mathbf{r}$. Are these flow fields compatible with incompressibility?
3. Show that (1.8) implies $\nabla \cdot \mathbf{v} = 0$. Show the reciprocal (more difficult). How can we express v_r and v_θ as a function of the stream function ψ in plane polar coordinates? Same question for a flow field with two components, v_r and v_z in cylindrical coordinates (meridional motion).
4. Retrieve the equation of continuity from (1.10).
5. Show that if all the components of the rate-of-strain tensor are zero then the velocity field is the sum of a rigid rotation and a translation.
6. Show that the evolution of the kinetic energy of a viscous fluid inside a fixed volume V, not submitted to any force field, is

$$\frac{dE_c}{dt} = -\int_{(V)} \mathcal{D}dV = -\frac{\mu}{2}\int_{(V)} (c_{ij}c_{ij})dV \tag{1.90}$$

where the second equality is valid only for incompressible fluids.
7. Give a demonstration of the relation (1.80) between uniaxial and biaxial viscosities.
8. Show the equivalence of the two forms of the viscous force in (1.41) and (1.42).

Further Reading

The fundamentals of fluid mechanics may be found in many books. The reader may find an interesting presentation in Batchelor (1967), Faber (1995), Landau and Lifchitz (1971, 1989), Paterson (1983), Ryhming (1991). For a presentation in the general framework of the mechanics of continuous media one may consult Sedov (1975). As for non-Newtonian fluids, the monograph of Barnes et al. (1989) is a good introduction. The Lagrangian formalism is discussed in papers like Friedman and Schutz (1978) while the Lie derivative is developed in the book of Schutz (1980). Some notes about the history of the discovery of heat convection may be found in Chandrasekhar (1961).

References

Barnes, H., Hutton, J., & Walters, K. (1989). *An introduction to rheology*. Amsterdam: Elsevier.

Batchelor, G. K. (1967). *An introduction to fluid dynamics*. Cambridge: Cambridge University Press.

Chandrasekhar, S. (1961). *Hydrodynamic and hydromagnetic stability*. Oxford: Clarendon Press.

Faber, T. (1995). *Fluid dynamics for physicists*. Cambridge: Cambridge University Press.

Friedman, J., & Schutz, B. (1978). Lagrangian perturbation theory of nonrelativistic fluids. *Astrophysical Journal, 221*, 937–957.

Goldstein, S. (1938, 1965). *Modern developments in fluid dynamics*. Oxford, Dover: Clarendon Press.

Guyon, E., Hulin, J.-P., Petit, L., & Mitescu, C. (2001). *Physical hydrodynamics*. Oxford: Oxford University Press.

Landau, L., & Lifchitz, E. (1971–1989). *Mécanique des fluides*. Moscow: Mir.

Lighthill, J. (1978). *Waves in fluids*. Cambridge: Cambridge University Press.

Paterson, A. (1983). *First course in fluid mechanics*. Cambridge: Cambridge University Press.

Ryhming, I. (1985, 1991). *Dynamique des fluides*. Lausanne: Presses Polytech. Univ. Romandes.

Schutz, B. (1980). *Geometrical methods of mathematical physics*. Cambridge: Cambridge University Press.

Sedov, L. (1975). *Mécanique des milieux continus*. Moscow: Mir.

Tabeling, P. (2004). Phénomènes de glissement à l'interface liquide solide. *Comptes Rendus Physique, 5*, 531–537.

Chapter 2
The Static of Fluids

The equilibrium of a fluid is certainly the most simple fluid "flow". However, not moving is not that easy for a fluid and we shall learn here, among other things, which conditions need to be satisfied for a fluid to remain in equilibrium.

2.1   The Equations of Static

If we let $\mathbf{v} = \mathbf{0}$, $\frac{\partial}{\partial t} = 0$ and $\sigma_{ij} = -p\delta_{ij}$ in (1.25) and (1.29), we find that mechanical and thermal equilibrium are governed by:

$$-\nabla P + \mathbf{f} = \mathbf{0} \tag{2.1}$$

$$\nabla \cdot (\chi \nabla T) + \mathcal{Q} = 0 \tag{2.2}$$

where $\mathbf{f}$ is an applied volumic force field and $\mathcal{Q}$ a heat source density. We immediately note that if $\mathbf{f}$ is zero then pressure is uniform.

The first important result from the above equations is that a static solution exists if, and only if, the external force can be derived from a potential. Thus, we may set $\mathbf{f} = -\nabla \phi_{ext}$ and solve for the pressure

$$P + \phi_{ext} = \text{Cst} .$$

This solution shows that isobars are identical to equipotential surfaces. We now know that if $\mathbf{f}$ is not the gradient of a potential no static solution exists. The fluid flows.

Equation (2.2) gives the temperature field. If the thermal conductivity is constant or a smooth function of the space coordinates, this equation has a solution.

© Springer International Publishing Switzerland 2015
M. Rieutord, *Fluid Dynamics*, Graduate Texts in Physics,
DOI 10.1007/978-3-319-09351-2_2

In most cases, $\mathbf{f}$ is proportional to the density ρ. Equations (2.1) and (2.2) need then to be completed by the equation of state:

$$P \equiv P(\rho, T)$$

The solution of the problem may be quite difficult, all the more that in general

$$\chi \equiv \chi(\rho, T)$$

2.2   Equilibrium in a Gravitational Field

The most common problem of fluid statics is certainly the one of a fluid at rest in a gravitational field. In this case

$$\mathbf{f} = -\rho \nabla \phi_g = \rho \mathbf{g}$$

where ϕ_g is the gravitational potential. The equation of mechanical equilibrium is then

$$\nabla P + \rho \nabla \phi_g = 0 \qquad (2.3)$$

which implies

$$\nabla \times (\frac{1}{\rho} \nabla P) = 0 \quad \Longleftrightarrow \quad \nabla \rho \times \nabla P = 0$$

This identity shows that isochore surfaces (i.e. surfaces where ρ is constant) need to be identical to isobar surfaces for a static solution to exist. This condition leads to

$$P \equiv P(\rho)$$

where we recognize the case of a barotropic fluid.

The foregoing result shows that a fluid in static equilibrium is necessarily barotropic. Now, we also note that

$$\frac{1}{\rho} \nabla P + \nabla \phi_g = 0$$

but because $P \equiv P(\rho)$, then $\rho^{-1} \nabla P = \nabla \int dP / \rho$,[1]

---

[1] We should observe that $\nabla \int dP/\rho = \left(\frac{d}{dP} \int \frac{dP}{\rho(P)} \right) \nabla P = \frac{1}{\rho} \nabla P$.

$$\implies \nabla(\int \frac{dP}{\rho} + \phi_g) = 0$$

$$\Longleftrightarrow \quad \int \frac{dP}{\rho} + \phi_g = \text{Cst} \tag{2.4}$$

A relation which determines the isobaric surfaces as a function of equipotentials.

2.2.1 Pascal Theorem

If we consider a fluid of constant density in a uniform gravity field, $\phi_g = gz$, the equation of mechanical equilibrium gives the relation

$$P + \rho g z = \text{Cst} \tag{2.5}$$

also known as Pascal[2] theorem. This relation shows that, in such a case, pressure only depends on the altitude z. We also see from this result that, in fluids at rest in a uniform gravity field, the difference of pressure between two points is just $\rho g h$, where h is the difference in their altitude.

A very direct application of this theorem is the barometer. For instance, the mercury barometer (see Fig. 2.1) is based on the fact that a column of mercury 76 cm high imposes a pressure difference similar to the atmospheric pressure at sea level.

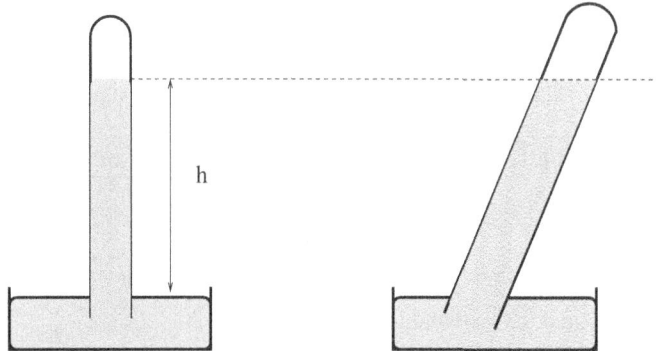

Fig. 2.1 The principle of a mercury barometer: the density of mercury is 1.36×10^4 kg/m^3 so that $\rho g h$ equals the atmospheric pressure (101,325 Pa) for h=76 cm. The void left by mercury is filled with mercury vapour but its pressure at room temperature is only 0.16 Pa, which is negligible compared to atmospheric pressure

---

[2]Blaise Pascal (1623–1662) was a French scientist and writer. As far as Physics is concerned, he is famous for his work on fluid's equilibria, *de l'Equilibre des liqueurs* and *de la Pesanteur de l'air* (the weight of air).

2.2.2   Atmospheres

Planetary atmospheres are a first application of the equilibria of fluids. The static solution is of course an approximation of an atmosphere. The Earth atmosphere is well known to be in constant evolution, with winds, clouds, etc. However, its mean vertical profile is not far from the static equilibrium. Here, we shall restrict ourselves to two very simple examples of atmosphere models: the isothermal and the isentropic ones. The latter will be compared to the actual Earth atmosphere.

2.2.2.1   The Isothermal Atmosphere

In some circumstances it is useful to simplify a model of atmosphere by assuming it being of constant temperature. Using the equation of state of ideal gases $P = \mathcal{R}_*\rho T$, which we combine with (2.4), we find the pressure profile

$$P(z) = P_0 e^{-z/z_0}$$

where $z_0 = \mathcal{R}_* T/g$ is called *the scale height* of the atmosphere. This expression shows that pressure, and hence density, decrease exponentially in an isothermal atmosphere. From the expression of z_0, we also see that the extension of such an atmosphere increases with temperature.

2.2.2.2   The Isentropic Atmosphere

The Earth atmosphere is far from being isothermal; everyone hiking in mountains has noticed that air temperature decreases with altitude. This is because the atmosphere of our planet is not very far from an isentropic state as we shall see now.

   Thermodynamics gives a relation between the differential of enthalpy, entropy and pressure, namely

$$dh = Tds + dP/\rho \,.$$

For an isentropic fluid, $ds = 0$ and thus

$$dh = dP/\rho$$

This relation implies a similar one on all the partial derivatives so that we also have $\nabla h = \nabla P/\rho$. Mechanical equilibrium reads $\nabla P = \rho \mathbf{g}$, hence

$$\nabla h = \mathbf{g} \tag{2.6}$$

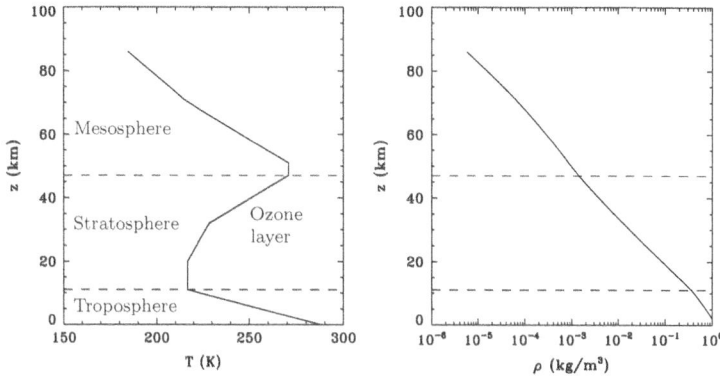

Fig. 2.2 Temperature and density profiles for the standard Earth atmosphere

This equation shows that the enthalpy gradient is just the local gravity. If the gas is ideal, then $h = c_p T$ and

$$\nabla T = \frac{\mathbf{g}}{c_p} \qquad (2.7)$$

which demonstrates that the temperature gradient is, like gravity, constant and directed towards the ground. This means that temperature decreases with altitude.

Now, (2.7) can be easily solved since $\mathbf{g} = -g\mathbf{e}_z$. We find

$$T = T_0(1 - z/z_0) \qquad (2.8)$$

where we introduced the ground temperature T_0 and, as before, the scale height which is now $z_0 = c_p T_0/g$. This quantity is only slightly different from the isothermal case if we take $T = T_0$. Pressure and density are derived from the relation $P^{1-\gamma} T^{\gamma} = \mathrm{Cst}$ valid for an isentropic ideal gas. They read

$$P = P_0(1 - z/z_0)^{\gamma/(\gamma-1)} \qquad (2.9)$$

$$\rho = \rho_0(1 - z/z_0)^{1/(\gamma-1)} \qquad (2.10)$$

These expressions show that the isentropic atmosphere has a finite height, given by z_0, unlike the isothermal atmosphere which is infinite. If we take standard values for the parameters, namely $T_0 = 289\,\mathrm{K}$, $g = 9.81\,\mathrm{m/s^2}$ and $c_p = 7/2\mathcal{R}$, we find $z_0 \simeq 30\,\mathrm{km}$. In fact, the atmosphere of the Earth is much more extended because isentropy is only approached in the troposphere (see Fig. 2.2 and the box on the standard atmosphere).

The gradient of temperature is found to be $-g/c_p = -9.8\,\mathrm{K/km}$, which represents a faster decrease than the actual atmosphere, which is close to $-6.5\,\mathrm{K/km}$. This comes from the simplifications that we adopted: in our model, the atmosphere is dry and in an isentropic state: there is no heat exchange between the fluid elements. Water vapour and heat exchanges reduce the temperature drop.

The standard atmosphere

The standard atmosphere has been defined for the needs in aeronautics and corresponds approximatively to the annual mean at a latitude of 40 degrees in North America. This atmosphere is defined up to an altitude of 86 km and is constructed with the temperature gradients defined in each layer of the model. Air is assumed to be an ideal gas with a mole mass of 28,9644 g, and located in a uniform gravity field with $g = 9.80665$ m/s^2.

Table 2.1 The standard atmosphere

Layers	Altitudes in km	∇T in K/km
Troposphere	0 – 11	−6.5
Stratosphere	11 – 20	0
	20 – 32	+1
	32 – 47	+2.8
Mesosphere	47 – 51	0
	51 – 71	−2.8
	71 – 86	−2.0

On ground (altitude $z = 0$ m), temperature is 15 °C (288.15 K) and pressure is 101325 Pa. Temperature decreases as 6.5 K/km up till 11 km, which is the upper limit of the *troposphere*. There, the temperature is 216.65 K (−56.5 °C). At this altitude the *stratosphere* begins and the temperature is first approximatively constant: this is the *tropopause*. The stratosphere contains two other layers like the famous ozone layer (20-32 km), and extends up to 47 km. Beyond and up to 86 km, one finds the mesosphere also divided into three layers (see Table 2.1). We should note that in the stratosphere, temperature increases and reaches a maximum of −2.5 °C near 50 km. This heating is essentially due to the absorption of solar UV radiation by the ozone molecules.

Beyond 86 km, we find the *thermosphere* where temperature increases again but density is so low that some part of the gas is always ionized. We touch here the *ionosphere* which extends up to 400 km, but this latter boundary is highly variable and rather fuzzy.

2.2.3   A Stratified Liquid Between Two Horizontal Plates

We now consider the equilibrium of a liquid inserted between two horizontal metallic plates. Such a device is used to study thermal convection in the laboratory (see Chap. 7). Here we shall describe the situation when the equilibrium of the fluid is stable and no convection occurs. To simplify we imagine that the two metallic plates are defined by the planes $z = 0$ and $z = d$, infinite horizontally. We also surmise that the metallic plates are perfect heat conductors and therefore impose the temperature to the fluid at these two heights. We denote these temperatures by T_b and T_t (bottom and top).

Liquids are weakly compressible; we introduced with (1.60) their simplified equation of state which we now use. Hence,

$$\rho = \rho_0(1 - \alpha(T - T_0))$$

where $\alpha > 0$ is the dilation coefficient which we assume to be constant. The thermal conductivity χ of the liquid is also assumed to be constant. With these assumptions the equations of mechanical and thermal equilibrium read:

$$\begin{cases} -\nabla P + \rho \mathbf{g} = \mathbf{0} \\ \nabla \cdot (\chi \nabla T) = 0 \end{cases} \quad \Rightarrow \quad \begin{cases} -\dfrac{dP}{dz} = \rho g \\ \dfrac{d^2 T}{dz^2} = 0 \end{cases}$$

where $\mathbf{g} = -g\mathbf{e}_z$ is the gravity. These equations can be easily solved and give the temperature, density and pressure profiles:

$$T(z) = T_b + (T_t - T_b)z/d$$

$$\rho(z) = \rho_b(1 - \alpha(T_t - T_b)z/d)$$

$$P(z) = P_b - \rho_b g z + \rho_b g \alpha (T_t - T_b)z^2/2d$$

The remarkable property of this system is that the temperature increase (or decreases) linearly with the altitude z. The stable situation corresponds to the increasing temperature. In this case light fluid is above dense fluid. The opposite case is obtained with a top plate cooler than the bottom one. As we shall see in Chap. 7, such a situation may be unstable if the temperature drop is strong enough. In such a case thermal convection takes place.

2.2.4 Rotating Self-gravitating Fluids ♣

Newton was the first to wonder about the shape of a rotating self-gravitating fluid. He was indeed interested in the shape of the Earth. This problem has then been tackled by the most renown mathematicians and physicists like Laplace, Jacobi, Riemann, Poincaré, Cartan, Chandrasekhar among the most famous. Recently, these results have been used in the theoretical approach of the dynamics of elliptical galaxies which may be viewed as a fluid of stars (see Binney and Tremaine, 1987).

Here we shall focus on the simplest of these kinds of problem: that of a fluid of constant density, self-gravitating and rotating uniformly like a solid body.

We first assume that the shape of such a system is that of an axisymmetric oblate ellipsoid and we look for the expression of its flatness as a function of its total mass M and angular velocity Ω. We shall verify afterwards that our assumption is indeed consistent with the solution.

It may be shown that the gravitational potential inside an ellipsoid of uniform density is given by

$$\Phi(r, z) = -\pi G \rho (I a^2 - A_1 s^2 - A_3 z^2)$$

where a is the equatorial radius and also the semi-major axis of a meridional section. (s, φ, z) are the cylindrical coordinates. We denote by e the eccentricity of this meridional section. Constants I, A_1 and A_3 are defined by

$$I = 2 \frac{\sqrt{1 - e^2}}{e} \arcsin e$$

$$A_1 = \frac{\sqrt{1 - e^2}}{e^2} \left(\frac{\arcsin e}{e} - \sqrt{1 - e^2} \right), \quad A_3 = 2 \frac{\sqrt{1 - e^2}}{e^2} \left(\frac{1}{\sqrt{1 - e^2}} - \frac{\arcsin e}{e} \right)$$

In a rotating frame the momentum equation reads:

$$-\nabla P - \rho \nabla \Phi - \rho \nabla \phi_c = 0$$

where $\phi_c = -\frac{1}{2} \Omega^2 s^2$ is the centrifugal potential. This equation shows that inside the body $P + \rho \Phi + \rho \phi_c$ is a constant. Since the pressure is vanishing at the surface, we have at this place

$$\Phi + \phi_c = \text{Cst} \quad \Longleftrightarrow \quad \pi G \rho (A_1 s^2 + A_3 z^2) - \frac{1}{2} \Omega^2 s^2 = \text{Cst}$$

which can be transformed into

$$\frac{s^2}{\pi G \rho A_3} + \frac{z^2}{\pi G \rho A_1 - \Omega^2/2} = \text{Cst}$$

This equation describes the surface of the fluid. Since we assumed it to be an ellipsoid, we write it

$$\frac{s^2}{a^2} + \frac{z^2}{b^2} = 1$$

where a and b are the semi-major and semi-minor axis of the meridional ellipse, respectively. By simple identification, we get the relations

$$\begin{cases} a^2 = \text{Cst} \times \pi G \rho A_3 \\ b^2 = \text{Cst} \times (\pi G \rho A_1 - \Omega^2/2) \end{cases} \tag{2.11}$$

Taking the ratio of these quantities (to eliminate the constant) and remembering that $a^2 = b^2 + c^2$ in an ellipse, where c is the distance between the center and a focus, while $c = ae$, we find

$$\frac{\Omega^2}{2\pi G \rho} = A_1 - A_3(1 - e^2) \tag{2.12}$$

Using the expression of A_1 and A_3, one may notice that the eccentricity (or the flatness) of the ellipsoid depends only on the ratio Ω^2/ρ.

The volume of an ellipsoid is $\frac{4\pi}{3} abc$, where a, b and c are the three semi-major axis of the ellipses defining this volume. Because density is constant, the volume is easily related to the mass and (2.12) may be rewritten as:

$$\frac{2\Omega^2 a^3}{3GM} = \frac{\arcsin e}{e^3}(3 - 2e^2) - \frac{3\sqrt{1 - e^2}}{e^2} \tag{2.13}$$

This equations gives the eccentricity as a function of rotation for a given density. It needs a numerical solution. However, by plotting the right-hand side as a function of e, like in Fig. 2.3, we immediately see that the solution is not unique: For each ratio Ω^2/ρ two eccentricities are possible. A low one and a high one. The latter is in fact always that of an unstable configuration.

As Newton did at his time, we now focus on the case of slow rotation and therefore on small eccentricities. An expansion of the right-hand side of (2.13) yields the relation

$$\frac{\Omega^2 a^3}{GM} = \frac{4\varepsilon}{5}$$

where we used the flatness instead of the eccentricity. The flatness is defined as

$$\varepsilon = \frac{a - b}{a} = 1 - \sqrt{1 - e^2} \simeq e^2/2$$

Observing that the surface gravity of the sphere is $g = GM/a^2$, we find the expression of ε, namely

$$\varepsilon = \frac{5\Omega^2 a}{4g}$$

Applying this formula to the case of the Earth, where $M = 5.974 \times 10^{24}$ kg, $a = 6.378 \times 10^6$ m, $g = 9.8$ m/s^2 and $\Omega = 2\pi/24$ h, we obtain

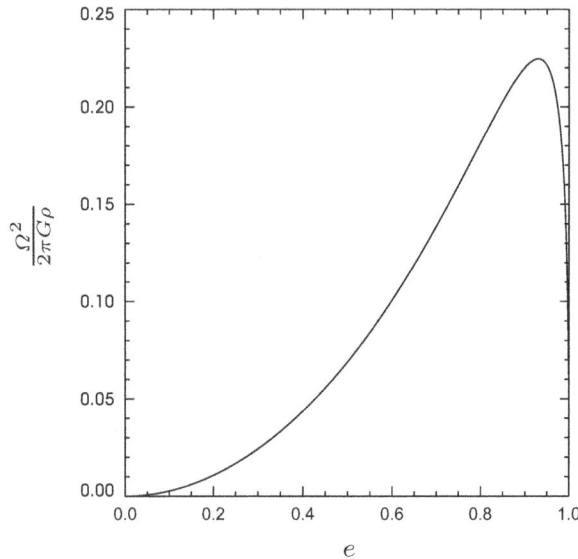

Fig. 2.3 The curve gives the value of the eccentricity of a MacLaurin ellipsoid when Ω^2/ρ, is known. The maximum, reached at $e = 0.929956$, shows that beyond some critical angular velocity (such that $\frac{\Omega^2}{2\pi G\rho} > 0.2246657$) no solution exists. In fact, an analysis of the stability of the configurations demonstrates that all solutions with $e \geq 0.9529$ are unstable, but if $0.81267 \leq e \leq 0.9529$ stable solutions exist only for an inviscid fluid. For rotations which give an eccentricity larger than 0.81267, stable solutions for a viscous fluid are triaxial Jacobi ellipsoids

$$\varepsilon_{\text{Earth}} = \frac{1}{232}$$

which is only slightly larger than the actual flatness $\varepsilon_{\text{Terre}} = 1/298$. The difference comes from the fact that the Earth is not homogeneous: central parts are much denser that the outer ones (the core of the Earth is essentially composed of iron, with a mean density of $10,500\,\text{kg/m}^3$ whereas the mantle is made of silicates and has a mean density of $\rho \sim 4,550\,\text{kg/m}^3$). This central condensation of the mass makes the shape of the Earth closer to that of a sphere.

2.3   Some Properties of the Resultant Pressure Force

When fluids are in equilibrium, one of the local body forces is the pressure gradient. This mathematical expression of the pressure force, which thus derives from a potential, implies some simple properties when it is integrated over a given volume.

2.3.1   Archimedes Theorem

Let us consider a solid fully immersed in a fluid that is *in equilibrium* in a uniform gravity field **g**. We wish to compute the resultant of pressure forces exerted on its surface. By definition this is simply

$$\mathbf{F}_{res} = -\int_{(S)} P d\mathbf{S}$$

where the differential element $d\mathbf{S}$ is oriented towards the exterior of the solid. To evaluate this integral, we may observe that we can substitute to the solid an equivalent volume of fluid without changing the equilibrium of the fluid around the solid. Indeed, there exists an equilibrium distribution of pressure inside the volume occupied by the solid that perfectly matches the outer distribution of pressures. It is obtained by a mere continuation of the isobar surfaces inside (S) (see Fig. 2.4). Then, using the theorem of divergence (see (12.8)), the foregoing surface integral can be transformed into a volume integral, like

$$\mathbf{F}_{res} = -\int_{(V)} \nabla P \, dV$$

Then, using the equation of mechanical equilibrium (2.3), we obtain

$$\mathbf{F}_{res} = -\mathbf{g} \int_{(V)} \rho dV = -M_f \mathbf{g}$$

where M_f is the mass of the fluid substituted to the solid. Archimedes theorem can now be stated:

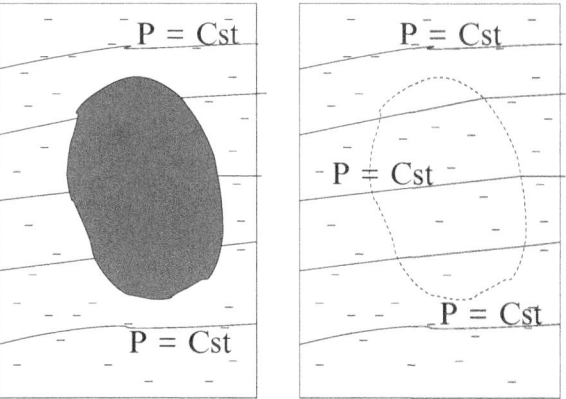

Fig. 2.4 Two equilibria of the fluid: with and without the solid

The resultant of pressure forces exerted on a volume V immersed in a fluid at equilibrium is equal and opposed to the weight of the displaced fluid.

This theorem can be applied in many situations. Note that ρ need not be constant. However, we see that it is crucial that the solid is completely surrounded by a fluid in mechanical equilibrium. This is because pressure needs to be continuous at the surface of the solid; some example where this is not the case are given in exercises.

2.3.2 The Centre of Buoyancy

A practical problem when considering the resultant of pressure forces is to know where to apply it. This is by definition the *centre of buoyancy*. When the buoyancy force is applied to it, it gives the same torque with respect to any point. In mathematical words, we need first the expression of the torque of pressure force with respect to an arbitrary point O:

$$-\int_{(S)} \mathbf{r} \times P d\mathbf{S}$$

where $\mathbf{r} = \overrightarrow{OM}$, M being the current point. Let us play with this integral using (12.9) and (12.39); we rewrite it as

$$\int_{(V)} \mathbf{\nabla} \times (P\mathbf{r})dV = \int_{(V)} \mathbf{\nabla}P \times \mathbf{r}dV = \int_{(V)} \rho \mathbf{g} \times \mathbf{r}dV = \mathbf{g} \times \int_{(V)} \rho \mathbf{r}dV$$

where we now see the appearance of a new point C_b, defined as

$$O\mathbf{C_b} = \frac{1}{M_f} \int_{(V)} \rho \mathbf{r}dV$$

We thus find that

$$\int_{(S)} -\overrightarrow{OM} \times P d\mathbf{S} = \overrightarrow{OC_b} \times (-M_f \mathbf{g}) \qquad (2.14)$$

which means that the torque exerted by pressure forces is the same as the one exerted by the resultant of pressure forces applied to the barycentre of the displaced fluid. Two remarks are now in order:

- The torque of the buoyancy force is not modified if we apply this force on a point different than C_b provided that the new point is on a line defined by $\mathbf{g}$ and C_b.
- The point where the buoyancy force is applied exist only if the fluid is in equilibrium and if the pressure varies continuously around the solid (otherwise (2.14) is not valid).

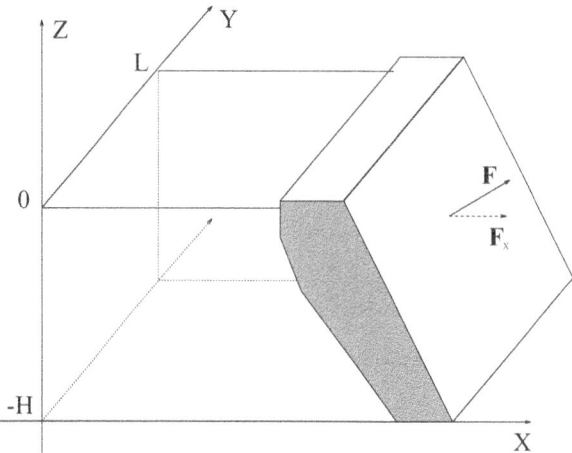

Fig. 2.5 Push on a dam

2.3.3 The Total Pressure on a Wall

The resultant pressure force exerted on a wall may easily be computed if one notices that the projection of the element $d\mathbf{S}$ on a plane whose normal is $\mathbf{e}_i$, is just $dS_i = d\mathbf{S} \cdot \mathbf{e}_i$. Hence,

$$F_i = \mathbf{e}_i \cdot \int_{(S)} P d\mathbf{S} = \int_{(S_i)} P dS_i$$

where the integral is computed on the projected surface (S_i). If this surface is a rectangle of width L and height H, like in Fig. 2.5, and pressure is only a function of z, we find that

$$F_x = \int_{-H}^{0} (P_{atm} - \rho g z) L dz = LH (P_{atm} + \rho g H/2) = LH P(-H/2)$$

for an incompressible fluid.

2.4 Equilibria with Surface Tension

In Chap. 1 we pointed out that surface tension is a source of normal stress at the surface of liquids. This stress is at the origin of some specific figures of equilibrium that we shall investigate in broad lines (we refer the reader to more specialized work for a detailed discussion, e.g. de Gennes et al. 2004).

2.4.1   Some Specific Figures of Equilibrium

2.4.1.1   The Soap Bubble

This is certainly the most simple fluid equilibrium which involves surface tension. There, only pressure opposes to surface tension. Neglecting any effect of gravity, the equilibrium of a liquid in the thin film which makes a soap bubble is given by

$$P_{air} - \frac{2\gamma}{R} = P_{liq}, \qquad P_{liq} - \frac{2\gamma}{R'} = P_{atm}$$

where P_{air} is the air pressure inside the bubble. Because the envelope is very thin, $R \simeq R'$ and

$$P_{int} \approx P_{atm} + \frac{4\gamma}{R},$$

a formula which permits the measurement of surface tension of some liquid–gas interfaces.

2.4.1.2   The Catenoid

Let us imagine now a liquid film where pressure is the same on each side of the film. In such a situation

$$\Delta P = 0 = \gamma \left(\frac{1}{R_1} + \frac{1}{R_2} \right) \qquad \Longleftrightarrow \qquad \frac{1}{R_1} + \frac{1}{R_2} = 0$$

This equation defines a surface called *the catenoid* which is such that the sum of its radii of curvature is always zero; one radius is always negative (see Fig. 2.6).

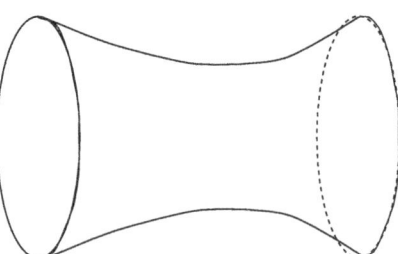

Fig. 2.6 The catenoid

2.4.2   Equilibrium of Liquid Wetting a Solid

The most spectacular effects of surface tension are certainly those associated with the wetting of solids. For instance, water raises in glass tube while mercury goes down (see Fig. 2.7). These different behaviours are the consequence of both the surface tension and the wetting properties of the solid by the liquid. These properties may be condensed in a single quantity ϑ, called the *wetting or contact angle*, in Young theory.[3] His theory assumes that the contact line gas–liquid–solid results from the equilibrium of three surface tensions: liquid–gas, liquid–solid and solid–gas. The angle between the liquid–gas and solid–liquid surfaces is called the contact angle ϑ (see Fig. 2.8). This theory gives a simple approach to very complex phenomena.

The equilibrium of the contact line yields *Young formula*:

$$\gamma_{\ell g} \cos \vartheta + \gamma_{\ell s} = \gamma_{sg} \tag{2.15}$$

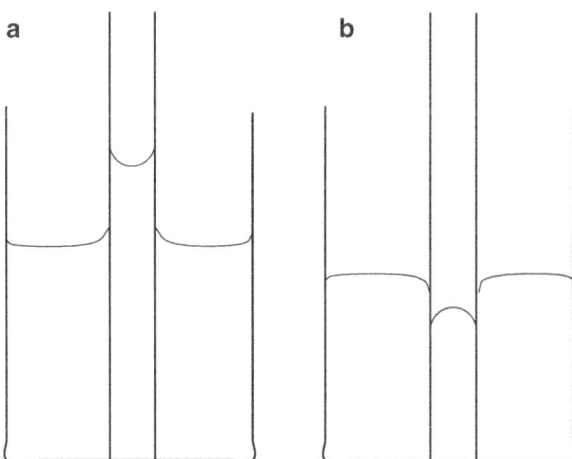

Fig. 2.7 *Upward* or *downward* displacement of a liquid due to the joint action of wetting and surface tension

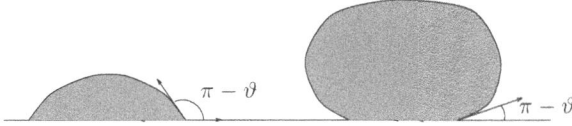

Fig. 2.8 The contact angle is ϑ but for the sake of clarity we show $\pi - \vartheta$

---

[3]Thomas Young (1773–1829) is well-known for his work in interferometry but he also studied the surface tension of liquids and the wetting of solids in 1805.

If $(\gamma_{sg} - \gamma_{\ell s})/\gamma_{\ell g} \simeq 1$, the contact angle is very small and the liquids wets the solid; if, on the contrary, $(\gamma_{sg} - \gamma_{\ell s})/\gamma_{\ell g} \simeq -1$, the contact angle is close to 180° and the liquid only weakly wets the solid. These two extreme cases are shown in Fig. 2.8. Now, what happens if $(\gamma_{sg} - \gamma_{\ell s})/\gamma_{\ell g} > 1$? Actually, no equilibrium is possible and the liquid spreads completely until it makes a very thin film: this is *total wetting*.

2.4.2.1   Jurin's Formula

Many of us have experienced the raise of water in a thin glass tube. This is a joint effect of surface tension and wetting. The contact angle imposes a negative curvature to the water's surface and thus a depression in the water inside the tube. Water thus raises.

We may easily determine this elevation of the liquid inside the tube if we assume that the meniscus has the shape of a spherical cap. Let r be the radius of the tube and ϑ the contact angle, then the radius of the spherical cap is $R = r/\cos\vartheta$. We infer the pressure difference between the liquid and the gas:

$$P_L = P_G - \frac{2\gamma\cos\vartheta}{r}$$

and the height of the raise

$$h = \frac{2\gamma\cos\vartheta}{\rho g r}. \qquad (2.16)$$

This is Jurin's formula.[4] We should stress here that this formula is an approximation valid for small values of the radius only. It is not valid for large radii since the meniscus is no longer spherical.

Jurin's formula shows that capillary rise is maximum for a total wetting ($\vartheta = 0$) but may be negative for a pair of liquid–solid such that $\cos\vartheta < 0$. For instance, water, whose surface tension is $\gamma = 0.0728\,\mathrm{J/m^2}$ at 20 °C may rise or sink by 15 mm in a tube of 1 mm radius.

2.5   Exercises

1. *About buoyancy*

   (a) An ice cube floats in a glass of water. When the ice melts, what does the level of water in the glass do?

---

[4]J. Jurin (1684–1750) was an English physician and physicist.

(b) Same question if the ice cube contains a piece of metal inside (but still floats)?

(c) And with a piece of cork?

(d) Explain why a balloon filled with some light gas (less dense than ambient air) that starts to rise, will reach a well-defined altitude while a submarine, which starts to sink, sinks to the bottom of the sea.

(e) In a car, a child holds a balloon filled with helium at the end of a string. When the car starts, how does the balloon move?

2. We consider a container filled with two immiscible liquids (oil and water for instance) and in a uniform gravity field.

(a) The two fluids being at rest, how do they settle in the container?

(b) What is the shape of the curve $P(z)$, the pressure as a function of the altitude z ($z = 0$ being the bottom of the container)?

(c) Oil and water densities are respectively $\rho_{oil} = 600\,\text{kg/m}^3$ and $\rho_{water} = 1{,}000\,\text{kg/m}^3$. A wooden sphere of density $\rho_{wood} = 900\,\text{kg/m}^3$ is left in this mixture; where is the equilibrium position of the sphere and what is the fraction of its volume inside water?

3. We consider a U-tube filled with water up to 10 cm from its bottom. The cross section of the tube is $1\,\text{cm}^2$. We then add $2\,\text{cm}^3$ of oil in one of the branches of the tube ($\rho_{oil}/\rho_{water} = 0.6$).

(a) At which height is the free surface of the oil?

(b) At which height is the interface oil–water?

(c) What is the height of water in the other branch?

4. A wooden sphere of density ρ and radius R is closing a circular hole of radius r at the bottom of a basin filled with water as shown in the figure below.

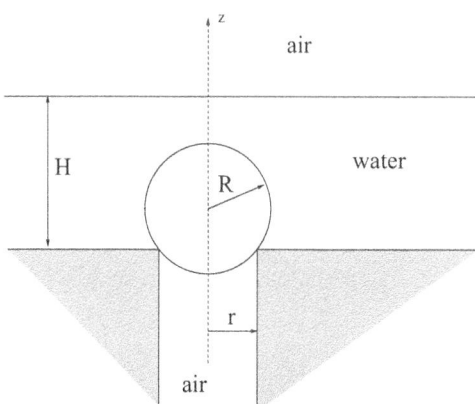

(a) Determine the force exerted by the sphere on the bottom of the basin.

(b) Give a numerical value using $\rho_{water} = 1{,}000\,\text{kg/m}^3$, $\rho = 850\,\text{kg/m}^3$, $H = 0.7\,\text{m}$, $R = 0.2\,\text{m}$, $r = 0.1\,\text{m}$, $g = 9.8\,\text{m/s}^2$.

(c) If the level of water is tunable, is there a value of this level which is such that the sphere rises to the surface before its top emerges?

5. We wish to compute the flight altitude of a balloon filled with hydrogen and left in the atmosphere assumed isentropic. Let M_b be the mass of the balloon (the nacelle and the envelope), V_b its volume assumed to be fixed and M_H its mass of hydrogen. We recall that $\rho_{air} = \rho_0(1 - z/z_0)^{\gamma/(\gamma-1)}$ for the isentropic atmosphere

(a) Which condition needs to be verified for the balloon to fly?

(b) If this condition is fulfilled, find the altitude of the flying balloon.

6. We now assume that the envelope of the balloon is opened in its lower part. At take-off, a fraction of the volume of the envelope is filled with hydrogen which is in thermal equilibrium with the surrounding air. The volume of the envelope is assumed constant.

(a) What can we say about the pressure of hydrogen in the balloon?

(b) Show that the mass of hydrogen must exceed some critical value so that the balloon takes off?

(c) Explain why the balloon reaches a well-defined altitude and give its expression.

7. Compute the temperature gradient at the equator of Jupiter assuming that its atmosphere is isentropic. The chemical composition is 85 % of molecular hydrogen and 15 % of helium. Jupiter's mass is $1.9 \times 10^{27}\,\text{kg}$, its radius 71,492 km and its rotation period 9.84 h.

8. A funnel is made of a tube with a very small cross section connected to a cone of aperture angle α. The funnel is put on a plane and filled with a liquid of density ρ as shown below.

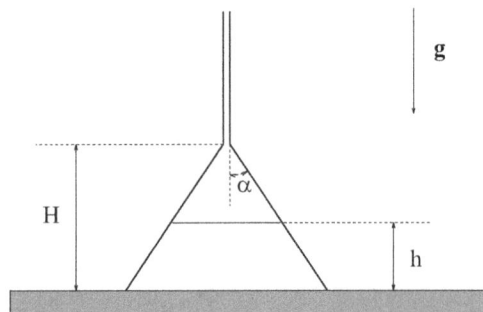

(a) Compute the vertical component of the resultant of pressure forces as a function of α, of the height h of the liquid inside the funnel, H, ρ and g the gravity.

(b) We suppose the funnel is filled up to height H; show that the funnel must have a minimum mass M_e to be equilibrium. Express this mass as a function of the mass of the liquid M_l. What does happen if the mass of the fluid is larger?

9. *A polytropic model for the Sun*

(a) We assume that a star is a ball of gas in hydrostatic equilibrium. We recall that pressure $P(r)$ and gravity $g(r)$ at a distance r from the centre verify:

$$g(r) = \frac{GM(r)}{r^2} \quad \text{and} \quad \frac{dP}{dr} = -\rho g$$

where $\rho(r)$ is the density at r and $M(r)$ is the mass inside the sphere of radius r. We also assume that the gas verifies a polytropic equation of state, namely

$$P = K\rho^{1+1/n}$$

where K is a constant and n is the polytropic index of the gas. Setting $\rho = \rho_c \theta^n$, with ρ_c being the central density and θ a non-dimensional function that varies between 0 (at the surface) and 1 at the centre, show that θ obeys the following differential equation

$$\frac{1}{\xi^2}\frac{d}{d\xi}\left(\xi^2\frac{d\theta}{d\xi}\right) + \theta^n = 0 \qquad (2.17)$$

called Emden equation, where $\xi = r/r_0$ with

$$r_0 = \sqrt{\frac{(n+1)K}{4\pi G \rho_c^{1-1/n}}}$$

(b) Show that pressure may be written

$$P = P_c\theta^n$$

(c) Show that if mass and radius of the star are known then we may deduce its central density with

$$\rho_c = -\frac{\xi_1}{3\theta_1'}\langle\rho\rangle$$

where ξ_1 is the first root of function θ and θ_1' is the value of the derivative of this function at ξ_1. $\langle\rho\rangle$ is the mean density of the star (its mass divided by its volume).

(d) Show that central pressure reads

$$P_c = \frac{4\pi G \rho_c^2 r_0^2}{n + 1}$$

(e) We now model the Sun by a polytrope of index $n = 3.37$. The numerical solution of Emden equation gives $\xi_1 = 8.686$ and $-\frac{\xi_1}{3\theta_1'} = 113.77$. Since the mass of the Sun is 2×10^{30} kg and its radius is 696×10^6 m, deduce the central density and pressure of the Sun according to this model.

(f) To derive the central temperature, we now assume that the solar plasma is an ideal gas. This gas is a mixture of protons, helium ions and electrons (other elements are neglected). We suppose that the mass fraction of helium is $Y = 28\%$. Show that the mole mass of this mixture is

$$\mathcal{M} = \frac{4}{8 - 5Y}$$

grams per mole. Deduce the central temperature of the Sun according to that model. Compare with the values obtained from more realistic models: $\rho_c = 1.62 \; 10^5$ kg/m^3, $P_c = 2.5 \; 10^{16}$ Pa, $T_c = 1.57 \; 10^7$ K.

Further Reading

For a deeper insight in the problems of wetting and capillarity, we refer the reader to de Gennes et al. (2004).

References

Binney, J., & Tremaine, S. (1987). *Galactic dynamics*. Princeton: Princeton University Press.
de Gennes, P.-G., Brochart-Wyart, F., & Quéré, D. (2004). *Capillarity and wetting phenomena: Drops, bubbles, pearls, waves*. Berlin: Springer.

Chapter 3
Flows of Perfect Fluids

3.1 Equations of Motions

In the first chapter we introduced the perfect fluid as a fluid that does not conduct
heat and for which the fluid elements interact only through pressure. We then derived
the equations of motion of such a fluid:

$$\frac{\partial \rho}{\partial t} + \nabla \cdot \rho \mathbf{v} = 0 \qquad (1.16)$$

$$\rho \frac{D\mathbf{v}}{Dt} = -\nabla P \qquad (1.33)$$

$$\frac{Ds}{Dt} = 0 \qquad (3.1)$$

These equations express mass, momentum and energy conservation, respectively.
The momentum equation is also called Euler's equation and the third equation shows
that the motion of fluid particles takes place at constant entropy. In other words
a particle of perfect fluid only sustains reversible adiabatic transformations in the
course of its motion.[1]

---

[1]On condition, of course, that the functions are continuous, i.e. that the fluid particles do not cross
a shock wave.

© Springer International Publishing Switzerland 2015
M. Rieutord, *Fluid Dynamics*, Graduate Texts in Physics,
DOI 10.1007/978-3-319-09351-2_3

3.1.1 Other Forms of Euler's Equation

Euler's equation (1.33) can be rewritten in several forms. Firstly, using the vector relation $(\mathbf{v} \cdot \nabla)\mathbf{v} = (\nabla \times \mathbf{v}) \times \mathbf{v} + \nabla \frac{1}{2}v^2$, we obtain Lamb's form:

$$\frac{\partial \mathbf{v}}{\partial t} = \mathbf{v} \times (\nabla \times \mathbf{v}) - \frac{1}{\rho}\nabla P - \nabla \frac{1}{2}v^2 \tag{3.2}$$

But Crocco's form is often more interesting. Let us introduce the enthalpy h, the total derivative of which is connected to that of pressure and entropy by

$$dh = Tds + \frac{1}{\rho}dP$$

This expression relates the differential forms of the three functions (pressure, enthalpy and entropy). It also relates the partial derivatives and therefore the gradients. Thus we can write:

$$\nabla h = T\nabla s + \frac{1}{\rho}\nabla P$$

which leads to Crocco's equation:

$$\frac{\partial \mathbf{v}}{\partial t} = \mathbf{v} \times \nabla \times \mathbf{v} + T\nabla s - \nabla(h + \frac{1}{2}v^2) \tag{3.3}$$

The quantity $h + \frac{1}{2}v^2$ is sometimes called the *total enthalpy.*

3.2 Some Properties of Perfect Fluid Motions

The form of equations (3.1) and (3.3) confers certain conservation properties on the motion of a perfect fluid and we shall study the simplest aspects of these. These properties are summarized by two theorems (Bernoulli and Kelvin) which express the conservation of mechanical energy and of angular momentum.

3.2.1 Bernoulli's Theorem

3.2.1.1 Statement and Proof

Let us consider a steady flow. It is governed by the equations:

$$\nabla \cdot \rho\mathbf{v} = 0 \tag{3.4}$$

$$\mathbf{v} \times (\nabla \times \mathbf{v}) + T\nabla s - \nabla(h + \frac{1}{2}v^2) = \mathbf{0} \qquad (3.5)$$

$$\mathbf{v} \cdot \nabla s = 0 \qquad (3.6)$$

where we dismissed all the time derivatives as required by steadiness. The last equation shows that entropy is constant along the streamlines. If we now project the momentum equation (3.5) onto the vector $\mathbf{v}$, we obtain

$$\mathbf{v} \cdot \nabla \left(\frac{1}{2}v^2 + h \right) = 0$$

so that

$$\frac{1}{2}v^2 + h = \text{Cst} \qquad (3.7)$$

along a streamline.

This result constitutes *Bernoulli's Theorem* in its fundamental form. It may be generalized to the case where the fluid flow is driven by a potential force $\mathbf{f} = -\rho\nabla\phi$. In this case

$$\frac{1}{2}v^2 + h + \phi = \text{Cst} \qquad (3.8)$$

along a streamline. This theorem simply expresses the conservation of mechanical energy per unit mass along a streamline. We notice that in this expression, enthalpy plays the role of a potential energy. If the fluid is incompressible (3.8) leads to

$$\frac{1}{2}\rho v^2 + P + \rho\phi = \text{Cst} \qquad (3.9)$$

and pressure plays the role of a potential. The quantity $\frac{1}{2}\rho v^2$ is called the *dynamic pressure*.

If the fluid is an ideal gas,

$$h = \frac{\gamma}{\gamma - 1} \frac{P}{\rho}$$

and (3.8) now reads

$$\frac{1}{2}v^2 + \frac{\gamma}{\gamma - 1} \frac{P}{\rho} + \phi = \text{Cst} \qquad (3.10)$$

also called *Saint-Venant's relation*.

Finally, it should be noted that the constant in (3.7) or (3.8) is *specific to each streamline* (see exercises).

3.2.2   The Pressure Field

The steady Euler's equation

$$\rho \mathbf{v} \cdot \nabla \mathbf{v} = -\nabla P \qquad (3.11)$$

leads to an interesting property of the pressure field associated with steady flows. Let us consider a streamline. We denote by s the curvilinear abscissa of a point on this curve and by $\mathbf{e}_s$ the tangent vector in s. We immediately see that $\mathbf{v} \cdot \nabla \equiv v \partial / \partial s$, therefore

$$(\mathbf{v} \cdot \nabla)\mathbf{v} = v \frac{\partial (v \mathbf{e}_s)}{\partial s} = v \frac{\partial v}{\partial s} \mathbf{e}_s + v^2 \frac{\partial \mathbf{e}_s}{\partial s} . \qquad (3.12)$$

Now

$$\frac{\partial \mathbf{e}_s}{\partial s} = \mathbf{n}/R_s ,$$

where R_s is the radius of curvature of the streamline at s and $\mathbf{n}$ a unit vector perpendicular to $\mathbf{e}_s$ (see Sect. 12.3). If one projects (3.11) on $\mathbf{e}_s$, one obtains

$$\frac{\partial P}{\partial s} = -\rho \frac{\partial}{\partial s} \left(\frac{v^2}{2} \right)$$

which leads to Bernoulli's theorem as we have seen above. However, if we project (3.11) along $\mathbf{n}$, we have

$$\frac{\partial P}{\partial n} = \frac{\rho v^2}{R_s} \qquad (3.13)$$

where n is the coordinate along $\mathbf{n}$. This equation expresses the equilibrium that exists between the local centrifugal force $\frac{\rho v^2}{R_s}$ and the normal component of the pressure gradient when the flow is steady. This equation also shows that *the pressure does not vary in the direction perpendicular to a streamline if the streamline is straight (infinite radius of curvature).*

Finally, we note that the relation (3.13) also applies to an unsteady flow because the term $\frac{\partial \mathbf{v}}{\partial t}$ does not have a component along $\mathbf{n}$; in this case it is necessary to replace the streamlines by the trajectories of fluid particles and R_s is the radius of curvature of such a trajectory.

3.2.3 Two Examples Using Bernoulli's Theorem

Waterfalls have been used for a very long time as a source of energy. In this example we calculate the maximum power available from a waterfall of height H having a volume flux q. We assume that water is an incompressible perfect fluid and that the flow is steady. Along a streamline we have, after Bernoulli's theorem:

$$\frac{1}{2}\mathbf{v}^2 + \frac{P}{\rho} + gz = \text{Cst} \tag{3.14}$$

We suppose that the origin of z is at the foot of the waterfall and that the water arrives at the entrance of the fall with a vanishing velocity (originating in a lake, for example).

By applying (3.14) along a streamline lying on the surface of the water, one can obtain the velocity of water at the foot of the waterfall:

$$\frac{1}{2}\mathbf{v}^2 + \frac{P_{atm}}{\rho} + 0 = 0 + \frac{P_{atm}}{\rho} + gH$$

where P_{atm} is the atmospheric pressure. We get

$$v = \sqrt{2gH} \tag{3.15}$$

also called *Torricelli's law*. This relation shows that the velocity at the foot of the waterfall is that of a free particle falling from a height H. The available power here is simply the flux of kinetic energy:

$$P_u = q \times \frac{1}{2}\rho v^2 = q\rho gH$$

For a height H of $10\,\text{m}$ and a flow rate q of $10\,\text{m}^3/\text{s}$, the available power is around $10^6\,\text{W}$. This is of course a theoretical limit and the study of a realistic case must take losses into account. Nevertheless the performance of hydraulic installations is high (actually higher than 90 %) and the preceding calculation provides a good order of magnitude.

Experts in hydraulics often rewrite (3.14) in the form

$$\frac{\mathbf{v}^2}{2g} + \frac{P}{\rho g} + z = H; \tag{3.16}$$

In this expression where the terms are all homogeneous to a length, permitting an immediate graphical representation (Fig. 3.1), the constant H represents the *hydraulic head* or *load* and

$$h = \frac{P}{\rho g} + z$$

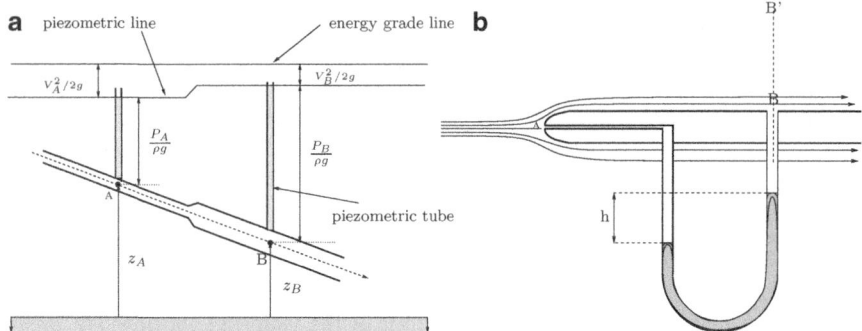

Fig. 3.1 (**a**) A representation of the hydraulic *grade line* in a pipe for a perfect fluid. With a real fluid the energy *grade line* would be inclined towards the downstream side since $H_A > H_B$. (**b**) The Pitot tube

the *piezometric height.*

In a real fluid, the head (or energy) line and the piezometric (or hydraulic grade) line are inclined towards the downstream side and the difference

$$H_A - H_B = \left(\frac{v_A^2}{2g} + \frac{P_A}{\rho g} + z_A \right) - \left(\frac{v_B^2}{2g} + \frac{P_B}{\rho g} + z_B \right)$$

on the load line represents the *head loss* between points A and B. The head loss measures the loss of mechanical energy of the flow.

Finally, we note that the power lost (or received) by a flow between two points is proportional to the product of the mass flux and the difference of load ΔH between the two points.

Another simple application of Bernoulli's Theorem is that of an apparatus called the *Pitot tube*,[2] permitting the measurement of velocity within a flow. This apparatus is sketched out in Fig. 3.1b. The principle of the device consists in estimating the difference in pressure between the stagnation point A and a point B along the tube. One admits that the holes for measuring pressure do not disturb the flow, and that the difference of elevation of the measurement points is negligible. If we consider the streamline ending in A, Bernoulli's theorem says that

$$P_A = P_\infty + \frac{1}{2}\rho v^2$$

where P_∞ is the pressure at infinity. The pressure in B is however the same as P_∞. We may see that by considering the streamline that passes through B: noting that the velocity in B is the same as at infinity (the fluid is inviscid), Bernoulli's theorem says

---

[2]H. Pitot (1695–1771) was a French physicist who invented this device around 1732 in order to measure the velocity of water in a river or the speed of a ship.

that pressure must also be the same as at infinity. Hence $P_B = P_\infty$. However, we may also note that pressure in B is also the same as the pressure along the straight line BB' because all the streamlines are straight lines there (see Sect. 3.2.2). Hence, far enough from the Pitot tube, we find streamlines along which the pressure is uniform (like the velocity) and equal to P_∞. This line of argument is interesting because it applies to real fluids also, and shows that the Pitot tube may measure the velocity even if a slight viscosity of the fluid modifies the flow in the neighbourhood of the solid, as it actually does. So we can write

$$P_A = P_B + \frac{1}{2}\rho v^2 \quad \text{and} \quad v = \sqrt{2(P_A - P_B)/\rho}$$

If the difference in pressure is measured by a U-shaped tube, $P_A - P_B = (\rho_\ell - \rho_f)gh$ where ρ_ℓ and ρ_f are respectively the densities of the liquid and the fluid that one supposes obviously non-mixable (for example, air–water, water–mercury, etc.).

3.2.4 Kelvin's Theorem

3.2.4.1 Statement

Let (C) be a contour moving with the fluid not intersecting any surface of discontinuity: if the fluid is barotropic and subject solely to forces deriving from a potential, then the circulation of the velocity along this curve is constant.

3.2.4.2 Proof

The circulation Γ along a contour (C) moving with the fluid (i.e. made of fluid particles, see Fig. 3.2) is defined as

$$\Gamma(t) = \oint_{C(t)} \mathbf{v}(\mathbf{x}, t) \cdot d\mathbf{l}$$

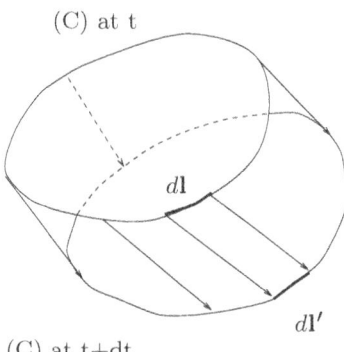

(C) at t

$d\mathbf{l}$

$d\mathbf{l}'$

Fig. 3.2 Example of a
contour moving with the fluid   (C) at t+dt

We calculate the derivative of this quantity with respect to time with the help of the relation (1.12) :

$$\frac{d\Gamma}{dt} = \oint_{C(t)} \left(\frac{Dv_i}{Dt} dl_i + v_i \partial_j v_i dl_j \right)$$

$$= \oint_{C(t)} \frac{Dv_i}{Dt} dl_i + \oint_{C(t)} \nabla(v^2/2) \cdot d\mathbf{l}$$

whence

$$\frac{d\Gamma}{dt} = \oint_{C(t)} \frac{D\mathbf{v}}{Dt} \cdot d\mathbf{l} \tag{3.17}$$

because the second integral always vanishes. We thus obtain

$$\frac{d\Gamma}{dt} = -\oint_{C(t)} \left(\frac{1}{\rho} \nabla P + \nabla \phi \right) \cdot d\mathbf{l} = -\oint_{C(t)} \left(\frac{1}{\rho} \nabla P \right) \cdot d\mathbf{l}$$

Since the fluid is barotropic $P \equiv P(\rho)$ and

$$\frac{1}{\rho} \nabla P = \nabla \int \frac{dP}{\rho}$$

where $h' = \int \frac{dP}{\rho}$ is a quantity that we can identify as the specific enthalpy if the fluid is isentropic. Finally

$$\frac{d\Gamma}{dt} = -\oint_{C(t)} \nabla h' \cdot d\mathbf{l} = 0$$

and thus

$$\Gamma(t) = \oint_{C(t)} \mathbf{v}(\mathbf{x}, t) \cdot d\mathbf{l} = \text{Cst} \tag{3.18}$$

3.2.4.3 Interpretation

Following Stokes' theorem, this result (3.18) can also be written as

$$\int_{(S(t))} \nabla \times \mathbf{v} \cdot d\mathbf{S} = \text{Cst} \tag{3.19}$$

where $S(t)$ is the surface delineated by the contour $C(t)$. The flux of vorticity across a surface moving with the fluid is constant.

If we consider an infinitesimal cylinder of fluid based on a contour $C(t)$, the angular momentum of this fluid particle is

$$\mathbf{L} = I(\frac{1}{2}\nabla \times \mathbf{v}) \propto mS(\frac{1}{2}\nabla \times \mathbf{v})$$

where we have used the fact that the moment of inertia I is proportional to the base S of the cylinder and that $\frac{1}{2}\nabla \times \mathbf{v}$ is nothing but the local rotation of the fluid element (see Chap. 1). Kelvin's theorem (3.18) implies the constancy of $S\frac{1}{2}\nabla \times \mathbf{v}$ and thus the constancy of the angular momentum $\mathbf{L}$ of the fluid particle of mass m.

Kelvin's theorem shows that in the motion of an inviscid fluid, the angular momentum of the fluid particles is conserved.

3.2.5 Influence of Compressibility

Bernoulli's theorem also allows the determination of the circumstances in which the compressibility of a gas has either a negligible or important role.

To see this, we need considering the flow of an ideal gas and Saint-Venant's relation. We apply it to a streamline that connects points far upstream where the velocity of the fluid is V_∞, the pressure P_∞ and the density ρ_∞, to a stagnation point on a solid surface where the pressure and density are respectively P_m and ρ_m. Then

$$\frac{1}{2}\mathbf{v}_\infty^2 + \frac{\gamma}{\gamma - 1}\frac{P_\infty}{\rho_\infty} = \frac{\gamma}{\gamma - 1}\frac{P_m}{\rho_m} \qquad (3.20)$$

We shall see in Chap. 5 that $\frac{\gamma P}{\rho}$ is simply the square of the local sound speed. The ideal gas flowing as a perfect fluid, fluid elements evolve isentropically and therefore $P \propto \rho^\gamma$. From this relation and (3.20), we deduce the expression of the density at the stagnation point as a function of that far upstream. One obtains

$$\rho_m = \rho_\infty \left(1 + \left(\frac{\gamma - 1}{2}\right)\frac{v_\infty^2}{c_\infty^2}\right)^{\frac{1}{\gamma - 1}} \qquad (3.21)$$

This expression shows that, at low velocity, the changes in density induced by the flow are of the order of v_∞^2/c_∞^2, which is the Mach number of the flow squared. From this particular case, we actually obtain a general result, namely that one can consider a fluid as incompressible as long as its velocity is very small in comparison with the sound speed. For example, the air flow around a car moving at 100 km/h causes variations of density less than a percent, which are therefore negligible in first approximation.

3.3 Irrotational Flows

3.3.1 Definition and Basic Properties

A flow is called *irrotational* if

$$\nabla \times \mathbf{v} = \mathbf{0}$$

or, equivalently, if there exists a function Φ such that

$$\mathbf{v} = \nabla \Phi.$$

This type of flow is also called a *potential flow* and Φ is the *velocity potential*.

Let us consider the case of irrotational flows of perfect fluids, whose motion is driven by a force field derived from a potential ϕ_{ext}. We look for the equations satisfied by the velocity potential Φ. Euler's equation is transformed in the following way:

$$\rho \frac{D\mathbf{v}}{Dt} = -\nabla P - \rho \nabla \phi_{ext}$$

$$\Longleftrightarrow \quad \nabla \left(\frac{\partial \Phi}{\partial t} + \frac{1}{2} \mathbf{v}^2 \right) = -\frac{1}{\rho} \nabla P - \nabla \phi_{ext}$$

We note that in order for this equation to make sense we require that

$$\nabla \times \frac{1}{\rho} \nabla P = \mathbf{0} \Longleftrightarrow \nabla \rho \times \nabla P = \mathbf{0},$$

namely that $P \equiv P(\rho)$, as has been seen in the previous chapter. So we can introduce h' such that $\nabla h' = \frac{1}{\rho} \nabla P$. Hence,

$$\nabla \left(\frac{\partial \Phi}{\partial t} + \frac{1}{2} \mathbf{v}^2 + h' + \phi_{ext} \right) = \mathbf{0}$$

or

$$\frac{\partial \Phi}{\partial t} + \frac{1}{2} \mathbf{v}^2 + h' + \phi_{ext} = \text{Cst} \tag{3.22}$$

We note the similarity of this expression with that obtained for Bernoulli's Theorem, but we must pay attention to the fact that in this new equation the *constant is the same in all the volume occupied by the fluid and thus identical for all streamlines*. Moreover, the expression is also valid for unsteady flows.

To (3.22), we add the equation of continuity

$$\frac{\partial \rho}{\partial t} + \nabla \cdot (\rho \nabla \Phi) = 0$$

This last equation takes a special form for incompressible fluids where $\rho = \mathrm{Cst}$, since

$$\Delta \Phi = 0 \qquad\qquad (3.23)$$

is simply Laplace's equation.

We observe that the potential Φ is defined to within a function of time: since Φ and $\Phi + f(t)$ give the same velocity field.

3.3.2 Role of Topology for an Irrotational Flow

Topology plays a very important role in irrotational flows. Let us first take an illustrative example. We consider a fluid which occupies all space except a cylinder of infinite length with a radius a centered on the axis O_z. The motion of fluid around the cylinder is given by its velocity field

$$\mathbf{v} = \mathbf{\Omega} \times \frac{a^2 \mathbf{e}_s}{s} = \frac{\Omega a^2}{s} \mathbf{e}_\varphi$$

which is derived from the potential $\Phi = a^2 \Omega \varphi$ (s, φ, z are the cylindrical coordinates). One will note that this potential possesses a special property: it is not single valued; at a given point, φ can take an infinite number of values like $\varphi + 2n\pi$. The immediate consequence of this property is that the circulation Γ along a closed curve can take many values depending on the chosen curve. In fact, if the curve does not enclose the cylinder $\Gamma = 0$. If, on the other hand, it encloses it n times $\Gamma = 2n\pi\Omega \neq 0$.

This example illustrates the effect of topology on circulation. The space occupied by the fluid here is *doubly connected*: there exist two irreducible paths[3] to connect two points in this space.

Double connectivity implies that the solutions to Laplace's equation are entirely defined only when the circulation around the regions not belonging to the fluid space is given.

---

[3]That are paths which cannot be reduced from one to the other by a continuous deformation within the space occupied by the fluid or, equivalently, the surface bounded by the two paths does not belong entirely to this space.

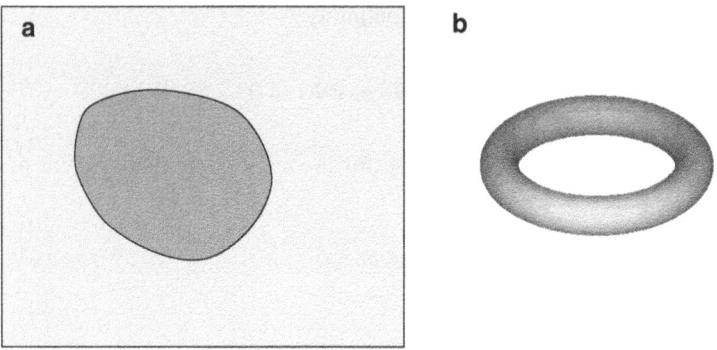

Fig. 3.3 Examples of doubly connected domains: in two-dimensions (**a**) any obstacle creates a doubly connected region; in three-dimensions a toroid (**b**) or an obstacle which is infinite in one dimension implies double connectivity

Two examples of doubly connected spaces are shown in Fig. 3.3. One may note that the presence of an obstacle in a two-dimensional flow renders the space occupied by the fluid doubly connected.

3.3.3 Lagrange's Theorem

If the flow of a barotropic fluid subjected to forces deriving from a potential is irrotational at time t_0 then it is (irrotational) at all other times.

In order to prove this theorem we shall suppose the volume occupied by the fluid to be simply connected. According to Kelvin's theorem,

$$\oint_{C(t)} \mathbf{v} \cdot d\mathbf{l} = \text{Cst}$$

at any time. But at t_0

$$\forall C \qquad \oint_C \mathbf{v} \cdot d\mathbf{l} = \oint_C \nabla \Phi \cdot d\mathbf{l} = 0$$

The equality is true for any curve (C) and, from Kelvin's theorem, at all time t. We have therefore

$$\oint_C \mathbf{v} \cdot d\mathbf{l} = \iint_S \nabla \times \mathbf{v} \cdot d\mathbf{S} = 0$$

for any surface S and any time t, thus

$$\Longleftrightarrow \nabla \times \mathbf{v} = \mathbf{0}, \quad \forall t \quad \text{or} \quad \mathbf{v} = \nabla \Phi, \quad \forall t$$

This result is important because it justifies the irrotationality of a large number of flows: in particular if an inviscid fluid is initially at rest and is set in motion by the action of a force deriving from a potential, one can state that the flow will be irrotational because $\mathbf{v} = \mathbf{0}$ is an irrotational flow.

3.3.4 Theorem of Minimum Kinetic Energy

For an incompressible flow of a perfect fluid, the irrotational solution is unique and is that of minimum kinetic energy.

The uniqueness (to within an additive constant) of the solution follows from Laplace's equation satisfied by the potential Φ. The solution is unique when the boundary conditions are specified. As for Lagrange's theorem, we consider only the case where the fluid occupies a simply connected space. If $\mathbf{n}$ is the outward normal at the surface bounding the fluid, the flux of $\mathbf{v}$ across the surface is zero and the potential therefore satisfies

$$\mathbf{n} \cdot \nabla \Phi = 0$$

on it. This boundary condition is called Neumann's boundary condition. Together with Laplace's equation it defines a unique solution for $\mathbf{v}$ (for Φ the solution is defined up to an additive constant). We now show that this solution is that of minimum energy. For this purpose we consider an irrotational flow $\mathbf{v} = \nabla \Phi$ such that $\nabla \cdot \mathbf{v} = 0$ as well as another flow $\mathbf{v}'$ such that $\nabla \cdot \mathbf{v}' = 0$ but which is not necessarily potential. The kinetic energies associated with each of these flows are:

$$E_c = \frac{1}{2} \rho \int_V \mathbf{v}^2 dV \quad \text{and} \quad E'_c = \frac{1}{2} \rho \int_V \mathbf{v}'^2 dV$$

Their difference is

$$E'_c - E_c = \frac{1}{2} \rho \int_V (\mathbf{v}'^2 - \mathbf{v}^2) dV$$

however

$$\mathbf{v}'^2 - \mathbf{v}^2 = (\mathbf{v}' - \mathbf{v})^2 + 2\mathbf{v} \cdot (\mathbf{v}' - \mathbf{v})$$

therefore

$$E'_c - E_c = \frac{1}{2} \rho \int_V (\mathbf{v}' - \mathbf{v})^2 dV + \rho \int_V \mathbf{v} \cdot (\mathbf{v}' - \mathbf{v}) dV$$

but

$$\int_V \mathbf{v} \cdot (\mathbf{v}' - \mathbf{v}) dV = \int_V (\mathbf{v}' - \mathbf{v}) \cdot \nabla \Phi \, dV = \int_V \nabla \cdot (\Phi(\mathbf{v}' - \mathbf{v})) dV = \int_{(S)} \Phi(\mathbf{v}' - \mathbf{v}) \cdot d\mathbf{S} = 0$$

because $\mathbf{v}$ and $\mathbf{v}'$ both satisfy the boundary condition $\mathbf{v} \cdot \mathbf{n} = \mathbf{v}' \cdot \mathbf{n} = 0$. We find the result

$$E'_c - E_c = \frac{1}{2} \rho \int_V (\mathbf{v}' - \mathbf{v})^2 dV \geq 0$$

This theorem is also due to Kelvin.

3.3.5   Electrostatic Analogy

Laplace's equation is encountered in numerous problems in Physics, in particular in electrostatics where it gives the variations of electrostatic potential in the absence of a charge density. Nevertheless, it is not the electric field that one uses as an analog of the velocity field, but rather a quantity which is proportional to it, like the current density $\mathbf{j}$. Ohm's law states that in a conductive medium, $\mathbf{j} = \sigma \mathbf{E}$, σ being the conductivity assumed constant. In making this analogy we actually substitute the flow of fluid for a flow of charges. The "obstacles" are thus the insulated regions. The situation is easily summed up in the following table:

Fields	$\mathbf{v}$	$\mathbf{j} = \sigma \mathbf{E}$
Equations	$\nabla \times \mathbf{v} = \mathbf{0} \Longleftrightarrow \mathbf{v} = \nabla \Phi$ $\nabla \cdot \mathbf{v} = 0$ $\Delta \Phi = 0$	$\nabla \times \mathbf{E} = \mathbf{0} \Longleftrightarrow \mathbf{j} = \nabla \phi_j$ $\nabla \cdot \mathbf{E} = 0$ $\Delta \phi_j = 0$
Boundary conditions	$\mathbf{v} = \mathbf{0}$ at infinity $\mathbf{v} \cdot \mathbf{n} = 0$ at the surface of the obstacle	$\mathbf{j} = \mathbf{0}$ at infinity $\mathbf{j} \cdot \mathbf{n} = 0$ at the surface of the insulated region

This is the *direct analogy*. We shall later encounter the inverse analogy where the analog of electrostatic potential is the stream function.

3.3.6 Plane Irrotational Flow of an Incompressible Fluid

3.3.6.1 Equation for the Stream Function

We have seen in Sect. 1.3.7 that a two-dimensional flow can be described with the help of a scalar function called the stream function ψ. If the velocity is derived from a potential then ψ also satisfies Laplace's equation. Indeed, $\nabla \times \mathbf{v} = \mathbf{0}$ implies that

$$\frac{\partial v_y}{\partial x} - \frac{\partial v_x}{\partial y} = 0$$

while $v_x = \partial \psi / \partial y$ and $v_y = -\partial \psi / \partial x$, therefore

$$\Delta \psi = 0 \tag{3.24}$$

It may then be shown (see exercise) that the streamlines ($\psi = $ Cst) are orthogonal to the "equipotentials of velocity" ($\phi = $ Cst).

3.3.6.2 Inverse Analogy

In view of the preceding relation we can make an analogy between the electrostatic potential and the stream function since they both satisfy the same equation. The two functions will be identical if they satisfy the same boundary conditions. For the velocity, these are simply $\psi = $ Cst along the boundaries and thus for the electrostatic potential we will require that $\phi_e = $ Cst along the bodies and these will be identified to perfect conductors (this is indeed the inverse of the preceding analogy!).

Fields	$\mathbf{v}$ ψ	$\mathbf{E} \times \mathbf{e}_z$ ϕ_e
Equations	$\nabla \times \mathbf{v} = \mathbf{0}, \mathbf{v} = \nabla \times (\psi \mathbf{e}_z)$ $\Downarrow$ $\Delta \psi = 0$ $\Uparrow$ $\nabla \cdot \mathbf{v} = 0, \mathbf{v} = \nabla \Phi$	$\nabla \times (\mathbf{E} \times \mathbf{e}_z) = \mathbf{0}$ $\Uparrow$ $\Delta \phi_e = 0$ $\Uparrow$ $\nabla \cdot \mathbf{E} = 0, \mathbf{E} = \nabla \phi_e$
Boundary conditions	$\mathbf{v} = \mathbf{0}$ at infinity $\psi = $ Cst on an obstacle	$\mathbf{E} = \mathbf{0}$ at infinity $\phi_e = $ Cst on a conductor

3.3.6.3   The Complex Potential

The existence of two harmonic functions[4] describing the flow allows the study of two-dimensional irrotational incompressible flows in a very thorough manner, thanks to the complex potential. We give here only the broad lines of this approach and refer the reader to the classical works for more details (see for example Batchelor 1967).

We thus introduce the complex function

$$f = \phi + i\psi \tag{3.25}$$

called the *complex potential*. Besides Laplace's equation, this function satisfies

$$\frac{\partial f}{\partial x} + i\frac{\partial f}{\partial y} = 0 \tag{3.26}$$

because

$$\begin{cases} \dfrac{\partial \phi}{\partial x} = \dfrac{\partial \psi}{\partial y} = v_x \\[3mm] \dfrac{\partial \phi}{\partial y} = -\dfrac{\partial \psi}{\partial x} = v_y \end{cases}$$

Equation (3.26) is also called Cauchy's conditions. It implies that $f \equiv f(x + iy)$, thus f is only a function of the complex variable $z = x + iy$.

We then introduce the *complex velocity* defined by

$$w = \frac{df}{dz} = \frac{\partial f}{\partial x} = v_x - iv_y$$

The interest in introducing the complex potential rests essentially in the ability to use *conformal transformation*. This type of transformation is defined by an analytical function G with non-zero derivative in a domain of the complex plane, which associates with each point z of the first domain a point z' of the image domain, such that

$$z' = G(z)$$

This transformation is called conformal because it conserves angles.

Let us seek the equation for the streamlines (the curves $\psi = $ Cst) in the image plane. $\psi = $ Cst is the equation of streamlines in the original plane, thus

---

[4] A harmonic function is a solution of Laplace's equation.

$\psi(G^{-1}(z')) = $ Cst is the equation in the image plane. $\psi \circ G^{-1}$ is the new stream function. More generally, if $F(z)$ is the complex potential of the flow $F \circ G^{-1}$ is the complex potential in the image plane. We derive from this, the new complex velocity:

$$w' = \frac{dF \circ G^{-1}}{dz'} = \frac{w(z)}{G'(z)} \tag{3.27}$$

In order to illustrate the power of this transformation, we shall use the example of Joukovski's transformation, namely

$$G(z') = z' + R^2/z' \tag{3.28}$$

This function is indeed analytic throughout the plane except at the origin.

We now consider a uniform flow past a flat plate represented by a segment of length $4R$ on the x axis. The velocity is simply $\mathbf{v} = V_0 \mathbf{e}_x$ and the associated complex potential is

$$f(z) = V_0 z$$

Let z be the transform of a system of coordinates z' by the conformal transformation G, so that $z = G(z')$. Substituting this in the above equation we have

$$f(G(z')) = V_0 G(z')$$

a new complex potential which is $f \circ G$; but since f is simply the identity (to within a multiplicative constant), G is in the new potential.

In choosing the Joukovsky's transformation for G, we can seek new streamlines and, in particular, the new shape of the obstacle in the (x', y') plane. For that purpose it suffices to take the imaginary part of (3.28)

$$\psi = \text{Im}(z' + R^2/z') = \text{Im}(r' e^{i\theta'} + R^2/r' e^{-i\theta'}) = \frac{r'^2 - R^2}{r'} \sin \theta'$$

which gives the new streamlines. Among them we find those bounding the obstacle: here it consists of the circle $r' = R$, the inverse transform of the line $\text{Im}(z) = 0$.

The inverse of Joukovski's transformation therefore takes us from the (trivial) flow past a flat plate to that past a circle (less obvious). Thus, we determine very easily the flow past more or less complicated forms. For example, starting from the flow past a circle, by shifting a direct Joukovsky transformation we obtain the flow past a wing profile, also called a Joukovsky profile (see Fig. 3.4).

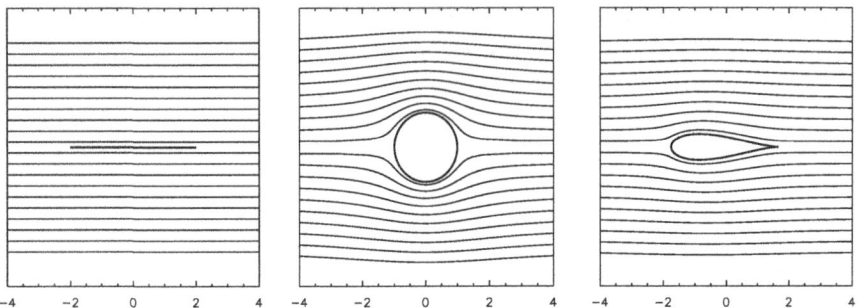

Fig. 3.4 Illustration of possible transformations of a flow past a flat plate. In this example we have first applied an inverse Joukovski transformation which has produced the flow past a circle; then, by application of the slightly shifted Joukovski transformation ($z' = z + c + (1 + c)^2/(z + c)$) one obtains the flow past a wing profile (note that if $c = 0$ the flow past the flat plate is recovered; here $c = -0.17$)

3.3.7 Forces Exerted by a Perfect Fluid

3.3.7.1 d'Alembert's Paradox

Statement:

The steady irrotational flow of an inviscid incompressible fluid around a solid body does not exert any force on it.

Proof:

We assume that the volume occupied by the fluid is simply connected. The solid is supposed to have a constant velocity $\mathbf{V}_s$. The potential satisfies Laplace's equation and the boundary conditions

$$\mathbf{n} \cdot \nabla\phi = \mathbf{n} \cdot \mathbf{V}_s \quad \text{on} \quad S \tag{3.29}$$

$$\phi = \mathcal{O}(1/r^2) \quad \text{if} \quad r \to \infty$$

The second boundary condition results from the properties of the solutions of Laplace's equation (see the mathematical supplement). The force exerted on the solid is just the sum of pressure forces

$$\mathbf{F} = -\int_{(S)} P d\mathbf{S}$$

Using (3.22) we write

$$P = P_\infty - \frac{1}{2}\rho v^2 - \rho \frac{\partial \phi}{\partial t}$$

where P_∞ is the pressure at infinity assumed constant. We calculate first of all the term $\partial \phi / \partial t$ while remarking that in a region attached to the solid $\phi \equiv \phi(x', y', z')$ where

$$x' = x - V_s t, \quad y' = y, \quad z' = z$$

$$\frac{\partial \phi}{\partial t} = -V_s \frac{\partial \phi}{\partial x} = -\mathbf{V}_s \cdot \nabla \phi$$

thus

$$\mathbf{F} = \frac{1}{2}\rho \int_{(S)} v^2 d\mathbf{S} - \rho \int_{(S)} (\mathbf{V}_s \cdot \mathbf{v}) d\mathbf{S} \tag{3.30}$$

Now we examine each component of each of these integrals. In particular,

$$\int_{(S)} v^2 dS_i = \int_{(S \cup S_\infty)} v^2 dS_i$$

where we have introduced a surface S_∞ at infinity which closes the volume of fluid. This is possible and interesting since $\lim_{r \to \infty} v = 0$. We have

$$\frac{1}{2}\int_{(S \cup S_\infty)} v^2 dS_i = \int_{(V)} \frac{1}{2}\partial_i v^2 dV = \int_{(V)} (\mathbf{v} \times \nabla \times \mathbf{v} + \mathbf{v} \cdot \nabla \mathbf{v})_i \, dV = \int_{(V)} v_j \partial_j v_i dV$$

$$= \int_{(V)} \partial_j (v_j v_i) dV = \int_{(S \cup S_\infty)} v_i v_j dS_j$$

$$= \int_{(S)} v_i v_j dS_j = \int_{(S)} v_i V_{sj} dS_j = V_{sj} \int_{(S)} v_i dS_j$$

where we used the boundary conditions (3.29). The second integral in (3.30) also reads

$$V_{sj} \int_{(S)} v_j dS_i \; .$$

Finally

$$F_i = -\rho V_{sj} \left(\int_{(S)} (v_j dS_i - v_i dS_j) \right) = -\rho V_{sj} \left(\int_{(V)} (\partial_i v_j - \partial_j v_i) dV \right)$$

$$= -\rho V_{sj} \left(\int_{(V)} (\partial_i \partial_j \phi - \partial_j \partial_i \phi) dV \right) = 0$$

whence the result.

This shows that a solid body moving in an inviscid fluid is not subjected to any force from the fluid if its motion is uniform. Viscosity is therefore paradoxically an essential element to insure, via the circulation that it induces, the lift of a wing, for example.

3.3.7.2 Case Where the Obstacle is Accelerated

The case of an accelerated body is quite different from the foregoing one and is worth discussing. In a referential attached to the accelerating solid, the flow is now unsteady and subject to an entrainment inertial force but the velocity potential still satisfies $\Delta\Phi = 0$. Therefore the dependence of Φ with respect to time comes from the boundary conditions at infinity where the velocity will be supposedly uniform and of the form $-U(t)\mathbf{e}_z$. One can show from this that the potential of the velocities can be written $\Phi = U(t) f(\mathbf{r})$. The force which is applied to the solid is still the result of the pressure forces, that is

$$\mathbf{F} = -\int_{(S)} P d\mathbf{S}$$

Noting that the entrainment inertial force $(-\rho \mathbf{a}_e = -\rho \nabla \phi_e)$ is derived from a potential, the momentum equation (3.22) reads

$$\frac{\partial \Phi}{\partial t} + \frac{1}{2} \mathbf{v}^2 + \frac{P}{\rho} + \phi_e = \text{Cst}$$

which leads to the following expression for the force exerted on the solid:

$$\mathbf{F} = \int_{(S)} \left(\rho \phi_e + \rho \frac{\partial \Phi}{\partial t} \right) d\mathbf{S} + \int_{(S)} \frac{1}{2} \rho \mathbf{v}^2 d\mathbf{S} \qquad (3.31)$$

where we have separated the term of kinetic energy since it is zero as we shall see now. Indeed,

$$\int_{(S)} \mathbf{v}^2 d\mathbf{S} = \int_{(S \cup S_\infty)} \mathbf{v}^2 d\mathbf{S}$$

because in enclosing the volume by a sphere of infinite radius, the integral remains unchanged since $\mathbf{v} = U(t)\mathbf{e}_z + \mathcal{O}(1/r^3)$. From the calculations of the preceding paragraph, the foregoing integral also reads

$$\int_{(S \cup S_\infty)} \mathbf{v}\,\mathbf{v}\cdot d\mathbf{S}$$

This integral is zero because of the boundary conditions on the solid and because of the form of the velocity at infinity. Finally, the expression for the force is

$$\mathbf{F} = \rho \int_{(S)} \left(\frac{\partial \Phi}{\partial t} + \phi_e \right) d\mathbf{S}$$

so that

$$\mathbf{F} = \rho \dot{U}(t) \int_{(S)} (f + z)\, d\mathbf{S} \tag{3.32}$$

where we have made use of $\phi_e = \dot{U}(t)z$ assuming a motion along the z-axis.

This integral is non-zero in general. This expression therefore shows that a solid having accelerated motion amidst the fluid, even if inviscid, sustains a force from the fluid. This force is at the origin of all swimming strokes: propulsion in the water is, in fact, efficient only if the solid accelerates with respect to the fluid. For this reason the motion of the fins of a fish is in perpetual acceleration (oscillating motion).

As an example we may calculate the force sustained by a sphere accelerated within of a perfect fluid with constant density. To determine the function f in (3.32) we must solve Laplace's equation in this particular case. In three dimensions and in this geometry we use the expansion of the solution in Legendre's polynomials.

$$\Phi = -U(t)r\cos\theta + \sum_{\ell=0}^{+\infty} \frac{A_\ell(t)}{r^{\ell+1}} P_\ell(\cos\theta) \tag{3.33}$$

where we have taken into account the boundary condition at infinity and the fact that the flow is axisymmetric with respect to the z-axis. The boundary conditions on the sphere, assumed to have radius R, give the functions $A_\ell(t)$. At $r = R$, $\mathbf{v}\cdot\mathbf{e}_r = 0$ so that

$$\left(\frac{\partial \Phi}{\partial r} \right)_{r=R} = 0 = -U(t)\cos\theta + \sum_{\ell=0}^{+\infty} -\frac{(\ell+1)A_\ell(t)}{R^{\ell+2}} P_\ell(\cos\theta)$$

This expression shows[5] that the A_ℓ are all zero except A_1, and

$$A_1(t) = -R^3 U(t)/2 \tag{3.34}$$

Finally, from (3.33)

$$\Phi(r, \theta, t) = U(t) \left(-r - \frac{R^3}{2r^2} \right) \cos \theta$$

and from (3.32)

$$\mathbf{F} = -\rho \int_{(S)} \dot{U}(t) R/2 \cos \theta \, d\mathbf{S} \quad \Longleftrightarrow \quad \mathbf{F} = -\frac{2\pi}{3} R^3 \rho \dot{U}(t) \mathbf{e}_z$$

The factor $\frac{2\pi}{3} R^3 \rho$ is a mass. It is often called the *added mass* because if we exert a force upon the sphere, the latter reacts as if its mass had increased by this quantity (which is equal in this case to half of the mass of the displaced fluid).

3.3.7.3 Drag and Lift of Two-Dimensional Flows

In the foregoing example we assumed that the volume occupied by the fluid was simply connected and therefore its flow was without circulation. In two dimensions, however, the presence of an obstacle makes the fluid "volume" automatically doubly connected and therefore, even if the flow is irrotational, one can have circulation along certain contours.

We shall now consider the same problem as in Sect. 3.3.7.1 but in two dimensions. We assume that the curves surrounding the obstacle possess a circulation Γ. Let us consider a region attached to the solid and assume that the velocity is uniform at infinity:

$$\mathbf{v}_\infty = V_0 \mathbf{e}_x$$

The solution that we are looking for is a solution of Laplace's equation which satisfies $\mathbf{n} \cdot \nabla \Phi = 0$ on the contour of the solid and $\nabla \Phi = V_0 \mathbf{e}_x$ at infinity. The general solution of this type of problem is:

$$\Phi = V_0 r \cos \theta + \frac{\Gamma \theta}{2\pi} + \sum_{n=0}^{\infty} A_n \frac{e^{in\theta}}{r^n}$$

---

[5]We just need to project the equation on Legendre's polynomials and to use their orthogonality with respect to the scalar product $\int_0^\pi P_\ell(\cos \theta) P_k(\cos \theta) d \cos \theta \propto \delta_{\ell k}$.

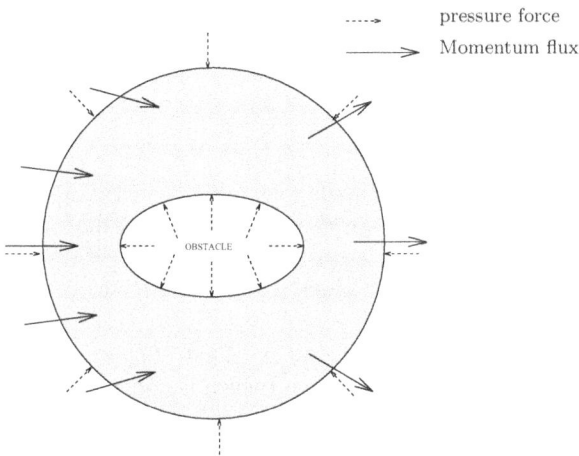

pressure force

Momentum flux

Fig. 3.5 At equilibrium, the sum of the forces and momentum flux is zero, hence $\mathbf{F}_{\text{obst/fluid}} + \mathbf{F}_{\text{press}} + \mathbf{F}_{\text{mom. flux}} = \mathbf{0}$. The force applied to the solid is $-\mathbf{F}_{\text{obst/fluid}} = \mathbf{F}_{\text{press}} + \mathbf{F}_{\text{mom. flux}}$

where the sum represents the multipolar terms that must be added in order to account for the precise shape of the solid.

The associated velocity field is

$$\mathbf{v} = \boldsymbol{\nabla}\Phi = V_0 \cos\theta \mathbf{e}_r + (\frac{\Gamma}{2\pi r} - V_0 \sin\theta)\mathbf{e}_\theta + \cdots$$

If we wish to find the force which is exerted on the solid, a simple method consists in writing the balance of forces and momentum flux that are exerted on a circle surrounding the obstacle at a distance R (see Fig. 3.5). The momentum flux on entry is given by

$$-\int_0^{2\pi} \rho \mathbf{v} v_r R d\theta = -\frac{\Gamma V_0 \rho}{2}\mathbf{e}_y \tag{3.35}$$

while the resultant of the pressure forces is

$$\mathbf{F}_p = -\int_0^{2\pi} P\mathbf{e}_r R d\theta$$

which we calculate using Bernoulli's theorem for an irrotational flow. Equation (3.22) yields

$$P = P_0 - \frac{1}{2}\rho v^2 \quad \text{and} \quad v^2 = V_0^2 - \frac{\Gamma}{\pi r}V_0 \sin\theta + \cdots$$

where the dots represent the multipolar terms. The calculation of the integral does not present any difficulty; we find that

$$\mathbf{F}_p = -\frac{\Gamma V_0 \rho}{2}\mathbf{e}_y \tag{3.36}$$

When we let R tend to infinity, the multipolar terms contribution vanishes and only one term remains. Finally, adding (3.35) and (3.36) we find the total force

$$\mathbf{F} = -\Gamma V_0 \rho \mathbf{e}_y = -\rho \mathbf{\Gamma} \times \mathbf{V}_0 \tag{3.37}$$

where $\mathbf{\Gamma} = \Gamma \mathbf{e}_z$. The force just found is called *Magnus' Force*. We see that depending on the sense of the circulation (which is connected to the shape or to the sense of rotation of the body when there is viscosity), the force is directed either upwards or downwards. It is this same force which is responsible for the trajectory of ping-pong balls or tennis balls when they are sliced, and for the lift on wings. Formulae (3.35)–(3.37) are obtained in a two-dimensional space so that the forces are actually forces per unit length. Equation (3.37) leads to the true Magnus force exerted on cylinder of length L by a simple multiplication by L, namely $\mathbf{F} = -\rho L \mathbf{\Gamma} \times \mathbf{V}_0$.

We further note that this force is perpendicular to the motion, consequently there is no resistance to the forward motion *or drag force*.

We could stop here and say that we need the effects of viscosity to calculate the circulation and therefore the lift. Quite surprisingly, this calculation is not necessary for the following reason: when we take into account the effects of viscosity we superimpose upon the irrotational flow the boundary layer corrections which allows the complete solution to verify all the boundary conditions (see next chapter). Actually, we may easily realize (see appendix at the end of this chapter) that the irrotational flow around the profile of a wing has a singularity in the velocity (which becomes infinite) at the trailing edge if the circulation is not adapted. The real flow (with viscosity), which should have for limit this singular irrotational flow, would be very unstable. The problem resolves itself when we observe that for a given circulation, this singularity disappears. This particular value of Γ is that which brings the second stagnation point[6] to the trailing edge (see Fig. 3.6). This condition is usually called *Kutta's Condition*.[7] For a wing profile where the angle of attack is α, we find (see appendix) that:

$$\Gamma = \pi \ell V_0 \sin \alpha \tag{3.38}$$

where ℓ is the wing chord (i.e. it's width).

---

[6]Point on the solid where the fluid's velocity is zero.

[7]This condition was also found independently by Joukovski in 1906 and is also called sometimes Joukovski's Condition.

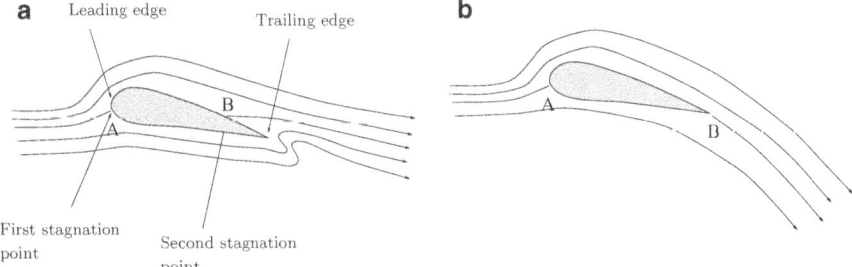

Fig. 3.6 (**a**) A and B are the two stagnation points. On this figure, the circulation is zero and the second stagnation point is located upstream of the trailing edge where the velocity has a singularity. In (**b**) the circulation is such that the trailing edge and the second stagnation point coincide; the velocity is finite everywhere

3.4 Flows with Vorticity

After the irrotational flows, the following step takes us naturally towards flows that own vorticity. These flows are more complex than the preceding ones because the distribution of vorticity is affected by the flow that the vorticity produces. The problem therefore becomes largely nonlinear (we no longer have the equation of the velocity potential $\Delta \Phi = 0$) and consequently only a small number of problems have analytical solutions. We now present the most classic examples.

3.4.1 The Dynamics of Vorticity

In all what follows we call $\boldsymbol{\omega} = \nabla \times \mathbf{v}$ the vorticity. The equation of this quantity is obtained by taking the curl of Euler's equation (1.33) which is made explicit using the following vector equality

$$\nabla \times (\mathbf{v} \cdot \nabla \mathbf{v}) = \nabla \times (\boldsymbol{\omega} \times \mathbf{v}) = (\mathbf{v} \cdot \nabla)\boldsymbol{\omega} - (\boldsymbol{\omega} \cdot \nabla)\mathbf{v} + (\nabla \cdot \mathbf{v})\boldsymbol{\omega}$$

We thus find that the vorticity satisfies:

$$\frac{D\boldsymbol{\omega}}{Dt} = (\boldsymbol{\omega} \cdot \nabla)\mathbf{v} - (\nabla \cdot \mathbf{v})\boldsymbol{\omega} + \frac{1}{\rho^2}\nabla \rho \times \nabla P \qquad (3.39)$$

This equation calls for several comments. In the first place, we note that the variations of $\boldsymbol{\omega}$ in a fluid particle result from three different sources:

1. $(\boldsymbol{\omega} \cdot \nabla)\mathbf{v}$ which is a term of *stretching-pivoting*: in order to understand its effect, we take the following simple example where $\boldsymbol{\omega}$ is parallel to $\mathbf{e}_z$ and $\mathbf{v}$ represents a shear along z (see Fig. 3.7). The equation $\frac{D\boldsymbol{\omega}}{Dt} = (\boldsymbol{\omega} \cdot \nabla)\mathbf{v}$ becomes $\frac{D\boldsymbol{\omega}}{Dt} = \omega \frac{\partial \mathbf{v}}{\partial z}$.

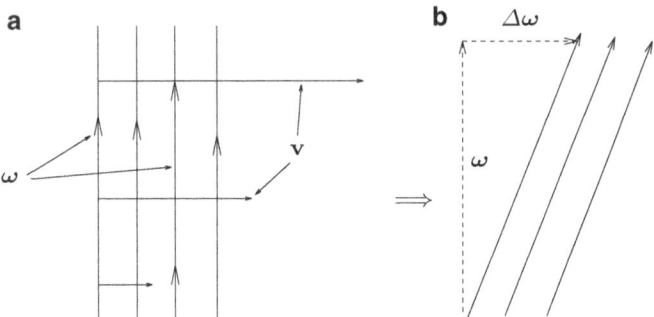

Fig. 3.7 Evolution of the vorticity field subject to a shear flow: from (a) to (b) the vorticity gains a component along the velocity field

It shows that, following each fluid particle, vorticity is created parallel to **v** as Fig. 3.7b shows. We will find again such a term when we analyse the evolution of the magnetic field in a fluid with electrical conductivity (Chap. 10).

2. $-(\nabla \cdot \mathbf{v})\,\boldsymbol{\omega}$. We have seen in Chap. 1 the physical meaning of $\nabla \cdot \mathbf{v}$; it represents the volume variations of the fluid elements. This term thus translates the variation in vorticity associated with these variations of volume: if the particle contracts its vorticity increases. Vorticity is created in the same direction and in proportion to the existing one.

3. $\frac{1}{\rho^2}\nabla\rho \times \nabla P$ is the baroclinic torque. This term does not exist (we noted it many times) if $P \equiv P(\rho)$. When it is present, the fluid elements can acquire vorticity, and thus angular momentum, because the pressure force then exerts a torque on them (see Fig. 3.8).

Let us now come back to the barotropic case where $P \equiv P(\rho)$. Equation (3.39) simplifies into

$$\frac{D\boldsymbol{\omega}}{Dt} = (\boldsymbol{\omega} \cdot \nabla)\mathbf{v} - \boldsymbol{\omega}\nabla \cdot \mathbf{v} \tag{3.40}$$

This equation shows that if initially, $\boldsymbol{\omega} = \mathbf{0}$ then $\boldsymbol{\omega}$ remains zero: vorticity cannot be created. This result is, of course, another version of Lagrange's Theorem (see Sect. 3.3.3).

In two dimensions, equation (3.40) takes a very remarkable form if the fluid is incompressible. Indeed, in this case the right-hand side is zero and

$$\frac{D\omega}{Dt} = 0 \tag{3.41}$$

where $\omega = \omega_z$ is the only non-zero component of $\boldsymbol{\omega}$.

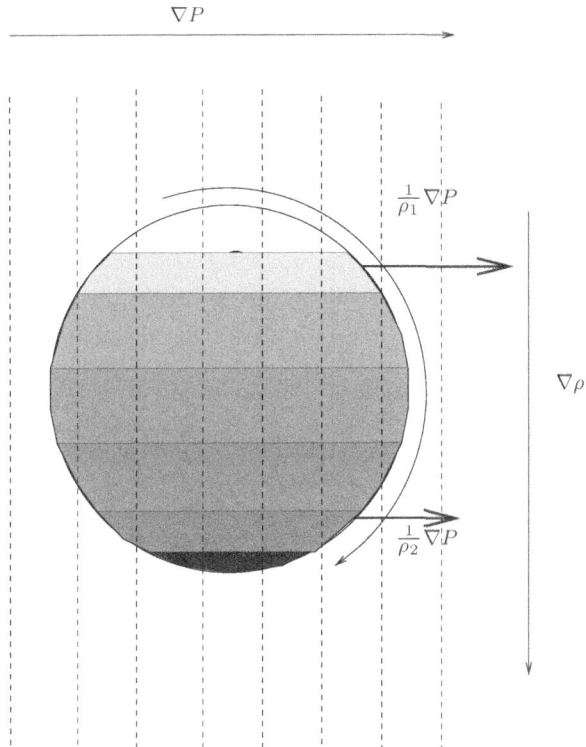

Fig. 3.8 Generation of vorticity by baroclinicity. Density increases towards the *bottom* of the sphere, thus the pressure force per unit mass ($\frac{1}{\rho_1}\nabla P$) is larger than $\frac{1}{\rho_2}\nabla P$. The resulting specific pressure force thus exerts a torque on the fluid element

This equation shows that in this case ω is a Lagrangian invariant. It implies Kelvin's theorem, but also

$$\frac{D\omega^n}{Dt} = 0 \qquad \Longleftrightarrow \qquad \int_{(S)} \omega^n dS = \text{Cst} \tag{3.42}$$

for all n, S corresponding to a surface advected by the fluid. We shall return to this equation when we study turbulence in two dimensions.

3.4.2 Flow Generated by a Distribution of Vorticity: Analogy with Magnetism

Let's imagine that the distribution of vorticity is given in the space occupied by the fluid. It is then easy to find the distribution of the associated velocity; it is sufficient to solve the equation for $\mathbf{v}$

$$\nabla \times \mathbf{v} = \boldsymbol{\omega}$$

where $\boldsymbol{\omega}$ is given. This equation, which is linear, strongly resembles Ampère's equation:

$$\nabla \times \mathbf{B} = \mu_0 \mathbf{j}$$

where $\mathbf{B}$ is the magnetic field, $\mathbf{j}$ the volumic current density and μ_0 the permittivity of vacuum. Ampère's equation can be solved quasi-analytically, but for this we must use the vector potential $\mathbf{A}$ such that $\mathbf{B} = \nabla \times \mathbf{A}$. The transposition of these results to fluid mechanics demands therefore $\nabla \cdot \mathbf{v} = 0$, that is to say, that we need to restrict ourselves to incompressible fluids. In such a case, just as we solve Ampère's equation with

$$\mathbf{A}(\mathbf{r}) = \frac{\mu_0}{4\pi} \int_{(V)} \frac{\mathbf{j}(\mathbf{r}')}{\|\mathbf{r} - \mathbf{r}'\|} dx'dy'dz'$$

and

$$\mathbf{B} = -\frac{\mu_0}{4\pi} \int_{(V)} \frac{(\mathbf{r} - \mathbf{r}') \times \mathbf{j}(\mathbf{r}')}{\|\mathbf{r} - \mathbf{r}'\|^3} dx'dy'dz'$$

which is Biot and Savart's law, we have for the velocity field:

$$\mathbf{v}(\mathbf{r}) = -\frac{1}{4\pi} \int_{(V)} \frac{(\mathbf{r} - \mathbf{r}') \times \boldsymbol{\omega}(\mathbf{r}')}{\|\mathbf{r} - \mathbf{r}'\|^3} d^3\mathbf{r}' \qquad (3.43)$$

Contrary to the magnetic case, this solution is not the end of the problem because the velocity field thus created modifies the vorticity field by way of the equation (3.40). Problems therefore have simple solutions if the distribution of vorticity is invariant by the advection that it generates. In particular, we can look for a necessary condition for steady flows to be possible. From Euler's equation, assuming that $\mathbf{v}$ is independent of time, we get

$$\boldsymbol{\omega} \times \mathbf{v} = -\nabla q, \qquad q = \frac{1}{2}v^2 + \int \frac{dP}{\rho} \qquad (3.44)$$

where we assumed the fluid to be barotropic. According to this equation

$$\mathbf{v} \cdot \nabla q = \boldsymbol{\omega} \cdot \nabla q = 0 ,$$

which means that the flow lines and the vorticity lines are on the surfaces $q = \text{Cst}$. If the flow is two-dimensional, the velocity is expressed with a stream function ψ

such that

$$\mathbf{v} = \nabla \times (\psi \mathbf{e}_z), \qquad \boldsymbol{\omega} = \nabla \times \nabla \times (\psi \mathbf{e}_z) = -\Delta \psi \mathbf{e}_z.$$

We can then transform (3.44) into

$$\Delta \psi \nabla \psi = \nabla q \quad \Longrightarrow \quad \nabla \Delta \psi \times \nabla \psi = \mathbf{0}$$

which shows that

$$\Delta \psi = F(\psi) \tag{3.45}$$

where F is an arbitrary function. We are now going to tackle some examples in this category.

3.4.3   Examples of Vortex Flows

3.4.3.1   Vortex Sheets

The first example of vortex flows is also the simplest; it concerns the *shear layer* also called the *vortex sheet*: it corresponds to a simple discontinuity in the tangential component of the velocity field, as shown in Fig. 3.9a. It is easy to see that a contour, such as that drawn in Fig. 3.9a, has a circulation; if the length of the longer side is L, the circulation is given by $\Gamma = (V_2 - V_1)L$.

We shall see in Chap. 6 that such a sheet is always unstable. This instability produces individualized vortices such as the vortex ring when the vortex sheet rolls up as indicated in the sketch of Fig. 3.9b under the impulsive motion of the piston.

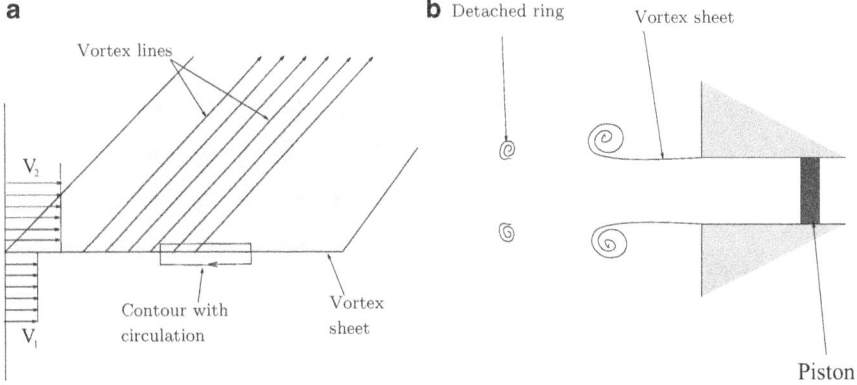

Fig. 3.9 (**a**) Vortex sheet. (**b**) Schematic view of the formation of a vortex ring from a vortex sheet

3.4.3.2 Rankine's Vortex

This is the most simple of the vortex flows. It is made up of a cylindrical kernel in which the vorticity is uniform, and out of which the flow is irrotational. The associated velocity field is then

$$
\begin{cases}
\boldsymbol{\omega} = \omega \mathbf{e}_z & s \le a & \Longrightarrow & \mathbf{v} = \tfrac{1}{2}\boldsymbol{\omega} \times \mathbf{r} & s \le a \\[2mm]
\boldsymbol{\omega} = \mathbf{0} & s > a & \Longrightarrow & \mathbf{v} = \frac{\omega a^2}{2s}\mathbf{e}_\varphi & s > a
\end{cases}
\tag{3.46}
$$

where a is the radius of the cylinder and (s, φ, z) are the cylindrical coordinates. We observe that the velocity field is purely azimuthal (only the component along $\mathbf{e}_\varphi$ is non-zero) and therefore the distribution of vorticity does not change with time. The velocity field on the outside of the core has been chosen such that the velocity is continuous at $r = a$.

Rankine's vortex is a very simplified model of the flow generated by a cyclone. We easily show that the pressure passes through a minimum in the centre of such a vortex (see exercises).

3.4.3.3 Hill's Vortex

Another exact solution of Euler's stationary equation consists in distributing the vorticity within a sphere in the following manner:

$$
\boldsymbol{\omega} = \frac{\omega r \sin \theta}{a}\mathbf{e}_\varphi \quad \text{if} \quad r \le a, \qquad \boldsymbol{\omega} = \mathbf{0} \quad \text{if} \quad r > a
$$

where (r, θ, φ) are the spherical coordinates. We thus formulate Hill's vortex which moves at constant velocity without being deformed (see Fig. 3.10). We can explain this property by first examining the velocity field of this vortex.

The components v_r and v_θ of the velocity field obey the two following equations:

$$
\begin{cases}
\dfrac{1}{r}\dfrac{\partial}{\partial r}(r v_\theta) - \dfrac{1}{r}\dfrac{\partial v_r}{\partial \theta} = \dfrac{\omega}{a}r \sin \theta \\[4mm]
\dfrac{1}{r^2}\dfrac{\partial}{\partial r}(r^2 v_r) + \dfrac{1}{r \sin \theta}\dfrac{\partial \sin \theta v_\theta}{\partial \theta} = 0
\end{cases}
\tag{3.47}
$$

which express respectively $\nabla \times \mathbf{v} = \boldsymbol{\omega}$ and $\nabla \cdot \mathbf{v} = 0$. We are looking for a solution to this system in the form:

$$
v_r = f(r) \cos \theta \quad \text{and} \quad v_\theta = g(r) \sin \theta
$$

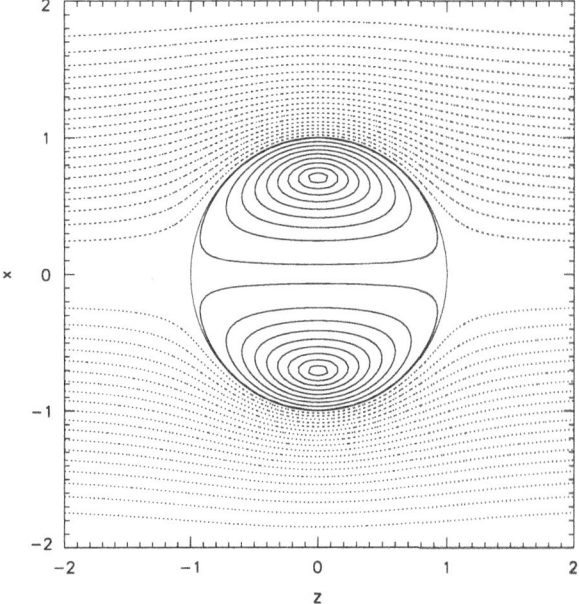

Fig. 3.10 Meridian streamlines associated with Hill's vortex. The *dotted lines* represent the irrotational flow

The equation of continuity yields:

$$g(r) = -\frac{1}{2r}\frac{d}{dr}(r^2 f)$$

The other equation gives the equation verified by f:

$$\frac{d^2}{dr^2}(r^2 f) - 2f = -2\omega r^2/a$$

the solution of which is:

$$f(r) = -\frac{\omega}{5a}r^2 + A + B/r^3$$

The two constants A and B are such that the velocity is regular at the centre of the sphere (so that $B = 0$) and that the radial velocity vanishes at $r = a$. Thus we get:

$$v_r = \frac{\omega}{5a}(a^2 - r^2)\cos\theta \qquad \text{and} \qquad v_\theta = \frac{\omega}{5a}(2r^2 - a^2)\sin\theta \qquad \text{for} \quad r \leq a$$

$$(3.48)$$

We note that on the bounding sphere $v_\theta = \omega a/5 \sin\theta \neq 0$. Outside the sphere the flow is irrotational and the constants of integration must be adjusted such that the velocity field be continuous on the sphere and regular at infinity. The velocity potential being solution of Laplace's equation we find that

$$\Phi(r, \theta) = (A'r + B'/r^2)\cos\theta$$

The boundary conditions $v_r(a) = 0$ and $v_\theta(a) = \omega a/5 \sin\theta$ allow the calculation of A' and B' and we thus infer the velocity field:

$$v_r = \frac{2\omega a}{15}\left(-1 + \left(\frac{a}{r}\right)^3\right)\cos\theta \quad \text{and} \quad v_\theta = \frac{2\omega a}{15}\left(1 + \frac{1}{2}\left(\frac{a}{r}\right)^3\right)\sin\theta \quad \text{for} \quad r > a$$

(3.49)

The remarkable feature in these expressions is the existence of a non-zero velocity at infinity. This velocity represents the velocity of the vortex with respect to the fluid at infinity; it is uniform and along the vortex axis. Its magnitude is:

$$V = \frac{2\omega a}{15} \tag{3.50}$$

The equations for the velocity field also provide the expression for the stream function inside and outside the vortex. For an axisymmetric flow, one notes that:

$$v_r = \frac{1}{r^2 \sin\theta}\frac{\partial\psi}{\partial\theta} \quad \text{and} \quad v_\theta = -\frac{1}{r \sin\theta}\frac{\partial\psi}{\partial r}$$

whence, the following two expressions:

$$\psi = \frac{\omega r^2}{10a}(a^2 - r^2)\sin^2\theta \quad \text{if } r \leq a \quad \text{and} \quad \psi = \frac{\omega a r^2}{15}\left(-1 + \left(\frac{a}{r}\right)^3\right)\sin^2\theta \quad \text{if } r > a$$

These two stream functions give the shape of the streamlines shown in Fig. 3.10.

3.4.3.4 The Vortex Ring

The vortex ring is a spectacular figure of a fluid motion usually known as the smoke ring (see Fig. 3.11). In fact this is a vortex filament that is closed on itself and forms a circular ring, hence the name. Around it, the flow is irrotational and can be calculated with the formula (3.43). The ring being axisymmetric, the velocity is the same at all of its points and thus its motion is a uniform translation. The exact calculation of its velocity can be performed if one assumes a finite interior

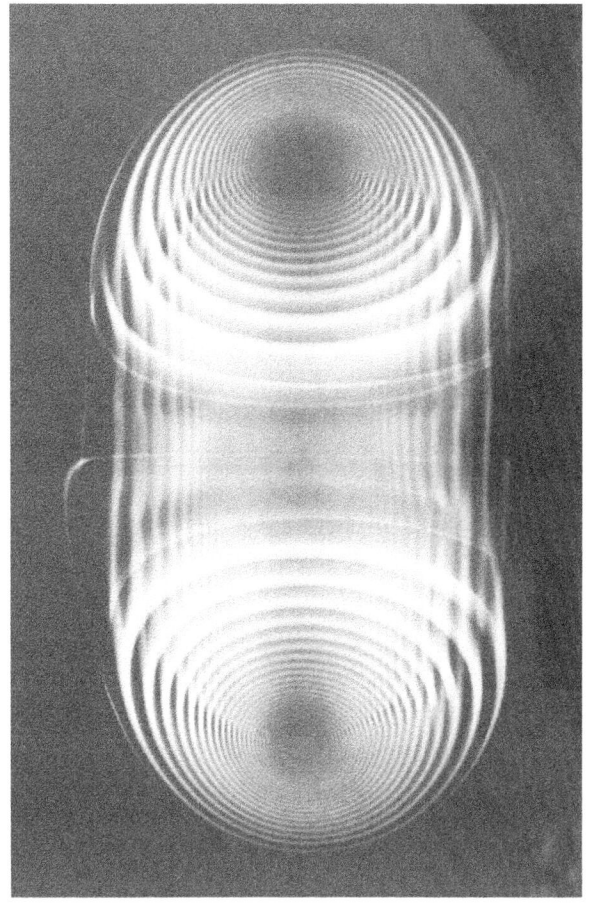

radius, but is quite lengthy and we shall limit ourselves to deriving an approximate expression of it. The velocity induced by the filament is, according to (3.43),

$$\mathbf{v} = \frac{1}{4\pi} \int_{(V)} \frac{\mathbf{r} \times \boldsymbol{\omega}}{r^3} dV \tag{3.51}$$

where we have located the origin of the coordinate system on the filament (see Fig. 3.12). The ring is assumed to be a torus of major radius R and minor radius a, with $a \ll R$. One can note that

$$r(\theta) = 2R \sin \theta, \qquad \boldsymbol{\omega}(\theta) = \omega(\cos \theta \mathbf{e}_r + \sin \theta \mathbf{e}_\theta)$$

and

$$dV = \pi a^2 R d\theta'$$

Fig. 3.12 Sketch of the vortex ring. Note that with this representation the equation of the *circle* is $r = 2R \sin \theta$

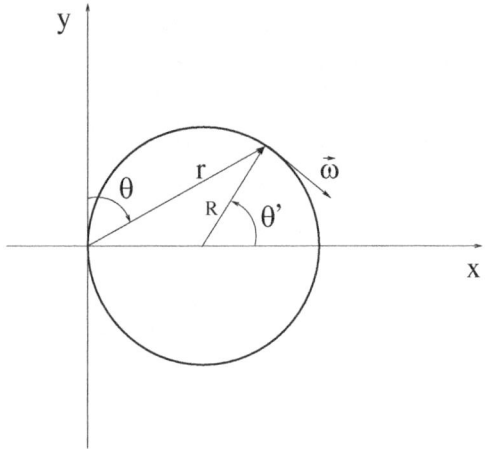

where θ' is the angle measured from the centre of the torus so that $\theta' = \pi - 2\theta$. Hence,

$$\mathbf{v} = \frac{\omega a^2}{8R} \int_0^\pi \frac{d\theta}{\sin \theta} \, \mathbf{e}_z$$

If one recalls that

$$\int \frac{d\theta}{\sin \theta} = \ln \tan \theta/2$$

it appears that the integral diverges at 0 and π. In fact, we have not accounted in this calculation for the fact that the core section is finite and that this effect is important for the points near the origin. An exact integration would involve elliptic integrals which are cumbersome to deal with. We thus simply estimate the order of magnitude of the integral by assuming that the integration domains is $[\varepsilon, \pi - \varepsilon]$ with $\varepsilon \approx a/R$. One finds

$$\mathbf{v} \approx \frac{\Gamma}{4\pi R} \ln(2R/a)\mathbf{e}_z$$

while the exact formula is:

$$\mathbf{v} = \frac{\Gamma}{4\pi R} \left(\ln \frac{8R}{a} - \frac{1}{4} \right) \mathbf{e}_z \qquad (3.52)$$

These two expressions have the same asymptotic behavior as $R \to \infty$ or $a \to 0$. Our derivation indicates that the logarithmic singularity is due to the regions that are the closest to the calculation point.

3.5   Problems

1. *Streamlines and velocity equipotentials*
   Show that for an irrotational plane flow of an incompressible fluid, the streamlines are orthogonal to the potential lines.
2. *Flow in a narrowing duct*

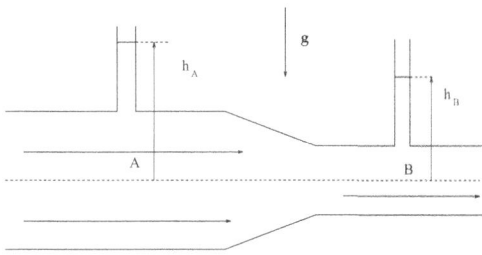

   The flow is assumed steady and horizontal between points A and B. Show that along the z-axis, hydrostatic equilibrium is satisfied. Derive from this equilibrium the relation between P_A and h_A. Calculate the difference $h_A - h_B$ in terms of V_A and V_B, assuming an incompressible fluid. What relation holds between V_A and V_B and the cross sections of the pipe S_A and S_B?
3. *Rankine's vortex*
   Let $\mathbf{v}$ be the velocity field of a fluid of constant density ρ:

$$\begin{cases} \mathbf{v} = s\Omega \mathbf{e}_\varphi & s < a \\ \mathbf{v} = \Omega a^2 \mathbf{e}_\varphi / s & s \geq a \end{cases}$$

   where (s, φ, z) are the cylindrical coordinates.

   (a) Show that the flow is irrotational outside the cylinder of radius a.
   (b) Give the expression for the pressure in each of the subdomains. At infinity, $P = P_\infty$.
   (c) What can be said about the quantity $\frac{1}{2}v^2 + P/\rho$ in each of the subdomains? What can be concluded?
   (d) Calculate the minimum pressure at the centre of a storm with winds blowing at a maximum velocity of 50 m/s (180 km/h).
   (e) If the vortex is located over the ocean, find from the previous results the shape of the ocean surface.

4. *Purge of a tank*

(a) Lets consider a water tank (assumed to be an inviscid, incompressible fluid), of cross section S with initial level h_0. A valve of cross section s ($s \ll S$) located at the bottom of the tank is open.

   i. Show that the flow is irrotational.
   ii. Assuming quasi-steady flow, derive the differential equation governing $h(t)$, and solve it. Find the time it will take to empty the tank.
   iii. Show "a posteriori" that the time derivatives are indeed negligible.

(b) One now adds to the reservoir a horizontal pipe of length ℓ, and of very small cross section compared to that of the tank. The tank is filled to level h_0 which is kept constant with time. The fluid is initially at rest. At $t = 0$ the valve at B (see figure) is opened.

   i. Derive the equation of motion of the fluid in the pipe. One assumes that the pressure in the exit jet is equal to the atmospheric pressure; solve the differential equation governing the exit velocity. Let's denote by $v_\infty = \sqrt{2gh_0}$.
   ii. The city water utility pressure is 6 bars; if the length of the connecting pipe from the main pipe to the sink is 10 m, what is the transient time when you open the tap?

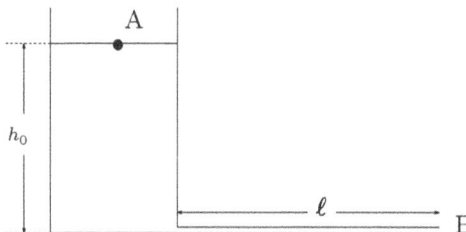

5. A U-tube contains an *incompressible* fluid subject to the gravity field $\mathbf{g} = -g\mathbf{e}_z$. The tube diameter is constant and very small compared to its length. The fluid level at equilibrium is $z = h_0$ and the free surface is at atmospheric pressure. ρ is the fluid density.

(a) We are interested in the small oscillations of the fluid height about the average value h_0; these oscillations occur for example when the tube is slightly shaken. The fluid is assumed *perfect*. Explain why the fluid motion is necessarily *irrotational*. What can be said about the velocity inside the tube?

(b) If Φ_A and Φ_B are the values of the velocity potential at the first and second free surfaces of the fluid, L the length of the wetted part of the tube and V the fluid velocity, show that

$$\Phi_A - \Phi_B = LV$$

(c) Derive from this the differential equation governing the time dependent height perturbation δh of the fluid in one of the branches of the tube.

6. *Motion of a liquid near an air bubble*

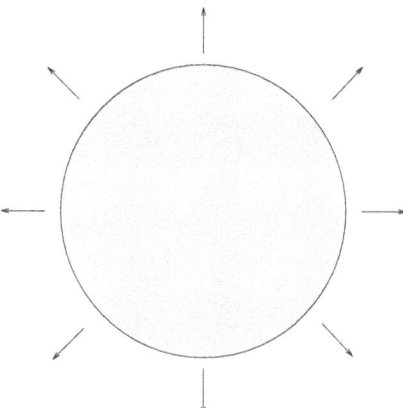

We assume that the liquid has a radial motion: $\mathbf{v} = v(r,t)\mathbf{e}_r$.

(a) Show that the liquid's flow is irrotational.
(b) Derive the expression for $v(r,t)$ in terms of the bubble radius $R(t)$.
(c) We assume that the air inside the bubble is an ideal gas which follows an isentropic transformation when the bubble radius varies. Neglecting the air flow, give the expression of the pressure inside the bubble in terms of the radius.
(d) Give the evolution equation of $R(t)$ (let P_0 be the value of the pressure at infinity and R_0 the radius of the bubble when $p = P_0$).
(e) If one supposes that the bubble radius oscillates slightly about the equilibrium value R_0, derive the expression for $R(t)$. What is the frequency f of such small oscillations?
(f) Numerical application: calculate f for $R_0 = 1\,\mathrm{mm}$ and $R_0 = 5\,\mathrm{mm}$; we give $\gamma_{\mathrm{air}} = 1.4$, $\rho_{\mathrm{water}} = 10^3\,\mathrm{kg/m^3}$, $P_0 = 10^5\,\mathrm{Pa}$.

7. Show that the potential vorticity of an inviscid compressible fluid, defined by ω/ρ, is governed by the equation

$$\mathcal{H}\left(\frac{\omega}{\rho}\right) = \mathbf{0} \tag{3.53}$$

where $\mathcal{H}$ is the "Helmholtzian" defined by

$$\mathcal{H}(\mathbf{a}) = \frac{\partial \mathbf{a}}{\partial t} + \mathbf{v} \cdot \nabla \mathbf{a} - (\mathbf{a} \cdot \nabla)\mathbf{v} \tag{3.54}$$

Appendix: Flow Past a Plane at Incidence

When we discussed the complex potential, we remarked that the Joukovski transformation can be used to transform a circle into a flat plate. In a previous example,
we used the Joukovski transformation to find the flow past a circle from the (trivial)
solution of the flow past a flat plate when the velocity is parallel to it.

Now we can do the opposite. Indeed, it is less obvious to find the solution of the
flow with circulation Γ past a flat plate at incidence α with respect to the flow
at infinity. Conversely, if we consider the circle, we have seen that the velocity
potential is:

$$\Phi = VRe(z + \frac{R^2}{z}) = V\left(r\cos\theta + \frac{R^2\cos\theta}{r}\right)$$

It is easy to add circulation to this flow since a potential vortex will still satisfy
the boundary conditions; it is also possible to rotate the incoming flow velocity by
an angle α with respect to the axes. With these changes, the velocity potential now
reads:

$$\Phi = V\left(r\cos(\theta - \alpha) + \frac{\Gamma(\theta - \alpha)}{2\pi V} + \frac{R^2\cos(\theta - \alpha)}{r}\right)$$

This is nothing but the real part of the complex velocity potential:

$$F(z) = Vze^{-i\alpha} + \frac{\Gamma}{2i\pi}\ln z + VR^2e^{i\alpha}/z$$

where we have overlooked the constants. From this expression, one obtains the
complex velocity in the image plane:

$$w' = \left(Ve^{-i\alpha} + \frac{\Gamma}{2i\pi z} - \frac{VR^2e^{i\alpha}}{z^2}\right)\left(1 - \frac{R^2}{z^2}\right)^{-1} \tag{3.55}$$

This expression is particular in the sense that it provides the velocity components
in the image plane (where the obstacle is a flat plate at incidence) in terms of the
coordinates in the initial plane (where the obstacle is a circle). To obtain w' in terms
of z', it would be necessary to invert the relation $z' = z + R^2/z$ and to substitute the
result into (3.55). But our goal is somewhat different: we only wish to examine the
singularities in the flow past the obstacle and find the condition for Γ to eliminate
them.

The points with z' on the flat plate correspond to z being on the circle, that is $z = Re^{i\theta}$, $\theta \in [0, 2\pi]$. Along the flat plate w' is given by:

$$w'(\theta) = \left(Ve^{-i\alpha} + \frac{\Gamma e^{-i\theta}}{2i\pi R} - Ve^{i(\alpha - 2\theta)} \right) \left(1 - e^{-2i\theta} \right)^{-1}$$

$$= \left(Ve^{-i(\alpha - \theta)} + \frac{\Gamma}{2i\pi R} - Ve^{i(\alpha - \theta)} \right) \left(e^{i\theta} - e^{-i\theta} \right)^{-1}$$

$$= \left(V\sin(\theta - \alpha) - \frac{\Gamma}{4\pi R} \right) \Big/ \sin\theta$$

This expression is singular at $\theta = 0$ and $\theta = \pi$ if Γ is arbitrary. The trailing edge corresponds to $\theta = \pi$; we see that the singularity disappears if the circulation is chosen such that:

$$\Gamma = 4\pi R V \sin\alpha$$

One recovers here the expression (3.38) remembering that the plate length is $4R$. One notices that the flow at the leading edge is also singular, but this singularity can be eliminated by rounding the profile as shown in Fig. 3.4c.

Further Reading

The theory of irrotational flows is often well developed in standard textbooks; one can refer to Batchelor (1967). With regards to the dynamics of vorticity, further developments to the notes presented here will be found in Saffman *Vortex dynamics* (1992) or in Ting and Klein *Viscous vortical flows* (1991). On the properties of the Euler equation, extended material is proposed in Zeytounian *Mécanique des fluides fondamentale* (1991).

References

Batchelor, G. K. (1967). *An introduction to fluid dynamics*. Cambridge: Cambridge University Press.
Magarvey, R. & MacLatchy, C. (1964) The formation and structure of vortex rings *Can. J. Phys.* **42**, 678–683
Saffman, P. (1992). *Vortex dynamics*. Cambridge: Cambridge University Press.
Ting, L., & Klein, R. (1991). *Viscous vortical flows. Lecture Notes in Physics*. Berlin: Springer.
Zeytounian, R. (1991). *Mécanique des fluides fondamentale. Lecture Notes in Physics*. Berlin: Springer.

Chapter 4
Flows of Incompressible Viscous Fluids

As it was shown in Sect. 3.2.5, the density variations in a fluid flow decrease with the square of the Mach number (the ratio of the fluid velocity to the sound speed). Hence, for many fluid flows, and especially for those of liquids, incompressibility is an excellent approximation. Moreover, it simplifies very much the equations of motion. This simplification provides us with the easiest context to study the effects of viscosity that we have neglected until now.

Thus, in this chapter we study the effects of viscosity using solely incompressible fluids. We first discuss the laws of similarity, which appear thanks to viscosity, then we deal with two limits: that of flows with a strong viscous force and that of flows with a slight viscous effect. Next, we review some classical solutions of Navier equation, and we end the chapter with a short study of forces exerted on solids by viscous fluid flows.

4.1 Some General Properties

4.1.1 The Equations of Motion

We have seen in the first chapter that the flow of an incompressible fluid with constant viscosity is governed by only two equations: the equation of continuity and the Navier–Stokes equation, namely

$$\begin{cases} \nabla \cdot \mathbf{v} = 0 \\ \rho \dfrac{D\mathbf{v}}{Dt} = -\nabla P + \mu \Delta \mathbf{v} \end{cases} \tag{4.1}$$

© Springer International Publishing Switzerland 2015
M. Rieutord, *Fluid Dynamics*, Graduate Texts in Physics,
DOI 10.1007/978-3-319-09351-2_4

Indeed, the third equation, the energy equation, uncouples completely from the two others. An important consequence of this uncoupling is that the pressure does not have a dynamic role. It doesn't drive the flow but is driven by the flow. This property follows from the fact that the velocity field is entirely determined by

$$\begin{cases} \nabla \cdot \mathbf{v} = 0 \\ \nabla \times \left(\dfrac{D\mathbf{v}}{Dt} - \nu \Delta \mathbf{v} \right) = \mathbf{0} \end{cases} \tag{4.2}$$

and the boundary conditions

$$\mathbf{v} = \mathbf{v}_s \quad \text{on} \quad (S)$$

where S is the boundary (supposedly solid with a velocity $\mathbf{v}_s$) which delimits the fluid. The pressure is thus entirely determined, up to a constant, by

$$\nabla P = \mu \Delta \mathbf{v} - \rho \frac{D\mathbf{v}}{Dt}$$

Another important property of this system of equations is that, if the viscosity of the fluid is large enough, the solution is unique. "Large enough" means larger than some critical value below which the system has several solutions. Physically, this shows up with the raise of an instability in the original solution. We shall discuss this point later on, but presently we just note that this phenomenon is a consequence of the nonlinear character of the equations.

4.1.2 Law of Similarity

A fluid flow always involves the *dynamic time scale*, which is the typical time that it takes for a fluid particle to cover the distance L, namely

$$T_d = \frac{L}{V} \tag{4.3}$$

where V is the typical velocity of the fluid. If the fluid is viscous then another time scale comes about; this is

$$T_v = \frac{L^2}{\nu} \tag{4.4}$$

also called *the viscous diffusion time*. The origin of this definition is the following: if we consider a very slow flow, the quadratic term $\mathbf{v} \cdot \nabla \mathbf{v}$ in the Navier–Stokes equation is very small compared to other terms. Neglecting this term and taking the curl of the momentum equation, we get the linearized equation of the vorticity $\boldsymbol{\omega} = \boldsymbol{\nabla} \times \mathbf{v}$, namely

$$\frac{\partial \boldsymbol{\omega}}{\partial t} = \nu \Delta \boldsymbol{\omega} \tag{4.5}$$

This is the diffusion equation (see Sect. 12.6.4 for a presentation of its basic properties). Schematically, if $\boldsymbol{\omega}$ varies on a length scale L, then $\Delta \boldsymbol{\omega} \sim \boldsymbol{\omega}/L^2$ and $\partial \boldsymbol{\omega}/\partial t \sim \nu \boldsymbol{\omega}/L^2$ which shows that $\boldsymbol{\omega}$ evolves on a time scale of order L^2/ν.

The dynamic and viscous time scales are compared through the non-dimensional ratio

$$Re = \frac{T_v}{T_d} \sim \frac{VL}{\nu} \tag{4.6}$$

also called the *Reynolds number*. It characterizes the ratio between two transport velocities: the macroscopic (dynamic) transport and the microscopic (diffusive) transport. This non-dimensional number is the only parameter intervening in the equations of motion of an incompressible viscous fluid. Indeed, if we make the following change in the variables:

$$\mathbf{v} = V\mathbf{u}, \qquad P = \rho V^2 p, \qquad \mathbf{r} = L\mathbf{x} \qquad \text{and} \qquad t = \frac{L}{V}\tau$$

then, $\mathbf{u}$, p, $\mathbf{x}$, τ represent respectively the non-dimensional velocity, pressure, spatial coordinates and time. The equations of motion read

$$\begin{cases} \boldsymbol{\nabla} \cdot \mathbf{u} = 0 \\ \dfrac{\partial \mathbf{u}}{\partial \tau} + \mathbf{u} \cdot \boldsymbol{\nabla} \mathbf{u} = -\boldsymbol{\nabla} p + \dfrac{1}{Re} \Delta \mathbf{u} \end{cases} \tag{4.7}$$

Save for the parameters that may be added in the boundary conditions, the solution $\mathbf{u}$ depends on just one quantity, which is the Reynolds number. All the flows having the same Reynolds number are identical up to a constant scale factor: they are said to be *similar*. This conclusion is true only if the solution is unique, that is to say, if the viscosity is large enough or if the Reynolds number is small enough.

Let us consider a simple example of the use of the similarity between flows. A solid represented by a cube of 1 cm side moves in air at a speed of 1 cm/s. The air flow is exactly the same as one around a cube of 1 m side moving at 0.1 mm/s. A practical application of the similarity relation is the use of reduced models to study some complex flows.

4.1.3 Discussion

System (4.7) gives us the first example of the flow equations written with non-dimensional variables. The use of non-dimensional variables is the rule in Fluid Mechanics. Thus doing, we are able to compare the various scales that intervene in a fluid flow. The foregoing example is very simple, but as we progress, we shall see that many non-dimensional numbers come into play. These numbers are crucial to compare the flows to each others and eventually evaluate the difficulties to compute them.

Finally, let us observe that perfect fluids correspond to the limit of infinite Reynolds numbers. However, this limit is *singular* because, as viscosity vanishes, second order derivatives disappear from the equations thus making some boundary conditions unmatched. This singularity is at the origin of the boundary layers, which appear when the Reynolds number is very large (see Sect. 4.3).

We shall further explore the dynamics of viscous fluids with the help of two limiting cases: the one of very viscous fluids and the one of nearly inviscid fluids. In other words, we shall study the two limits: the very small and the very large Reynolds numbers. We begin with the first case, which is the easiest one.

4.2 Creeping Flows

Creeping flows are all the flows for which the inertia of the fluid is negligible. Their Reynolds number is therefore very small compared to unity.

Examples of such flows come from the very viscous fluids (magma, for instance) or from the flows with very small scales (lubrication, microfluidic,...).

4.2.1 Stokes' Equation

We consider the momentum equation in (4.7) and multiply it by the Reynolds number while carrying out the substitution $p \rightarrow p/\text{Re}$. We get

$$\text{Re}\frac{\partial \mathbf{u}}{\partial \tau} + \text{Re}\, \mathbf{u} \cdot \nabla \mathbf{u} = -\nabla p + \Delta \mathbf{u}$$

Setting $\text{Re} = 0$, we get *Stokes' Equation* :

$$-\nabla p + \Delta \mathbf{u} = \mathbf{0} \qquad\qquad (4.8)$$

We may observe that, by taking the limit Re=0, we eliminated the time derivative of the velocity. Does this mean that all flows with very small Reynolds number are stationary? Not quite, of course! It means that if the appropriate time scale is L/V

then the temporal variations are truly negligible and the flow is steady. But we can easily envision a flow where there is a time dependent forcing. In this case, the time scale of this forcing is a new independent parameter which controls the amplitude of the term $\partial \mathbf{v}/\partial t$.

Stokes equation can take two other equivalent forms for an incompressible fluid:

$$\nabla p + \nabla \times \boldsymbol{\omega} = \mathbf{0} \qquad \text{or} \qquad \Delta \boldsymbol{\omega} = \mathbf{0} \tag{4.9}$$

where $\boldsymbol{\omega} = \nabla \times \mathbf{v}$ is the vorticity. We also note, by taking the divergence of (4.8), that the pressure verifies Laplace's equation:

$$\Delta p = 0$$

The essential property of these equations is their linear character. The solutions thus own all the properties associated with linearity. For instance, an interesting property is *the reversibility* : if $\mathbf{v}$ is a solution then $-\mathbf{v}$ is also a solution. Any fluid particle goes back to its initial position if the forcing is reversed (the nonlinear terms break this symmetry).

Another important consequence of the linearity of Stokes' equation is the uniqueness of the solution for a given set of boundary conditions. Below, we demonstrate this property by showing that the solutions obey a variational principle. The unicity of the solutions also resolves the problem of stability: the solutions are always stable.

Finally, note a third possible form of Stokes' equation:

$$\mathbf{Div}[\sigma] = \mathbf{0} \qquad \text{or} \qquad \partial_j \sigma_{ij} = 0 \tag{4.10}$$

where $[\sigma]$ is the stress tensor. This form is more general than the previous ones since it does not make use of the explicit form of the stress tensor.

4.2.2   Variational Principle ♣

Equation (4.10) can be obtained with the help of a variational principle, such as the least action principle. This means that the solutions of (4.10) render extremum a functional of the velocity field defined on the space occupied by the fluid. This functional is just the viscous dissipation.

In order to show this result, we shall consider a Newtonian fluid inside a given volume, limited by a surface (S), where the velocity is given as on a solid wall. On this surface the variations of the velocity $\delta \mathbf{v}$ vanish. The dissipation in this volume is given by:

$$D = \int_{(V)} \left(\frac{\mu}{2} c_{ij} c_{ij} + \zeta (\nabla \cdot \mathbf{v})^2 \right) dV$$

where we do not assume incompressibility. A variation of D associated with the variations of $\mathbf{v}$ is easily obtained by a functional derivation of the integral with respect to the velocity field:

$$\delta D = \int_{(V)} \{\mu c_{ij}\delta c_{ij} + 2\zeta(\nabla \cdot \mathbf{v})(\nabla \cdot \delta\mathbf{v})\} \, dV \tag{4.11}$$

According to the rheological law of Newtonian fluids $\mu c_{ij} = \sigma_{ij}^v - \zeta\nabla \cdot \mathbf{v}\,\delta_{ij}$ where $[\sigma^v]$ is the viscous stress tensor. Moreover, $\delta c_{ii} = 0$; hence, the foregoing expression may be rewritten as

$$\delta D = \int_{(V)} \{\sigma_{ij}\delta c_{ij} + 2\zeta(\nabla \cdot \mathbf{v})\nabla \cdot \delta\mathbf{v}\} \, dV$$

The expression (1.40) of c_{ij} and the symmetry of the viscous stress tensor allows us to simplify the preceding expression. Thus

$$\delta D = 2\int_{(V)} \sigma_{ij}\partial_i\delta v_j \, dV$$

Using the equation of motion (4.10) together with the divergence theorem, we finally obtain

$$\delta D = 2\int_{(S)} \sigma_{ij}\delta v_j \, dS_i = 0 \tag{4.12}$$

This last integral is zero because of the boundary conditions imposed on $\delta\mathbf{v}$. The dissipation is therefore at an extremum for the velocity field verifying Stokes' equation. We now show that this extremum is a minimum. For this, we observe that the dissipation is a linear function of the squared gradient of velocity. Symbolically, we can write that

$$D(v) = \mathcal{L}\left[(\partial v)^2\right]$$

where $\mathcal{L}$ is a linear operator. We have shown that

$$\delta D = 2\mathcal{L}\left((\partial v)(\partial\delta v)\right) = 0$$

Now we make the difference between the dissipation associated with the field v solution of the equations and that associated with the field $v + \delta v$ where δv is a variation. We have:

$$\begin{aligned}
D(v + \delta v) - D(v) &= \mathcal{L}\left([\partial(v + \delta v)]^2\right) - \mathcal{L}\left((\partial v)^2\right) \\
&= \mathcal{L}\left((\partial v)^2 + (\partial\delta v)^2 + 2(\partial v)(\partial\delta v)\right) - \mathcal{L}\left((\partial v)^2\right) \\
&= \mathcal{L}\left((\partial\delta v)^2\right) = D(\delta v) \geq 0
\end{aligned}$$

This result shows that whatever the variation of v made around the true solution, the dissipation increases. This quantity is therefore a minimum for the true solution.

Let us now show that this implies the uniqueness of the solution. For that, it is convenient to take two solutions and show that they are in fact identical. Let v_1 and v_2 be two such solutions; their dissipation being at the minimum is therefore identical. From the foregoing results, we necessarily have

$$\mathcal{L}\left([\partial(v_1 - v_2)]^2\right) = 0$$

The operator $\mathcal{L}$ being only an integration, we infer from the preceding equation that the integrant is zero everywhere within the fluid $\partial(v_1 - v_2) = 0$. In fact $\partial(v_1 - v_2)$ symbolizes all the components of the shear tensor and the divergence; we therefore have:

$$c_{ij}(\mathbf{v}_1 - \mathbf{v}_2) = 0 \quad \text{and} \quad \nabla \cdot (\mathbf{v}_1 - \mathbf{v}_2) = 0$$

From these two equations we derive a third one, namely

$$s_{ij}(\mathbf{v}_1 - \mathbf{v}_2) = 0$$

which means that the symmetric part of the velocity gradient tensor is zero. We have seen in Chap. 1 that this implies that $\mathbf{v}_1 - \mathbf{v}_2$ is the combination of a solid body rotation and a translation. But $\mathbf{v}_1$ and $\mathbf{v}_2$ satisfy the same boundary conditions, thus, in general, the rotation and the translation are both zero. Thus, $\mathbf{v}_1$ and $\mathbf{v}_2$ are identical.

The preceding results are not valid when the Reynolds number is large. In this case the solution is not unique and does not produce a minimum of dissipation.

4.2.3 Flow Around a Sphere

As a first example we shall consider the flow of a viscous fluid around a sphere moving slowly. We assume that the fluid fills the whole space and that the flow is steady. The Reynolds number, based on the velocity of the sphere and its diameter, is very small compared to unity so that we can use Stokes' equation. Returning to the dimensional variables, we have

$$\begin{cases} \nabla \cdot \mathbf{v} = 0 \\ 0 = -\nabla P + \mu \Delta \mathbf{v} \end{cases} \tag{4.13}$$

Using a reference frame whose origin is at the centre of the sphere, the boundary conditions are

$$\mathbf{v} = \mathbf{0} \quad \text{at} \quad r = R \quad \text{and} \quad \mathbf{v} \to U \mathbf{e}_z \quad \text{when} \quad r \to \infty$$

Note that $-U\mathbf{e}_z$ is the velocity of the sphere in a rest frame. We use the spherical coordinates. Since the flow is axisymmetric around the z-axis, the velocity and the pressure only depend on r and θ. Moreover, $\mathbf{v}$ has no component along $\mathbf{e}_\varphi$.

In order to solve system (4.13) we expand the functions on the basis of Legendre's polynomials in the following manner:

$$
\begin{cases}
v_r = \sum_\ell u_\ell(r) P_\ell(\cos\theta) \\
v_\theta = \sum_\ell v_\ell(r) \frac{dP_\ell}{d\theta} \\
P = \sum_\ell p_\ell(r) P_\ell(\cos\theta)
\end{cases}
\tag{4.14}
$$

Legendre's polynomials satisfy the differential equation

$$
\frac{1}{\sin\theta} \frac{d}{d\theta}\left(\sin\theta \frac{dP_\ell}{d\theta}\right) + l(l+1)P_\ell = 0
$$

Using the equation of continuity, we derive the relation between $u_\ell(r)$ and $v_\ell(r)$

$$
\ell(\ell+1)v_\ell(r) = \frac{1}{r}\frac{dr^2 u_\ell}{dr}
$$

We also find that

$$
\nabla \times \mathbf{v} = -\sum_\ell \frac{\Delta_\ell(r u_\ell)}{\ell(\ell+1)} P_\ell(\cos\theta)\mathbf{e}_\varphi,
$$

and

$$
\Delta\mathbf{v} = -\nabla \times (\nabla \times \mathbf{v}) = \frac{\Delta_\ell(r u_\ell)}{r} P_\ell(\cos\theta)\mathbf{e}_r + \frac{1}{r}\frac{d}{dr}\left(\frac{r\Delta_\ell(r u_\ell)}{\ell(\ell+1)}\right)\frac{dP_\ell}{d\theta}\mathbf{e}_\theta
$$

Δ_ℓ being the operator

$$
\Delta_\ell = \frac{1}{r}\frac{d^2}{dr^2}r - \frac{\ell(\ell+1)}{r^2}
$$

As Legendre's polynomials form an orthogonal basis as well as their derivative, we easily find that

$$
\begin{cases}
\dfrac{dp_\ell}{dr} = \mu\Delta_\ell(r u_\ell)/r \\[2mm]
p_\ell(r) = \mu\dfrac{d}{dr}\left(\dfrac{r\Delta_\ell(r u_\ell)}{\ell(\ell+1)}\right)
\end{cases}
\tag{4.15}
$$

which yields

$$\Delta_\ell \Delta_\ell (r u_\ell) = 0 \tag{4.16}$$

The general solution of this equation is in the form of the powers of r; namely

$$u_\ell(r) = A r^{\ell-1} + B r^{\ell+1} + C r^{-\ell} + D r^{-\ell-2}$$

In order to find these four constants, we need four boundary conditions; $v_r(R) = v_\theta(R) = 0$ imply

$$u_\ell(R) = v_\ell(R) = 0, \qquad \forall \ell$$

while at infinity we have

$$\mathbf{v} \to U \mathbf{e}_z = U(\cos\theta \mathbf{e}_r - \sin\theta \mathbf{e}_\theta), \qquad \text{as} \quad r \to \infty$$

One may verify that this last condition leads to

$$\lim_{r\to\infty} u_1(r) = U \qquad \text{and} \qquad \lim_{r\to\infty} u_{\ell\neq1}(r) = 0$$

The boundary conditions implies that all the coefficients, except those of u_1, are zero. More explicitly, we have

$$u_1(r) = A + B r^2 + \frac{C}{r} + \frac{D}{r^3} \tag{4.17}$$

Using the conditions at infinity we find $A = U$ and $B = 0$, while with those on the sphere it turns out that

$$C = -3UR/2, \qquad D = UR^3/2$$

Since $P_1(\cos\theta) = \cos\theta$, we finally obtain

$$\begin{cases} v_r = U \cos\theta \left(1 - \frac{3R}{2r} + \frac{R^3}{2r^3}\right) \\[2ex] v_\theta = -U \sin\theta \left(1 - \frac{3R}{4r} - \frac{R^3}{4r^3}\right) \\[2ex] p = -\frac{3\mu UR}{2r^2} \cos\theta \end{cases} \tag{4.18}$$

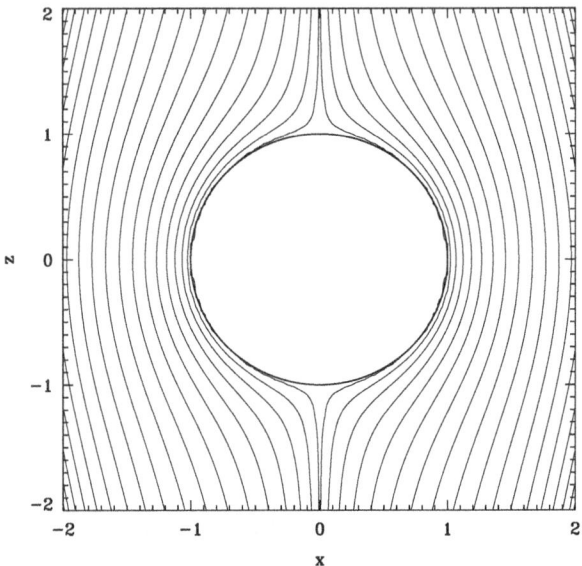

Fig. 4.1 Streamlines in the meridional plane of the Stokes' flow around a sphere in uniform motion in a viscous fluid

The velocity field can be expressed with a stream function ψ describing the streamlines in the meridional plane as shown in Fig. 4.1. The expression of the stream function is

$$\psi = \frac{1}{2} U r^2 \sin^2 \theta \left(1 - \frac{3R}{2r} + \frac{R^3}{2r^3} \right)$$

since

$$v_r = \frac{1}{r^2 \sin \theta} \frac{\partial \psi}{\partial \theta} \qquad \text{and} \qquad v_\theta = -\frac{1}{r \sin \theta} \frac{\partial \psi}{\partial r}$$

From the expression of the velocity, we also infer the expression of the vorticity field:

$$\nabla \times \mathbf{v} = -\frac{3UR}{2r^2} \cos \theta \, \mathbf{e}_\varphi, \tag{4.19}$$

which will be used later on to compute the drag force exerted on the sphere.

4.2.4   Oseen's Equation

Stokes' equation has been derived for vanishing Reynolds numbers. However, in any experiment, even if the Reynolds number is very small, it is finite. This makes Stokes' equation invalid far from the solid. This strange property comes from the nature of the solutions of Stokes equation, which is a kind of Laplace equation. The solutions of Stokes's equation that vanish at infinity are power laws of the distance (like (4.17) for instance). For these solutions, the length scale characterizing the velocity variations grows with r. Indeed, a typical length scale of the velocity field V is $L = (d \ln V/dr)^{-1}$, thus if $V \sim 1/r^n$ then $L \sim r/n$. The consequence of this growing scale is that nonlinear terms decrease more slowly than viscous ones as they contain only first order derivatives. The distance by which nonlinear terms overtake the linear one can be guessed from an order of magnitude estimate

$$(\mathbf{u} \cdot \nabla)\mathbf{u} \sim \frac{1}{\text{Re}}\Delta\mathbf{u} \implies \frac{1}{L} \sim \frac{1}{L^2\text{Re}} \implies L \sim \text{Re}^{-1}$$

where the Reynolds number is computed from the dimensions of the object. The foregoing result shows that this critical length goes to infinity as the Reynolds number vanishes.

Hence, the computation of flows extending to distance larger than Re^{-1} must take into account the corrections imposed by nonlinear terms. For instance, if we wish to compute the flow around a solid body moving at constant speed in a very viscous fluid, we may set $\mathbf{u} = \mathbf{U}_\infty + \delta\mathbf{u}$ ($\mathbf{U}_\infty$ is the fluid velocity at infinity in a frame attached to the solid) and first solve Stokes equation. However, at distances larger than Re^{-1}, corrections from nonlinear terms are important. These are taken into account by keeping the leading order of these terms. Hence, in these regions, one has to solve

$$(\mathbf{U}_\infty \cdot \nabla)\delta\mathbf{u} = -\nabla p + \frac{1}{\text{Re}}\Delta\delta\mathbf{u} \tag{4.20}$$

which is *Oseen's equation*. Although this equation seems more complete than Stokes one, it is not valid close to the solid as the flow is not close to a uniform velocity field. Thus, Oseen's equation is useful to complement Stokes' equation when the fluid's domain is larger than Re^{-1}; in this case the solution of both equations must be matched together, which may be delicate. Of course, if the domain is not that large, Stokes' equation is sufficient.

4.2.5   The Lubrication Layer

We end this section with the study of another type of flows at very small Reynolds number, namely the case of the lubrication layer, which was analysed for the first

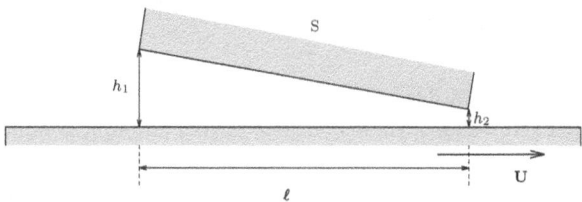

Fig. 4.2 Schematic view of a lubrication layer. In the reference frame of the solid S the ground moves to the right at speed U entraining the fluid to flow in the same direction

time by Reynolds in 1886. The flow in a lubrication layer has numerous applications, especially in *tribology* (the study of friction). The lubricating effect of a thin fluid layer between two solids is shown by an experiment of everyday life: that of a sheet of paper, which glides practically without friction on a smooth floor. A thin layer of air forms between the floor and the paper, making an air cushion, which reduces drastically the friction. We may also observe that in the same conditions of incidence and velocity, but far above the floor, the sheet of paper has not a sufficient lift to compensate its weight, and falls.

In order to understand the fundamentals of lubrication, we shall consider the simple system illustrated in Fig. 4.2. A solid with a length ℓ glides with a velocity U above a fixed solid plane. An incompressible viscous fluid flows between the two solids, forming the lubrication layer. The Reynolds number of this flow, based on the thickness of the fluid film, is supposedly very small:

$$\frac{Uh}{\nu} \ll 1$$

We assume the contact surface of the solid S to be plane and slightly inclined as in Fig. 4.2. As shown, the thickness h varies linearly with the abscissa; we thus set

$$h(x) = h_1 + (h_2 - h_1)x/\ell \quad \text{and} \quad \alpha = (h_1 - h_2)/\ell \ll 1 \ .$$

In order to analyse the flow, we use a frame attached to the moving solid. The boundary conditions are therefore

$$\mathbf{v} = U\mathbf{e}_x \quad \text{at} \quad z = 0 \quad \text{and} \quad \mathbf{v} = \mathbf{0} \quad \text{at} \quad z = h(x)$$

The flow is stationary and with a very small Reynolds number. It therefore satisfies Stokes' equation (4.13). Using the x-component of the momentum equation, we get

$$-\frac{\partial P}{\partial x} + \mu \frac{\partial^2 v_x}{\partial z^2} = 0, \tag{4.21}$$

where we neglected the x-dependence of the velocity field. Terms coming from this dependence are $\mathcal{O}(\alpha)$ or smaller, thus (4.21) is just the zeroth order in α. At this same order we see that $G_p = \partial P/\partial x$ is independent of x. Solving for the z-dependence gives the following form of v_x:

$$v_x = U\frac{h-z}{h} + \frac{G_p}{2\mu}(z-h)z \tag{4.22}$$

This solution is the superimposition of two exact solutions of the Navier–Stokes equation: Couette's flow

$$\mathbf{u} = U(h-z)/h \; \mathbf{e}_x$$

and Poiseuille's flow

$$\mathbf{u} = G_p z(z-h)/2\mu \; \mathbf{e}_x$$

that we shall discuss in Sect. 4.4.1.

In the expression of the velocity, G_p is an unknown. However, it may be related to the volume flux Q which is the same at any x since the fluid is incompressible. Neglecting the third dimension, we have

$$Q = \int_0^h v_x dz = \frac{Uh}{2} - \frac{G_p h^3}{12\mu}$$

which leads to

$$G_p = \frac{dP}{dx} = \frac{12\mu}{h^3}\left(\frac{Uh}{2} - Q\right) \tag{4.23}$$

If the solid S is completely immersed in the same fluid (as the sheet of paper), the pressure on the two ends is identical. Rewriting (4.23) as

$$\frac{G_p}{\mu} = \frac{1}{\mu}\frac{dP}{dx} = \frac{1}{\mu}\frac{dh}{dx}\frac{dP}{dh} = -\frac{\alpha}{\mu}\frac{dP}{dh} = \frac{6U}{h^2} - \frac{12Q}{h^3}$$

and integrating between h_1 and h_2, we can express the volume flux as a function of the parameters of the problem. We find

$$Q = U\frac{h_1 h_2}{h_1 + h_2} \tag{4.24}$$

We can then give the expression of the pressure field in the domain $[0, \ell]$:

$$P(x) - P_o = \frac{6\mu U(h_1 - h(x))(h(x) - h_2)}{\alpha(h_1 + h_2)h(x)^2} \tag{4.25}$$

where P_o is the pressure at the ends of the solid. Using this expression, in which we insert $h(x) = h_1 - \alpha x$, we can observe that

$$\delta P = P(x) - P_o \propto x(\ell - x)/(h_1 - \alpha x)^2,$$

showing that the pressure reaches a maximum in the neighborhood of $\ell/2$. The maximum value of the pressure may be expressed as a function of the parameters of the problem, namely

$$\delta P_{max} = \frac{6\mu U \alpha \ell}{(h_1 + h_2)^3} \tag{4.26}$$

This result shows that the pressure strongly increases when the thickness of the fluid layer vanishes. Furthermore, the total pressure force can be derived from $F_l = \int_0^\ell \delta P dx$. After little algebra,[1] we find

$$F_p = \frac{6\mu U}{\alpha^2} \left\{ \ln\left(\frac{h_1}{h_2}\right) - 2\frac{h_1 - h_2}{h_1 + h_2} \right\} \tag{4.27}$$

It is interesting to compare this lift force to the total shear stress exerted upon the moving solid, which is just the drag force. By using a similar calculation, we find

$$F_d = \int_0^\ell \mu \frac{\partial v_x}{\partial z} dx = \frac{2\mu U}{\alpha} \left\{ 3\frac{h_1 - h_2}{h_1 + h_2} - \ln\left(\frac{h_1}{h_2}\right) \right\} \tag{4.28}$$

so that

$$\frac{F_{drag}}{F_{lift}} = \frac{\alpha}{3} \frac{\left\{ 3\frac{h_1-h_2}{h_1+h_2} - \ln\left(\frac{h_1}{h_2}\right) \right\}}{\left\{ \ln\left(\frac{h_1}{h_2}\right) - 2\frac{h_1-h_2}{h_1+h_2} \right\}} \tag{4.29}$$

Since $\alpha \ll 1$, the horizontal force is very much smaller than the vertical one. This is why a large mass can be moved effortlessly with bearings. An example is given in exercise.

---

[1] The reader may note that after an integration by part

$$\int_0^\ell \frac{x(\ell - x)}{(h_1 - \alpha x)^2} dx = -\frac{1}{\alpha} \int_0^\ell \frac{\ell - 2x}{h_1 - \alpha x} dx$$

while

$$\int_0^\ell \frac{\ell - 2x}{h_1 - \alpha x} dx = \int_0^\ell \left\{ \frac{\ell - 2h_1/\alpha}{h_1 - \alpha x} + \frac{2}{\alpha} \right\} dx$$

4.3 Boundary Layer Theory

In the foregoing section we considered flows with very low Reynolds numbers, fully dominated by the viscous force. Now, we shall examine the opposite limit, that of laminar flows at large Reynolds numbers.

4.3.1 Perfect Fluids and Viscous Fluids

Contrary to the low Reynolds number flows, the present ones are not always stable. Beyond some critical Reynolds number, several solutions are possible. However, this critical value may be large compared to unity. Thus it is often the case that the flow is stable even if $\mathrm{Re} \gg 1$. This is the typical situation that we shall investigate now.

A convenient example to bear in mind is the one of a flow around a solid body. Let us think of a car or an airplane moving at constant speed. In the frame attached to the solid, the fluid shows a uniform velocity field in the far distance of the body. We assume that the Reynolds number, based on the typical size of the object, is large compared to unity. Such a set-up was already discussed in the case of perfect fluids (see Sect. 3.3.7). There we argued that the flow was irrotational, so that there exist ϕ such that $\mathbf{v} = \nabla\phi$. It is interesting to note that such kind of solution is almost acceptable for a viscous fluid. Indeed, for an incompressible fluid ($\nabla \cdot \mathbf{v} = 0$), the viscous force associated with a potential flow is zero, since:

$$\Delta\nabla\phi = \nabla(\nabla \cdot \mathbf{v}) - \nabla \times \nabla \times \nabla\phi = \mathbf{0}$$

However, this solution is not fully acceptable since it does not meet the no-slip boundary conditions on the solid. The fluid sticks to the solid and we can surmise that close enough to it, the viscous force dominates over the other forces, which is clearly not possible if the flow remains irrotational.

To specify what is meant by "close enough", we have to go back to the momentum equation (4.7). If the viscous force is important in some region of the flow, then, in this place

$$\Delta\mathbf{u} \gtrsim \mathcal{O}(\mathrm{Re})$$

since other terms are supposedly of order unity. Such an inequality can be realized in only two ways: either $\mathbf{u}$ is very large compared to unity or its spatial variations are very rapid. The first possibility can be eliminated thereof since close to the solid the velocity cannot grow much as it vanishes on the boundary. The second possibility is

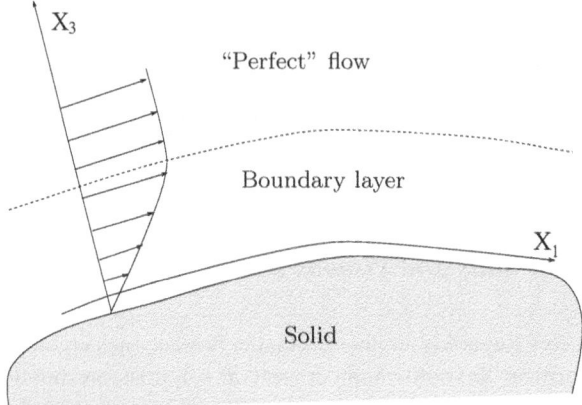

Fig. 4.3 Geometry of a boundary layer flow

therefore the right one. The function $\mathbf{u}$ varies very rapidly in the vicinity of the solid. If ε is the characteristic scale of the velocity field, the viscous force is dominating if

$$\frac{1}{\mathrm{Re}} \Delta \mathbf{u} \sim \frac{1}{\mathrm{Re}} \frac{u}{\varepsilon^2} \sim \mathcal{O}(1) \qquad \Longleftrightarrow \qquad \varepsilon = \mathcal{O}(1/\sqrt{\mathrm{Re}})$$

since we assumed that $\mathbf{u} \sim \mathcal{O}(1)$.[2]

We thus find that around a solid body in a large Reynolds number flow, there is a very thin layer of thickness $\varepsilon = 1/\sqrt{\mathrm{Re}} \ll 1$ where the viscous force may control the flow. This is the *boundary layer*. At a distance of a few times ε, the viscous force is usually negligible and the fluid behaves as if perfect. This example shows us that high Reynolds number flows may be split into regions with very different dynamics: (*i*) the boundary layers controlled by viscosity (or other diffusion processes in more general situations) and (*ii*) the "remaining" where Euler's equation is sufficient to describe the flow. This is schematically illustrated in Fig. 4.3.

Before closing this heuristic introduction to boundary layers, let us mention that regions where viscous force is important are not systematically attached to boundaries. It turns out that in some cases strong shear layers occur in the middle of the fluid. They are no longer boundary layers but *detached (shear) layers*. Let us also underline that the technique of splitting the fluid domain into various subdomains is not specific to fluid mechanics but is in fact a way of obtaining an approximate solution of (partial) differential equations that are too difficult to be solved analytically.[3]

---

[2] Here is a very simple example of such a property: $\sin(x/\varepsilon)$ is $\mathcal{O}(1)$ but $\frac{d^2}{dx^2} \sin(x/\varepsilon)$ is $\mathcal{O}(1/\varepsilon^2)$.

[3] An example where we determine the solution of a differential equation using boundary layer theory is given in Sect. 12.4.

4.3.2   Method of Resolution

The preceding discussion has shown that a small parameter $\varepsilon \ll 1$ is naturally introduced in this problem. A way of taking advantage of this peculiarity is to expand the solutions into powers of ε and write:

$$\mathbf{u} = \mathbf{u}_0 + \varepsilon\,\mathbf{u}_1 + \varepsilon^2\,\mathbf{u}_2 + \cdots \tag{4.30}$$

Furthermore, we can take advantage of the partition of the flow into the boundary layer and the inviscid outer flow. In the boundary layer viscosity is important and the full equation need to be solved. However, the shape of the flow is rather simple as it is parallel to the boundaries. On the other hand, the outer flow does not need viscosity, which may be neglected at zeroth order. Hence, in both domains the solution can be simplified. The strategy is therefore obvious: in each domain solutions are expanded according to (4.30). Each order is solved in each domains and the solutions are matched together. This technique is called asymptotic matching. The final result is an asymptotic solution valid up to some higher order correction in ε^n. We shall now detail all these steps.

4.3.3   Flow Outside the Boundary Layer

Outside the boundary layer, the derivatives are all of order unity. Thus using the expansion (4.30) and identifying each order in ε, we get

$$\begin{cases} \mathbf{u}_0 \cdot \nabla \mathbf{u}_0 = -\nabla p_0 \\ \nabla \cdot \mathbf{u}_0 = 0 \\ \mathbf{u}_0 \cdot \mathbf{n} = 0 \quad \text{on } S \end{cases} \tag{4.31}$$

at zeroth order, and

$$\begin{cases} \mathbf{u}_0 \cdot \nabla \mathbf{u}_1 + \mathbf{u}_1 \cdot \nabla \mathbf{u}_0 = -\nabla p_1 \\ \nabla \cdot \mathbf{u}_1 = 0 \end{cases} \tag{4.32}$$

at first order. We do not write the boundary conditions yet; they need further discussion and will be introduced at the end of the next subsection. Note that at second or higher orders, viscous terms need to be taken into account, even if we are outside the boundary layer.

In the simple case where the zeroth order velocity field is irrotational, equation may be simplified

$$\begin{cases} \Delta \Phi_0 = 0 \\ \mathbf{n} \cdot \nabla \Phi_0 = 0 \quad \text{on } S \end{cases} \tag{4.33}$$

4.3.4 Flow Inside the Boundary Layer

The foregoing equations give solutions that are not valid close to the boundaries of the fluid. There, the viscous force is important. Viscous terms make the equations of higher order, but, as noted above, the geometry of the flow is simpler, as almost parallel to the boundary. To take advantage of this property, it is natural to introduce three curvilinear coordinates (x_1, x_2, x_3) where $x_3 = \mathrm{Cst}$ is the equation of the boundary. In order to simplify the following discussion, we shall assume that the bounding surface of the fluid is just the $z = 0$-plane. Plain cartesian coordinates are thus sufficient.

As in the "inviscid domain", we expand the unknowns in powers of the small parameter ε. We use the tilde to denote boundary layer quantities; hence, the boundary layer flow is expressed:

$$\tilde{u} = \tilde{u}_0 + \varepsilon\,\tilde{u}_1 + \varepsilon^2\,\tilde{u}_2 + \cdots$$

Boundary layer quantities are characterized by their rapid variations in the direction perpendicular to the boundary. Partial derivatives should therefore be ordered as

$$\partial/\partial x,\ \ \partial/\partial y \ll \partial/\partial z \tag{4.34}$$

This inequality can be made more quantitative since we know the thickness of the boundary layers, namely ε. Thus, a typical boundary layer function $\tilde{f}$ reads

$$\tilde{f} \equiv \tilde{f}\left(x, y, \frac{z}{\varepsilon}\right)$$

One usually introduces the *stretched coordinate* $\tilde{z} = z/\varepsilon$, so that $\tilde{f} \equiv \tilde{f}(x, y, \tilde{z})$. With this new coordinate, the inviscid region, which is at $z = \mathcal{O}(1)$, is now rejected at infinity, since for a fixed z, $\tilde{z} \to \infty$ as $\varepsilon \to 0$.

The thickness of the boundary layer, and thus the stretched coordinate, is such that $\frac{\partial \tilde{f}}{\partial \tilde{z}} = \mathcal{O}(1)$; hence, normal variations of the boundary layer functions, namely $\partial_z \tilde{f}$, are all of order ε^{-1}. Besides, the horizontal variations of the fields (velocity and pressure) are controlled by those in the perfect domains. Indeed, the solutions in the boundary layer match those of the perfect domain at each point on the bounding surface, thus horizontal variations in the boundary layer are the same as those just outside of it. In the perfect domain, all the scales are of order unity, therefore horizontal gradient in the boundary layer are also of order unity; thus inequalities (4.34) mean

$$\partial_x \tilde{f} \equiv \mathcal{O}(1), \qquad \partial_y \tilde{f} \equiv \mathcal{O}(1), \qquad \text{and} \qquad \partial_z \tilde{f} \equiv \mathcal{O}(\varepsilon^{-1})$$

Let us now consider the equation of mass conservation. Using (4.30) together with $\nabla \cdot \mathbf{v} = 0$, we find that the lowest order is $\mathcal{O}(\varepsilon^{-1})$. It yields

$$\frac{\partial \tilde{u}_{0,z}}{\partial \tilde{z}} = 0 \ .$$

This equation implies that

$$\tilde{u}_{0,z} = 0 \tag{4.35}$$

since the velocity is zero on $\tilde{z} = 0$. This result shows an important property of boundary layers: the component of the velocity that is perpendicular to the layer is much smaller than the one parallel to it; it is at least of the next order in ε.

The following order of the equation of continuity reads

$$\frac{\partial \tilde{u}_{0,x}}{\partial x} + \frac{\partial \tilde{u}_{1,z}}{\partial \tilde{z}} = 0 \tag{4.36}$$

The Navier–Stokes equation develops in the same way and the first terms, of zeroth order, yield the two equations

$$\begin{cases} \tilde{u}_{0,x}\dfrac{\partial \tilde{u}_{0,x}}{\partial x} + \tilde{u}_{1,z}\dfrac{\partial \tilde{u}_{0,x}}{\partial \tilde{z}} = -\dfrac{\partial \tilde{p}_0}{\partial x} + \dfrac{\partial^2 \tilde{u}_{0,x}}{\partial \tilde{z}^2} \\[2ex] 0 = -\dfrac{\partial \tilde{p}_0}{\partial \tilde{z}} \end{cases} \tag{4.37}$$

(4.36) and (4.37) are known as *Prandtl's equations* of the boundary layer. These equations show that the pressure does not depend on the coordinate $\tilde{z}$; in other words, it is determined, like $\tilde{u}_{0,z}$, by the flow outside the boundary layer. This implies that in (4.37) $\frac{\partial \tilde{p}_0}{\partial x}$ is given by the "perfect fluid flow", so that

$$\frac{\partial \tilde{p}_0}{\partial x} = \frac{\partial p_0}{\partial x}$$

Prandtl's equations can be rearranged in the following way. We derive $\tilde{u}_{1,z}$ from (4.37), and we substitute it into (4.36). Thus

$$\tilde{u}_{1,z} = \frac{-\tilde{u}_{0,x}\frac{\partial \tilde{u}_{0,x}}{\partial x} - \frac{\partial \tilde{p}_0}{\partial x} + \frac{\partial^2 \tilde{u}_{0,x}}{\partial \tilde{z}^2}}{\frac{\partial \tilde{u}_{0,x}}{\partial \tilde{z}}} \tag{4.38}$$

and

$$\frac{\partial \tilde{u}_{0,x}}{\partial x} + \frac{\partial}{\partial \tilde{z}}\left(\frac{-\tilde{u}_{0,x}\frac{\partial \tilde{u}_{0,x}}{\partial x} - \frac{\partial \tilde{p}_0}{\partial x} + \frac{\partial^2 \tilde{u}_{0,x}}{\partial \tilde{z}^2}}{\frac{\partial \tilde{u}_{0,x}}{\partial \tilde{z}}} \right) = 0 \tag{4.39}$$

This last expression shows that the boundary layer equations are nonlinear and of the third order. Three boundary conditions are thus necessary to determine the solution. These are:

- $\tilde{u}_{0,x} = 0$ at $\tilde{z} = 0$,
- $\tilde{u}_{0,x} \longrightarrow u_{0,x}$ when $\tilde{z} \longrightarrow +\infty$,
- $\tilde{u}_{1,z} = 0$ at $\tilde{z} = 0$.

We observe that the limit value of $\tilde{u}_{1,z}$ when $\tilde{z} \to +\infty$ is not specified. It cannot be since $\tilde{u}_{1,z}$ obeys a first order equation (4.36). In general, $\lim_{\tilde{z}\to+\infty} \tilde{u}_{1,z} \neq 0$ so that *there exist a mass flux between the boundary layer and the perfect fluid domain.* The value of $\tilde{u}_{1,z}$ plays in this way the role of the boundary value for the first order terms in the perfect domain. Thus, system (4.32) is completed by the boundary condition

$$u_{1,z}(z = 0) = \lim_{\tilde{z}\to+\infty} \tilde{u}_{1,z} \tag{4.40}$$

4.3.5 Separation of the Boundary Layer

We may observe that (4.38), which gives $\tilde{u}_{1,z}$, becomes singular if, at some point on the boundary layer,

$$\frac{\partial \tilde{u}_{0,x}}{\partial \tilde{z}} = 0 \tag{4.41}$$

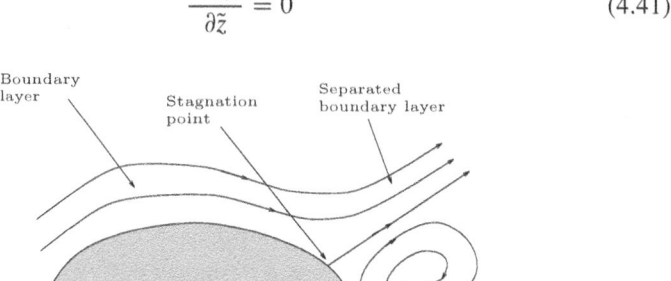

When approaching such a point, the vertical variations of $\tilde{u}_{0,x}$ are on an increasingly larger scale, in other words the boundary layer becomes thicker up to the point of being "infinite". One says that there is a *separation of the boundary layer*. Let us note that the boundary layer becomes infinitely thick with respect to the coordinate $\tilde{z}$. It does not mean that the boundary layer is overrunning all the fluid domain, but simply that its true scale is no longer ε but a much greater scale. For example, if the thickness is $\varepsilon^{\frac{1}{2}}$ and that it is developed in spite of everything in powers of ε, the thickness will be $\mathcal{O}(1/\varepsilon^{\frac{1}{2}})$ in the coordinate $\tilde{z}$ and therefore infinite when ε is vanishing. Thus, at the point of separation, $\tilde{u}_{1,z}$ diverges but in the neighbourhood of such a point $\tilde{u}_{1,z}$ is no longer of order unity and the expansion in powers of ε is no longer valid.

Equation (4.41) determines the position of the point of separation when solving Prandtl's equation. In fact, it is not necessary to solve this equation in order to know the position of this point. Indeed, if $\frac{\partial \tilde{u}_{0,x}}{\partial \tilde{z}} = 0$, then, using boundary conditions, $\tilde{u}_{0,x} = 0$. As the tangential velocity in the boundary layer is also the tangential velocity of the perfect fluid on the solid, we find that the separation of the boundary layer occurs close to a (downstream) stagnation point of the perfect fluid flow (see figure).

4.3.6   Example of the Laminar Boundary Layer: Blasius' Equation

We shall now illustrate the foregoing general theory with a very classical example which is Blasius flow. This is the boundary layer flow generated by a thin horizontal plate parallel to the flow at infinity (see Fig. 4.4). Far from the plate the pressure is

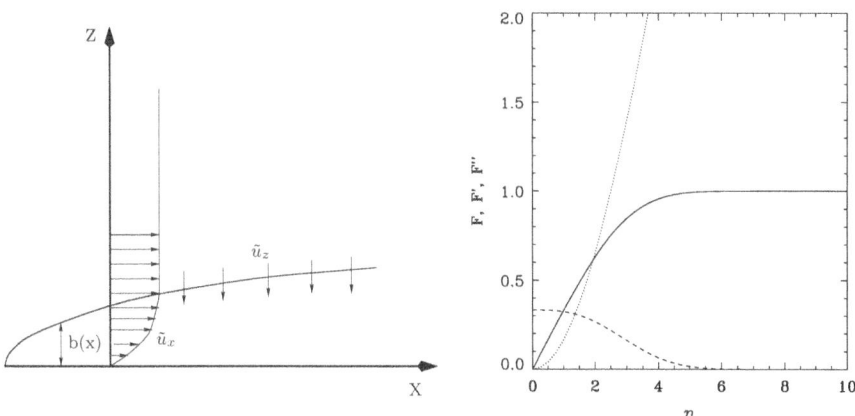

Fig. 4.4 *Left*: Shape of the boundary layer on a plate in a uniform flow. *Right*: View of the functions F (*dotted line*), f (*solid line*) and f' (*dashed line*)

uniform. Equations (4.36) and (4.37) thus reduce to

$$
\begin{cases}
\tilde{u}_{0,x} \dfrac{\partial \tilde{u}_{0,x}}{\partial x} + \tilde{u}_{1,z} \dfrac{\partial \tilde{u}_{0,x}}{\partial \tilde{z}} = \dfrac{\partial^2 \tilde{u}_{0,x}}{\partial \tilde{z}^2} \\[2mm]
\dfrac{\partial \tilde{u}_{0,x}}{\partial x} + \dfrac{\partial \tilde{u}_{1,z}}{\partial \tilde{z}} = 0
\end{cases}
\tag{4.42}
$$

Despite their apparent complexity, these equations admit analytical solutions when one imposes self-similarity. Indeed, we may set

$$
\tilde{u}_{0,x} = U f \left(\frac{\tilde{z}}{b(x)} \right) \quad \text{and} \quad \tilde{u}_{1,z} = V(x) g \left(\frac{\tilde{z}}{b(x)} \right)
\tag{4.43}
$$

where U is the velocity at infinity. Such a velocity profile is said to be *self-similar*, because for all x, its shape is identical and given by f. Then, we introduce the similarity variable $\eta = \tilde{z}/b(x)$ and rewrite the equations. The equation of mass conservation (4.42b) gives

$$
V(x) g'(\eta) = U \eta f'(\eta) b'(x)
\tag{4.44}
$$

The primed functions designate the derivatives. This equation shows that if self-similar solutions exist then $V(x)/b'(x)$ is a constant. Dimensionally, this constant is a velocity that can be set to U without loss of generality.

Turning to the momentum equation, we find

$$
f'' = -U b(x) b'(x) \eta f f' + b(x) V(x) g f'
\tag{4.45}
$$

As before, this equation admits self-similar solutions if $b(x)b'(x)$ and $b(x)V(x)$ are constants. We therefore set

$$
b(x) = \sqrt{Lx} \quad \Longrightarrow \quad bb' = L/2 \quad \text{and} \quad b(x)V(x) = UL/2
$$

$b(x)$ is the thickness of the boundary layer. The preceding expressions show that this quantity grows like the square root of the distance to the leading edge of the plate. We shall return further on to the physical interpretations of this result.

Finally, (4.45) is rewritten as

$$
f'' = \frac{UL}{2} \left(g f' - \eta f f' \right)
$$

where U and L are dimensionless constants, which represent a velocity and a length scale respectively. We use them as the velocity scale and length scale, which is equivalent to setting $U = L = 1$. By using (4.44), which we now write $g'(\eta) = \eta f'(\eta)$, we thus deduce a first form of Blasius' equation :

$$\left(\frac{f''}{f'}\right)' = -\frac{1}{2}f \quad \Longleftrightarrow \quad f'' = -\frac{1}{2}f' \int_0^\eta f(y)dy \tag{4.46}$$

Another classical form of Blasius' equation may be found by introducing a stream function like $F = \int f d\eta$. Equation (4.46) yields then

$$2F''' + FF'' = 0 \tag{4.47}$$

This equation is completed by three boundary conditions:

- $F'(0) = 0$, $(v_x = 0$ on the plate),
- $F'(\tilde{z} \to +\infty) = 1$, (velocity is constant at infinity)
- $F(0) = 0$, $(v_z = 0$ on the plate).

The functions f or F need to be determined numerically[4] and are shown in Fig. 4.4.

The solution of Blasius' equation allows us to show two general phenomena of boundary layers: The flow in a boundary layer vanishes exponentially in the outer region and there is a flux of matter between the boundary layer and the rest of the fluid. This is the so-called boundary layer pumping. We can demonstrate this last point by recapitulating the asymptotic form of f' for the large values of η. In this case, $f' \sim \exp(-\eta^2/4)$ and we get the component of the velocity v_z by way of the equation of mass conservation $g' = \eta f'$, which we integrate taking into account that $\lim_{\eta \to \infty} g = 0$. Hence,

$$g \sim -2e^{-\eta^2/4}$$

---

[4]We can get an idea of the shape of the function $f(\eta)$ by considering the asymptotic limits $\eta \sim 0$ and $\eta \to \infty$.

Near the origin, (4.46) and the boundary conditions impose that $f(0) = f''(0) = f'''(0) = 0$; hence, a Taylor expansion yields

$$f(\eta) \approx a\eta - \frac{a^2}{48}\eta^4 + \mathcal{O}(\eta^7)$$

where $a = f'(0) \simeq 0.332058$ (this value is determined by the boundary condition $f(\infty) = 1$). This expression shows that the profile of the velocity is almost linear just before reaching the asymptotic value where $f(\eta) \simeq 1$. In this region ($\eta \gg 1$), the function f verifies approximately $f'' = -\eta f'/2$ whose solution for f' is Gauss function and thus for f the error function:

$$f'(\eta) = Ae^{-\eta^2/4} \quad \Longrightarrow \quad f(\eta) \sim \text{erf}(\eta) \quad \eta \to \infty$$

so that v_z is negative far from the boundary; everything happens as if the boundary layer "breathes" the exterior fluid.

4.4 Some Classic Examples

We continue our tour of flows with incompressible viscous fluids by a short review of the very classic examples, which are either very simple solutions of the Navier–Stokes equation or just very common flows.

4.4.1 Poiseuille's Flow

4.4.1.1 Stationary Regime

One of the simplest cases of steady flows is that of a viscous fluid in a very long cylindrical pipe. In this case the velocity has just one component that is parallel to the pipe axis and which we identify to the z-axis. We also assume that the flow is axisymmetric. These two symmetries imply that the velocity field may be written as $\mathbf{v} = v(r, z)\mathbf{e}_z$. Using mass conservation, we find that $\partial v/\partial z = 0$ so that $\mathbf{v} = v(r)\mathbf{e}_z$. This velocity field belongs to the class of *plane-parallel shear flows*: it has just one component, which varies in a direction perpendicular to it. As a consequence, the velocity gradient is orthogonal to the velocity itself and thus the nonlinear term $(\mathbf{v} \cdot \nabla)\mathbf{v}$ is zero. The momentum equation reads

$$-\nabla p + \mu \Delta \mathbf{v} = \mathbf{0}$$

and we find Stokes equation again. We note, however, that in this case the Reynolds number is not necessarily small. If we project this equation along $\mathbf{e}_r$ and $\mathbf{e}_z$ we find that:

$$\frac{\partial p}{\partial r} = 0 \qquad \text{and} \qquad \frac{\partial p}{\partial z} = \frac{\mu}{r}\frac{\partial}{\partial r}\left(r\frac{\partial v}{\partial r}\right)$$

The pressure is therefore independent of r and if we differentiate the second equation with respect to z, we see that the pressure gradient is necessarily constant. We call this gradient G_p and integrate the equation of v, which leads to

$$v(r) = -\frac{G_p}{4\mu}(R^2 - r^2) \tag{4.48}$$

where the constants of integration have been chosen so that $v(0)$ is finite and $v(R) = 0$. Solution (4.48) is *Poiseuille's flow*.[5] The velocity profile is parabolic. We see in this expression that the flow is in the opposite direction to the pressure gradient. The fluid flows from the high pressures toward the low pressures.

Such a velocity profile may also be found for the laminar flow of a viscous fluid between two infinite flat plates staying at a distance d from each other. If a pressure gradient G_p is set up along, say, the x-axis, then

$$v_x(z) = -\frac{G_p}{\mu} z(d - z)$$

which is also a parabolic profile with a maximum velocity of $z = d/2$.

4.4.1.2 Transients to a Poiseuille's Flow

We shall now briefly examine the way the Poiseuille flow sets up. For this, we consider two situations. The steady flow inside a pipe but close to the inlet, and the transient flow occurring in an infinitely long pipe when a pressure gradient is abruptly set up.

• When a viscous fluid enters a pipe, the Poiseuille flow is not immediately set up, especially if the Reynolds number is large. Indeed, at large Reynolds numbers, a boundary layer appears. Such a layer is very similar to the one described by Blasius' equation. It thickens as the square root of the distance to the entrance as illustrated in Fig. 4.5. When the thickness of the layer has reached

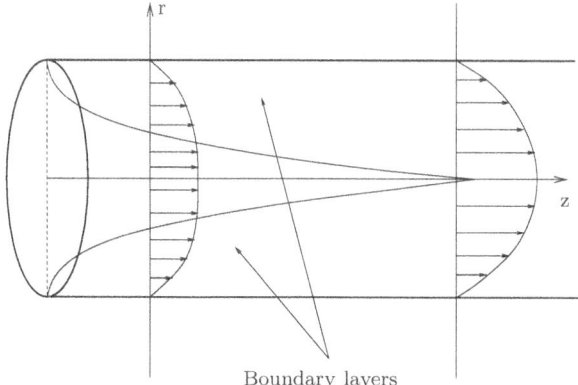

Fig. 4.5 Boundary layers at the inlet of a cylindrical pipe

---

[5]Sometimes called the Hagen–Poiseuille flow. Hagen studied it in 1839 and Poiseuille in 1840.

the radius of the pipe, Poiseuille flow is almost established. Using the results of Blasius boundary layer, we can estimate the distance from the entrance at which Poiseuille flow appears. The thickness of the layer is given by

$$e = \frac{D}{\sqrt{\text{Re}}} \sqrt{z}$$

In this expression, D is the diameter of the pipe. From the boundary layer theory, we know that the boundary layer thickness scales like $D/\sqrt{\text{Re}}$, so that using the result of the Blasius flow we find the above expression (which only gives an order of magnitude). We thus see that Poiseuille flow appears at a distance from the entrance which is typically $\text{Re}/2$ times the diameter.

- We now consider the case of an infinite pipe in which a pressure gradient is suddenly set up. Such a situation occurs when one rapidly opens a tap or a sluice gate. In this case, the velocity field evolves according to

$$\frac{\partial v}{\partial \tau} = -G_p + \text{Re}^{-1} \frac{1}{r} \frac{\partial}{\partial r} \left(r \frac{\partial v}{\partial r} \right)$$

If we solve this equation numerically, we find a result similar to that of Fig. 4.6. In this figure we clearly see the boundary layers at early times and their progressive diffusion towards the interior up until the formation of the parabolic profile.

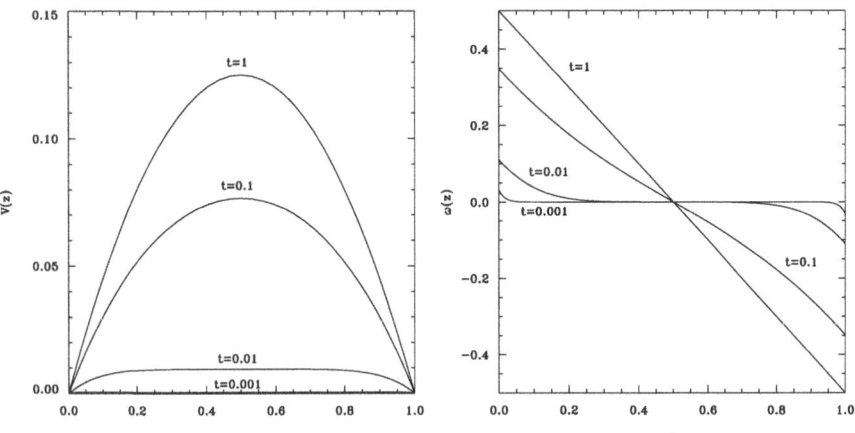

Fig. 4.6 Time evolution of the velocity and vorticity during a transient leading to the Poiseuille flow between two plates. The equation $\frac{\partial v}{\partial \tau} = -\frac{\partial p}{\partial x} + \frac{\partial^2 v}{\partial z^2}$ with $\partial p / \partial x = -1$ has been integrated numerically, giving the velocity while vorticity is $\omega_y = \partial v / \partial z$

4.4.1.3 Generation of Vorticity

The set-up of Poiseuille flow in the foregoing examples comes from a single phenomenon: the diffusion of vorticity from the walls. The case of the time evolution is very clear: just after the pressure gradient is set up, at $t = 0.001$, vorticity is zero everywhere except near the bounding planes. As time passes, it diffuses slowly to the interior until the steady state is reached. This shows the key role played by the no-slip boundary conditions in the generation of vorticity. These conditions prevent the flow from remaining irrotational.

In the other example, vorticity also diffuses from the walls, but it is simultaneously advected by the main stream. These two effects combine, and give birth to the square root law that we met in analysing the Blasius layer. Indeed, the diffusion of a quantity proceeds with the square root of time (see the discussion of the diffusion equation in the maths complements) : if δ is the distance to the wall at which the vorticity takes a given value, then $\delta \propto \sqrt{t}$. But t is such that $z = Vt$; thus $\delta \propto \sqrt{z/V}$. We thus find again the square root law of Blasius boundary layer. It is a consequence of advection and diffusion acting simultaneously.

4.4.2 Head Loss in a Pipe

When we studied the motion of perfect fluids, we introduced the notion of head losses which we connected to energy dissipation. We are now in a position to estimate these losses in some simple cases.

4.4.2.1 Regular Head Losses

We begin with the case of the Poiseuille's flow of an incompressible fluid. We can easily calculate the head loss between two points separated by a distance L and belonging to the same streamline. Since the kinetic energy is constant along each streamline (this is mandatory because of mass conservation and incompressibility), the loss of mechanical energy $P + \frac{1}{2}\rho v^2$ comes from the pressure gradient G_p. Thus, over a distance L, the loss is $G_p L$; the head therefore decreases linearly in the downstream direction. One says that the head loss is *regular*.

More quantitatively, if the volume flux in the pipe is Q and the volumic mechanical energy is $E_m(x)$, x being the coordinate along the streamline, the power dissipated between the two points is just:

$$D = (E_m(x) - E_m(x + L))Q$$

Since $E_m = \frac{1}{2}\rho v^2 + P$ and since the velocity does not depend on x, we find that $D = LG_pQ$. We may verify that this expression also comes out of a direct calculation of the viscous dissipation, and that D does not depend on the velocity profile inside the pipe, but just on the pressure gradient.

4.4.2.2 Singular Head Losses

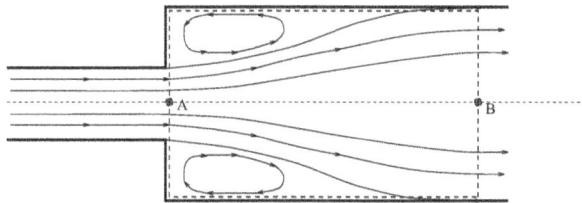

Let us now imagine the case of a pipe flow at the place where the pipe's cross section abruptly increases, just as shown in the above figure. Such a change in the pipe, provokes the separation of streamlines from the wall and gives birth to a jet. Further downstream, this jet reconnects to the wall of the pipe. In between, we find a region of "dead water" where recirculation vortices stand. The flow is quite complicated there, but an evaluation of the losses and gains of momentum, between the upstream and downstream sides, allows us to find out the head loss due to this singularity of the cross section. Such kinds of head losses are called *singular*. Other examples of singularities are pipe junctions, pipe bends, etc.

To understand the effect of the abrupt change of pipe section, we consider a fixed control surface (shown with dashed lines in the figure). The difference between the in and out momentum flux is compensated by the pressure difference in A and B. In B the pressure is uniform in the cross section since streamlines are all parallel to the pipe boundaries (see Sect. 3.2.2), however in A this is less obvious. In fact it is almost uniform there also. The reason is similar as for B: in a cross section of the jet, pressure does not vary because of its almost parallel streamlines; it is equal to the one just outside it. Thus, provided the pressure is constant in the dead water region (this is approximate of course), we may assume that the pressure is constant all over the section in A. Thus we write

$$(P_A - P_B)S_B = \rho V_B^2 S_B - \rho V_A^2 S_A$$

But since the volume flux is conserved, $V_A S_A = V_B S_B$, and thus we get

$$P_A - P_B = \rho V_B(V_B - V_A) \tag{4.49}$$

This result is sometimes called *Bélanger's Theorem*. We show now that some energy is lost in the crossing of the enlargement; for this we calculate the difference

$$\Delta = (P_A + \frac{1}{2}\rho V_A^2) - (P_B + \frac{1}{2}\rho V_B^2) \ .$$

From the preceding formula we get

$$\Delta = \frac{1}{2}\rho(V_A - V_B)^2 > 0 \qquad\qquad (4.50)$$

This difference is thus always positive: there is always a head loss due to the sudden change of cross section.

We examined here the case of an abrupt enlargement of the cross section; in the opposite case of a cross section narrowing abruptly, some head loss also exists but not as large. This is because no jet forms, so that recirculating vortices are much smaller.

4.4.3 Flows Around Solids

Flows around solids constitute a wide class of flows with numerous applications. They are sometimes called *external flows* to underline the differences with pipe flows which are therefore *internal flows*. We shall describe these kinds of flows with the vorticity field, considering examples with increasingly high Reynolds numbers.

When the Reynolds number is small compared to unity, vorticity fills the whole space, although decreasing like $1/r^2$ as shown by (4.19) in the case of the sphere. When the Reynolds number is large compared to unity, we have seen that it is confined inside the boundary layer. However, this confinement is not complete: the boundary layer always separate somewhere on the downstream side of the solid and forms *the wake*. Thus, far from the solid, the vorticity may be found in the wake only.

When we discussed Stokes' equation in Sect. 4.2.1, we noticed the symmetry of the solutions between upstream and downstream sides. We observed that the nonlinear terms break this symmetry. Here, we see that this symmetry breaking actually occurs through the raise of the wake on the downstream side.

The shape of a wake much depends on the Reynolds number. In Fig. 4.7, we see that the wake consists of two recirculating vortices. When Re $\gtrsim$ 50, these vortices separate from the solid and form a *vortex streak* also called von Kármán streak (see Fig. 4.8). Finally, if Re $\gtrsim$ 1000, the wake becomes turbulent (e.g. Fig. 4.9).

Fig. 4.7 A glimpse at unsteady recirculating vortices behind a cylinder at Re∼ 330 (photo of the author)

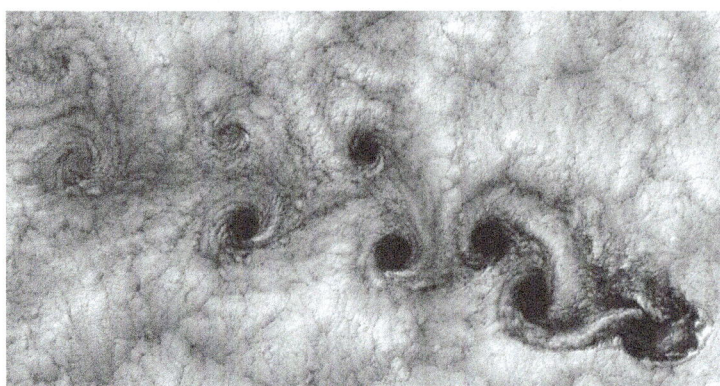

Fig. 4.8 Vortex street in the wake of Juan Fernandez islands imprinted in the clouds (Landsat 7 image, NASA)

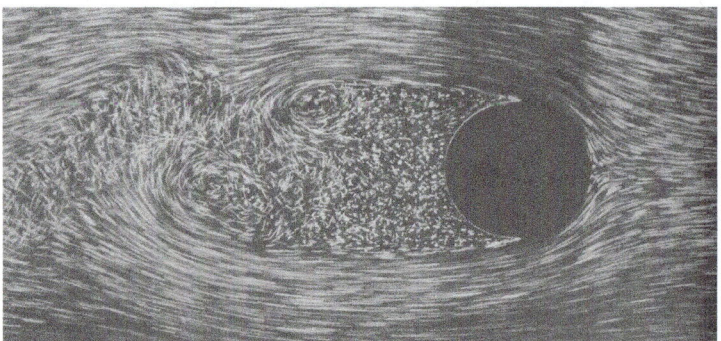

Fig. 4.9 Turbulent wake behind a cylinder at Re=2000 (ONERA photograph in Werlé and Gallon 1972).

The computation of such flows is solely possible with a direct numerical resolution of the Navier–Stokes equation (together with mass conservation) and only when the Reynolds number is not too large (presently less than a few thousands). Such flows are extremely complex.

4.5 Forces Exerted on a Solid

When the flows are known, the stresses they exert on the boundaries can be computed. We therefore continue our investigations on viscous fluids by a look at forces that they may apply on solid bodies.

4.5.1 General Expression of the Total Force

The total force exerted on a solid by a viscous fluid flowing around it is the sum of all the stresses applied to its surface, namely

$$\mathbf{F} = \int_{(S)} [\sigma] d\mathbf{S}$$

where the surface element $d\mathbf{S}$ is directed outside the solid.

To illustrate this formula we take the simple example of Poiseuille flow for which we have an explicit expression of $\mathbf{v}$. In this case, the expression of $[\sigma]$ is

$$[\sigma] = \begin{pmatrix} -p & 0 & \mu\frac{\partial v_z}{\partial r} \\ 0 & -p & 0 \\ \mu\frac{\partial v_z}{\partial r} & 0 & -p \end{pmatrix}$$

now $d\mathbf{S} = -2\pi R dz \mathbf{e}_r$ since the surface must be oriented towards the fluid which exerts the stress. The resultant force is therefore

$$\mathbf{F} = -\int \begin{vmatrix} -p \\ 0 \\ \mu\left(\frac{\partial v_r}{\partial z}\right)_R \end{vmatrix} dS = -\mu 2\pi RL \left(\frac{\partial v_z}{\partial r}\right)_{r=R} \mathbf{e}_z$$

where L is the length of the tube and R its radius. But $\left(\frac{\partial v_z}{\partial r}\right)_R = G_p R/2\mu$ so that the fluid entrains the tube with a force $\mathbf{F} = -\pi R^2 G_p L \mathbf{e}_z$ in the same direction as the flow. We observe that viscosity has disappeared from the expression of the force which means that this force can be obtained without knowing the details of the flow, simply by an integral balance (see exercises).

4.5.2 Coefficient of Drag and Lift

The expression of the force exerted on a solid is rarely accessible by direct calculation, in particular when the Reynolds number is large compared to unity. It is then necessary to resort to numerical calculation and/or to modelling (if there is turbulence). However, even if we totally ignore the form of the flow it is always possible to connect this force, may be just dimensionally, to the fundamental quantities of the flow. This is why one introduces non-dimensional coefficients which concentrate our ignorance about the flow.

When the Reynolds number is large, the pressure field due to the inertia of the fluid is the main source of stresses exerted on a solid moving in a fluid. When we studied the motion of perfect fluids, we saw that pressure at an upstream stagnation point was $\frac{1}{2}\rho v^2$ (also called dynamic pressure); multiplying this pressure by a surface typical of the solid, we get an order of magnitude for the force. This may be expressed in the following way:

$$\mathbf{F} = \frac{1}{2}\rho V^2 \begin{vmatrix} S_x C_x \\ S_y C_y \\ S_z C_z \end{vmatrix}$$

where ρ is the density of the fluid, V the velocity of the solid, and S_x, S_y and S_z are the surfaces projected on planes perpendicular to each axis (see Fig. 4.10).

If the solid moves along Ox, C_x is called *drag coefficient*, C_z *lift coefficient*, while C_y, rarely utilized, could be called *coefficient of lateral lift*.

These coefficients depend on the shape of the body and on the Reynolds number: a well-shaped body has a smaller C_x than an ill-shaped one. When the Reynolds number is very large ($\gtrsim 10^6$), these coefficients are almost constant.

The dependence on the shape of a body is most easily shown with the case of a wing. In this example, the coefficients much depend on the incidence of the wing: for small values the lift increases with the sine of this angle. But if this angle exceeds some critical value, the lift drops abruptly: the boundary layer separates from the wing near the leading edge. Streamlines are no longer curved and the resulting pressure drop, which is responsible of the lift, disappears. In addition, the drag strongly raises: this is known as *the wing stall*. This situation is illustrated in Fig. 4.11.

4.5.3 Example: Stokes' Force

To conclude this section, we examine the case of the sphere in uniform motion in a viscous fluid when the Reynolds number is very small. The solution that we obtained in Sect. 4.2.3 allows us to calculate the expression of the resultant force. This force

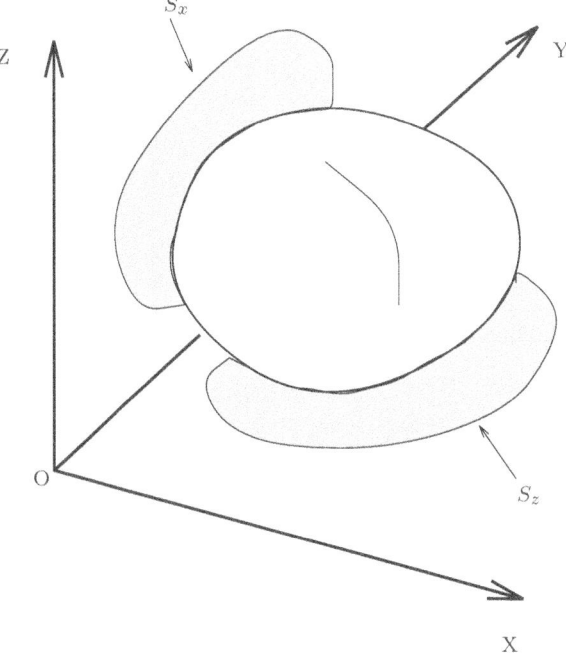

Fig. 4.10 Schematic views of the various projections of a solid on a plane

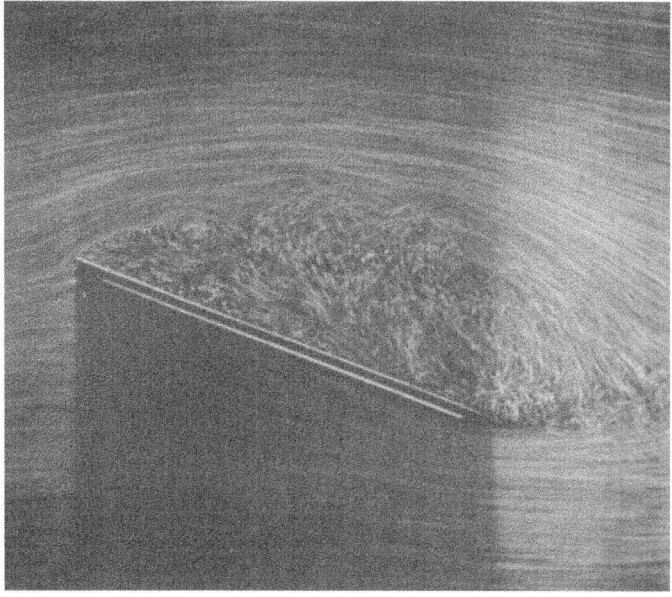

Fig. 4.11 Flow around an inclined plate at stall (© ONERA photograph, Werlé 1974)

expresses as

$$\mathbf{F} = \int_{(S)} [\sigma]\mathbf{e}_r dS = \int_{4\pi} (\sigma_{rr} \cos \theta - \sigma_{r\theta} \sin \theta) R^2 \sin \theta d\theta d\phi \mathbf{e}_z$$

where we have used the symmetry of the flow around Oz. The boundary conditions on the sphere allow us to write:

$$\sigma_{rr}(R) = -p(R) \quad \text{and} \quad \sigma_{r\theta}(R) = \mu \left(\frac{\partial v_\theta}{\partial r} \right)_{r=R}$$

By using the solution (4.18) and after evaluation of the integrals we get

$$F_z = 6\pi \mu R U \tag{4.51}$$

which is the expression of *Stokes' force*. This formula may be used in various ways, but an interesting one is the measurement of the dynamic viscosity of Newtonian fluids. Indeed, if we let a ball falling in a viscous fluid, provided its radius is small enough, its velocity quickly reaches a constant value. This value results from the balance between the weight (minus the buoyancy force) and the viscous friction (Stokes force). The result is that dynamic viscosity is given by

$$\mu = (m - m_f)g/(6\pi R V)$$

where V is the velocity of the falling ball, m is its mass and m_f the mass of the displaced fluid. g is the local gravity. Since Stokes' formula is valid only at very low Reynolds number, it is necessary to check that this condition is verified once the viscosity is determined. For instance, a small glass ball, weighing 0.02 g, left in glycerin, falls with a constant speed of 1 cm/s. This corresponds to a Reynolds number $\sim$0.04. However, if the same experiment is made with water, we would expect, from Stokes formula, a final velocity of 10 m/s and a Reynolds number of 20,000, which is certainly not consistent with the use of Stokes equation.

From Stokes formula, we may also compute the drag coefficient of a sphere at low Reynolds numbers. We find

$$C_x = \frac{6\pi \mu R U}{\frac{1}{2}\rho U^2 \pi R^2} = \frac{24}{\text{Re}}$$

using a Reynolds number based on the diameter of the sphere.

Thus, the C_x coefficient decreases like the inverse of the Reynolds number. At infinite Reynolds number it is zero which is reminiscent of d'Alembert's paradox, but in this limit, again, Stokes' formula is not valid!

The variations of the sphere's C_x with Reynolds number has been well studied experimentally. Figure 4.12 reproduces the curve derived from experiments like in Fig. 4.13. We see the decrease in 1/Re for the small numbers, then a plateau and

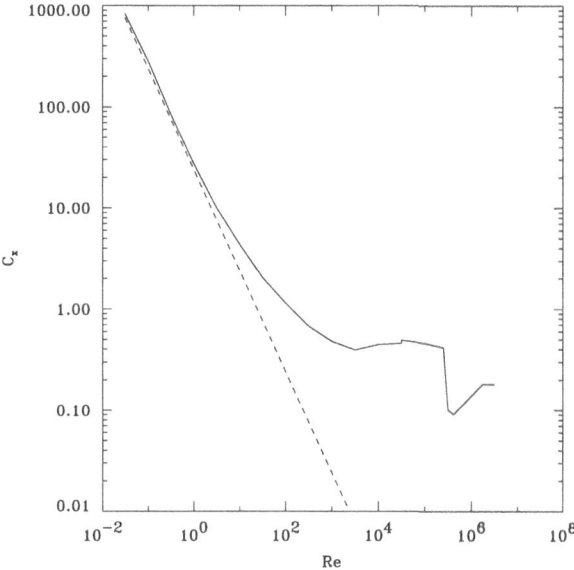

Fig. 4.12 Variation of the C_x coefficient of a sphere with the Reynolds number (*solid line*). The *dashed line* shows the law $C_x = 24/\text{Re}$ valid at small Reynolds numbers

Fig. 4.13 View of the turbulent wake of a sphere at $\text{Re} = 15{,}000$ (© ONERA photograph, H. Werlé)

an abrupt jump which correspond to the transition of the boundary layers towards turbulence. It is usually admitted that beyond this value C_x remains constant.

4.6   Exercises

1. Find the expression of the force exerted by a Poiseuille flow in a cylindrical pipe using an integral balance. L is the length of the pipe, R its radius and G_p the applied constant pressure gradient.
2. *Lubrication layer:* show that if the solid inclination is very small, i.e. that $h_1 = h_2(1 + \varepsilon)$, then

$$F_{\text{drag}} \simeq \frac{h_2}{3\ell} F_{\text{lift}}$$

   Compute the force needed to push a solid of 10^3 kg at constant speed on an oil film 1 mm thick if the length of the contact surface is 1 m (the contact surfaces are assumed to be perfect planes).
3. *Flow of a viscous fluid on a slope*

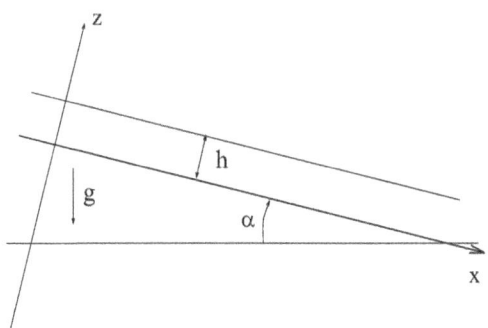

The fluid layer, as shown in the above figure, meets free-slip boundary conditions on the top plane and no-slip ones on the bottom plane. The planes make an angle α with the horizontal. The fluid is incompressible and of kinematic viscosity ν. The flow is steady.

(a) Determine the velocity profile assuming that $\mathbf{v} = V(z)\mathbf{e}_x$.
(b) Find the volume flux through a cross section S.

4. *The Taylor–Couette flow or cylindrical Couette flow*
   We consider a viscous fluid contained between two rotating cylinders of radii R_1 and R_2, respectively. Their angular velocity is Ω_1 and Ω_2.

(a) We look for a solution like $\mathbf{v} = v(s)\mathbf{e}_\varphi$. What are the symmetries of this solution? Show that $v(s)$ is the solution of a linear equation. Solve it and give the solution verifying the boundary conditions.

(b) What is the torque exerted by the interior cylinder on the outer one.

(c) How can we measure the viscosity of a fluid with such a device?

5. *Falkner–Skan equation*: We take the Prandtl equations of the laminar boundary layer (4.37), but we look for solutions more general than the Blasius ones. We set $\tilde{u}_{0,x} = U(x)f(\eta)$ and $\tilde{u}_{1,z} = V(x)g(\eta)$, where η is still the self-similarity variable and $U(x)$ the x-component of the velocity at infinity.

(a) Give the expression of $\partial p_0/\partial x$ as a function of $U(x)$.

(b) Show that the existence of such solutions implies that $U(x)$, $V(x)$ and $b(x)$ verify:

$$2Ubb' = c_1, \qquad U'b^2 = c_2 \qquad \text{and} \qquad Vb = c_3$$

where c_1, c_2, c_3 are constants.

(c) Derive the general form of $U(x)$ and $b(x)$.

(d) Show that if one chooses $c_1/2 + c_2 = 1$ (why is that always possible?), then $F = \int f d\eta$ verifies Falkner–Skan equation:

$$F''' + FF'' - \frac{2m}{m+1}(F'^2 - 1) = 0 \qquad (4.52)$$

where m is a constant to be related to c_1, c_2, c_3.

(e) For which value of m is the boundary layer thickness constant? How can we find the Blasius equation again?

Further Reading

The matter of this chapter belongs to the very base of Fluid Mechanics and therefore may be found in all the books aimed at introducing Fluid Mechanics; for instance, Batchelor (1967), Faber (1995), Guyon et al. (2001), Landau and Lifchitz (1971–1989), Paterson (1983), Ryhming (1985, 1991).

References

Batchelor, G. K. (1967). *An introduction to fluid dynamics*. Cambridge: Cambridge University Press.
Faber, T. (1995). *Fluid dynamics for physicists*. Cambridge: Cambridge University Press.
Guyon, E., Hulin, J.-P., Petit, L., & Mitescu, C. (2001). *Physical hydrodynamics*. Oxford: Oxford University Press.

Landau, L., & Lifchitz, E. (1971–1989). *Mécanique des fluides*. Moscow: Mir.
Paterson, A. (1983). *First course in fluid mechanics*. Cambridge: Cambridge University Press.
Ryhming, I. (1985, 1991). *Dynamique des fluides*. Lausanne: Presses Polytech. Univ. Romandes.
Werlé, H. & Gallon, M. (1972). *Aéronaut. Astronaut.*, **34**, 21–33

Chapter 5
Waves in Fluids

5.1 Ideas on Disturbances

Disturbances play an important role in Physics and notably in Fluid Mechanics. Indeed all flows in Nature are constantly subjected to perturbations of various origin: thermal noise, variations of boundary conditions, etc. If the flow is stable, these disturbances are always damped: otherwise, some of them grow, up to the disappearance of the initial flow replaced by, perhaps, more stable one. The study of the stability of a flow therefore begins with the study of perturbations. However, before addressing the case of flow stability in the next chapter, we shall first concentrate on the simplest manifestation of disturbances, namely *the waves*. Their existence is indeed the first evidence that an equilibrium (or a steady state) has been slightly perturbed.

5.1.1 Equation of a Disturbance

Let us begin with the simple case of a disturbance affecting the steady flow $\mathbf{V}$ of an incompressible fluid. The fluid is in a domain D delimited by a solid wall ∂D on which $\mathbf{V} = \mathbf{0}$. The motion is generated by a force $\mathbf{f}$. The equations of the original steady flow are simply

$$\begin{cases} \rho \mathbf{V} \cdot \nabla \mathbf{V} = -\nabla P + \mu \Delta \mathbf{V} + \mathbf{f} \\ \\ \nabla \cdot \mathbf{V} = 0 \\ \\ \mathbf{V} = \mathbf{0} \quad \text{on} \quad \partial D \end{cases} \qquad (5.1)$$

© Springer International Publishing Switzerland 2015   149
M. Rieutord, *Fluid Dynamics*, Graduate Texts in Physics,
DOI 10.1007/978-3-319-09351-2_5

Given that $(\delta P, \delta \mathbf{v})$ is a disturbance of this flow, the total field $(P + \delta P, \mathbf{V} + \delta \mathbf{v})$ must also be a solution to the equations of motion

$$
\begin{cases}
\rho\left(\frac{\partial(\mathbf{V}+\delta\mathbf{v})}{\partial t} + (\mathbf{V}+\delta\mathbf{v})\cdot\nabla(\mathbf{V}+\delta\mathbf{v})\right) = -\nabla(P+\delta P) + \mu\Delta(\mathbf{V}+\delta\mathbf{v}) + \mathbf{f} \\[2mm]
\nabla\cdot(\mathbf{V}+\delta\mathbf{v}) = 0 \\[2mm]
\mathbf{V}+\delta\mathbf{v} = \mathbf{0} \qquad \text{on} \quad \partial D
\end{cases}
$$

$$(5.2)$$

We develop these terms and subtract (5.1), which leads to the following equations for the disturbances:

$$
\begin{cases}
\rho\left(\frac{\partial\delta\mathbf{v}}{\partial t} + \overbrace{\mathbf{V}\cdot\nabla\delta\mathbf{v}}^{\text{I}} + \overbrace{\delta\mathbf{v}\cdot\nabla\mathbf{V}}^{\text{II}} + \delta\mathbf{v}\cdot\nabla\delta\mathbf{v}\right) = -\nabla\delta P + \mu\Delta\delta\mathbf{v} \\[2mm]
\nabla\cdot\delta\mathbf{v} = 0 \\[2mm]
\delta\mathbf{v} = \mathbf{0} \qquad \text{on} \qquad \partial D
\end{cases}
$$

$$(5.3)$$

Terms I and II show that the disturbances do not obey the same equations as the original flow. Their evolution is indeed a function of the flow on which they appear and this dependence is the result of the nonlinearities of the original equations.

5.1.2   Analysis of an Infinitesimal Disturbance

The study of disturbances is done in successive steps. The first of these consists in analyzing the case of disturbances whose amplitude is infinitesimal: indeed, for these disturbances the equations are linear and therefore easy to resolve (in theory!). Two situations can thus occur: either we are seeking the evolution of disturbances in a homogeneous region of the flow (whose properties are independent of the local coordinates) and we make an analysis using plane waves, also called *local analysis*, or we are facing important spatial variations and we must do a *global analysis* (this is the case if the boundary conditions play a role).

5.1.2.1   Local Analysis

Local analysis is the easiest one because the form of the disturbances is known in advance. Let us consider the system (5.3); by linearizing it, we have

$$\begin{cases} \rho \left(\dfrac{\partial \delta \mathbf{v}}{\partial t} + \mathbf{V} \cdot \nabla \delta \mathbf{v} + \delta \mathbf{v} \cdot \nabla \mathbf{V} \right) = -\nabla \delta P + \nu \Delta \delta \mathbf{v} \\[2ex] \nabla \cdot \delta \mathbf{v} = 0 \\[2ex] \delta \mathbf{v} = \mathbf{0} \quad \text{on} \quad \partial D \end{cases} \tag{5.4}$$

for which we seek a solution of the plane wave type, namely

$$\delta \mathbf{v} = \delta \mathbf{v}_0 \, e^{i(\omega t + \mathbf{k} \cdot \mathbf{r})} \quad \text{and} \quad \delta P = \delta P_0 \, e^{i(\omega t + \mathbf{k} \cdot \mathbf{r})} \tag{5.5}$$

where $\mathbf{k}$ is the wave vector. Observing that the operators ∇ and $\frac{\partial}{\partial t}$ are transformed in the following manner:

$$\nabla \to i \mathbf{k} \quad \text{and} \quad \frac{\partial}{\partial t} \to i \omega \,,$$

we immediately get the system

$$\begin{cases} i(\omega + \mathbf{k} \cdot \mathbf{V}) \delta \mathbf{v}_0 + (\delta \mathbf{v}_0 \cdot \nabla) \mathbf{V} = -\delta P_0 i \mathbf{k} - \nu k^2 \mathbf{v}_0 \\[2ex] \mathbf{k} \cdot \delta \mathbf{v}_0 = 0 \end{cases} \tag{5.6}$$

The solution (5.5) and the relation (5.6) that follows are valid only if $\mathbf{V}$ and $\nabla \mathbf{V}$ are almost constant. This is obviously not the case in general: but if we limit ourselves to a small area of the flow, then these quantities are almost constants and the derivation that we did does make sense. This is the reason for which it is called local analysis. It is valid only if the wavelength λ of the disturbance is very small compared to the scale of the variations of $\mathbf{V}$ or $\nabla \mathbf{V}$; in other words

$$\lambda \ll \text{Min} \left[\frac{\|\mathbf{V}\|}{\|\nabla \mathbf{V}\|}, \frac{\|\nabla \mathbf{V}\|}{\|\nabla^2 \mathbf{V}\|} \right] \tag{5.7}$$

Let us continue our analysis and give a matrix form to relation (5.6), namely

$$\begin{bmatrix} D_{xx} & D_{xy} & D_{xz} & ik_x \\ D_{yx} & D_{yy} & D_{yz} & ik_y \\ D_{zx} & D_{zy} & D_{zz} & ik_z \\ k_x & k_y & k_z & 0 \end{bmatrix} \begin{pmatrix} \delta v_{0,x} \\ \delta v_{0,y} \\ \delta v_{0,z} \\ \delta P_0 \end{pmatrix} = \mathbf{0} \tag{5.8}$$

This system has a non-zero solution if the determinant of the matrix is vanishing. Since each component D_{ij} of this matrix is a function of $\mathbf{k}$ and ω, we finally get the *dispersion relation* of the waves:

$$\det[D] = \mathcal{D}(\omega, \mathbf{k}) = 0 \tag{5.9}$$

We note that the dispersion relation is an implicit relation between ω and $\mathbf{k}$. We shall see in the next chapter that this form of the relation has important consequences as far as stability is concerned.

5.1.2.2 Global Analysis

When one cannot neglect the boundary conditions or the heterogeneities of the disturbed system, one cannot impose the plane wave form to the disturbances. Their partial differential equations (5.4) need to be solved directly.

If the solution $\mathbf{V}$, which we analyse, is stationary, we can look for disturbances in the form

$$\delta \mathbf{v}(\mathbf{r}, t) = \mathbf{v}_\lambda(\mathbf{r}) e^{\lambda t} \tag{5.10}$$

Such a solution is called an *eigenmode* of the system. It is associated with the eigenvalue λ, which is a complex number in general. The search for the eigenmodes of a system is also called *modal analysis*.

If we note that the system (5.4) may be written

$$\begin{cases} \mathcal{L}(\delta \mathbf{v}) = \lambda \delta \mathbf{v} & \text{in} \quad D \\ \delta \mathbf{v} = 0 & \text{on} \quad \partial D \end{cases} \tag{5.11}$$

the search of the eigenmodes is equivalent to finding the eigenfunctions of the operator $\mathcal{L}$ verifying the boundary conditions. Simultaneously, we determine the associated eigenvalues which give the point spectrum (the set of eigenvalues) of the operator $\mathcal{L}$. If the operator is compact[1] then the spectrum is discrete and each eigenvalue can be identified by a triplet of quantum numbers (ℓ, m, n).

The resolution of such a problem is difficult in general and must be carried out numerically. In the examples that we shall consider, we shall combine the local analysis and the global analysis so as to reduce the partial differential equation to ordinary differential equations. This is possible when the system owns symmetries.

5.1.3 Disturbances with Finite Amplitude

When the amplitude of the disturbances cannot be neglected, the problem becomes very complicated because of the nonlinearities of the equations. Several strategies are then possible.

---

[1] To say it in a simple way, compact (linear) operators are like matrices of finite dimension although the space on which they act is a space of functions and therefore of infinite dimension.

- The amplitudes are finite but small: we can develop the solution into powers of the amplitude.
- Several, very different, scales intervene in the problem and we are able to make a multi-scale expansion of the solution.

An example of each of these strategies will be given in Chap. 7 using the case of thermal convection.

5.1.4 Waves and Instabilities

In the following we analyse the simple case of disturbances that are neither amplified nor damped (or very little). They are waves which freely propagate in the fluid. When an amplification appears, ones speaks rather of an instability, the study of which is postponed to the following chapter.

5.2 Sound

5.2.1 Equation of Propagation

Sound waves are the simplest and the most frequent of the disturbances which propagate in a fluid. In order to study them, we assume that the undisturbed fluid is at rest, i.e. $\mathbf{V} = \mathbf{0}$. With regard to (5.3), we must take into account the compressibility of the fluid: sound does not exist in an incompressible environment!

We make the following expansion:

$$\begin{cases} P = P_0 + \delta P \\ T = T_0 + \delta T \\ \rho = \rho_0 + \delta\rho \\ \mathbf{v} = \mathbf{0} + \delta\mathbf{v} \end{cases} \tag{5.12}$$

We assume moreover that the fluid be *perfect* and initially at constant pressure, density and temperature. By neglecting all the nonlinear terms we get

$$\frac{\partial \delta\rho}{\partial t} + \rho_0 \mathbf{\nabla} \cdot \delta\mathbf{v} = 0 \tag{5.13}$$

from the equation of continuity,

$$\rho_0 \frac{\partial \delta\mathbf{v}}{\partial t} = -\mathbf{\nabla}\delta P \tag{5.14}$$

from the equation of momentum and

$$\frac{\partial \delta s}{\partial t} = 0 \tag{5.15}$$

from the equation of entropy. This last equation can be used immediately. Indeed, this equation implies $\delta s = f(x, y, z)$; but at $t = 0$ (or before the disturbance starts) $\delta s = 0$, and therefore the disturbance stays isentropic. We can thus write a relation between the fluctuations of ρ and P:

$$\delta P = \left(\frac{\partial P}{\partial \rho}\right)_s \delta \rho \tag{5.16}$$

where the partial derivative of the pressure is taken at constant entropy. Now, if we take the time derivative of (5.13) and combine it with (5.14) and (5.16), we get the following wave equation:

$$\Delta \delta \rho - \frac{1}{c_s^2} \frac{\partial^2 \delta \rho}{\partial t^2} = 0 \quad \text{with} \quad c_s^2 = \left(\frac{\partial P}{\partial \rho}\right)_s \tag{5.17}$$

c_s is naturally identified with the speed of propagation of the disturbance. We easily verify by exercise that δP and $\delta \mathbf{v}$ obey the same wave equation.

If the gas is ideal, the speed of sound can be expressed as

$$c_s^2 = \frac{\gamma P_0}{\rho_0} = \gamma R_* T_0 \tag{5.18}$$

where γ and R_* are defined in Sect. 1.7.1 This equation shows that, for an ideal gas, c_s depends only on temperature. Let us calculate an order of magnitude of the speed of sound in the air at 300 K. For $\gamma = 1.4$, $R_* = 8.314/0.029$ J/kg and $T_0 = 300$ K, we find $c_s = 347$ m/s. We observe that sound propagates faster in hot gases and with small molecular mass. In hydrogen at 300 K, $M = 0.002$ kg/mole, the sound speed is $c_s = 1321$ m/s thus almost four times faster than in the air.

The sound speed is of the same order of magnitude as the rms[2] velocity of the molecules of the gas. Pressure is indeed due to collisions between molecules and pressure disturbances cannot go much faster than the molecules themselves!

5.2.2 The Dispersion Relation

In the medium that we have chosen the sound waves have a very simple dispersion relation; assuming

$$(\delta P, \delta \rho, \delta \mathbf{v}) \propto \exp(i \omega t - i \mathbf{k} \cdot \mathbf{r})$$

---

[2] For "root mean square"; this is the typical dispersion of molecules velocities in a gas.

we easily obtain *the dispersion relation*

$$\omega^2 = c_s^2 k^2 \tag{5.19}$$

which shows that the waves are not dispersives since the phase velocity ω/k is independent of k.

Let us now consider the orientation of the velocity field associated with the wave, and the wavevector $\mathbf{k}$. We take back (5.14) which we transform into

$$\rho_0 i \omega \delta \mathbf{v} = i \mathbf{k} \delta P \tag{5.20}$$

This last relation shows that the velocity vector is parallel to the wave vector. One says that the wave is *longitudinal*.

5.2.3   Examples of Acoustic Modes in Wind Instruments

The study of sound waves naturally leads to the vast domain of acoustics. We shall just outline the subject by examining the acoustic oscillations associated with wind instruments.

5.2.3.1   The Flute

The flute is one of the oldest instruments and its principle is one of the simplest. It is based on the oscillation of an air column in a cylindrical pipe. In order to study this oscillation, we neglect the viscosity of the air and assume its motion to be one-dimensional. Taking the axis of the tube parallel to Ox, with the origin at one extremity and the other at $x = L$, we write that the velocity, the pressure, etc. are the superimposition of two plane waves propagating in opposite directions, namely

$$v_x = A e^{i(\omega t - kx)} + B e^{i(\omega t + kx)}$$

$$\delta p = A' e^{i(\omega t - kx)} + B' e^{i(\omega t + kx)}$$

At the extremities of the tube, the pressure is fixed (it is the atmospheric pressure), so that the pressure disturbance vanishes there.[3] These two boundary conditions allow us to write

$$\delta p = 0 \quad \text{at} \quad x = 0 \qquad \Longrightarrow \qquad A' = -B'$$

$$\delta p = 0 \quad \text{at} \quad x = L \qquad \Longrightarrow \qquad A' e^{-ikL} + B' e^{ikL} = 0$$

---

[3]This is an idealization of course. In reality the fluctuations of pressure do not exactly vanish, but their amplitude is very small compared to the one inside the tube.

from which comes the relation

$$\sin kL = 0 \quad \Longleftrightarrow \quad k = \frac{n\pi}{L}, \quad n \in \mathbb{N} \tag{5.21}$$

(5.21) gives the wavelength of the acoustic modes of a fluid in a cylindrical cavity open at both ends, namely

$$\lambda = \frac{2L}{n}$$

The reader has certainly observed that we just determined an eigenmode of the air column since we took the boundary conditions into account. This example shows the utility of the local analysis, which, in some cases, is easily extended to the global one.

The frequency F of these modes is immediately obtained from the dispersion relation $\omega = kc_s$ with $\omega = 2\pi F$. We have therefore

$$F_n = \frac{nc_s}{2L} \tag{5.22}$$

Let us apply this result to a flute in which the lowest note ($n = 1$) is the C at 261.6 Hz. Its length should be (if $c_s = 347$ m/s) L= 66.3 cm, to be compared to the length of a modern transverse flute in C which is 67 cm. We also note that the next harmonic, $n = 2$, vibrates at a frequency exactly two times higher than the fundamental $n = 1$. Hence, if the player is able to excite the second harmonic, a new set of notes with a frequency twice higher, i.e. at the next octave of the fundamental, is available.

5.2.3.2 The Clarinet

The clarinet is another interesting instrument because it uses different boundary conditions: one of the extremities is closed and we must set the velocity to zero there.[4] We have

$$\delta p = 0 \quad \text{at} \quad x = 0 \quad \Longrightarrow \quad A' = -B'$$
$$\delta \mathbf{v} = \mathbf{0} \quad \text{at} \quad x = L \quad \Longrightarrow \quad Ae^{-ikL} + Be^{ikL} = 0$$

Now, additional relations between A, A', B and B' are necessary. These relations are given by (5.20), thus

$$A' = \rho_0 c_s A \quad \text{and} \quad B' = -\rho_0 c_s B$$

---

[4]This doesn't imply that $\delta p = 0$ because (5.20) doesn't apply to a superimposition of plane waves.

which implies that $A = B$. Moreover,

$$\cos kL = 0 \qquad \Longleftrightarrow \qquad k = \frac{(2n+1)\pi}{2L}, \quad n \in \mathbb{N} \qquad (5.23)$$

The frequencies of the different harmonics are thus

$$F_0 = \frac{c_s}{4L}, \qquad F_1 = \frac{3c_s}{4L}, \quad \dots, \qquad F_n = \frac{(2n+1)c_s}{4L}, \dots$$

Let us apply this result to a real instrument like the B-flat clarinet. The fundamental is the D at 146.8 Hz. The calculation of its theoretical length is 59 cm to be compared to the real length of 60 cm. Contrary to the flute, the first harmonic ($n = 1$) is not the next octave, but a frequency that is three times the fundamental one ($n = 0$), namely the A at 440 Hz (for the player this is an octave plus a fifth or a perfect twelfth).

Because of this dispersion relation of the modes, this instrument is necessarily more complex to make. Other examples, like the oboe and the bassoon are studied in the exercises.

5.3   Surface Waves

A second category of very common waves in our environment is that of surface gravity waves, namely all the waves which agitate the surface of water planes. Contrary to sound waves, these waves are very dispersive, i.e. their phase and group velocity strongly depends on the wavelength. We would have a hard time making music if sound waves behaved that way!

5.3.1   Surface Gravity Waves

In order to understand the way in which these waves propagate, we must return to the boundary conditions ruling a free surface. We have seen in Chap. 1 (Sect. 1.8.1.2) that the surface obeys the kinematic condition

$$\frac{\partial S}{\partial t} + \mathbf{v} \cdot \nabla S = 0 \qquad (5.24)$$

where $S = \text{Cst}$ is the equation of the given surface. To this condition we add the dynamic one, which imposes the continuity of the stress when we cross the surface. As we neglect viscosity, this last condition amounts to the continuity of the pressure. We shall also neglect the effects of surface tension which will be examined separately.

In order to treat this problem we make several simplifying hypotheses: we first assume that the fluid is incompressible and that its motion is vorticity free. This latter assumption means that the flow is driven by forces derived from a potential. To be more precise, we consider the case of an interface between air and water and simultaneously treat the motion of the two fluids. The equations which govern these motions are (3.22) and (3.23):

$$\begin{cases} \Delta \Phi_w = 0 \\ \dfrac{\partial \Phi_w}{\partial t} + \dfrac{1}{2} \mathbf{v}_w^2 + \dfrac{P_w}{\rho_w} + gz = \text{Cst} \end{cases} \tag{5.25}$$

$$\begin{cases} \Delta \Phi_a = 0 \\ \dfrac{\partial \Phi_a}{\partial t} + \dfrac{1}{2} \mathbf{v}_a^2 + \dfrac{P_a}{\rho_a} + gz = \text{Cst} \end{cases} \tag{5.26}$$

where the indices a and w refer to air and water respectively. We first look for one-dimensional solutions that propagate in the x-direction:

$$\Phi = \Phi(z)e^{ikx - i\omega t} \tag{5.27}$$

Laplace's equation then implies

$$\frac{\partial^2 \Phi}{\partial z^2} - k^2 \Phi = 0 \qquad \Longrightarrow \qquad \Phi = Ae^{kz} + Be^{-kz} \tag{5.28}$$

5.3.1.1 In Deep Water

We assume, as a first step, that the air occupies the upper half-space $z \geq 0$ while the water occupies the lower half-space $z \leq 0$. In this case

$$\Phi_a = A_a e^{-kz} \qquad \text{and} \qquad \Phi_w = A_w e^{kz} \tag{5.29}$$

Let us now consider the boundary conditions. The surface verifies an equation of the form

$$S(\mathbf{r}, t) = z - z_s(x, t) = 0$$

from which we derive that $\nabla S = \mathbf{e}_z - \dfrac{\partial z_s}{\partial x}\mathbf{e}_x$.

If we now assume that the amplitudes of motions are small, we find from (1.63), that

$$v_z = \frac{\partial z_s}{\partial t} \qquad \text{at} \quad z = 0 \tag{5.30}$$

neglecting second order terms. Finally, on the surface the velocity potential verifies

$$\frac{\partial \Phi}{\partial z} = \frac{\partial z_s}{\partial t} \quad \text{at} \quad z = 0 \tag{5.31}$$

as given by (5.24) for small amplitude motions. We apply this relation to the air and the water and thus

$$k\Phi_w(0) = -i\omega z_s = -k\Phi_a(0) \tag{5.32}$$

which shows incidentally that $A_w = -A_a$. We now use the linearized equation of dynamics and the second boundary condition (continuity of pressure). At $z = 0$ we have

$$\begin{cases} -i\omega \rho_w \Phi_w + \rho_w g z_s + \delta P_w = 0 \\ -i\omega \rho_a \Phi_a + \rho_a g z_s + \delta P_a = 0 \end{cases} \tag{5.33}$$

by subtracting these two equations and by using (5.32) together with the fact that $\delta P_w = \delta P_a$, we get the sought-after dispersion relation:

$$\omega^2 = \frac{\rho_w - \rho_a}{\rho_w + \rho_a} gk \tag{5.34}$$

If we neglect the influence of the air we simply have:

$$\omega = \sqrt{gk} \tag{5.35}$$

We easily derive from this expression the phase and group velocities, namely

$$v_\varphi = \frac{\omega}{k} = \sqrt{\frac{\rho_w - \rho_a}{\rho_w + \rho_a} \frac{g}{k}}, \qquad v_g = \frac{\partial \omega}{\partial k} = \frac{1}{2} v_\varphi \tag{5.36}$$

These two relations show that the waves are dispersive: the waves with long wavelength are the fastest.

5.3.1.2 In Shallow Water

If the depth of the water is not infinite (and especially if it is smaller than the wavelength of the waves), the dispersion relation is much simplified.

Taking the bottom into account, which we assume to be flat and located in $z = -H$, we have to modify the solution (5.28) so that the boundary condition:

$$v_z = 0 \quad \text{at} \quad z = -H$$

is verified. We easily find that it implies that

$$\Phi_w(z) = 2A_w e^{-kH} \cosh[k(z+H)] = A'_w \cosh[k(z+H)]$$

Using this relation at the surface, it turns out that

$$A'_w k \sinh(kH) = -i\omega z_s = -kA_a \tag{5.37}$$

which replaces (5.32). The equation of pressure allows us, after minor calculations, to find the new dispersion relation

$$\omega^2 = \frac{(\rho_e - \rho_a)\tanh(kH)}{\rho_e + \rho_a \tanh(kH)} gk. \tag{5.38}$$

If we neglect the density of the air, it simplifies into

$$\omega^2 = gk \tanh(kH) \tag{5.39}$$

We find again the foregoing relation, (5.35), when $\lambda \ll H$ because then $kH \gg 1$ and $\tanh(kH) \sim 1$. On the other hand, if we take the opposite case where $\lambda \gg H$, that is in the case of *a shallow layer*, then $\tanh(kH) \sim kH$ and the dispersion relation (5.39) becomes

$$\omega^2 = gHk^2 \tag{5.40}$$

The phase velocity is

$$v_\phi = \sqrt{gH}$$

identical to the group velocity: *the waves are no longer dispersive.*
Some example of the use of these results may be found in the list of exercises.

5.3.2 Capillary Waves

When discussing the case of surface gravity waves, we voluntarily ignored the role of surface tension. We may wonder whether this simplification was justified or not. In order to evaluate the effects of this new phenomenon, we just need to modify the dynamic boundary conditions. Indeed, now

$$P_{water} = P_{air} + \gamma \left(\frac{1}{R_1} + \frac{1}{R_2} \right)$$

As before we linearize these equations and simplify the problem to two dimensions, thus $R_2 = \infty$. With the linear approximation we have

$$\frac{1}{R} = -\frac{\partial^2 z_s}{\partial x^2}$$

(see 12.12 in the complements of Mathematics). This equation allows us to write

$$P_{water} = P_{air} - \gamma \frac{\partial^2 z_s}{\partial x^2} \tag{5.41}$$

The relations (5.33) are thus transformed into

$$-i\omega(\rho_w \Phi_w - \rho_a \Phi_a) + (\rho_w - \rho_a)g z_s + \gamma k^2 z_s = 0 \tag{5.42}$$

from which we derive the following dispersion relation:

$$\omega^2 = \frac{\rho_w - \rho_a}{\rho_w + \rho_a} gk + \frac{\gamma k^3}{\rho_w + \rho_a} \tag{5.43}$$

which is also written under the form

$$\omega^2 = gk + \frac{\gamma}{\rho} k^3 \tag{5.44}$$

when we neglect the density of the air. In this last relation, we have of course assumed the depth of the liquid to be infinite (the case of finite depth is proposed as an exercise). It shows that the effects of surface tension are expected at short wavelengths. They dominate if

$$\frac{\gamma}{\rho} k^3 > gk \qquad \Longleftrightarrow \qquad \lambda < \lambda_t = 2\pi \sqrt{\frac{\gamma}{\rho g}}$$

For water, we find that $\lambda_t = 1.7$ cm. We may observe that when the surface tension dominates, the waves are also dispersive.

5.4 Internal Gravity Waves

Gravity waves or internal gravity waves (in order to distinguish them from surface gravity waves) are present in all the fluids that are stably stratified by gravitation (see Chap. 2 sect. 2.2.3 for a presentation of a stratified fluid). This type of situation is encountered frequently in our environment. For example, in a lake where the cold water is found at the bottom and the warm water, lighter, close to the surface. Such a situation is stable. All disturbances of this equilibrium give birth to waves which are the internal waves of gravity. For such waves the restoring force is the buoyancy force which has a privileged (vertical) direction. Hence, these waves propagate anisotropically.

In order to get more familiar with these waves, we consider the following idealized situation: a quasi-incompressible fluid (such as water) is in equilibrium under the effect of gravitation. Its temperature is supposed to increase in a linear manner with z (see Chap. 2, sect. 2.2.3). We suppose moreover that the variations of density associated with the variations of temperature are negligible except those generating the buoyancy force (this is *the Boussinesq approximation* that we shall thoroughly describe in Chap. 7).

Neglecting the effects of diffusion (viscosity and thermal conduction) and the nonlinear terms, the equations of disturbances are now

$$
\begin{cases}
\dfrac{\partial \delta \mathbf{v}}{\partial t} = -\dfrac{1}{\rho}\nabla \delta p + \dfrac{\delta \rho}{\rho}\mathbf{g} \\[2ex]
\dfrac{\partial \delta T}{\partial t} + \delta \mathbf{v} \cdot \nabla T_0 = 0 \\[2ex]
\nabla \cdot \delta \mathbf{v} = 0
\end{cases}
\tag{5.45}
$$

T_0 is the temperature of the equilibrium configuration that we assume to vary linearly with z. We set

$$
T_0 = T_{00} + \beta z \qquad \text{with} \quad \beta > 0
$$

If the fluid is a liquid (see 1.60),

$$
\frac{\delta \rho}{\rho} = -\alpha \delta T
$$

where α is the dilation coefficient of the fluid.

We are looking for a solution in the form of plane waves, thus setting

$$
\delta f = f_0 \exp(i\omega t - \mathbf{k} \cdot \mathbf{r}),
$$

we transform (5.45) into

$$
\begin{cases}
i\omega \delta \mathbf{v}_0 = i\mathbf{k}p/\rho + \alpha \delta T_0 g \mathbf{e}_z \\
i\omega \delta T_0 + \beta \delta v_{0z} = 0 \\
\mathbf{k} \cdot \delta \mathbf{v}_0 = 0
\end{cases}
\tag{5.46}
$$

Taking the dot product of the first equation with $\mathbf{k}$, we find that $ipk^2 = -\alpha \rho \delta T_0 g k_z$. We combine this equation with $i\omega \delta v_z = ik_z p/\rho + \alpha \delta T_0 g$ and $i\omega \delta T_0 + \beta \delta v_{0z} = 0$, in order to finally obtain the following dispersion relation:

$$\omega^2 = N^2 \frac{k_x^2 + k_y^2}{k^2} \tag{5.47}$$

where we have set $N = \sqrt{\alpha \beta g}$. N is a frequency, called *the Brunt-Väisälä frequency*. Its interpretation is simple: it is the frequency of the oscillations of a fluid element when it is slightly moved from its position of equilibrium. Often, it is written in an equivalent manner as

$$N^2 = -\frac{g}{\rho} \frac{d\rho}{dz}$$

We note that if the density gradient is of the opposite sign (density increases with height), this frequency is imaginary. This situation corresponds to an instability (the Rayleigh-Taylor instability—Sect. 6.3.2.1- or thermal convection—Chap. 7).

The dispersion relation (5.47) shows that the waves are anisotropic. If θ is the angle between the wavevector $\mathbf{k}$ and $\mathbf{e}_z$, then (5.47) reads

$$\omega^2 = N^2 \sin^2 \theta$$

This relation clearly shows that the frequency of the wave depends on the direction of propagation and never exceeds N. In particular, such waves do not propagate vertically. We can calculate the group velocity

$$\mathbf{v}_g = \nabla_k \, \omega = N \begin{vmatrix} k_z^2/k^3 \\ 0 \\ -k_z k_s/k^3 \end{vmatrix} = N(k^2 \mathbf{e}_s - k_s \mathbf{k})/k^3 = N \frac{\mathbf{k} \times (\mathbf{e}_s \times \mathbf{k})}{k^3} \tag{5.48}$$

where we used cylindrical coordinates ($\mathbf{e}_s$ is the radial unit vector). From this relation, it turns out that $\mathbf{v}_g \cdot \mathbf{k} = 0$: *Energy propagates perpendicularly to the phase*. Such a property, stemming from the anisotropy of the background, is also shared by inertial waves (see Chap. 8).

We note that these waves are transversal, namely $\mathbf{k} \cdot \delta \mathbf{v} = 0$. This property is the consequence of mass conservation $\nabla \cdot \mathbf{v} = 0$ and is shared by all the waves propagating in an incompressible fluid.

5.5 Waves Associated with Discontinuities

Until now we neglected the nonlinear terms in the equations of disturbances. We just considered waves of infinitesimal amplitude. However, is this approximation always relevant? To answer this question we need to estimate the relative importance of nonlinear terms to linear ones. Linear and nonlinear terms are not unique, forcing us to be more specific. We shall therefore take the pressure term $-\nabla \delta P$ as typical of the linear part and the inertial one, $\rho(\delta \mathbf{v} \cdot \nabla)\delta \mathbf{v}$, typical of nonlinear ones. Hence, nonlinear effects are important when

$$\nabla \delta P \sim \rho(\delta \mathbf{v} \cdot \nabla)\delta \mathbf{v}$$

If the characteristic scale of motion is L (namely the wavelength), then the foregoing criterion becomes

$$\delta P \sim \rho(\delta \mathbf{v})^2$$

saying that the kinetic energy of the fluctuations is of the order of the pressure disturbances. If we consider sound waves, from (5.20) it turns out that

$$\omega \rho \delta v = k \delta P \implies \rho v_\phi \delta v \sim \delta P,$$

where v_ϕ is the phase velocity. The criterion is now

$$\delta v \sim v_\phi \tag{5.49}$$

Nonlinear effects are therefore important when the velocity of the fluid, associated with the passing wave, is of the same order as the wave velocity. In fact, writing (5.49), we introduced a dimensionless number

$$M = \frac{V}{V_\phi} \tag{5.50}$$

which is just a *Mach number*. The most famous of these numbers is the ratio of the fluid velocity and the speed of sound. This is the number which is referred to when one speaks about the Mach number without any precision. The foregoing discussion shows that it may be defined for any type of waves.

5.5.1 Propagation of a Disturbance as a Function of the Mach Number

The difference between a flow with $M < 1$ and a flow with $M > 1$ is not just quantitative: it is also qualitative. The propagation of perturbations is very different in these two cases.

Let us consider a source of low amplitude waves (sound waves or gravity waves for instance) moving at a speed V while emitting waves that propagate with a phase velocity c in the fluid. Using a reference frame attached to the source, the space filled by the fluid appears very differently when the Mach number is changed. If this number is smaller than unity, waves can reach any point in this space; in the opposite case they are confined to the *Mach cone* (see Fig. 5.1). The transition $M = 1$ defines

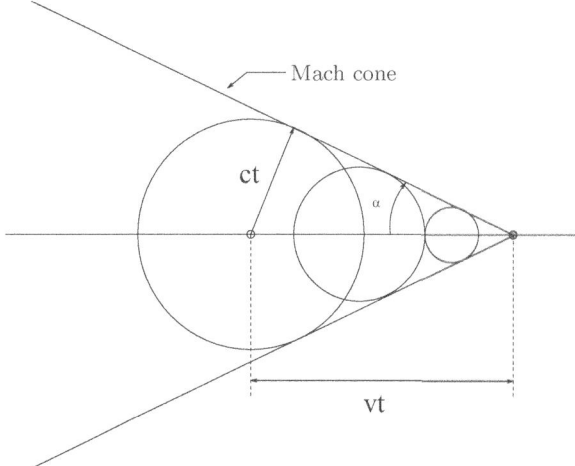

Fig. 5.1 The Mach cone formed by a source of periodic perturbations moving with a supersonic speed

the raise of a partition of space. Now, if we consider the example of a plane flying at a supersonic speed, air disturbances are of finite amplitude and usually produce a discontinuity.

These discontinuities are the consequence of the nonlinear evolution of the waves. In the case of a supersonic flight, the discontinuity is just the supersonic "bang", and in other words the shock wave. We shall see below that shock waves are part of a more general phenomenon which gathers all the waves resulting from a discontinuity. The other common example is the one of crunching water waves.

5.5.2   Equations for a Finite-Amplitude Sound Wave

The first step needed to study shock waves, is to write down the equations governing the evolution of a finite-amplitude sound wave. To simplify the matter, we restrict our discussion to the one-dimensional case. Although much simplified, this model represents fairly well the formation and propagation of a shock wave in a shock tube as we shall see below.

The equations of momentum and mass conservation are :

$$
\begin{cases}
\dfrac{\partial \rho}{\partial t} + \dfrac{\partial \rho u}{\partial x} = 0 \\[2ex]
\dfrac{\partial u}{\partial t} + u \dfrac{\partial u}{\partial x} = -\dfrac{1}{\rho}\dfrac{\partial p}{\partial x}
\end{cases}
\tag{5.51}
$$

In addition, we assume that the gas is isentropic so that $p \propto \rho^\gamma$. The sound speed $\sqrt{\gamma P / \rho}$ is a convenient variable. If we note that

$$\frac{d\rho}{\rho} = \frac{2}{\gamma - 1} \frac{dc}{c} \quad \text{and} \quad \frac{dp}{p} = \frac{2\gamma}{\gamma - 1} \frac{dc}{c}$$

then, (5.51) may be written as

$$\begin{cases} \left(\frac{\partial}{\partial t} + u \frac{\partial}{\partial x} \right) \frac{2c}{\gamma - 1} + c \frac{\partial u}{\partial x} = 0 \\[3mm] \left(\frac{\partial}{\partial t} + u \frac{\partial}{\partial x} \right) u + \frac{2c}{\gamma - 1} \frac{\partial c}{\partial x} = 0 \end{cases} \tag{5.52}$$

which may be mixed to yield:

$$\begin{cases} \left(\frac{\partial}{\partial t} + (u + c) \frac{\partial}{\partial x} \right) r = 0 \\[3mm] \left(\frac{\partial}{\partial t} + (u - c) \frac{\partial}{\partial x} \right) s = 0 \end{cases} \tag{5.53}$$

where we introduced

$$\begin{cases} r = \frac{u}{2} + \frac{c}{\gamma - 1} \\[3mm] s = \frac{u}{2} - \frac{c}{\gamma - 1} \end{cases} \tag{5.54}$$

called the *Riemann invariants*.

5.5.3   The Equations of Characteristics

Equation (5.53) are nonlinear and may be difficult to solve. Fortunately, these are a little simpler and often called *quasi-linear equations*. They can be solved qualitatively at least. For this purpose, we use the theory of characteristics. The reader who may not be familiar with this approach of partial differential equations, may have a look to Sect. 12.6.2 first.

The first result of characteristics theory to be used is the following. If r and s are solutions of (5.53), then r is constant along the characteristic curves of equation

$$\frac{dx}{dt} = u + c \tag{5.55}$$

Fig. 5.2 Schematic view of a shock tube

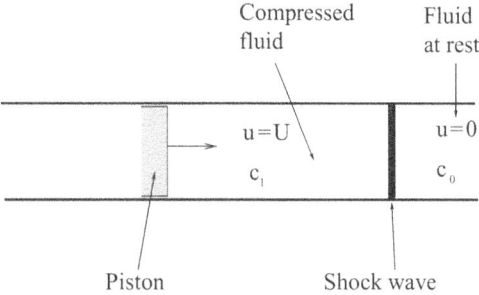

while s is constant on the other characteristic curves

$$\frac{dx}{dt} = u - c \tag{5.56}$$

If the (5.53) were linear, the characteristic curves could be determined directly from (5.55) and (5.56); the initial conditions completely specify the solution. Here, u and c are unknown; nevertheless, we shall see that one can determine the shape of characteristics and understand the evolution of the solutions.

5.5.4   Example: The Compression Wave

We consider the following system: a piston inside a tube (of infinite length) starts at $t = 0$ and reach a constant velocity after a time t_a. The set-up is schematically drawn in Fig. 5.2.

Initially, the fluid is at rest: $u = 0$ and $c = c_0$ on the $t = 0$ line (see Fig. 5.3). The characteristics of s have a slope $\frac{dt}{dx} = -1/c_0$ on the x-axis (the $t = 0$ line), which they cut (see Fig. 5.3). But $s(x, 0) = -c_0/(\gamma - 1)$ is constant on the axis at $t = 0$ and therefore *it is constant everywhere in a region of the (x, t) plane bounded by the piston trajectory.* Using the definition of s on the piston where the gas velocity is U, we find that

$$c = c_0 + \frac{\gamma - 1}{2}U > c_0 \tag{5.57}$$

Since the gas is isentropic and ideal $c \propto \rho^{(\gamma-1)/2}$, density increases when one gets closer to the piston.

In the same region we can write the other Riemann invariant as

$$r = s_0 + \frac{2c}{\gamma - 1} \tag{5.58}$$

Fig. 5.3 Characteristic lines
in the (x, t) plane. The *solid
lines* show r-characteristics
and the *dotted lines* show the
s-ones

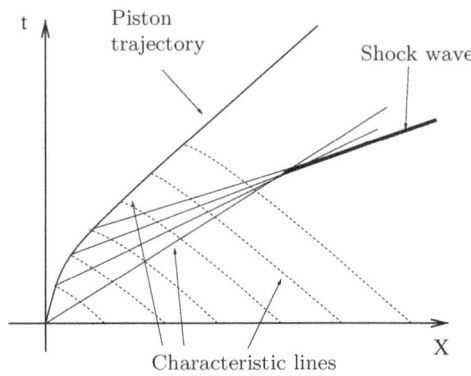

where $s_0 = -c_0/(\gamma - 1)$. Thus, we find that *characteristics associated with r are straight lines*. Indeed, on its characteristic, r is constant and therefore c is constant from (5.58). Thus, u is also constant from the definition of r, (5.54). Hence, along an r-characteristic, $u + c$ is constant showing that these curves are straight lines.

Let us now consider characteristics emitted by the piston. They are straight lines verifying:

$$\frac{dt}{dx} = \frac{1}{c_0 + \frac{\gamma+1}{2}U}$$

The slope of these lines decreases with time since U increases. Consequently, the straight lines will cross somewhere. The function r is then no longer single valued and a discontinuity appears: the shock forms.

We may estimate the time by which the shock has formed. It is given by the point where the characteristic emitted at $t = 0$ (with a slope $1/c_0$) crosses the one when the piston reaches its asymptotic velocity after a time t_a. The slope is then $1/(c_0 + (\gamma + 1)U/2)$. We find that in this case the shock forms after a time t_c such that

$$t_c \sim \frac{2c_0 t_a}{(\gamma + 1)U}$$

Qualitatively, the formation of the discontinuity may be understood in the following manner: The acceleration of the piston increases the density in its vicinity. Sound waves move more rapidly in this denser region. A shock appears when the sound waves emitted in the compressed region overtake the ones emitted at $t = 0$, leading to a steepening of the wave front (see Fig. 5.4).

We would infer from the last formula that a shock forms whatever the conditions. This not the case of course. Indeed, our discussion neglects the dissipative effect as well as the finite length of the tube. If we still assume the infinite length of the tube, we may say that the shock will appear only if t_c is short enough, shorter that

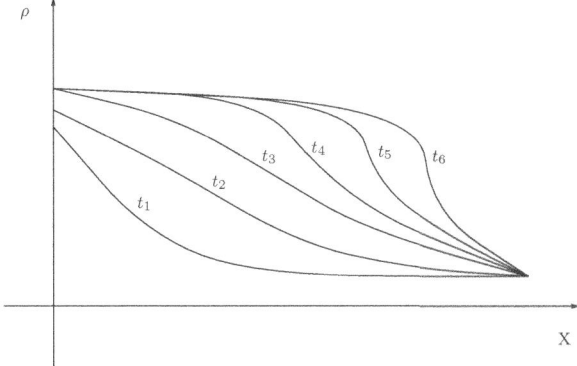

Fig. 5.4 Schematic evolution of density during the formation of a shock: note the steepening of the wave front. Similar shapes may be found for the pressure and temperature

$t_d \sim d^2/v$ where v is the kinematic viscosity of the fluid. Inequality $t_c < t_d$ implies that

$$\frac{U}{t_a} \gtrsim \frac{c_0 v}{d^2}$$

showing that the piston acceleration must be large enough. Let us give a numerical example to fix ideas: if we take a cylindrical tube with a diameter of 3 cm, filled with air, then $\frac{U}{t_a} \gtrsim 5$ m/s^2. Thus, the piston needs to accelerate about half that of terrestrial gravity.

5.5.5 Interface and Jump Conditions

When the shock is formed, it may be described as a pure discontinuity. Indeed, its thickness is only a few mean-free paths which may be neglected macroscopically. However, not all the variables are discontinuous. For instance, the mass flux must be the same on each side of the shock. Thus, in a frame attached to the shock

$$\rho_1 \mathbf{v}_1 \cdot \mathbf{n} = \rho_2 \mathbf{v}_2 \cdot \mathbf{n} \tag{5.59}$$

where indices 1 and 2 refer to the upstream and downstream quantities. Let us first give some precisions about the up- and downstream regions. The flow goes from the upstream to the downstream, of course. The upstream region is the one of low pressure and supersonic velocity whereas the downstream one is of high pressure and subsonic velocity. The supersonic region may sometimes be qualified as "before" the shock since the shock wave may be seen as propagating, supersonically, in the low pressure region. The foregoing case of the shock tube is clear in this respect: the shock wave propagates in a fluid at rest. However, standing on the shock wave, we would see low pressure supersonic air rushing into the high pressure subsonic region!

The next interface condition which must be met at a discontinuity demands the balance of forces. More precisely, the flux of momentum through the shock wave must compensate the pressure jump. Hence,

$$p_1\mathbf{n} + \rho_1\mathbf{v}_1\,\mathbf{v}_1 \cdot \mathbf{n} = p_2\mathbf{n} + \rho_2\mathbf{v}_2\,\mathbf{v}_2 \cdot \mathbf{n}$$

Finally, the energy flux must also be continuous, thus

$$\frac{1}{2}v_1^2 + h_1 = \frac{1}{2}v_2^2 + h_2$$

where h is the enthalpy of the fluid. Let us demonstrate this latter relation, where the reader might wonder why the enthalpy comes into play. For this purpose we consider a cylindrical control surface whose generatrix lines are parallel to the flow and the ends of which are on each side of the shock. The energy flux entering the cylinder is just

$$(\frac{1}{2}v^2 + e)\rho\mathbf{v} \cdot \mathbf{n}\Big|_1$$

In a steady state, it differs from the outgoing flux by the power of forces applied on this volume. In our case we just need to consider pressure forces and their power $-p\mathbf{v} \cdot \mathbf{n}$. Thus,

$$(\frac{1}{2}v^2 + e)\rho\mathbf{v} \cdot \mathbf{n}\Big|_1 - (\frac{1}{2}v^2 + e)\rho\mathbf{v} \cdot \mathbf{n}\Big|_2 = -p\mathbf{v} \cdot \mathbf{n}|_1 + p\mathbf{v} \cdot \mathbf{n}|_2$$

After some rearrangements,

$$(\frac{1}{2}v^2 + e + \frac{p}{\rho})\rho\mathbf{v} \cdot \mathbf{n}\Big|_1 = (\frac{1}{2}v^2 + e + \frac{p}{\rho})\rho\mathbf{v} \cdot \mathbf{n}\Big|_2$$

Taking into account mass conservation, we find that

$$\frac{1}{2}v^2 + e + \frac{p}{\rho}$$

must be continuous; we note that $h = e + p/\rho$ is just the enthalpy.

Actually, the demonstration could be far shorter if we used Bernoulli theorem (3.7), which shows that

$$\frac{1}{2}v^2 + h$$

is constant along a streamline.

5.5.6 Relations Between Upstream and Downstream Quantities in an Orthogonal Shock

The foregoing relations are much simpler when the velocity field is orthogonal to the shock wave. The shock is said to be *normal* to underline the difference with the general case of an *oblique* shock. The conditions are now

$$\begin{cases} \rho_1 v_1 = \rho_2 v_2 \\ p_1 + \rho_1 v_1^2 = p_2 + \rho_2 v_2^2 \\ h_1 + v_1^2/2 = h_2 + v_2^2/2 \end{cases} \tag{5.60}$$

They may be used to rewrite the downstream quantities (index 2) as a function of the upstream ones. Using the upstream M_1 and downstream M_2 Mach numbers, one may show (see the demonstration at the end of the chapter) that

$$v_2 = v_1 \frac{(\gamma - 1)M_1^2 + 2}{(\gamma + 1)M_1^2} \tag{5.61}$$

$$p_2 = \rho_1 \frac{(\gamma + 1)M_1^2}{(\gamma - 1)M_1^2 + 2} \tag{5.62}$$

$$M_2 = \frac{M_1}{\sqrt{(v_1/v_2)^2 + (\gamma - 1)((v_1/v_2)^2 - 1)M_1^2/2}} \tag{5.63}$$

$$p_2/p_1 = 1 + \frac{2\gamma}{\gamma + 1}(M_1^2 - 1) \tag{5.64}$$

These relations allow us to determine the state of the fluid after crossing a shock wave. The upstream flow being supersonic, $M_1 > 1$, we immediately find the following inequalities:

$$v_2 < v_1, \qquad \rho_2 > \rho_1, \qquad p_2 > p_1$$

After crossing the shock wave the fluid slows down and is compressed. Pressure and density increase. Using (5.60c), which we rewrite

$$T_2 = T_1 + \frac{v_1^2 - v_2^2}{2c_p}$$

we also see that temperature increases. Obviously the downstream flow is subsonic and $M_2 < 1$. This inequality is not clear in (5.63), but becomes as such when this equation is transformed using the ratio of velocities (5.61). One then finds

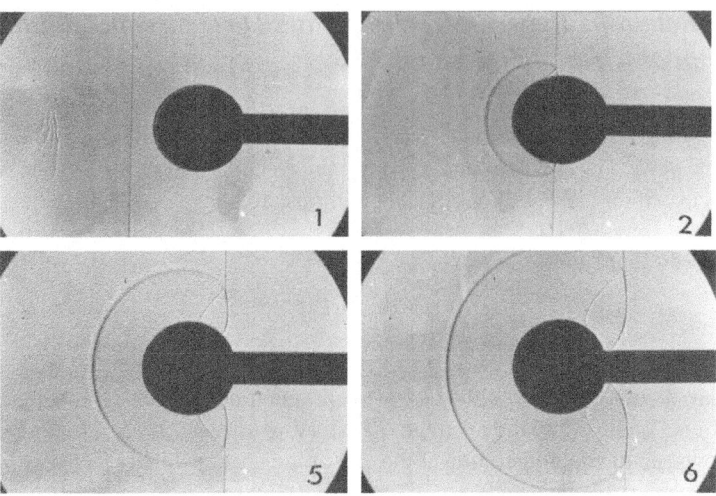

Fig. 5.5 Interaction of a plane shock wave with a cylinder. In (2), (5) and (6), we observe the evolution of the incident, reflected and refracted shock waves. We also observe (in 5 and 6) the appearance of a line joining the intersection point of the three waves and the cylinder; this is a "contact surface" where the pressure is continuous but where there is still an entropy and temperature jump

$$M_2 = \sqrt{\frac{(\gamma - 1)M_1^2 + 2}{2\gamma M_1^2 - \gamma + 1}} \tag{5.65}$$

showing the equivalence $M_1 > 1 \iff M_2 < 1$ (Fig. 5.5).

To summarize, when going through a shock wave, a supersonic flow becomes subsonic and pressure, density, temperature increase. The temperature rise is not only the consequence of compression, but also that of the strong dissipation which occurs within the shock wave. Macroscopically, the velocity gradient are infinite but the volume of the dissipative region is vanishing; one may expect that a finite dissipation implies an *increase of entropy*.

Recalling the entropy expression (1.59), we can derive the entropy jump:

$$s_2 - s_1 = c_v \ln(p_2/p_1) - c_p \ln(\rho_2/\rho_1) = c_v \ln\left[\frac{p_2}{p_1}\left(\frac{\rho_1}{\rho_2}\right)^{\gamma}\right]$$

We may show that the entropy jump is always positive as expected in a dissipative process. For this purpose, we first express the entropy jump as a function of M_1. It turns out that

$$\Delta s = c_v \ln\left[\left(1 + \frac{2\gamma}{\gamma + 1}(M_1^2 - 1)\right)\left(\frac{(\gamma - 1)M_1^2 + 2}{(\gamma + 1)M_1^2}\right)^{\gamma}\right]$$

Let us use a Taylor expansion of this function in a neighbourhood of the threshold $M_1 = 1$ up to the third order. Setting $\varepsilon = M_1^2 - 1 \ll 1$, we have

$$\Delta s = \Delta s(1) + \Delta s'(1)\varepsilon + \Delta s''(1)\frac{\varepsilon^2}{2} + \Delta s'''(1)\frac{\varepsilon^3}{6} + \mathcal{O}(\varepsilon^4)$$

One may show that $\Delta s(1) = \Delta s'(1) = \Delta s''(1) = 0$ and that

$$\Delta s'''(1) = 4c_v \frac{\gamma(\gamma - 1)}{(\gamma + 1)^2}$$

hence

$$\Delta s = c_v \frac{2\gamma(\gamma - 1)}{3(\gamma + 1)^2}(M_1^2 - 1)^3 + \mathcal{O}(\varepsilon^4) \tag{5.66}$$

This expression shows that the entropy of the fluid increases when passing the shock, as soon as $M_1 > 1$. However, the jump is very small if M_1 is not very different from unity. This leads to a classification of shocks into weak and strong ones (see below).

Let us now show that the entropy jump is an increasing function of the Mach number. Setting $m = M_1^2$, we compute ds/dm. We have

$$\frac{1}{c_v}\frac{ds}{dm} = \frac{d}{dm}\left[\ln\left(1 + \frac{2\gamma}{\gamma + 1}(m - 1)\right) + \gamma\ln\left(\frac{(\gamma - 1)m + 2}{(\gamma + 1)m}\right)\right]$$

$$\frac{1}{c_v}\frac{ds}{dm} = \frac{2\gamma}{\gamma + 1 + 2\gamma(m - 1)} - \frac{2\gamma}{(\gamma - 1)m^2 + 2m}$$

$$= \frac{2\gamma(\gamma - 1)(m - 1)^2}{m(\gamma + 1 + 2\gamma(m - 1))((\gamma - 1)m + 2)}$$

Thus $ds/dm > 0$ when $m > 1$. We also note that if M_1 goes to infinity, Δs also tends to infinity as $\ln M_1$.

5.5.7   Strong and Weak Shocks

The strength of a shock is usually measured by its pressure jump $(p_2 - p_1)/p_1$. We may note that this quantity is proportional to M_1^2 and is not bounded. This is not the case for the density or velocity jumps. When $M_1 \to \infty$

$$\frac{\rho_2}{\rho_1} \to \frac{\gamma + 1}{\gamma - 1} \quad \text{and} \quad \frac{v_2}{v_1} \to \frac{\gamma - 1}{\gamma + 1}$$

The weak shocks are defined as those where the variation of entropy is negligible. This distinction between weak and strong shocks is possible because of the very slow variation of entropy in the neighbourhood of $M_1 = 1$. If, for instance, $M_1 = 1.1$ then $\Delta s/s \sim 6 \times 10^{-5}$. If we neglect entropy variations less than 10%, shocks are weak as long as $M_1 \leq 1.46$.

5.5.8 Radiative Shocks

Some stars like cepheids or "RR Lyrae" show periodic variations of their luminosity (see Fig. 5.6). These variations are understood as the signature of radiative shock waves that propagate inside these stars as a result of the breaking-up of some acoustic waves. Radiative shock waves are much more powerful than the foregoing strong shocks. They are also called hypersonic shock waves because $M_1^2 \gg 1$. They occur in a low-density and hot medium. After the shock, matter is ionized by collisions but free electrons recombine with ions while emitting mainly ultraviolet radiations. Part of these photons propagate towards the upstream region and pre-heat the gas (see Fig. 5.7). This phenomenon makes the shock almost isothermal: the downstream gas cools efficiently by radiating photons after its compression. This is very different from the previous shocks where we assumed that the gas was evolving adiabatically after the shock. If we remember that pressure is proportional to density for an ideal isothermal gas, namely $P \propto \rho$, the adiabatic index would be $\gamma = 1$. Thus, the compression ratio $(\gamma + 1)/(\gamma - 1)$ may raise to infinity if the gas supports a quasi-isothermal compression. This ratio should be of order of M_1^2. In actual stellar models, typical shocks have a compression ratio of order 30. They are obviously in the hypersonic regime. Such shocks have been reproduced in the laboratory only very recently, thanks to the development of powerful lasers that can deposit a lot of energy in a very small volume.

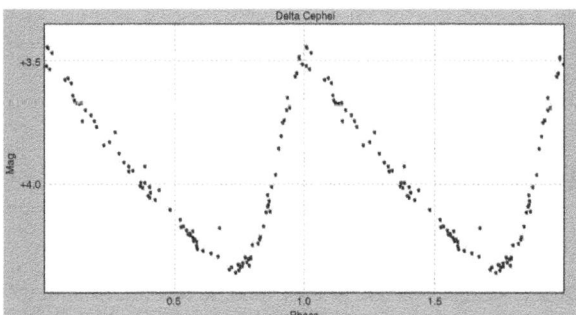

Fig. 5.6 Lumosity variations of the cepheid star δ Cephei. Its oscillation period is 5.37 days. The luminosity of the star varies by more than a factor two between minimum and maximum. This variation comes from an acoustic oscillation of the star. Its large amplitude leads to the formation of a radiative shock wave (source ThomasK Vbg)

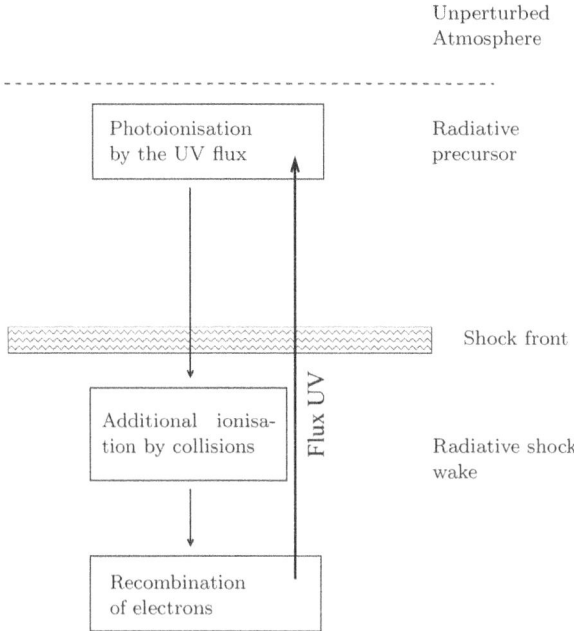

Fig. 5.7 Schematic view of a radiative shock in a star according to Gillet (2006)

5.5.9   The Hydraulic Jump

Another, very common type of discontinuity wave is the discontinuity of water depth in breaking water waves like in Fig. 5.8: this is the *hydraulic jump* (see the schematic view in Fig. 5.9). A connection can be easily made with sound waves if we consider waves propagating in shallow water. In this case, indeed, wave velocity is $\sqrt{gh}$ showing that in a similar way as in Fig. 5.4, the wave front steepens inevitably since the wave velocity increases like $\sqrt{h}$. In this case there is a direct analogy between depth h and temperature T of an ideal gas where the sound wave propagates.[5] However, the analogy, cannot be pushed too far, since gravity waves are necessarily two dimensional because of the incompressibility of the fluid. Another complication comes from the fact that these waves are naturally dispersive. The equality of phase and group velocity is only true asymptotically for wavelengths long compared to the depth. We shall see below that these dispersive effects can stop the steepening of the wave front and give rise to a solitary wave.

---

[5]Generally, we use the density as the analog of the depth [just compare (5.67) et (5.60a)], but the analog of the hydraulic jump is a shock wave in gas where $\gamma = 2$ in which case T and ρ are proportional.

Fig. 5.8 Hydraulic jumps from braking waves on a sand beach (Picture from the author)

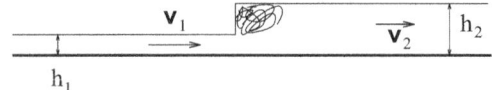

Fig. 5.9 A schematic view of a hydraulic jump

Let us examine the jump conditions of a hydraulic jump. Considering the fluid as incompressible, the conservation of the mass flux when crossing the hydraulic jump implies that:

$$v_1 h_1 = v_2 h_2 \tag{5.67}$$

in the most simple set-up where the velocity is assumed to be constant in the whole cross section of the flow.

The conservation of momentum leads to the same reasoning as for shock waves. The variation of the momentum flux must be compensated by the total pressure forces. Hence,

$$\rho(v_1^2 h_1 - v_2^2 h_2) + \int_0^{h_1} p(z)dz - \int_0^{h_2} p(z)dz = 0$$

Since we neglect vertical motions, the hydrostatic balance controls the z-dependence of the pressure. Thus,

$$p = \rho g(h - z)$$

where we assumed a zero pressure above the fluid. It yields

$$\int_0^{h_1} p(z)dz = \rho g \frac{h_1^2}{2} \qquad \text{and} \qquad \int_0^{h_2} p(z)dz = \rho g \frac{h_2^2}{2}$$

From these relations, a second jump condition connecting upstream and downstream quantities[6] can be derived:

$$v_1^2 h_1 + g h_1^2/2 = v_2^2 h_2 + g h_2^2/2 \qquad (5.68)$$

With (5.67), we find out the ratio between upstream and downstream depth:

$$\frac{h_2}{h_1} = \frac{\sqrt{1 + 8\mathrm{Fr}_1^2} - 1}{2} \qquad (5.69)$$

In this expression we introduced the Froude number:

$$\mathrm{Fr}_1 = \frac{v_1}{\sqrt{g h_1}}$$

This number quantify the ratio between the fluid velocity and the speed of waves. It is the analog of the Mach number for acoustic waves. When this number is larger than unity, the flow is *supercritical* or *torrential*. On the contrary, when Fr < 1, the flow is said to be *subcritical* or *fluvial*. One may show as an exercise, that if the flow is supercritical in the upstream region, it is subcritical in the downstream region. Hence, we have the following equivalence:

$$\mathrm{Fr}_1 \gtreqless 1 \quad \Longleftrightarrow \quad \mathrm{Fr}_2 \lesseqgtr 1 \qquad (5.70)$$

The first observation of a solitary wave

"I believe I shall best introduce this phænomenon by describing the circumstances of my own first acquaintance with it. I was observing the motion of a boat which was rapidly drawn along a narrow channel by a pair of horses, when the boat suddenly stopped - not so the mass of water in the channel which it had put in motion; it accumulated round the prow of the vessel in a state of violent agitation, then suddenly leaving it behind, rolled forward with great velocity, assuming the form of a large solitary elevation, a rounded, smooth and well-defined heap of water, which continued its course along the channel apparently without change of form or diminution of speed. I followed it on horseback, and overtook it still rolling on at a rate of some eight or nine miles an hour, preserving its original figure some thirty feet long and a foot to a foot and a half in height. Its height gradually diminished, and after a chase of one or two miles I lost it in the windings of the channel. Such, in the month of August 1834, was my first chance interview with that singular and beautiful phænomenon which I have called the Wave of Translation, a name which it now very generally bears;" From "Report on waves", *Rep. 14th Meet. Brit. Assoc. Adv. Sci., York*, 319–320 par S. Russell (1844).

---

[6]As for the shock waves, we use a frame attached to the discontinuity. Upstream and downstream regions are defined similarly.

As it may be guessed, the hydraulic jump is a dissipative structure: the hydraulic load decreases through a hydraulic jump. To show this, let us consider a streamline on the plane $z = 0$ and compare the energy per unit mass upstream and downstream. For this purpose we just need to check that

$$\frac{1}{2}v_1^2 + gh_1 \geq \frac{1}{2}v_2^2 + gh_2$$

Using expressions (5.86), demonstrated in the exercises, the preceding inequality implies

$$\left(\frac{h_2}{h_1} - \frac{h_1}{h_2}\right)\frac{h_1 + h_2}{4} + h_1 - h_2 \geq 0$$

$$\Longleftrightarrow (h_2 - h_1)^2 \geq 0$$

which is always true.

5.6 Solitary Waves 🍄

Nonlinear effects have not always dramatic consequences such as the formation of a discontinuity. The steepening of the wave front can indeed be compensated by some dispersion effects which tend to spread the wave packet. When this balance occurs, one may observe a *solitary wave* which is remarkable for its stability.

The first observation of a solitary wave was made by Scott Russell in 1834 (see box and Fig. 5.10) on a surface gravity wave. In Sect. 5.3.1 we saw that these waves are dispersive. It is just in the asymptotic case of long wavelengths compared to the depth, that dispersion disappears. This property allows a control of the effects of dispersion by tuning the ratio of the wavelength to the depth. Another possible small parameter is obviously the amplitude of the wave. We shall see that when these two possibly small parameters are linked through a simple relation, one obtains a new equation, first derived by Korteweg and de Vries in 1895, which governs the motion of solitary waves.

5.6.1 The Korteweg and de Vries Equation

We first set again the general equations governing surface waves, still neglecting the effects of viscosity. We concentrate on the propagation of a wave in a one-dimension water basin of depth h. The motion is assumed irrotational and the velocity potential verifies:

$$\begin{cases} \Delta \Phi = 0 \\[2mm] \dfrac{\partial \Phi}{\partial t} + \dfrac{1}{2}\mathbf{v}^2 + \dfrac{P}{\rho} + gz = \text{Cst} \end{cases} \tag{5.71}$$

This system is completed by the following boundary conditions:

$$v_z = \frac{\partial \Phi}{\partial z} = 0 \qquad \text{at} \quad z = 0$$

$$\left. \begin{array}{l} \dfrac{\partial z_s}{\partial t} + \mathbf{v} \cdot \nabla z_s = v_z \\[4mm] P = 0 \end{array} \right\} \qquad \text{at} \quad z = h + z_s(x,t)$$

where the pressure above the fluid has been set to zero. We first rewrite this system using non-dimensional variables. $\sqrt{gh}$ is a natural scale for the velocities and h for the lengths. Thus we set

$$\mathbf{v} = \sqrt{gh}\,\mathbf{u}, \qquad \Phi = h\sqrt{gh}\,\phi, \qquad t = \sqrt{h/g}\,\tau \qquad \text{and} \qquad z_s = h\,\zeta_s$$

We then get

$$\Delta \phi = 0 \tag{5.72}$$

Fig. 5.10 Repetition of Russell's observation of a solitary wave in Union Canal of Edinburgh during a conference at Heriot-Watt University (Nature, 3 August 1995)

$$\left.\begin{aligned}\frac{\partial \zeta_s}{\partial \tau} + \mathbf{u} \cdot \nabla \zeta_s &= \frac{\partial \phi}{\partial z}\\[2mm]\frac{\partial \phi}{\partial \tau} + \frac{1}{2}(\nabla \phi)^2 + \zeta_s &= 0\end{aligned}\right\} \qquad \text{at} \quad z = 1 + \zeta_s(x,t) \qquad (5.73)$$

$$\frac{\partial \phi}{\partial z} = 0 \qquad \text{at} \quad z = 0 \qquad (5.74)$$

where we substituted to the boundary condition $P = 0$ the equation of momentum taken on the surface, with the constant set to zero.

The next step is slightly delicate. We wish to introduce the small parameters and to consider the situations that are weakly nonlinear and weakly dispersive. Since the solitary wave seems to be a steady solution in some appropriate reference frame, we look for slowly evolving solutions in a new frame. We therefore introduce the following new form for the solutions:

$$\zeta_s \equiv \varepsilon \tilde{\zeta}_s(\tilde{\tau}, \tilde{x}) \qquad \text{and} \qquad \phi = \varepsilon^{1/2} \tilde{\phi}(\tilde{\tau}, \tilde{x})$$

where we set:

$$\tilde{x} = \varepsilon^{1/2}(x - \tau), \quad \tilde{\tau} = \varepsilon^{3/2}\tau$$

Theses new functions and new variables are sensitive to large scale or long time-scale variations only: x and τ need to vary a lot to yield significant variations on $\tilde{x}$ and $\tilde{\tau}$. With these new variables (5.72) now reads:

$$\frac{\partial^2 \tilde{\phi}}{\partial z^2} + \varepsilon \frac{\partial^2 \tilde{\phi}}{\partial \tilde{x}^2} = 0 \qquad (5.75)$$

while (5.73) yields

$$\frac{\partial \zeta_s}{\partial \tau} + \mathbf{u} \cdot \nabla \zeta_s = \frac{\partial \phi}{\partial z}$$

$$\Longleftrightarrow \qquad \frac{\partial \zeta_s}{\partial \tilde{\tau}} \frac{\partial \tilde{\tau}}{\partial \tau} + \frac{\partial \zeta_s}{\partial \tilde{x}} \frac{\partial \tilde{x}}{\partial \tau} + \frac{\partial \zeta_s}{\partial \tilde{x}} \frac{\partial \tilde{\phi}}{\partial \tilde{x}} \left(\frac{\partial \tilde{x}}{\partial x}\right)^2 = \frac{\partial \phi}{\partial z}$$

where we took into account that $\tilde{x}$ depends on τ. We now deduce:

$$\left.\begin{aligned}\varepsilon^2 \left(\frac{\partial \tilde{\zeta}_s}{\partial \tilde{\tau}} + \frac{\partial \tilde{\phi}}{\partial \tilde{x}} \frac{\partial \tilde{\zeta}_s}{\partial \tilde{x}}\right) - \varepsilon \frac{\partial \tilde{\zeta}_s}{\partial \tilde{x}} &= \frac{\partial \tilde{\phi}}{\partial z}\\[2mm]\varepsilon \frac{\partial \tilde{\phi}}{\partial \tilde{\tau}} - \frac{\partial \tilde{\phi}}{\partial \tilde{x}} + \frac{1}{2}\left(\frac{\partial \tilde{\phi}}{\partial z}\right)^2 + \frac{\varepsilon}{2}\left(\frac{\partial \tilde{\phi}}{\partial \tilde{x}}\right)^2 + \tilde{\zeta}_s &= 0\end{aligned}\right\} \qquad \text{at} \quad z = 1 + \varepsilon \zeta(x,t)$$

$$(5.76)$$

and

$$\frac{\partial \tilde{\phi}}{\partial z} = 0 \qquad \text{at} \quad z = 0 \tag{5.77}$$

We then make the expansion of $\tilde{\zeta}_s$ and $\tilde{\phi}$ in powers of ε:

$$\tilde{\phi} = \sum_{n=0}^{\infty} \phi_n(\tilde{x}, z, \tilde{\tau})\varepsilon^n \qquad \text{and} \qquad \tilde{\zeta}_s = \sum_{n=0}^{\infty} \zeta_n(\tilde{x}, \tilde{\tau})\varepsilon^n$$

The first step consists in solving (5.75) for $\tilde{\phi}$. The first orders give:

$$\frac{\partial^2 \phi_0}{\partial z^2} = 0, \qquad \frac{\partial^2 \phi_1}{\partial z^2} + \frac{\partial^2 \phi_0}{\partial \tilde{x}^2} = 0, \qquad \frac{\partial^2 \phi_2}{\partial z^2} + \frac{\partial^2 \phi_1}{\partial \tilde{x}^2} = 0, \cdots$$

Using the boundary conditions in $z = 0$, we easily find a new expression for $\tilde{\phi}$, namely:

$$\tilde{\phi}(\tilde{x}, z, \tilde{\tau}) = c_0 + \varepsilon(c_1 - c_0'' z^2/2) + \varepsilon^2(c_0^{(4)} z^4/24 - c_1'' z^2/2 + c_2) + \mathcal{O}(\varepsilon^3)$$

where c_0, c_1, c_2 are functions of $\tilde{x}$ and $\tilde{\tau}$. The primes indicate the derivatives with respect to $\tilde{x}$. The first boundary condition, taken at order ε, gives:

$$\frac{\partial \zeta_0}{\partial \tilde{x}} = c_0''$$

and at order ε^2:

$$\frac{\partial \zeta_0}{\partial \tilde{\tau}} + c_0' \frac{\partial \zeta_0}{\partial \tilde{x}} - \frac{\partial \zeta_1}{\partial \tilde{x}} = -c_0'' \zeta_0 + c_0^{(4)}/6 - c_1''$$

Concerning the second boundary condition, it gives

$$\zeta_0 = c_0' \qquad \text{and} \qquad \frac{\partial c_0}{\partial \tilde{\tau}} - c_1' + c_0^{(3)}/2 + c_0'^2/2 + \zeta_1 = 0$$

These equations are used to obtain the equation controlling ζ_0, which we rename ζ. We thus find:

$$\frac{\partial \zeta}{\partial \tilde{\tau}} + \frac{3}{2}\zeta \frac{\partial \zeta}{\partial \tilde{x}} + \frac{1}{6}\frac{\partial^3 \zeta}{\partial \tilde{x}^3} = 0 \tag{5.78}$$

known as Korteweg and de Vries equation (or also KdV equation).

5.6.2 The Solitary Wave

This equation is solved by looking for a solution of the form $\zeta(\tilde{x} - \tilde{\tau})$. Hence,

$$-\zeta' + \frac{3}{2}\zeta\zeta' + \frac{1}{6}\zeta''' = 0$$

which we integrate once, and get:

$$-\zeta + \frac{3}{4}\zeta^2 + \frac{1}{6}\zeta'' = A$$

The constant of integration A is set to zero as we are interested in solutions vanishing at infinity. Multiplying this equation by ζ' and integrating again, we find:

$$-\frac{1}{2}\zeta^2 + \frac{1}{4}\zeta^3 + \frac{1}{12}\zeta'^2 = 0$$

As before, the constant of integration has been set to zero. This equation can be solve analytically since the variables can be separated.

$$\int \frac{d\zeta}{\zeta\sqrt{1 - \zeta/2}} = \sqrt{6}(\tilde{x} - \tilde{\tau})$$

Despite of its look, the integration of the left-hand side is very easy if we set

$$\zeta = \frac{2}{\cosh^2 \vartheta} \; ;$$

we immediately find that

$$\vartheta = -\sqrt{\frac{3}{2}}(\tilde{x} - \tilde{\tau}) \, .$$

$1/\cosh \vartheta$ is called the hyperbolic secant, and noted *sech*. Back to the dimensional variables, we have

$$z_s = h(1 + \varepsilon\zeta)$$

$$= h + 2h\varepsilon \, \mathrm{sech}^2 \left[\sqrt{\frac{3\varepsilon}{2}} \, (x - (1 + \varepsilon)\tau) \right]$$

We now introduce the wave amplitude $a = 2h\varepsilon$, the dimensional length and time scale:

$$z_s = h + a \, \mathrm{sech}^2 \left[\sqrt{\frac{3a}{4h^3}} \left\{ x - \sqrt{gh}\left(1 + \frac{a}{2h}\right)t \right\} \right] \tag{5.79}$$

Fig. 5.11 This graph shows three solitary waves of amplitudes $a, a/2, a/4$ respectively

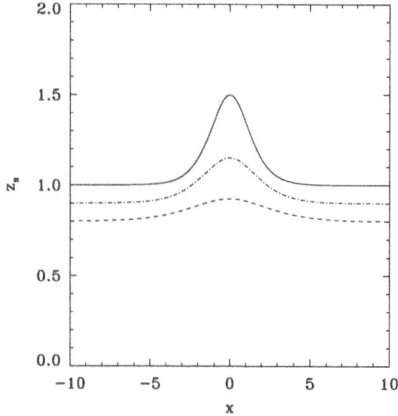

The solitary wave thus moves at a speed of:

$$c = \sqrt{gh}\left(1 + \frac{a}{2h}\right) \tag{5.80}$$

This velocity is only slightly different from the velocity of a gravity wave moving in shallow water if we take the total height $c \simeq \sqrt{g(h + a)}$.

We also observe that the horizontal scale of the wave (the width of the "bump") is given by

$$L = \sqrt{\frac{4h^3}{3a}} \tag{5.81}$$

We show in Fig. 5.11 the shape of the solitary wave for three amplitudes.

5.6.3 Elementary Analysis of the KdV Equation

The properties of the KdV equation are numerous and we could write a full book on it! Here, we content ourselves with an elementary analysis so as to appreciate the role of the various terms involved in this equation. Let us first focus on the linear term $\partial^3 \zeta / \partial x^3$. Eliminating nonlinear terms, the KdV equation reads

$$\frac{\partial \zeta}{\partial \tau} + \frac{1}{6}\frac{\partial^3 \zeta}{\partial x^3} = 0 \tag{5.82}$$

which we modify by changing of reference frame, namely $\zeta(x, \tau) = \zeta'(x - \tau, \tau)$. The equation for ζ' is

$$\frac{\partial \zeta'}{\partial \tau} + \frac{\partial \zeta'}{\partial x} + \frac{1}{6}\frac{\partial^3 \zeta'}{\partial x^3} = 0.$$

A solution of the plane wave type gives the following dispersion relation:

$$\omega = k - k^3/6$$

It can be compared to (5.39) which is the dispersion relation of gravity waves in a basin of finite depth. If we expand (5.39) for the long wavelengths, it turns out that

$$\omega^2 \simeq ghk^2\left(1 - \frac{k^2h^2}{3}\right) \qquad \Longleftrightarrow \qquad \omega \simeq \sqrt{ghk}\left(1 - \frac{k^2h^2}{6}\right)$$

which is exactly the foregoing relation up to some dimensional coefficients. We see that the dispersive term $\partial^3\zeta/\partial x^3$ comes from the finite depth of the fluid.

We may now have a look to the way this term contributes to the spreading of the wave packet. For this purpose, we use a frame attached to the waves and the relation (5.82).

Let $\zeta_0(x)$ be the shape of the wave packet initially. We suppose that this shape is a kind of bell curve such that its Fourier transform $\hat{\zeta}_0$ exists. Taking the Fourier transform of (5.82), it turns out that

$$\hat{\zeta}(k,\tau) = \hat{\zeta}_0 e^{ik^3\tau/6}$$

hence

$$\zeta(x,\tau) = \int_{-\infty}^{+\infty} \hat{\zeta}_0 e^{ikx+ik^3\tau/6}dk$$

This expression is nicer if we observe that $e^{ik^3/3}$ is the Fourier transform of Airy's function $Ai(x)$, i.e.

$$e^{ik^3/3} = \int_{-\infty}^{+\infty} Ai(z)e^{-ikz}dz$$

After some easy manipulations, it turns out

$$\zeta(x,\tau) = \left(\frac{2}{\tau}\right)^{1/3}\int_{-\infty}^{+\infty} \zeta_0(z)Ai\left[\frac{x-z}{(\tau/2)^{1/3}}\right]dz$$

If the initial wave packet is strongly peaked and may be assimilated to a Dirac peak, then

$$\zeta(x,\tau) = \left(\frac{2}{\tau}\right)^{1/3} Ai\left[\frac{x}{(\tau/2)^{1/3}}\right] \qquad (5.83)$$

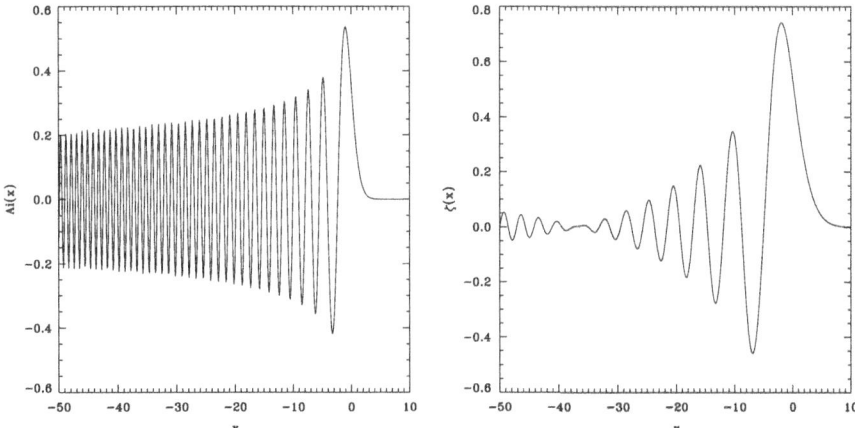

Fig. 5.12 (a) A plot of Airy's function. Schematically, this function behaves as $\cos x/(-x)^{1/4}$ when $x \to -\infty$ and as $e^{-2x^{3/2}/3}$ when $x \to +\infty$. (b) The convolution of Airy's function and a gaussian $\zeta_0(z) = e^{-10z^2}$

Airy's function and its convolution by a gaussian are shown in Fig. 5.12. With this figure, we note that energy is dispersed in the domain $]-\infty, 0]$ through a set of oscillations. The expression (5.83) shows the spreading of the wave packet: its width increases like $\tau^{1/3}$ (in fact the width of oscillations) whereas its amplitude decreases as $\tau^{-1/3}$.

Let us now examine the role of the nonlinear term. We leave aside the linear term and rewrite the KdV equation as

$$\frac{\partial \zeta}{\partial \tau} + \frac{3}{2} \zeta \frac{\partial \zeta}{\partial x} = 0 \qquad (5.84)$$

This equation is of the same type as the one verified by Riemann's invariant. Thus we write the equation of characteristics, namely

$$\frac{dx}{d\tau} = \frac{3}{2} \zeta$$

which are straight lines since ζ is constant on such a line. If, at initial time, ζ has a bell shape, the construction of characteristics issued from the wave front immediately shows that a discontinuity will appear after a finite time (see Fig. 5.13).

Equation (5.84) is in fact of the same type as a famous equation in Fluid Mechanics, namely

$$\frac{\partial u}{\partial t} + u \frac{\partial u}{\partial x} = v \frac{\partial^2 u}{\partial x^2} \qquad (5.85)$$

Fig. 5.13 A schematic view
of the formation of a
discontinuity through Burgers
equation

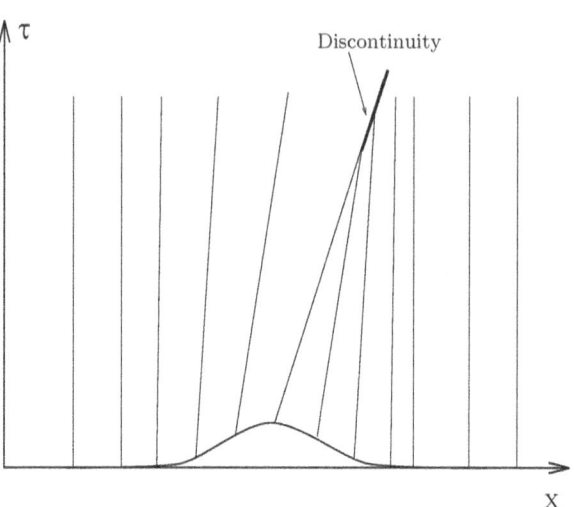

which is called *Burgers equation*. This equation is just the equation of momentum
of an incompressible fluid in which one would have neglected the pressure force.
When viscosity is neglected (and time and length are appropriately scaled), (5.85)
and (5.84) are equivalent. The foregoing reasoning shows that the solutions of
Burgers equation without viscosity always form discontinuities.

5.6.4 Examples

Similarly as the observation of Scott Russell, there exist some natural phenomena
where solitary waves appear. We shall mention two of them: the tidal bores and the
tsunamis.

The tidal bore is the wave that propagates in an estuary, shallow enough, when
the tide rises. This wave is usually first breaking and can be describe as a hydraulic
jump. Getting upstream this hydraulic jump decreases and may give birth to a train
of waves, which, like the wave observed by Russell, have a very long life time and
are also solutions of the KdV equation: these are cnoidal waves. Tidal bores are very
spectacular at the equinoxial high tides. In Europe, famous ones are in the Gironde
in France and in the river Severn in England (see Fig. 5.14).

Tsunamis ("thunderstorm" wave in Japanese) designate the tidal waves which
break on the coasts of the Pacific ocean (where they are the most common). Their
origin is generally related to an earthquake. The seismic wave gives momentum
to a large mass of water which may generate a solitary wave. Such a wave can
cross the Pacific ocean without much damping. For instance, it is well-known that
earthquakes occurring along the Alaska coast can generate a few hours later a
tsunami on the Hawaiian shores. The wave has an horizontal scale which may reach

Fig. 5.14 The hydraulic jump made by the tidal bore of river Severn (photographed by D. H. Peregrine, in *An Album of Fluid Motion*, van Dyke 1982).

a hundred nautical miles (180 km). In this case, taking into account the depth of the Pacific ocean (5 km), the amplitude may be estimated to 5 m from (5.81). Its velocity may also be computed; it is close to $\sqrt{gh}$, which gives 800 km/h. Thus it crosses half of the Pacific ocean (4,800 km) in 6 h. When it arrives on a shore the steepening of the wave front may generate a water wall up to 20 or 30 m high.

5.7 Exercises

1. The bassoon and the oboe are two instruments whose air column is conical. Using the fact that a cone is a part of a sphere, rewrite the equation of disturbances and show that the eigenmodes obey the same dispersion relation as those of the flute. Compute the length of a bassoon whose gravest note is at 58.27 Hz (third B flat). Compare the result to its real length of 295 cm.
2. What is the frequency variation of the fundamental mode of a flute when the air temperature varies from 10 to 30 °C. Compare it to the change of frequency in a half-tone interval. The variation of the length of the tube is neglected and we recall that an octave is divided into twelve equal half-tones (tempered scale).
3. In a harbour along the coast of the Atlantic ocean, waves arrive periodically with a period of 15 s. What is their wavelength? their phase velocity? How long would it take for them to cross the Atlantic ocean (4,800 km)? We assume that the ocean is infinitely deep.
4. How long does it take for a wave of very long wavelength to cross the Atlantic ocean whose width is 4,800 km and depth 5 km? We give $g = 9.81$ m/s^2. Show that the Atlantic ocean is a resonant cavity for the tides; what are the consequences?

5. Derive the dispersion relation of capillary waves in the shallow water approximation.
6. a) From the jump conditions of a hydraulic jump, show that upstream and downstream velocities are of the form:

$$v_1 = \sqrt{g\frac{h_2}{h_1}\left(\frac{h_1+h_2}{2}\right)} \quad \text{and} \quad v_2 = \sqrt{g\frac{h_1}{h_2}\left(\frac{h_1+h_2}{2}\right)} \qquad (5.86)$$

   b) Show that on each side of a hydraulic jump, Froude numbers are related by

$$\mathrm{Fr}_2 = \mathrm{Fr}_1 \left(\frac{\sqrt{1+8\mathrm{Fr}_1^2}-1}{2}\right)^{-3/2}$$

   Derive the equivalence (5.70).
7. Show that the following quantities

$$\int_{-\infty}^{+\infty} \zeta(x,\tau)dx \quad \text{and} \quad \int_{-\infty}^{+\infty} \zeta^2(x,\tau)dx$$

are conserved by the KdV equation. What is the physical interpretation of these conservation laws?

Appendix: Jump Conditions

We give here the demonstration of the relations (5.61)–(5.64) relating upstream and downstream quantities in a normal shock. Let us recall that the enthalpy of an ideal gas is:

$$h = \frac{\gamma}{\gamma-1}\frac{p}{\rho} = \frac{c^2}{\gamma-1} \qquad (5.87)$$

The energy relation can thus be written

$$c_1^2 + (\gamma-1)v_1^2/2 = x\gamma p_2/\rho_1 + x^2(\gamma-1)v_1^2/2$$

where we introduced $x = v_2/v_1$. The conservation of momentum (5.60b) reads now

$$p_2 = p_1 + \rho_1 v_1^2(1-x)$$

Combining the two foregoing equations, we find

$$(\gamma+1)x^2 - 2(\gamma+1/M_1^2)x + 2/M_1^2 + \gamma - 1 = 0$$

This second order equation necessarily has $x = 1$ as a solution (why?). Thus, factorizing it we get straightforwardly

$$(x - 1)((\gamma + 1)x - 2/M_1^2 - \gamma + 1) = 0$$

which gives the non-trivial solution sought after. From (5.60a), we derive (5.62), relating the densities.

The relation between upstream and downstream pressures comes from (5.60b) which we divide by p_1. Thus

$$\frac{p_2}{p_1} = 1 + \frac{\rho_1 v_1^2}{p_1}\left[1 - \frac{\rho_2}{\rho_1}\left(\frac{v_2}{v_1}\right)^2\right] = 1 + \frac{\gamma v_1^2}{c_1^2}\left(1 - \frac{v_2}{v_1}\right)$$

The desired expression is obtained using (5.61).

The relation on Mach numbers (5.63) comes from the equation on enthalpy. Using (5.87), we find

$$c_2^2 = c_1^2 + \frac{(\gamma - 1)}{2}(v_1^2 - v_2^2)$$

Dividing this expression by v_2^2 we get (5.63).

Further Reading

The monograph "Waves in Fluids" of Lighthill (1978), cannot be ignored, but, at a less ambitious level, general books on fluid mechanics may be useful. As far as shock waves are concerned the reader may consult the monograph of Courant and Friedrichs (1976), *Supersonic flow and shock waves*, while the study of solitary waves may be followed up with the introduction of Drazin and Johnson (1989) and the more mathematical approach of solitons by Newell (1985).

References

Courant, R. & Friedrichs, K. (1976). *Supersonic flow and shock waves*. New York: Springer.
Drazin, P. & Johnson, R. (1989). *Solitons: An introduction*. Cambridge: Cambridge University Press.
Gillet, D. (2006). Radiative shocks in stellar atmosphere: Structure and turbulence amplification. In M. Rieutord & B. Dubrulle (Eds.), *EAS Publications Series* (Vol. 21, pp. 297–324).
Lighthill, J. (1978). *Waves in fluids*. Cambridge: Cambridge University Press.
Newell, A. (1985). Solitons in mathematics and physics. *The Society for Industrial and Applied Mathematics, 39*, 422–443.
van Dyke, M. (1982). An Album of Fluid Motion. The Parabolic Press.

Chapter 6
Flows Instabilities

The study of the stability of flows is one of the cornerstones of Fluid Mechanics: the subject is so large that it would deserve a whole book to be reviewed. Leaving aside such an ambitious goal, we shall concentrate, in this chapter and the following one, on the fundamentals, although, here and there, making some excursions in more specialized topics.

The importance of instabilities, or stability questions, comes from their relation to turbulence and mixing. An unstable flow is a necessary path to a turbulent one. Turbulence is indeed a fundamental process in Fluid Mechanics because it controls in many circumstances the fluid transport properties. The conditions within which turbulence sets in, can be appreciated only when the questions of stability are settled. Often, this is not sufficient, but always necessary.

6.1 Local Analysis of Instabilities

When we discussed the equations of perturbations, we found that a simple way to understand their evolution was to consider them as plane waves and analyse their dispersion relation. Owing to the simplicity of the approach, we again start with this type of analysis.

6.1.1 Definitions

First of all, let us recall that the local analysis is only valid if the wavelength of disturbances is very small compared to the scales of the velocity field as given by expression (5.7).

© Springer International Publishing Switzerland 2015
M. Rieutord, *Fluid Dynamics*, Graduate Texts in Physics,
DOI 10.1007/978-3-319-09351-2_6

If, for some wavevectors $\mathbf{k}$ belonging to a subset of $\mathbb{R}^3$, the dispersion relation gives a frequency ω with a negative imaginary part, then there are waves whose amplitude grows exponentially with time. This is called the *absolute* or *temporal* instability. It is the most frequent case, but the opposite one also exists: if, for some real values of the frequency, the wavevector is complex, then we face a *spatial or convective instability*.

The existence of these two types of instabilities is tied to the implicit nature of the dispersion equation: $D(\omega, \mathbf{k}) = 0$. Two types of explicit solutions are thus possible:

$$\omega(\mathbf{k}) \qquad \text{or} \qquad \mathbf{k}(\omega)$$

The first are called "temporal branches" when $\mathbf{k} \in \mathbb{R}^3$, while the second ones define the "spatial branches" if $\omega \in \mathbb{R}$. For example, the dispersion relation

$$\omega + 2k - k^2 = 0$$

possesses one temporal branch $\omega = k^2 - 2k$, which is stable, and two spatial branches $k = 1 \pm \sqrt{1 + \omega}$ which can generate a spatial instability.

6.1.2 The Gravitational Instability

A simple example of an absolute instability comes from Astrophysics with the gravitational instability, which is at the origin of star formation. To make things as simple as possible, we consider an unbounded fluid of uniform temperature and pressure. We assume that it is an ideal gas of adiabatic index γ. The sound waves propagate with a velocity

$$c_s = \sqrt{\gamma P_0 / \rho_0}$$

where P_0 and ρ_0 are respectively the pressure and density of the undisturbed medium.

The linearized equations satisfied by the disturbances of the medium are:

$$
\begin{cases}
\dfrac{\partial \delta\rho}{\partial t} + \rho_0 \nabla \cdot \mathbf{v} = 0 \\[2mm]
\rho_0 \dfrac{\partial \mathbf{v}}{\partial t} = -\nabla \delta P - \rho_0 \nabla \delta\Phi \\[2mm]
\delta P = c_s^2 \delta\rho \\[2mm]
\Delta \delta\Phi = 4\pi G \delta\rho
\end{cases}
\tag{6.1}
$$

where we have taken into account the fluctuations in the gravitational field generated by the fluctuations of density (the last equation of the system). For plane wave solutions

$$\delta\rho = \delta\rho_0 e^{i(\omega t + \mathbf{k}\cdot\mathbf{r})}, \quad \mathbf{v} = \mathbf{v}_0 e^{i(\omega t + \mathbf{k}\cdot\mathbf{r})}, \text{ etc.} \tag{6.2}$$

we find the following dispersion relation:

$$\omega^2 = c_s^2 k^2 - 4\pi G\rho_0 \tag{6.3}$$

This relation clearly shows a temporal instability since all the perturbations with a wavenumber smaller than

$$k_J = \sqrt{\frac{4\pi G\rho_0}{c_s^2}} \tag{6.4}$$

are unstable and grow exponentially. The associated wavelength $\lambda_J = 2\pi/k_J$ is called *Jeans' length* and the associated criterion, *Jeans' criterion*. The dispersion relation (6.3) also shows that there is no spatial instability:

$$c_s^2 k^2 = \omega^2 + 4\pi G\rho_0 > 0, \quad \forall\omega \in \mathbb{R};$$

thus k is always real.

In order to fix ideas, let us calculate Jeans' length in the case of the Earth's atmosphere. We assume it to be a mass of air at $P_0 = 10^5$ Pa and $T_0 = 20\,°C$. Then, $c_s = 343$ m/s and $\rho_0 = 1.2$ kg/m^3 giving $\lambda_J = 6.8\,10^4$ km. The Earth's atmosphere, much smaller (in thickness) than this length, is not, therefore, in danger of gravitational collapse! On the other hand, an interstellar cloud of a hundred solar masses,[1] with a temperature of 50 K and a diameter of two light-years can be wiped out gravitationally (see exercises).

The example of the Earth's atmosphere is interesting as it points out the limits of the local analysis: if the dimensions of the fluid domain are smaller than the wavelength of the disturbances we are interested in, the local solutions are invalid because of boundary conditions.

6.1.3 Convective Instability

Such an instability is usually found in shear flows, for instance in a boundary layer. Perturbations are amplified in the downstream direction and may transform a laminar flow into a turbulent one (see Fig. 6.14 for the growth of a perturbation in

---

[1] A solar mass, symbolized by $M_\odot$, is equal to 2×10^{30} kg.

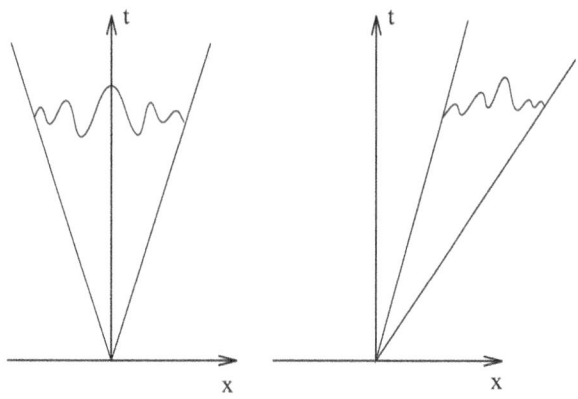

Fig. 6.1 Sketch of an absolute (*left*) and convective (*right*) instability in the (x, t)-plane

the downstream direction). This instability can also be regarded as the growth of an absolute instability advected by the background flow as shown in Fig. 6.1.

To further illustrate this mechanism, we shall consider a model problem based on the perturbations of Burgers flow.[2] The background flow is uniform and disturbances are only on the velocity field and one-dimensional; thus $\delta \mathbf{u} = \delta u(x, t)\mathbf{e}_x$ and

$$\frac{\partial \delta u}{\partial t} + U \frac{\partial \delta u}{\partial x} = \nu \frac{\partial^2 \delta u}{\partial x^2}$$

where ν is the kinematic viscosity. The dispersion relation of the Fourier modes is

$$i\omega + ikU = -\nu k^2$$

This relation immediately shows that the temporal branch is stable since, for a given k, the temporal dependence $\exp(i\omega t)$ leads to an exponential decay.

Let us now extract the spatial branches of this dispersion relation. We easily find that two branches exist, namely

$$k_\pm = \frac{iU}{2\nu}\left(1 \pm \sqrt{1 + \frac{4i\omega\nu}{U^2}}\right)$$

To discuss its properties, it is convenient to consider the limiting case of a small viscosity such that $\omega\nu \ll U^2$. Thus,

$$k_+ = -\frac{\omega}{U} + \frac{iU}{\nu} \qquad \text{and} \qquad k_- = \frac{\omega}{U} - \frac{2i\nu\omega^2}{U^3}$$

---

[2]Burgers equation is given by (5.85).

To see whether these branches correspond to growing or decaying perturbations, it is useful to write the resulting velocity field, namely

$$\mathbf{u}^+ = \mathbf{u}_0^+ e^{\frac{i\omega}{U}(x+Ut)+\frac{U}{\nu}x} \qquad \text{and} \qquad \mathbf{u}^- = \mathbf{u}_0^- e^{-\frac{i\omega}{U}(x-Ut)-\frac{2\nu\omega^2}{U^3}x}$$

These expressions show that a given phase of $\mathbf{u}^+$ propagate to negative values of x, therefore its amplitude decreases rapidly as $\exp(Ux/\nu)$. On the other hand the phase of $\mathbf{u}^-$ propagates to positive values of x, and its amplitude also decreases (but slower, as $\exp(-\nu\omega^2 x/U^3)$) as the perturbation moves to high values of x. In this example, we see that if some disturbance is forced at a given frequency ω at $x = 0$ say, it will propagate upstream and downstream, but both waves will be damped. Thus the flow is stable.

6.2   Linear Analysis of Global Instabilities

Although local analysis is very handy to get a first impression of the stability of a steady flow, it is often limited in its applications because of the boundary conditions. To deal with this constraint, we need moving to the global analysis, which is often difficult. A medium way is to take into account the boundary conditions only in one direction. Although still quite idealized, the resulting solutions are usually very instructive on the physics of the flow. With this approach, we shall investigate selected examples of instabilities, which will enlight us, at the same time, on the properties of rotating fluids, shear flows, etc.

6.2.1   Centrifugal Instability: Rayleigh's Criterion

Let us consider a perfect incompressible fluid filling the gap between two cylinders of radii R_1 and R_2. The fluid rotates with the prescribed angular velocity profile $\Omega(s)$. We wish to know the conditions to be met by this rotation law, for this flow to be stable or unstable. The original flow, $\mathbf{U} = s\Omega(s)\mathbf{e}_\varphi$, is a solution of Euler's equation and satisfies $\nabla \cdot \mathbf{U} = 0$.

To simplify the analysis, we assume that the cylinders are infinitely long. Thus, boundary conditions are only imposed in the radial direction and we are allowed to make a local analysis in the z-direction. We further restrict the disturbances to the axisymmetric ones; we thus write the perturbations of the velocity and pressure fields as

$$\mathbf{u}(s,z,t) = \mathbf{u}(s)e^{ikz+\lambda t}, \qquad P(s,z,t) = \rho p(s)e^{ikz+\lambda t} \tag{6.5}$$

where ρ is the fluid density.

6.2.1.1 Equations for the Perturbations

The momentum equation leads to the following equations for $p(s)$ and $\mathbf{u}(s)$:

$$\lambda \mathbf{u} + (\mathbf{U} \cdot \nabla)\mathbf{u} + (\mathbf{u} \cdot \nabla)\mathbf{U} = -\nabla p \qquad (6.6)$$

which we rewrite as

$$\lambda \mathbf{u} - 2\Omega(s)u_\varphi \mathbf{e}_s + (\Omega(s)u_s + (\mathbf{u} \cdot \nabla)U(s))\mathbf{e}_\varphi = -\nabla p . \qquad (6.7)$$

with cylindrical coordinates (s, φ, z).

Finally, taking mass conservation into account, we get the four following equations:

$$\begin{cases} \lambda u_s - 2\Omega(s)u_\varphi = -\dfrac{dp}{ds} \\[2mm] \lambda u_\varphi + \Phi_1(s)u_s = 0 \\[2mm] \lambda u_z = -ikp \\[2mm] \dfrac{1}{s}\dfrac{d}{ds}(su_s) + iku_z = 0 \end{cases} \qquad (6.8)$$

where we have introduced

$$\Phi_1(s) = \frac{1}{s}\frac{d(s^2\Omega)}{ds} .$$

The second equation gives the expression of u_φ as a function of u_s. The third and fourth ones relate p and u_s. Altogether they lead to a single equation for u_s now denoted u, namely

$$\frac{d}{ds}\left[\frac{1}{s}\frac{d(su)}{ds}\right] - k^2 u = \frac{k^2}{\lambda^2}\kappa^2(s)u \qquad (6.9)$$

where

$$\kappa^2(s) = 2\Omega\Phi_1 = \frac{1}{s^3}\frac{d(s^2\Omega)^2}{ds} \qquad (6.10)$$

is proportional to the radial derivative of the angular momentum $\ell = s^2\Omega$ of the fluid particles in the original flow. $\kappa(s)$ is called the *epicyclic frequency*. We shall comment later about its physical meaning.

Setting, $\Lambda = 1/\lambda^2$, the differential equation (6.9) has the general and interesting form:

$$\mathcal{L}u = \Lambda k^2 \Phi(s) u \tag{6.11}$$

Supplemented with the boundary conditions $u = 0$ at $s = R_1$ and $s = R_2$, this is the classical Sturm–Liouville problem in the theory of differential equations. We refer the reader to the maths complements in Chap. 12 to get acquainted with the basic properties of Sturm–Liouville problems and proceed to the consequences for our problem.

6.2.1.2 The Rayleigh Criterion

First of all, let us compare (6.11) with the general equation (5.11): it is of the same form. Thus, in order to determine the stability of the flow **U**, we "just" need to know the spectrum of the operator $k^{-2}\Phi(s)^{-1}\mathcal{L}$, which gives the set of allowed values of Λ. Usually, this is not an easy game; however, because of the Sturm–Liouville nature of the eigenvalue problem, the answer is straightforward. For such problems indeed, it may be shown that the eigenvalues are discrete, real and of the sign of $-k^2\Phi(s)$ if this function keeps the same sign in the interval of definition $[R_1, R_2]$. If Φ changes sign the eigenvalues are of both signs.

These properties of the Sturm–Liouville problems allow us to conclude on the stability of the flow. Indeed, if $\Phi(s) \geq 0$, all the eigenvalues Λ are negative, which means that all the eigenvalues λ are purely imaginary. Thus perturbations are just neutral; the flow is stable. On the other hand, if there exist an interval where Φ is negative, then there exist some eigenvalues Λ that are positive, implying the existence of real positive λ, and thus the existence of amplified disturbances making the flow unstable.

The foregoing result shows that the flow under consideration is unstable when, somewhere, the specific angular momentum ℓ decreases with r (making $\Phi < 0$). In this case, some axisymmetric disturbances grow exponentially. The opposite situation, where $\Phi(s) \geq 0$, does not mean that the flow is stable; it means that axisymmetric disturbances are not amplified, however, some non-axisymmetric ones could be growing.

We thus find *a sufficient condition* for an instability ($\Phi(s) < 0$ somewhere) or *a necessary condition* for stability ($\Phi(s) \geq 0$ everywhere). This criterion was discovered by Rayleigh and named after him.

6.2.1.3 The Rayleigh Criterion: A Heuristic Derivation

The foregoing argument is rather mathematical and little intuitive. However, the result may be explained on more physical grounds as follows. Let us consider two annular fluid elements of radii s_1 and s_2, with $s_1 < s_2$. Their angular momentum is respectively ℓ_1 and ℓ_2. Their total kinetic energy is

$$E_k = \frac{1}{2}\left(\frac{\ell_1^2}{s_1^2} + \frac{\ell_2^2}{s_2^2}\right)$$

Now let's suppose that the position of these two fluid elements is inverted: their angular momentum and mass are conserved, but the kinetic energy of the two elements is now

$$E_k' = \frac{1}{2}\left(\frac{\ell_1^2}{s_2^2} + \frac{\ell_2^2}{s_1^2}\right)$$

Making the difference between these two expressions, we find

$$E_k - E_k' = \frac{1}{2}\left(\ell_2^2 - \ell_1^2\right)\left(\frac{1}{s_2^2} - \frac{1}{s_1^2}\right)$$

If the angular momentum increases outwards then $\ell_2 > \ell_1$ and $E_k < E_k'$, therefore the change imposed is energetically unfavorable: the situation is stable. If, in the opposite case, the angular momentum decreases outwards then $\ell_2 < \ell_1$ and $E_k > E_k'$: some energy is released if the position of the fluid elements are interchanged. The system cannot stay in its initial configuration and will evolve towards a new state of lower energy.

6.2.2 Shear Instabilities of Parallel Flows

Parallel shear flows represent a vast category of flows that are very common in Nature and often at the origin of turbulence. A parallel shear flow is basically very simple: its velocity field is like:

$$\mathbf{V} = U(z)\mathbf{e}_x . \tag{6.12}$$

It has only one component, taken here in the x-direction, which is a function of only one coordinate normal to the direction of the flow, here z. We only consider steady flows. We note that, if the density of the fluid is independent of the coordinate in the velocity direction then, the equation of continuity is automatically satisfied. To further simplify, we restrict our discussion to the case of incompressible fluids.

The stability of parallel shear flows has an interesting property formulated by *Squire's theorem*: the most unstable disturbances of these flows are two-dimensional. This greatly simplifies the analysis of the stability of such flows. We shall therefore start by proving this theorem before presenting some famous examples of shear instabilities.

6.2.2.1 Squire's Theorem

Statement: *To every unstable disturbance of a parallel shear flow of an incompressible fluid there corresponds a more unstable two-dimensional disturbance.*
Proof: We begin by proving this theorem in the inviscid case. We assume that the perturbations are in the following form:

$$f(\mathbf{r}, t) = f(z)e^{ik_x x + ik_y y + \lambda t} \tag{6.13}$$

where the Fourier form is in the homogenous directions of the flow. The perturbations satisfy

$$\begin{cases} \dfrac{\partial \mathbf{v}}{\partial t} + U(z)\dfrac{\partial \mathbf{v}}{\partial x} + v_z U'(z)\mathbf{e}_x = -\nabla P \\[2mm] \nabla \cdot \mathbf{v} = 0 \end{cases} \tag{6.14}$$

After substitution by (6.13) and projection along the three axes, we find

$$\begin{cases} (\lambda + ik_x U)v_x + v_z U'(z) = -ik_x P \\ (\lambda + ik_x U)v_y = -ik_y P \\ (\lambda + ik_x U)v_z = -DP \\ Dv_z + ik_x v_x + ik_y v_y = 0 \end{cases} \tag{6.15}$$

where we have set $D = \frac{\partial}{\partial z}$. We now make *Squire's transformation* and set

$$\tilde{k} = \sqrt{k_x^2 + k_y^2}, \qquad \tilde{k}\tilde{v} = k_x v_x + k_y v_y, \qquad \tilde{P} = \frac{\tilde{k}}{k_x} P$$

The equations are now

$$\begin{cases} (\lambda + ik_x U)\tilde{v} + \dfrac{k_x}{\tilde{k}}v_z U'(z) = -i\tilde{k}P \\[3mm] (\lambda + ik_x U)v_z = -\dfrac{k_x}{\tilde{k}}D\tilde{P} \\[3mm] Dv_z + i\tilde{k}\tilde{v} = 0 \end{cases} \tag{6.16}$$

which can again be written as

$$\begin{cases} (\tilde{\lambda} + i\tilde{k}U)\tilde{v} + v_z U'(z) = -i\tilde{k}\tilde{P} \\ (\tilde{\lambda} + i\tilde{k}U)v_z = -D\tilde{P} \\ Dv_z + i\tilde{k}\tilde{v} = 0 \end{cases} \tag{6.17}$$

by introducing $\tilde{\lambda} = \lambda \frac{\tilde{k}}{k_x}$. Noting the similarity of (6.17) and (6.15) with $k_y = v_y = 0$, we conclude that, if the flow is unstable, namely if $Re(\lambda) > 0$, for every three-dimensional disturbance, we can construct a two-dimensional disturbance $(\tilde{v}, v_z, \tilde{P})$ that grows faster, since $Re(\tilde{\lambda}) \geq Re(\lambda)$.

The case with viscosity is treated in a similar manner. While observing that for the disturbances (6.13), the Laplacian is changed into $D^2 - \tilde{k}^2$, we rewrite (6.15) in the form

$$\begin{cases} [\lambda + ik_x U - \nu(D^2 - \tilde{k}^2)]v_x + v_z U'(z) = -ik_x P \\ [\lambda + ik_x U - \nu(D^2 - \tilde{k}^2)]v_y = -ik_y P \\ [\lambda + ik_x U - \nu(D^2 - \tilde{k}^2)]v_z = -DP \\ Dv_z + ik_x v_x + ik_y v_y = 0 \end{cases} \tag{6.18}$$

We apply Squire's transformation

$$\begin{cases} [\tilde{\lambda} + i\tilde{k}U - \tilde{\nu}(D^2 - \tilde{k}^2)]\tilde{v} + v_z U'(z) = -i\tilde{k}\tilde{P} \\ [\tilde{\lambda} + i\tilde{k}U - \tilde{\nu}(D^2 - \tilde{k}^2)]v_z = -D\tilde{P} \\ Dv_z + i\tilde{k}\tilde{v} = 0 \end{cases} \tag{6.19}$$

where $\tilde{\nu} = \nu \frac{\tilde{k}}{k_x} \geq \nu$. Thus, with every three-dimensional disturbances, we can associate a two-dimensional disturbance, for which the Reynolds number is smaller. Consequently, the critical Reynolds number, above which a given perturbation grows exponentially, can be decreased by applying Squire's transformation to that perturbation. Hence, the perturbations, which give the lowest critical Reynolds number of shear flows, are the two-dimensional ones.

6.2.3 Rayleigh's Equation

In order to complete our study of parallel shear flows, we now transform (6.15) into an ordinary differential equation for the stream function of the disturbances, since, thanks to Squire's theorem, we can restrict our study to two-dimensional perturbations only. Accordingly, we set

$$v_x = \frac{\partial \psi}{\partial z} = D\psi \qquad \text{and} \qquad v_z = -\frac{\partial \psi}{\partial x} = -ik\psi$$

where $k = k_x$ and $k_y = 0$. We then transform (6.15) into

$$(\lambda + ikU)k^2\psi = D\left[(\lambda + ikU)D\psi - ik\psi U'\right]$$

then finally into

$$(\lambda + ikU)(D^2 - k^2)\psi - ikU''\psi = 0 \qquad (6.20)$$

which is Rayleigh's equation.

6.2.3.1 Criteria of Stability

We can infer from Rayleigh's equation a necessary condition for instability, that is to say, a condition so that $Re(\lambda) > 0$ is possible. We return to (6.20) and assume that the fluid is bounded by two planes located in $z = a$ and $z = b$. By integrating the equation over this domain after multiplication it by ψ^*, the complex conjugate of ψ, we find

$$\int_a^b \psi^*(D^2 - k^2)\psi\, dz - ik \int_a^b \frac{U''|\psi|^2}{\lambda + ikU} dz = 0 \,.$$

Since $v_z = 0$ on each bounding plane, integration by parts yields

$$\int_a^b (|D\psi|^2 + k^2|\psi|^2) dz + ik \int_a^b \frac{U''|\psi|^2}{\lambda + ikU} dz = 0 \qquad (6.21)$$

The imaginary part of this equation leads to

$$Re(\lambda)k \int_a^b \frac{|\psi|^2 U''}{|\lambda + ikU|^2} dz = 0 \qquad (6.22)$$

which shows that a necessary condition for the existence of an instability is that the integral be zero. This condition implies that U'' changes sign at least once in the interval $[a, b]$. Reciprocally, this condition shows that if a velocity profile has no point of inflexion, then $Re(\lambda) = 0$ and the flow is stable with respect to infinitesimal disturbances.

This condition is evidently not sufficient: even if $Re(\lambda) \neq 0$, this quantity is not necessarily positive!

Rayleigh proved this result in 1880. In 1950 Fjørtoft found a more constraining version of it. He showed that a necessary condition for instability was that

$$U''(U - U_i) < 0$$

at some point in the flow where U_i is the velocity at the inflexion point. We propose the proof of this theorem as part of the exercises.

6.2.4 The Orr–Sommerfeld Equation

The Orr–Sommerfeld equation is the variant of Rayleigh's equation including viscosity. We obtain this equation after several manipulations of (6.19), by expressing v_x and v_z with the help of the stream function. Orr–Sommerfeld equation has the following form:

$$\left(\lambda + ikU - \nu(D^2 - k^2)\right)(D^2 - k^2)\psi = ikU''\psi \tag{6.23}$$

which we complete with the no-slip boundary conditions at the walls (planes $z = 0$ and $z = d$), namely

$$\psi = D\psi = 0 \quad \text{at} \quad z = 0 \quad \text{and} \quad z = d$$

We shall not discuss the solutions of this equation because it would bring us too far, and refer the interested reader to the book of Drazin and Reid (1981). We shall give a few comments only.

Shear flows, like boundary layers, jets, wakes, mixing layers, etc. are usually the seat of strong turbulence, which is a consequence of shear instabilities. The Orr–Sommerfeld equation offers a nice model to study these instabilities and its solutions have therefore numerous applications.

Many cases have been studied. The simplest ones are those at low Reynolds number, which can be investigated by perturbation methods on the diffusion equation. However, they are not the most interesting since applications usually require the other extreme: a very high Reynolds number. As we saw in Chap. 4, this implies the existence of boundary layers, but not only. Indeed, from Rayleigh equation, to which Orr–Sommerfeld reduces at infinite Reynolds number, we observe that something special must occur when

$$\lambda + ikU = 0$$

or when $c = \frac{\omega}{k} = -U$. This equality means that the phase velocity of the perturbations is equal and opposite to the fluid velocity; the phase perturbation stands still in the reference frame. At this place, the coefficient of the second derivatives of ψ vanishes. A singularity of the perturbed flow shows up: this is *a critical layer*. In such a layer, viscosity smooths out the singularity, which usually consists in a discontinuity of the parallel component of the velocity field (see Sect. 6.3.3 for instance). Critical layer are also called *detached shear layers*; their thickness, like the one of boundary layers, scales like some fractional power of the viscosity ($\nu^{1/3}$ and $\nu^{1/4}$ are the most common cases). They are important in the global dynamics of a fluid layer as they are strong dissipative structures.

Ending the chapter, we shall use Orr–Sommerfeld equation to introduce algebraic instabilities that represent another path to turbulent flows.

6.3   Some Examples of Famous Instabilities

6.3.1   Example: The Kelvin–Helmholtz Instability

The Kelvin–Helmholtz instability is a shear instability that appears when two fluid layers of different densities, slide one on the other.

In order to analyse this instability, we shall consider the setup of an air flow on top of a water plane. *The two fluids are assumed to be inviscid.* The air occupies the $z > 0$ half-space, while the water fills the remaining space. The air is assumed to be moving at a constant velocity $V \mathbf{e}_x$ with respect to the water. To be complete, we also take into account the surface tension γ between the two fluids. Thus, except for the air motion, the set-up is exactly the same as the one use in Sect. 5.3, when studying surface waves.

As in Sect. 5.3, we assume the perturbations of the velocity field to be irrotational, namely $\delta \mathbf{v} = \nabla \Phi_a$. We thus rewrite the second equation of (5.26) directly as:

$$\frac{\partial \Phi_a}{\partial t} + V \frac{\partial \Phi_a}{\partial x} + \frac{\delta P_a}{\rho_a} + g \delta z = cst \tag{6.24}$$

Since the potential Φ_a still satisfies Laplace's equation, (5.29) is always satisfied because we are still looking for solutions in the form of (5.27). On the other hand the boundary condition (5.30) is modified on the air side, indeed

$$v_{z,water} = \frac{\partial z_s}{\partial t} \qquad \text{and} \qquad v_{z,air} = \frac{\partial z_s}{\partial t} + V \frac{\partial z_s}{\partial x}$$

Fig. 6.2 The great red spot of Jupiter as viewed by the Galileo probe. Note the vortices around it. They come from the shear instabilities forced by this flow (Credit NASA)

(5.32) is therefore replaced by

$$k\Phi_{water}(0) = -i\omega z_s \quad \text{and} \quad k\Phi_a(0) = i(\omega - kV)z_s$$

Since we take surface tension into account we have

$$\delta P_w = \delta P_a + \gamma k^2 z_s$$

according to (5.40). Finally, we derive the following dispersion relation

$$\omega^2(\rho_w + \rho_a) - 2\omega\rho_a kV - (\rho_w - \rho_a)gk - \gamma k^3 + \rho_a k^2 V^2 = 0 \qquad (6.25)$$

The temporal branches can be easily extracted:

$$\omega_\pm = \frac{kV\rho_a \pm \sqrt{\Delta}}{\rho_w + \rho_a} \quad \text{and} \quad \Delta = (\rho_w + \rho_a)[k^3\gamma + (\rho_w - \rho_a)gk] - k^2 V^2 \rho_a \rho_w$$

$$(6.26)$$

The expression of $\omega_\pm$ shows that the instability arises when $\Delta < 0$, that is when

$$V^2 > \frac{\rho_w + \rho_a}{\rho_a \rho_w}\left[\gamma k + (\rho_w - \rho_a)\frac{g}{k}\right] .$$

Since the term in brackets has a minimum when $k = k_{min} = \sqrt{(\rho_w - \rho_a)g/\gamma}$, we see that the flow will be unstable if, and only if, the velocity V is greater than the critical velocity given by:

$$V_{crit} = \left(\frac{2}{\rho_w} + \frac{2}{\rho_a}\right)^{1/2}[\gamma g(\rho_w - \rho_a)]^{1/4} \qquad (6.27)$$

With typical values of a water-air interface, namely $\rho_w = 1{,}000\,\text{kg/m}^3$, $\rho_a = 1.2\,\text{kg/m}^3$, $g = 9.81\,\text{m/s}^2$ and $\gamma = 0.072\,\text{N/m}$, we find $V_{crit} = 6.4\,\text{m/s}$. The wavelength of the most unstable mode, namely that for which $k = k_{min}$, is $\lambda_{crit} = 1.7$ cm, which is the length where the capillary effects are of the same order of magnitude as those of gravity (see Sect. 5.3.2).

6.3.2 Instabilities Related to Kelvin–Helmholtz Instability

Formula (6.25) actually contains many interesting cases that we shall discuss now.

6.3.2.1 Rayleigh–Taylor Instability

If in (6.25) we set $V = 0$, we immediately find the dispersion relation of gravity or capillary waves (5.43). Now, let us assume that we manage to put the water above the air. This situation is likely unstable. In fact, since

$$\omega^2 = \frac{(\rho_a - \rho_w)gk + \gamma k^3}{\rho_e + \rho_a} \, , \tag{6.28}$$

we see that this is not necessarily the case. In order for the situation to be unstable, it is necessary that $k < \sqrt{(\rho_w - \rho_a)g/\gamma}$, namely that the perturbations with a wavelength greater than $\lambda_{crit} = 2\pi\sqrt{\gamma/(\rho_w - \rho_a)g}$ can grow.

The foregoing instability, which appears when a layer of fluid covers a layer of a less dense fluid in a gravitational field, is known as Rayleigh–Taylor instability. It usually occurs in Nature when a fluid layer is heated from the bottom. The instability then leads to a fluid flow known as thermal convection, which we shall study in detail in Chap. 7.

Now, the instability shown by (6.28) can be illustrated by a simple experiment. Taking a bottle filled with water, we turn it upside down, maintaining the cork on the orifice. Removing it delicately, we observe that if the diameter of the bottleneck is small enough,[3] the water remains in the bottle. If the diameter is too large, however, the stability of the equilibrium can be restored by increasing artificially the surface tension: a piece of paper laid on the interface will do the job.

Finally, let us note that if the surface tension is zero, for example if both fluids are gases, then the equilibrium is always unstable.

Figure 6.3 shows the development of Rayleigh–Taylor instability in a numerical simulation of a supernova explosion. This instability plays an important role in the mixing of elements yielded by this stellar explosion.

6.3.2.2 The Instability of the Mixing Layer

Another example that is easily derived from (6.25) is the one where the two fluids are identical. Thus, $\rho_a = \rho_w = \rho$.

The configuration thus obtained is the famous "vortex sheet" presented in Fig. 3.9 where the velocity sustains a discontinuity that usually develops into vortices (see Fig. 6.2). From (6.25) we see that such a configuration is unstable for all wavelengths since

$$\omega = (1 \pm i)kV/2 \, .$$

---

[3]We may expect that if the diameter of the bottleneck is smaller than 1.7 cm, the equilibrium is stable. However, we should keep in mind that the value was derived for pure water; impurities decrease the surface tension and lead to a smaller value of the critical wavelength.

Fig. 6.3 Growth of the Rayleigh–Taylor instability in the wake of the shock wave associated with a supernova explosion. The four quadrants show the concentration of helium, oxygen, nickel and silicium. The numerical simulation has been made by Joggerst et al. (2010)

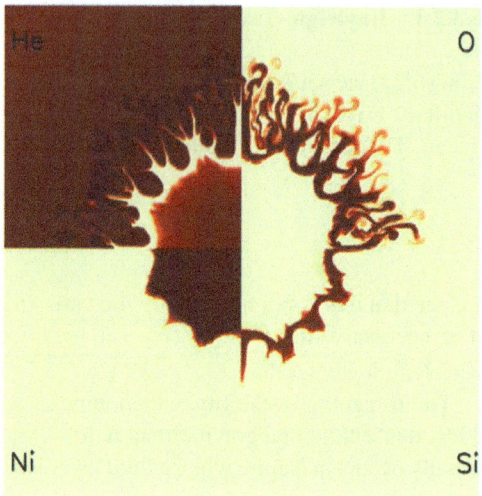

The growth rate increases with the wavenumber without bounds apparently. This dispersion relation comes from the discontinuity of the velocity field. In real systems, the discontinuity has some thickness (due to viscosity) and the growth rate reaches a maximum for perturbations with a wavelength similar to the thickness of the vortex sheet.

6.3.3 Disturbances of the Plane Couette Flow

The plane Couette flow is a shear flow for which the profile is linear:

$$U(z) = z/T \tag{6.29}$$

where T is a constant homogenous to a time. A discussion of the perturbations of this flow is interesting. Setting $\lambda = i\omega$ with $\omega \in \mathbb{R}$ and substituting (6.29) in Rayleigh's equation we find

$$(\omega + kU)(D^2 - k^2)\psi = 0 \tag{6.30}$$

If we assume that $(\omega + kU) \neq 0$, then ψ is given by

$$\psi = A \operatorname{sh}(kz + Q)$$

If the flow takes place between two planes situated at $z = 0$ and $z = d$, at which the disturbances vanish, then $\psi = 0$ throughout. Therefore, in order that the perturbations exist, it is necessary that $\omega + kz/T = 0$ in the interval $[0, d]$. In other words, the disturbances are such that

$$\omega \in [-kd/T, 0]$$

so that their spectrum is continuous.

The form of the solutions is always given by (6.30), but the solutions have a discontinuity in $z = z_c = -\omega T/k$. Actually, we have

$$\psi(z) = A \operatorname{sh} kz, \qquad \text{if} \quad 0 \le z \le z_c$$

$$\psi(z) = B \operatorname{sh} k(z - d), \qquad \text{if} \quad z_c \le z \le d$$

At $z = z_c$, ψ is continuous because v_z is continuous, which is imposed by mass conservation. Therefore, we have

$$B = A \frac{\operatorname{sh} kz_c}{\operatorname{sh} k(z_c - d)}$$

Let us now calculate $v_x = D\psi$ on each side of z_c. We easily verify that $v_x(z_c^-) \ne v_x(z_c^+)$. The component v_x is discontinuous at this point. This discontinuity illustrates a property of linear operators, which connects the existence of a continuous spectrum to that of discontinuous eigenfunctions.

This discontinuity of the perturbed v_x means that the plane Couette flow is likely unstable to finite-amplitude disturbances. Indeed, such a perturbation will contain a vortex sheet, which is always unstable. This inference has been actually verified experimentally and numerically.

6.3.4   Shear and Stratification

To conclude this section on famous unstable shear flows, we now study the case where the fluid is stably stratified in the vertical direction. In this way, we can examine the case where shear instabilities are opposed by a positive temperature gradient that inhibits vertical motions but allows the propagation of internal gravity waves. This situation is often met in natural systems, for instance a lake over which a wind is blowing. The wind entrains surface water and thus imposes some shear flow in the lake. But lake water is often stably stratified with cold (dense) water below (light) warmer water. Because of this stratification, shear flow instabilities may be inhibited, and thus the mixing of waters in the lake.

In order to study the evolution of disturbances in such a system, we return to (6.14) modified to take the buoyancy force into account and completed by the equation of temperature (5.45b).

Staying with the two-dimensional case and using the same notations as before, we now have:

$$\begin{cases} (\lambda + ikU)v_x + U'v_z = -ikP \\ (\lambda + ikU)v_z = -DP + \alpha gT \\ (\lambda + ikU)T + v_z\partial_z T_0 = 0 \\ Dv_z + ikv_x = 0 \end{cases} \quad (6.31)$$

where T is the temperature fluctuation, α the coefficient of thermal expansion (see 1.60) and T_0 is the background temperature profile of the fluid in equilibrium. We also introduce the Brunt–Väisälä frequency N such that $N^2 = \alpha g \partial_z T_0$ and the stream function ψ such that $v_x = D\psi$ and $v_z = -ik\psi$. We can then cast the preceding system into a single equation for ψ:

$$(\lambda + ikU)[D^2 - k^2]\psi - ikU''\psi = \frac{k^2 N^2}{\lambda + ikU}\psi \quad (6.32)$$

also called the *Taylor-Goldstein equation*. If we set the Brunt–Väisälä frequency to zero, we recover Rayleigh equation. As for this equation, we shall derive a criterion of stability when the flow is bounded by two horizontal plates. We could, as for Rayleigh's equation, multiply the equation by the conjugate of ψ and integrate z between the two boundaries. We would then get

$$\int_a^b (|d\psi|^2 + k^2|\psi|^2)dz + \int_a^b \frac{ikU''|\psi|^2}{\lambda + ikU}dz = -k^2 \int_a^b \frac{N^2|\psi|^2}{(\lambda + ikU)^2}dz$$

By requiring the cancellation of the imaginary part of this equation, we find the following necessary condition for the instability of the flow:

$$U'' = \frac{2(\lambda_I + kU)kN^2}{\lambda_R^2 + (\lambda_I + kU)^2}$$

where we set $\lambda = \lambda_R + i\lambda_I$. Unfortunately, this equation is not a criterion of the flow itself, unlike Rayleigh's one, since it depends on the eigenvalue. The way to obtain a true criterion on the flow was discovered by L. Howard in 1961. It consists in making use of the function

$$\chi = \frac{\psi}{\sqrt{\lambda + ikU}}$$

which obeys

$$D\left[(\lambda + ikU)D\chi\right] + \left[k^2\frac{U'^2/4 - N^2}{\lambda + ikU} - \frac{ikU''}{2} - k^2(\lambda + ikU)\right]\chi = 0 \quad (6.33)$$

which we can multiply by χ^* and integrate between a and b. Taking the real part of the result, we thus find

$$\lambda_R \int_a^b \left\{ |D\chi|^2 + k^2 |\chi|^2 - k^2 \frac{U'^2/4 - N^2}{|\lambda + ikU|^2} |\chi|^2 \right\} dz = 0$$

In this equation the integral can vanish if, and only if, $U'^2/4 - N^2 > 0$, or if

$$\text{Ri} = \frac{N^2}{U'^2} \leq \frac{1}{4} \qquad (6.34)$$

Ri is called *the Richardson number*. Equation (6.34) is generally called the *Richardson's criterion*. We see that it is a necessary condition for instability. For certain particular flows it is also sufficient. This criterion shows that when the stratification is sufficiently large, that is to say when the Brunt–Väisälä frequency is sufficiently high, the flow is stable.

This criterion, like Rayleigh's criterion, can be recovered on heuristic arguments, which allow a more physical understanding. To do this, we shall take two fluid elements respectively at z and $z + \delta z$. In order to exchange them, some work against the buoyancy force must be provided, namely

$$W = -g\delta\rho\delta z$$

The energy will be taken from the reservoir of kinetic energy, which stays in the original flow. We then make the following transformation:

$$
\begin{array}{ccc}
z + \delta z & \rho + \delta\rho \quad U + \delta U \\
z & \rho \qquad U
\end{array}
\longrightarrow
\begin{array}{c}
\rho \qquad U + \alpha\delta U \\
\rho + \delta\rho \quad U + (1-\alpha)\delta U
\end{array}
$$

where α is a free number between 0 and 1. We see that this transformation conserves the mass and momentum at first order. Let us now calculate the difference of kinetic energy δE_c between the initial and final states. We have

$$2\delta E_c = \rho U^2 + (\rho + \delta\rho)(U + \delta U)^2 - \rho(U + \alpha\delta U)^2 - (\rho + \delta\rho)(U + (1-\alpha)\delta U)^2$$

$$= 2\alpha(1 - \alpha)\rho\delta U^2 + 2\alpha U\delta\rho\delta U$$

We observe that if $\alpha < 1$ then $\alpha(1 - \alpha) \leq 1/4$, and the maximum is reached at $\alpha = 1/2$, so that

$$2\delta E_c \lesssim \frac{1}{2}\rho(\delta U)^2 + 2U\delta U\delta\rho . \qquad (6.35)$$

Because of these constraints, stability is insured if

$$\frac{1}{4}\rho\delta U^2 + U\delta U\delta\rho \leq -g\delta\rho\delta z$$

that is to say if the maximum variation of the kinetic energy is smaller than the work needed to exchange two fluid elements. Finally, there is stability if

$$\frac{1}{4}\left(\frac{dU}{dz}\right)^2 \leq -\frac{g}{\rho}\left(\frac{d\rho}{dz}\right) - \frac{U}{\rho}\frac{d\rho}{dz}\frac{dU}{dz}$$

We shall see later that in many circumstances, stratified flows can be computed using the Boussinesq approximation, which implies the neglect of ρ variations while maintaining constant the product $g\delta\rho$, (buoyancy force must not disappear!). Thus, the second term of the right-hand side is usually negligible compared to the first; in this way, we recover Richardson's criterion, which was discovered in 1920 (see Richardson 1920).

6.3.5   The Bénard-Marangoni Instability ♣

At the turn of the twentieth century Bénard (1874–1939) discovered that a thin film of liquid heated from below exhibits some vortical cellular motions. For almost 60 years, these fluid flows have been interpreted as the result of thermal convection, an instability driven by the buoyancy force (this is the subject of our next chapter). However, Pearson (1958) showed that when the fluid layer is very shallow, buoyancy effects are dominated by surface tension effects that are able, as we shall see, to destabilize the fluid at rest.[4]

To understand this phenomenon, we consider a fluid layer of thickness d, infinite in the x and y directions. In the z direction, i.e. across the layer, some temperature gradient is imposed, for instance by heating the bottom boundary. In the equilibrium situation, we thus have

$$T_{\mathrm{eq}} = T_0 + \beta z,$$

for the temperature field. We assume that the density variations are negligible altogether, thus perturbations of the velocity field $\mathbf{v}$, of the pressure field δp and of the temperature field δT, verify

---

[4]The name of Carlo Marangoni (Pavia 1840–Firenze 1925) is generally associated with this instability as he was the first physicist to describe fluid flows driven by surface tension gradients (with a paper in Annalen der Physik in 1871).

$$\begin{cases} \nabla \cdot \mathbf{v} = 0 \\[2mm] \dfrac{\partial \mathbf{v}}{\partial t} = -\dfrac{1}{\rho}\nabla \delta p + v \Delta \mathbf{v} \\[2mm] \dfrac{\partial \delta T}{\partial t} + \mathbf{v} \cdot \nabla T_{\text{eq}} = \kappa \Delta \delta T \end{cases} \qquad (6.36)$$

These equations need to be completed by boundary conditions. On the bottom plane, we impose no-slip boundary conditions for the velocity and a fixed temperature; thus

$$\mathbf{v} = \mathbf{0} \qquad \text{and} \qquad \delta T = 0 \quad \text{on} \quad z = 0$$

On the top boundary the fluid (a liquid) meets another fluid (a gas). Surface tension is therefore important, and above all its temperature dependence. Since the temperature fluctuations are assumed very small, a linear law is valid and sufficient; we take

$$\gamma(T) = \gamma_0(1 + \gamma_T T) \qquad (6.37)$$

where γ_0 and γ_T are given by the nature of the liquid-gas interface. Usually, $\gamma_T < 0$ since surface tension decreases with temperature.[5]

We also assume that the deformation of the interface is negligible (γ_0 is large enough), and neglect the effects of gas motion. In these circumstances, the boundary conditions on the liquid at the interface are that the vertical velocity of the liquid vanishes there and that no horizontal stress applies on this surface. From (1.70), it turns out that:

$$\mathbf{v} \cdot \mathbf{e}_z = 0 \qquad \text{and} \qquad ([\sigma_{\text{liq}}]\mathbf{e}_z - \nabla \gamma) \times \mathbf{e}_z = \mathbf{0} \qquad (6.39)$$

---

[5]Surface tension comes from the binding energy of molecules due to their mutual interactions in a liquid. We may expect that at the critical temperature, which is the temperature where the gas and liquid phases are undistinguishable, the surface tension disappears. This remark lead L. Eötvös (1848–1919) to propose that surface tension varies with temperature like

$$\gamma = k(T_c - T)/V^{2/3}$$

Here k is a universal constant for the liquids, V is the volume of one mole and T_c is the critical temperature. This law, which is known as Eötvös rule, is only approximate, but suggests that γ decreases linearly with temperature, as actually observed experimentally. For instance, the following fit

$$\gamma = 7.3\,10^{-2}\,[1 - 0.0023(T - 291)]\ \text{N/m} \qquad (6.38)$$

matches rather well the variations of surface tension of water in the range 273–373 K, as illustrated in Fig. 6.4.

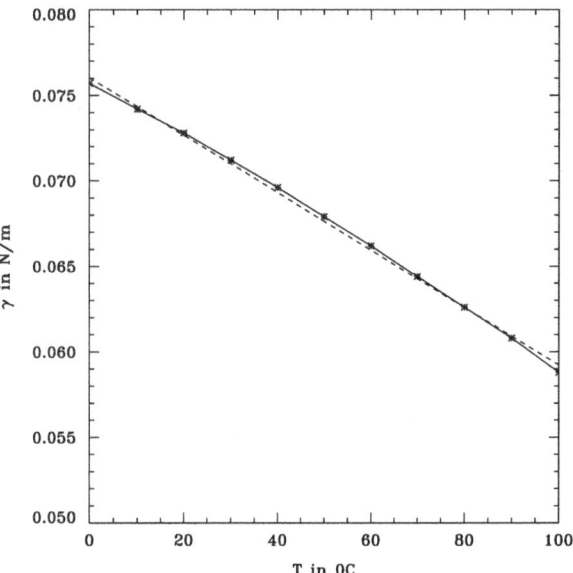

Fig. 6.4 Temperature variations of the surface tension of pure water; the *dashed line* shows the linear fit given in the text

One condition remains to be specified: that on the temperature at the interface. There, we should impose the general conditions between two conducting materials, namely (1.66), with the additional point that air is a transparent medium where energy may be carried out by radiation. Assuming that the liquid radiates like a black body into the gas, boundary conditions at the interface read:

$$T_l = T_g \qquad \text{and} \qquad -\chi_l \frac{\partial T_l}{\partial z} = \sigma T_l^4 - \chi_g \frac{\partial T_g}{\partial z}$$

where σ is Stefan constant. If we now consider the temperature perturbations around a steady state, these perturbations verify:

$$\delta T_l = \delta T_g \qquad \text{and} \qquad \frac{\partial \delta T_l}{\partial z} + q \delta T_l = \frac{\chi_g}{\chi_l} \frac{\partial \delta T_g}{\partial z}$$

Usually, the thermal conductivity of liquids is much higher than the one of gases (see Table 1.1) so that we can safely neglect the right-hand side of the second condition. $q = 4\sigma T_l^3 / \chi_l$ is a parameter which measures the efficiency with which the heat flux permeating the liquid is radiated. If the liquid is a good conductor then the gradient of temperature fluctuation must be small near the boundary. Hence, we shall take

$$\frac{\partial \delta T}{\partial z} + q \delta T = 0 \quad \text{on} \quad z = d \qquad (6.40)$$

as the boundary condition on the temperature of the liquid at the interface. The remaining condition $\delta T_l = \delta T_g$ is useful only in the case we are interested in the gas temperature fluctuations.

We have now prescribed all the equations and boundary conditions, which control the fate of perturbations. We shall rewrite them using non-dimensional variables. We choose the thickness of the layer as the length scale and d^2/κ as the time scale. The temperature scale is naturally given by $d|\beta|$. Furthermore, as we are making a global analysis of stability, we impose that disturbances evolve as $\exp(\lambda t)$; hence we write the equations of motion:

$$\begin{cases} \nabla \cdot \mathbf{u} = 0 \\ \lambda \mathbf{u} = -\nabla p + \mathcal{P} \Delta \mathbf{v} \\ \lambda T - u_z = \Delta T \end{cases} \qquad (6.41)$$

since we take $\beta < 0$. $\mathcal{P}$ is the Prandtl number of the liquid. Using the equation of continuity together with the $\mathbf{u} = \mathbf{0}$ conditions, we derive the following boundary conditions on the $z = 0$ plane:

$$u_z = \frac{\partial u_z}{\partial z} = 0 \qquad \text{and} \qquad T = 0 \quad \text{on} \quad z = 0 \qquad (6.42)$$

On the $z = 1$ plane, we should first make the stress condition (6.39) more explicit; it yields

$$\mu\left(\frac{\partial v_z}{\partial x} + \frac{\partial v_x}{\partial z}\right) - \gamma_0 \gamma_T \frac{\partial \delta T}{\partial x} = 0, \quad \mu\left(\frac{\partial v_z}{\partial y} + \frac{\partial v_y}{\partial z}\right) - \gamma_0 \gamma_T \frac{\partial \delta T}{\partial y} = 0 \quad \text{on } z = 1$$

These conditions are completed by $v_z = 0$. Using dimensionless variables, and mass conservation, the three top boundary conditions give

$$u_z = 0, \qquad \frac{\partial^2 u_z}{\partial z^2} - \mathrm{Ma}\left(\frac{\partial^2 T}{\partial x^2} + \frac{\partial^2 T}{\partial y^2}\right) = 0 \qquad (6.43)$$

where we introduced the *Marangoni number*:

$$\mathrm{Ma} = \frac{\gamma_0 |\gamma_T| |\beta| d^2}{\mu \kappa} \qquad (6.44)$$

Finally, the boundary condition on temperature at $z = 1$ reads

$$\frac{\partial T}{\partial z} + \mathrm{Bi}\, T = 0 \qquad (6.45)$$

where Bi is the Biot number[6]

$$Bi = \frac{4\sigma T_{eq}^3(d)d}{\chi}$$

The system (6.41) can be further reduced to two equations controlling the vertical velocity and the temperature fluctuations; namely:

$$\begin{cases} \lambda \Delta u = \mathcal{P} \Delta \Delta u \\ \lambda T = u + \Delta T \end{cases} \tag{6.46}$$

where $u \equiv u_z$. Since the fluid layer is infinite in the x and y directions, we may express the functions $f(x, y, z) = f(z) \exp(ik_x x + ik_y y)$ and set $k^2 = k_x^2 + k_y^2$. Thus,

$$\begin{cases} \mathcal{P}(D^2 - k^2)^2 u = \lambda(D^2 - k^2)u \\ (D^2 - k^2)T + u = \lambda T \end{cases} \tag{6.47}$$

where $D = \partial/\partial z$. This is a system of sixth order, which is completed by the six boundary conditions:

$$\begin{cases} u = Du = T = 0 & \text{at} \quad z = 0 \\ u = D^2 u + k^2 \text{Ma}\, T = DT + \text{Bi}\, T = 0 & \text{at} \quad z = 1 \end{cases} \tag{6.48}$$

The stability of the fluid layer is determined by the set of eigenvalues λ. It may be shown that the λ's are all real negative numbers when the Marangoni number is zero, hence the system is stable. When this number is increased, the real part of the least-damped mode vanishes for some critical value Ma_c of the Marangoni number. We assume that the associated eigenvalue remains real (the instability is assumed not to be oscillatory). Thus doing, when $\text{Ma} = \text{Ma}_c$, $\lambda = 0$, and we can determine the solutions at the threshold of instability.

The solution of $(D^2 - k^2)^2 u = 0$ verifying $u(0) = Du(0) = u(1) = 0$ is

$$u(z) = A\left[\sinh(kz) + (k \coth k - 1)z \sinh(kz) - kz \cosh(kh)\right]$$

---

[6]The Biot number is the ratio of two heat transfer coefficients. The heat transfer coefficient is a flux surface density divided by a temperature; for instance, χ_l/d is the heat transfer coefficient of the liquid layer, while σT^3 is that of the vacuum.

We also find that

$$T(z) = \frac{1}{4}\left\{\frac{3}{k}z\cosh(kz) + \frac{k\cosh k - \sinh k}{k\sinh k}\left(z^2\cosh(kz) - z\frac{\sinh(kz)}{k}\right) - z^2\sinh(kz)\right.$$

$$\left. - \frac{k^2\sinh^2 k + (\mathrm{Bi}+1)A(k)}{k^2\sinh k(k\cosh k + \mathrm{Bi}\sinh k)}\sinh(kz)\right\}$$

where $A(k) = k^2 + k\sinh k\cosh k + \sinh^2 k$. This solution verifies the boundary conditions $T(0) = 0$ and $DT(1) + \mathrm{Bi}T(1) = 0$. Using these two solutions we can express the Marangoni number as a function of the wavenumber k, as:

$$\mathrm{Ma}(k,\mathrm{Bi}) = \frac{8k(k - \sinh k\cosh k)(k\cosh k + \mathrm{Bi}\sinh k)}{k^3\cosh k - \sinh^3 k} \tag{6.49}$$

This function, plotted in Fig. 6.5 for various values of Bi, determines the minimum value of Ma beyond which the instability sets in. We note that, in the ideal case where $\mathrm{Bi} = 0$, the critical value of the Marangoni number is $\mathrm{Ma}_{\mathrm{crit}} = 79.607$ reached at a wavenumber of $k_{\mathrm{crit}} = 1.993$.

As we mentioned it at the beginning of this section, this instability has long been confused with the Rayleigh–Bénard instability, which is driven by the buoyancy

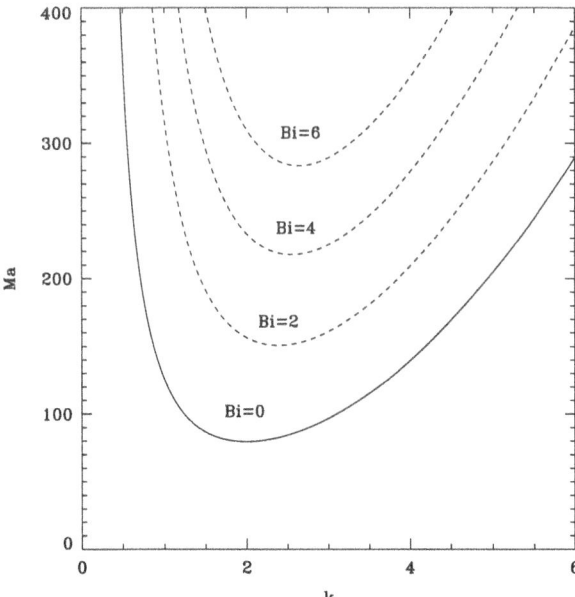

Fig. 6.5 Critical curves for Marangoni-Bénard instability for various values of the Biot number

force. However, as we shall see now, when the thickness of the layer is small enough, the surface tension instability dominates over the buoyancy driven one.

Anticipating on the following chapter, we note that the Rayleigh–Bénard instability is controlled by the Rayleigh number

$$\text{Ra} = \frac{\alpha|\beta|gd^4}{\nu\kappa}$$

which critical value, in similar conditions,[7] is 669.

The dependence of the Rayleigh number with the fourth power of the thickness of the layer, shows that for increasing values of d, the supercriticality of the Rayleigh–Bénard instability is growing faster than that of the tension driven instability, which grows only with the square of the thickness.

We may compute the thickness where the two instabilities are of similar strength. The critical thickness for the tension driven instability is

$$d_t = \left(\frac{\text{Ma}_{\text{crit}}\rho\nu\kappa}{\gamma_0|\gamma_T\beta|}\right)^{1/2}$$

whereas it is

$$d_b = \left(\frac{\text{Ra}_{\text{crit}}\nu\kappa}{\alpha g|\beta|}\right)^{1/4}$$

for the buoyancy driven one. For a given fluid under a similar temperature gradient, these two thicknesses are equal at:

$$d_{\text{bt}} = \sqrt{\frac{\text{Ra}_{\text{crit}}\gamma_0|\gamma_T|}{\text{Ma}_{\text{crit}}\alpha g\rho}}$$

As a numerical illustration, let us consider the case of pure water around $20\,^{\circ}\text{C}$. At this temperature $\alpha = 2.07 \times 10^{-4}\,\text{K}^{-1}$, and using the linear fit of the surface tension (6.38), we find that the critical thickness is 2.6 cm. Hence, a water layer a few millimeters thick is destabilized by surface tension when heated from below.

6.4 Waves Interaction ♠

Another way to tackle instabilities is to interpret their development as the consequence of the interaction of two waves with energies of opposite sign. The total energy of the system stays constant but the amplitude of the two waves can increase

---

[7]This means the same boundary conditions on the bottom plate and on the top plate, stress-free and fixed-flux conditions (this is for the case Bi = 0).

indefinitely (in a linear regime, of course!). This approach has been introduced in Fluids Mechanics by Cairns (1979), who adapted technics devised in plasma physics.

6.4.1   The Energy of a Wave

There is no universal definition of the energy of a wave. Following Cairns' work, we shall define it as the work needed to make its amplitude increase from zero to a given finite value A_0. We assume that the passing wave causes a small displacement of matter, which we denote by

$$\xi(x,t) = A(t)e^{i(\omega_0 t - k_0 x)} . \tag{6.50}$$

The associated pressure disturbance has a similar form. The function $A(t)$ is assumed to vary slowly: the amplitude of the wave increases very progressively. We express this "slowness" by claiming that

$$\frac{1}{A}\frac{dA}{dt} \ll \omega_0 \quad \Longrightarrow \quad \dot{\xi} \simeq i\omega_0 \xi .$$

In order to define the work done to raise the wave, we assume that the displacement (6.50) is the result of the action of the pressure forces, which act on both sides of a surface. As long as the wave is not established, the pressure on both sides differs; thus the work reads

$$W = \int_{-\infty}^{+\infty} (p_2 - p_1)\dot{\xi}\, dt$$

or, in complex notations,

$$W = \frac{1}{2}Re\left\{ \int_{-\infty}^{+\infty} (p_2 - p_1)^* \dot{\xi}\, dt \right\} = Re\left\{ \frac{i\omega_0}{2} \int_{-\infty}^{+\infty} (p_2 - p_1)^* \xi\, dt \right\}$$

In a linear problem, all quantities are proportional and therefore we can write

$$\begin{cases} p_1 = D_1(\omega, k_0)A(t)e^{i(\omega_0 t - k_0 x)} \\ p_2 = D_2(\omega, k_0)A(t)e^{i(\omega_0 t - k_0 x)} \end{cases} \tag{6.51}$$

let

$$(p_2 - p_1)^* = D(\omega, k_0)A(t)e^{-i(\omega_0 t - k_0 x)}$$

where $D(\omega, k) = D_2(\omega, k) - D_1(\omega, k)$. When the wave is established, $p_1 = p_2$ and $D(\omega, k) = 0$ is the dispersion relation of the waves system.

Let us calculate the Fourier Transform of $(p_2 - p_1)^*$; we have

$$\Delta \tilde{p}(\omega) = \int D(\omega, k_0) A(t) e^{-i(\omega_0 t - k_0 x)} e^{i\omega t} dt = D(\omega, k_0) e^{ik_0 x} \tilde{A}(\omega - \omega_0)$$

Since A varies slowly with t, $\tilde{A}(\omega - \omega_0)$ differs from zero only at low frequencies, that is to say for $\omega - \omega_0 \approx 0$. In the neighbourhood of ω_0, we have

$$D(\omega, k) = D(\omega_0, k_0) + (\omega - \omega_0) \left(\frac{\partial D}{\partial \omega} \right)_{\omega_0} + \cdots$$

with $D(\omega_0, k_0) = 0$, therefore

$$(p_2 - p_1)^*(t) = e^{ik_0 x} \int \tilde{A}(\omega - \omega_0) D(\omega, k) e^{-i\omega t} d\omega \; ;$$

taking into account our remark about A, this integral is approximated by

$$(p_2 - p_1)^*(t) = \left(\frac{\partial D}{\partial \omega} \right)_{\omega_0} e^{i(k_0 x - \omega_0 t)} \int (\omega - \omega_0) \tilde{A}(\omega - \omega_0) e^{-i(\omega - \omega_0)t} d\omega$$

$$= -i \left(\frac{\partial D}{\partial \omega} \right)_{\omega_0} e^{i(k_0 x - \omega_0 t)} \frac{dA^*}{dt}$$

From which we find that

$$W = \frac{\omega_0}{2} \left(\frac{\partial D}{\partial \omega} \right)_{\omega_0} \int_{-\infty}^{+\infty} Re(A \frac{dA^*}{dt}) dt = \frac{\omega_0}{4} \left(\frac{\partial D}{\partial \omega} \right)_{\omega_0} |A_0|^2$$

The energy of a wave is therefore defined by

$$\mathcal{E} = \frac{\omega_0}{4} \left(\frac{\partial D}{\partial \omega} \right)_{\omega_0} |A_0|^2 \tag{6.52}$$

6.4.2 Application to the Kelvin–Helmholtz Instability

We now apply the preceding calculations to the Kelvin–Helmholtz instability studied previously.

The dispersion relation (6.25) shows that two waves corresponding to $\omega_\pm$ are possible. We easily calculate their energy

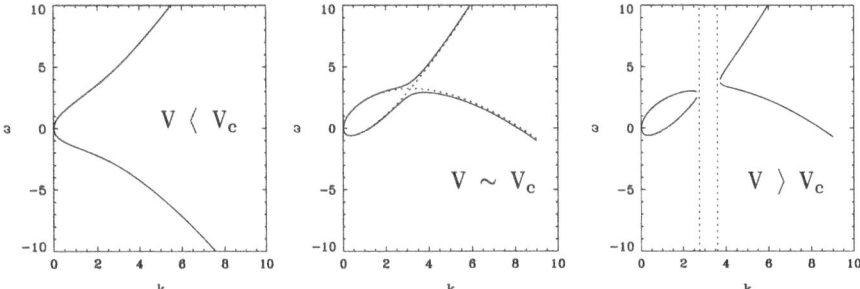

Fig. 6.6 Change of the dispersion relation with the background velocity in a set-up prone to Kelvin–Helmholtz instability. We see that as the background velocity gets close to the critical velocity, a negative energy branch narrows the positive energy one. In the third plot, where $V > V_c$, unstable modes have imaginary frequencies and their wavenumber belongs to the interval limited by the *dotted lines*

$$\mathcal{E}_\pm = \omega_\pm \left(\frac{\partial D}{\partial \omega} \right)_{\omega_\pm} = 2\omega_\pm (\omega_\pm - \rho_a kV) = \pm 2\omega_\pm \sqrt{\Delta}$$

It therefore follows that for every $k > 0$, $\omega_+ > 0$ and $\mathcal{E}_+ > 0$. If $V \ll V_{\mathrm{crit}}$ then $\omega_- < 0$ and $\mathcal{E}_- > 0$; the energy of the two waves are of the same sign. However, if $V \lesssim V_{\mathrm{crit}}$, there appears a band of wavenumbers k for which $\omega_- > 0$ and thus associated with negative energy waves. Moreover, there exists a wavenumber $(k \sim 3)$, such that the two waves are close to resonance, i.e. $\omega_+ \simeq \omega_-$. As illustrated in Fig. 6.6c, this resonance is at the origin of the band of unstable waves.

6.5   The Nonlinear Development of an Instability

Up to now we have studied the evolution of disturbances with infinitesimal amplitudes and noted their exponential growth in the case of instability. Obviously, this growth cannot continue indefinitely because the increasing amplitude inevitably leads to non-negligible nonlinear terms. Their role might simply be to trigger the damping of the instability and to insure a new equilibrium: this is the most simple case that we shall find again in thermal convection in Chap. 7. In general, the situation is more complex: for example, it often happens that a group of modes are unstable because of the set-up. The question we are faced with then is to know towards which solution the system is evolving: is it systematically towards the mode with the highest growth rate? or is it that the nonlinear terms will decide the choice of the final solution which, if it exists, should be stable? It is also possible that no stable solution exists. If the system is chaotic, it wanders indefinitely without ever returning to a point (in the phase space) previously visited.

The nonlinear development of instabilities is a vast field, which would deserve an entire book. The object of this section is thus more modest: we shall examine a few of the simplest cases from which we can shape our intuition about the possible developments of an instability.

6.5.1 Amplitude Equations

When we discussed global instabilities, we expressed the growth of disturbances in the form:

$$\frac{\partial \mathbf{u}}{\partial t} = \mathbf{L}(\mathbf{u})$$

by choosing a time dependence in $e^{\lambda t}$. We generalize this approach by writing

$$\mathbf{u}(\mathbf{r}, t) = A(t)\mathbf{u}_0(\mathbf{r}) \tag{6.53}$$

where $A(t)$ is the amplitude of the mode $\mathbf{u}_0$. If A is very small, we always have

$$\dot{A}(t)\mathbf{u}_0 = A(t)\mathbf{L}(\mathbf{u}_0)$$

but since $\mathbf{u}_0$ is an eigenmode, $\mathbf{L}(\mathbf{u}_0) = \lambda \mathbf{u}_0$, A thus evolves according to

$$\dot{A}(t) = \lambda A(t) \tag{6.54}$$

Such an equation is called an *amplitude equation*. This one is the simplest and its solution $A = A_0 e^{\lambda t}$ is already known to us.

Now, let us suppose that $\mathbf{u}$ is always in the form (6.53), but that its growth is determined by a nonlinear equation that we may write

$$\dot{A}(t) = f(A) \tag{6.55}$$

But for small amplitudes, we have

$$f(A) = f(0) + f'(0)A + \frac{f''(0)}{2}A^2 + \frac{f'''(0)}{6}A^3 + \cdots \tag{6.56}$$

Since A is the amplitude of a disturbance, $A = 0$ should be the equilibrium solution such that $f(A = 0) = 0$; therefore, $f(0) = 0$. Further identification shows that $f'(0) = \lambda$. Hence, we rewrite the preceding equation as:

$$\dot{A}(t) = \lambda A + \frac{f''(0)}{2}A^2 + \frac{f'''(0)}{6}A^3 + \cdots \tag{6.57}$$

We can still progress in the determination of the coefficients of the Taylor expansion of f by using the symmetries of the system. Imagine that the system is invariant in the symmetry $A \rightarrow -A$, i.e. if A is a solution, $-A$ is also a solution, then it is obvious that $f(-A) = -f(A)$ because of the linearity of ∂_t. In this way all the even derivatives of f are zero and (6.57) shortens to:

$$\dot{A}(t) = \lambda A + \frac{f'''(0)}{6} A^3 + \cdots \tag{6.58}$$

Setting $\mathcal{L} = -f'''(0)/6$, the preceding equation is known as *Landau equation*:

$$\dot{A}(t) = \lambda A - \mathcal{L} A^3 \tag{6.59}$$

and $\mathcal{L}$ is the *Landau constant* of the system(cf. Landau and Lifchitz 1971–1989, Sect. 27).

6.5.2   A Short Introduction to Bifurcations

Landau equation describes the behaviour of many systems in Physics, especially in Fluid Mechanics (we shall meet it again when discussing thermal convection in Chap. 7). Thus, it is worth a little study, which will also allow us to introduce the basic ideas of bifurcation theory. First of all, we shall assume that $\mathcal{L} > 0$.

Assuming $\mathcal{L} > 0$, (6.59) is easily solved: after dividing it by A^3, it is solved for $1/A^2$, which gives

$$A(t) = \frac{A_0}{\sqrt{(1 - A_0^2 \mathcal{L}/\lambda)e^{-2\lambda t} + A_0^2 \mathcal{L}/\lambda}}$$

where A_0 is the amplitude at $t = 0$. Figure 6.7 shows a plot of this solution.

By writing Landau equation in the form

$$\frac{dA}{d\tau} = (\lambda - \mathcal{L}A^2)A \,,$$

we observe that the solution saturates thanks to the term in A^3: the increasing amplitude causes a reduction of the effective growth rate $(\lambda - \mathcal{L}A^2)$. The final amplitude is such that $\lambda - \mathcal{L}A^2 = 0$, or

$$A = A_{eq} = \sqrt{\frac{\lambda}{\mathcal{L}}} \tag{6.60}$$

We see that this solution exists only if $\lambda > 0$. In the opposite case, $A \rightarrow 0$. This situation can be summed up by a *bifurcation diagram* (Fig. 6.8) that outlines the

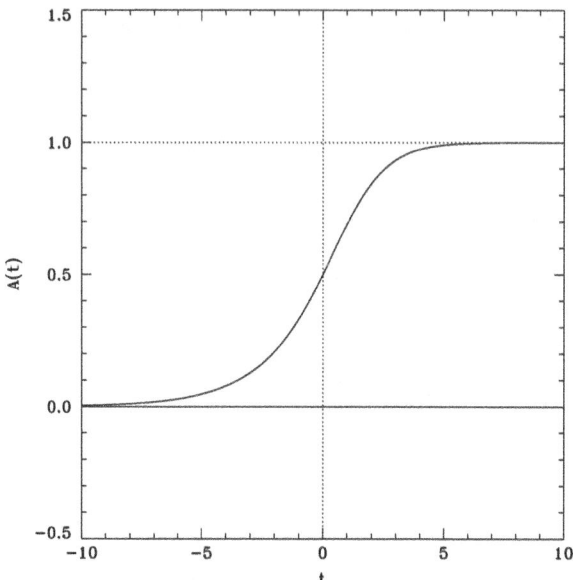

Fig. 6.7 The solution of Landau equation when $\lambda = 0.5$ and $\mathcal{L}/\lambda = 1$

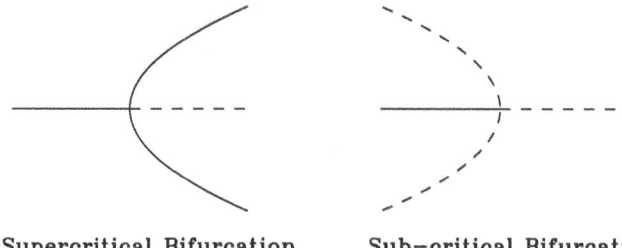

Supercritical Bifurcation   **Sub−critical Bifurcation**

Fig. 6.8 Bifurcation diagrams for Landau equation in the supercritical and sub-critical cases. *Solid lines* indicate stable branches while *dashed lines* are for unstable ones

equilibrium solutions,[8] namely the values of A such that $f(A) = 0$ when the control parameter λ is varied. The control parameter is also known as the *order parameter* in reference to phase transitions, where bifurcations are also playing an important role. In fluid flows, this parameter is usually a number like the Reynolds one.

We now return to Landau equation and its possible equilibrium solutions. $f(A) = 0$ leads to

$$\lambda A - \mathcal{L}A^3 = 0 \quad \Longrightarrow \quad A = 0 \quad \text{or} \quad A = \pm\sqrt{\frac{\lambda}{\mathcal{L}}}$$

---

[8]The equilibrium solutions are also called fixed points in the language of dynamic systems.

These three solutions constitute the different *branches* of the diagram. It is then necessary to examine their stability. For this, we perturb Landau equation by writing $A = A_{eq} + \delta A$; thus,

$$\frac{d\delta A}{dt} = (\lambda - 3\mathcal{L}A_{eq}^2)\delta A$$

The stability of each branch is given by the sign of $\tau = \lambda - 3\mathcal{L}A_{eq}^2$. If $A_{eq} = 0$, $\tau = \lambda$: the branch is stable if $\lambda < 0$ and unstable if $\lambda > 0$. If $A_{eq} = \sqrt{\lambda/\mathcal{L}}$, the system is in the bifurcated state and its perturbations evolve according to

$$\frac{d\delta A}{dt} = -2\lambda\delta A \tag{6.61}$$

Therefore, when this branch exists (if $\lambda > 0$), it is stable ($-2\lambda < 0$).

The bifurcation controlled by Landau equation is called a *pitchfork bifurcation*. When $\mathcal{L} > 0$, it is *supercritical*. If λ passes from negative values to positive ones, the system bifurcates from a solution that has become unstable ($A_{eq} = 0$) towards a new stable solution ($A_{eq} = \sqrt{\lambda/\mathcal{L}}$). The bifurcation takes place at the critical value $\lambda = 0$.

In some systems, the critical value of λ is not zero but purely imaginary $\lambda = i\omega$: at the bifurcation point the system oscillates with a frequency ω. This kind of bifurcation is called a *Hopf bifurcation*. The behaviour of the system is very similar to the Landau one and we propose its study as an exercise.

Let us now return to Landau equation and consider the case where Landau constant is negative. In this case, non-zero equilibrium solutions exist only if $\lambda < 0$. The bifurcation is called sub-critical and we note from (6.61) that the bifurcated state is always unstable (Fig. 6.8). The evolution from these branches cannot be described by Landau equation (except in an initial phase where the amplitudes are small), because the nonlinear cubic term strengthens the instability rather than reducing it. We should then extend the development of f to the next order in amplitude, namely the one in A^5. This brings us to the consideration of a somewhat more complex system, where we can find a finite amplitude instability.

6.5.3 Finite Amplitudes Instabilities ♣

We shall now analyse a system having a sub-critical bifurcation at $\lambda = 0$ taking into account the A^5-term. The dynamics of the system is assumed to be controlled by the following equation:

$$\frac{dA}{dt} = \lambda A + 2\mathcal{L}A^3 - A^5 \tag{6.62}$$

where the coefficient of A^5 has been set to -1 for simplicity (its negative value is necessary for the instability to saturate). We could also, as we did for Landau equation, explicitly solve this equation but this is not really necessary because the drawing of the bifurcation diagram as well as the analysis of the stability of the different branches allows a good understanding of the dynamics of such a system.

The points of equilibrium are the five solutions that cancel out the right-hand side of (6.62), namely

$$A = 0 \quad \text{and} \quad A = \pm\sqrt{\mathcal{L} \pm \sqrt{\mathcal{L}^2 + \lambda}} = \pm A_\pm \qquad (6.63)$$

We thus find the axis $A = 0$ plus a fourth-degree curve (see Fig. 6.9). In order to find the stability of the different branches, we must determine the sign of the rate of growth

$$\tau = \lambda + 6\mathcal{L}A^2 - 5A^4$$

for each equilibrium solution. The case of $A = 0$ is immediate. If $A \neq 0$, we can use the equilibrium equation $\lambda + 2\mathcal{L}A^2 - A^4 = 0$ to eliminate λ; recalling the expressions of $A_\pm$ given by (6.63), it turns out that

$$\tau(A_\pm) = \mp 4A^2\sqrt{\mathcal{L}^2 + \lambda}$$

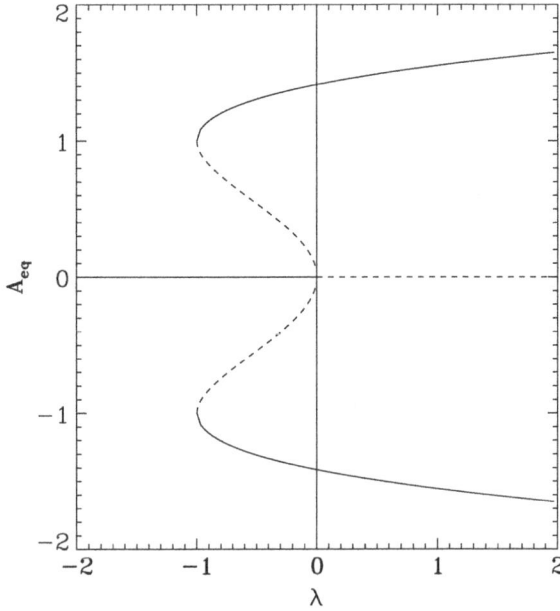

Fig. 6.9 Bifurcation diagram for a system endowed with a subcritical bifurcation obeying (6.62). *Dashed lines* indicate unstable branches, *solid* ones show stables branches (we set $\mathcal{L} = 1$)

then $\tau(A_+) < 0$ and A_+ is stable whereas A_- is unstable since $\tau(A_-) > 0$. The existence of the solutions for $A_\pm$ obviously depends on λ and we may verify that

- A_+ exists if $\lambda \geq -\mathcal{L}^2$,
- A_- exists if $-\mathcal{L}^2 \leq \lambda \leq 0$.

Several conclusions about the dynamics of the system can now be drawn. If $\lambda < -\mathcal{L}^2$, branch $A = 0$ is absolutely stable: whatever the disturbance might be, the system will return to this equilibrium. If $0 > \lambda > -\mathcal{L}^2$ three stable solutions are possible: 0 and $\pm A_+$. The system "will choose" according to initial conditions but henceforth we can note that if the solution $A = 0$ is disturbed strongly enough, we can make it bifurcate towards the other stable branches $\pm A_+$. Although stable with respect to infinitesimal perturbations the solution $A = 0$ is unstable with respect to disturbances of finite amplitude, provided that this amplitude is large enough (the same applies to the branch A_+). We can illustrate this property by noting that the equation of the dynamical system (6.62) can be written using a potential $V_\lambda(A)$ such that

$$\frac{dA}{dt} = -\frac{\partial V_\lambda(A)}{\partial A}$$

The diagram of $V_\lambda(A)$ for different values of λ shows the "valleys" of stabilities and the "peaks" of instability (see Fig. 6.10).

Finally, if $\lambda > 0$, $A = \pm A_+$ are the only stable solutions. Figures 6.9 and 6.10 summarize the properties of this system.

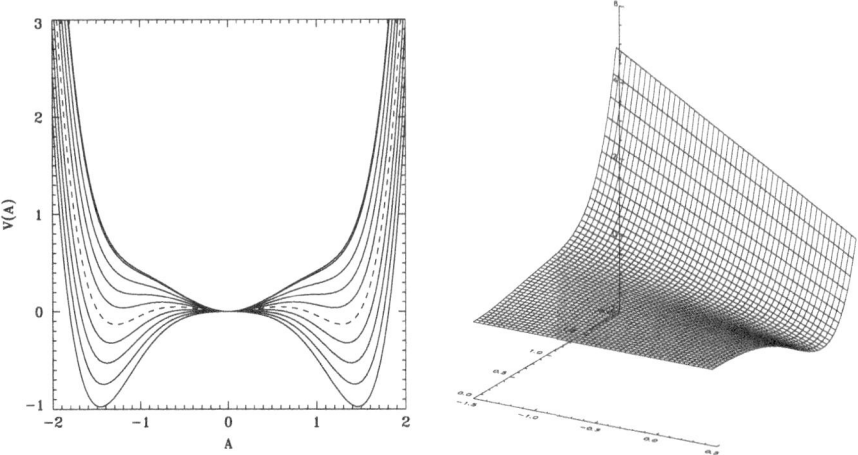

Fig. 6.10 *Left* The potential $V_\lambda(A) = \lambda A^2/2 - \mathcal{L}A^4/2 - A^6/6$ for various values of λ. The *dashed line* shows a value of λ such that $0 > \lambda > -\mathcal{L}^2$ where the potential has three local minima. *Right* A view of the surface $V(A, \lambda)$

6.6 Optimal Perturbations ♣

6.6.1 Introduction

In the foregoing sections we categorized flow disturbances into two families: those of infinitesimal amplitude which are controlled by a linear operator and those of finite amplitude which require the solving of a nonlinear problem. However, on the latter perturbation we discussed only the possible role of a finite amplitude in the context of amplitude equations. We may thus wonder if, like in dynamical systems of Sect. 6.5.3, there exist flows that are stable with respects to infinitesimal perturbations but unstable with respect to some finite amplitudes ones and also wonder how these finite amplitude perturbations are generated. These questions lead us to present a recent progress of Fluid Mechanics that bridges the gap between the two aforementioned categories of disturbances.

Let us first come back to small amplitude perturbations. We decided of the stability of the flow when these perturbations were exponentially damped, that is to say when their temporal evolution was controlled by the $e^{\lambda t}$-factor with $\mathrm{Re}(\lambda) < 0$, the instability criterion being the existence of a perturbation with a positive growth rate. This condition for instability is sufficient of course but not necessary as we shall see. We can indeed imagine the existence of other perturbations that are not described by the eigenvalues of the disturbances operator, like algebraically growing disturbances. One may even imagine situations where a flow is stable as far as perturbations like $f(\mathbf{r})e^{\lambda t}$ are concerned, but that would be transformed into an unstable one by some slowly growing disturbances. This is precisely what has been uncovered in the years 1980: some flows well known to develop turbulence but otherwise known to be stable with respect to small amplitude perturbations have been revealed as the seat of slowly growing perturbations that in the end completely destabilize them. These perturbations are now known as *optimal perturbations*: the linear analysis shows that they can be strongly amplified before disappearing, but during the course of their growth they might transform the original flow into another one that is exponentially unstable.

This scenario shows that finite amplitude disturbances may be spontaneously generated by some small amplitude noise. The existence of optimal perturbations explains why a flow like the cylindrical Poiseuille flow, which is stable linearly for any Reynolds number, shows turbulence bursts when this number is over $\sim 10^3$. In this section we shall introduce the reader to this new page of Fluid Dynamics, a page that has started being written 25 years ago.

6.6.2 Plane-Parallel Flows

Squire theorem told us that two-dimensional perturbations of plane-parallel flows were the most unstable. But three-dimensional ones have other properties, unnoticed for a long time, that might also efficiently control the stability of flows as we shall see.

Let us reconsider the disturbances that might affect a plane-parallel flow $\mathbf{v} = V(z)\mathbf{e}_x$ and let us assume that the fluid is bounded only in the z-direction. We are now considering perturbations of the same shape as (6.13) but we do not impose an exponential time dependance. Thus we write:

$$f(\mathbf{r}, t) = f(z, t)e^{ik_x x + ik_y y} \tag{6.64}$$

for the general form of the perturbations. System (6.18) has now the more general shape:

$$
\begin{cases}
[\partial_t + ik_x U - \nu(D^2 - k^2)]v_x + v_z U'(z) = -ik_x P \\
[\partial_t + ik_x U - \nu(D^2 - k^2)]v_y = -ik_y P \\
[\partial_t + ik_x U - \nu(D^2 - k^2)]v_z = -DP \\
Dv_z + ik_x v_x + ik_y v_y = 0
\end{cases}
\tag{6.65}
$$

where $k^2 = k_x^2 + k_y^2$ and $D = \partial_z$. Eliminating pressure and the components v_x and v_y we re-derive Orr–Sommerfeld equation for the vertical velocity v_z:

$$[\partial_t + ik_x U(z) - \nu(D^2 - k^2)](D^2 - k^2)v_z - ik_x U'' v_z = 0 \tag{6.66}$$

We may check that this new form of Orr–Sommerfeld equation gives back (6.23) which we derived previously. Since we now consider three-dimensional perturbations, it is necessary to complete it with an equation for the spanwise v_y component of the velocity. Following tradition, we write the equation verified by the vertical component of the vorticity $\omega_z = \partial_x v_y - \partial_y v_x$. Using the first two equations of (6.65), we easily find that

$$[\partial_t + ik_x U(z) - \nu(D^2 - k^2)]\omega_z + U'(z)ik_y v_z = 0 \tag{6.67}$$

also called *Squire equation*. These two equations form a coupled system whose coupling coefficient is proportional to k_y which represents the variation of perturbations in the third spanwise dimension.

Let us now write Squire and Orr–Sommerfeld equations in the following symbolic form:

$$\frac{\partial}{\partial t}\begin{pmatrix} \Delta v_z \\ \omega_z \end{pmatrix} + \begin{pmatrix} \mathcal{D}_4 & 0 \\ ik_y U' & \mathcal{D}_2 \end{pmatrix}\begin{pmatrix} v_z \\ \omega_z \end{pmatrix} = 0 \tag{6.68}$$

where we introduced the differential operators

$$\Delta = D^2 - k^2, \quad \mathcal{D}_2 = ik_x U(z) - \nu(D^2 - k^2), \quad \mathcal{D}_4 = \mathcal{D}_2(D^2 - k^2)$$

In order to understand the properties of the solutions of this system, it is useful to study a much simpler problem but which shares many of the properties of (6.68).

6.6.3 A Simplified Model

System (6.68) is a differential system where space and time coordinates are coupled. We shall uncouple these variables by forgetting space variations and focusing on time evolution. For that, we consider the following simple system:

$$\frac{d}{dt}\begin{pmatrix} x \\ y \end{pmatrix} = \begin{pmatrix} -\varepsilon & 0 \\ 1 & -2\varepsilon \end{pmatrix}\begin{pmatrix} x \\ y \end{pmatrix} \tag{6.69}$$

where ε is the model parameter. We look for the temporal evolution of $x(t)$ and $y(t)$ whose initial values are (x_0, y_0). The resolution of these two differential equations is straightforward and we find the general solution:

$$\begin{cases} x(t) = x_0 e^{-\varepsilon t} \\ y(t) = (y_0 - x_0/\varepsilon)e^{-2\varepsilon t} + \frac{x_0}{\varepsilon}e^{-\varepsilon t} \end{cases} \tag{6.70}$$

We might observe that at long times ($t \to +\infty$), these solutions vanish for any initial conditions. The short time evolution, that is when $\varepsilon t \ll 1$ is on the contrary sensitive to initial conditions. An expansion of the solution to first order in ε gives

$$x(t) = x_0(1 - \varepsilon t + \mathcal{O}(\varepsilon^2))$$

$$y(t) = x_0(t - \frac{3}{2}\varepsilon t^2 + \mathcal{O}(\varepsilon^2)) + y_0(1 - 2\varepsilon t + \mathcal{O}(\varepsilon^2))$$

These expressions show that $x(t)$ starts decreasing and this is indeed what says the general solution. However, this is not the case for $y(t)$. The first order expansion shows that if $x_0 \gg y_0$, the solution y first increases at a rate controlled by x_0. Obviously, if initial conditions are such that $x_0 \ll y_0$ then $y(t)$ also decreases.

Let us now consider initial conditions where $y(t)$ is increasing with time and search the time t_m where y is maximum. Using the general solution, we easily find that the maximum of y is reached at time

$$t_m = -\frac{1}{\varepsilon}\ln\left(\frac{1 + \varepsilon y_0/x_0}{2}\right)$$

as long as $\varepsilon y_0/x_0 > -1$. If we assume that $x_0 \sim y_0$ and $\varepsilon \ll 1$, we see that

$$t_m \simeq \frac{\ln 2}{\varepsilon}$$

so that the growing of y lasts longer when ε is smaller. y then reaches the amplitude

$$y_m = \frac{x_0}{4\varepsilon}$$

This amplitude is therefore the larger, the longer is the growth. This expression also shows that even if initial perturbations are small, they can be strongly amplified if they are optimally chosen. We shall note however that this condition is not very severe: in the simple case that we are studying we just need to avoid the case $x_0 \ll y_0$. Figures 6.11 and 6.12 illustrate the growth of the y component in the optimal case for various values of ε.

In general the amplification is measured by the energy gain, that is to say by a quadratic function of the amplitude. In our case this gain is simply

$$G(t) = \frac{x(t)^2 + y(t)^2}{x_0^2 + y_0^2} \tag{6.71}$$

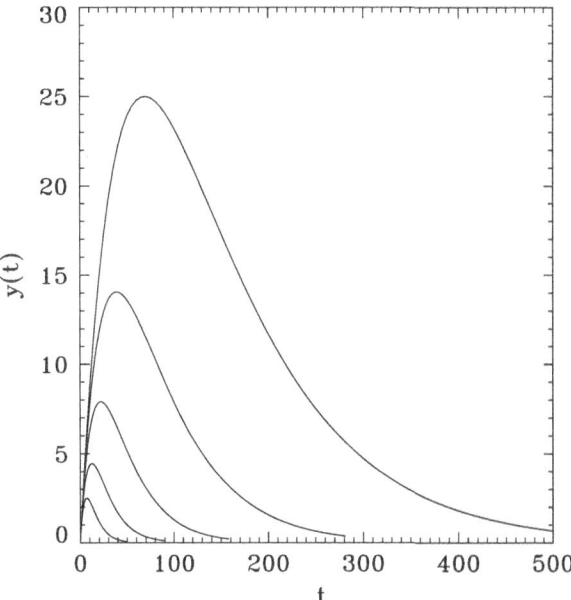

Fig. 6.11 Growth of $y(t)$ for the solution (6.70) for various values of ε with $0.1 \le \varepsilon \le 0.01$ when $x_0 = 1$ and $y_0 = 0$. In this case if $t < \ln 2/\varepsilon$ the function is strictly increasing with time. It reaches its maximum $1/4\varepsilon$ at $t = \ln 2/\varepsilon$

For our optimal perturbations verifying $x_0 \gg y_0$, the gain reaches a maximum close to

$$G_{\max} = \frac{1}{16\varepsilon^2}$$

6.6.4 Back to Fluids: Algebraic Instabilities

In view of the foregoing example it is interesting to reconsider system (6.68). We note that if perturbations are such that $k_x \ll 1$, $k_y = \mathcal{O}(1)$ and that ν is small, we qualitatively retrieve the foregoing system. The two diagonal operators are 'small' while the coupling term is of order unity. More rigorously, if $k_x/k_y \sim 1/\mathrm{Re} \ll 1$ we should expect that perturbations are amplified with a gain $\mathcal{O}(\mathrm{Re}^2)$. This is precisely what is found when one solves the full problem of disturbances verifying Orr–Sommerfeld and Squire equations. Table 6.1 illustrates the characteristics of optimal perturbations for a few classical plane-parallel flows. The analogy with the simple system is clear if we set $\varepsilon = 1/\mathrm{Re}$.

The foregoing example shows us the possible existence of perturbations with shear flows whose growth is algebraic. If $\mathrm{Re} \gg 1$, the growth is not limited neither in time neither in amplitude, but like exponentially growing disturbances the nonlinear terms will stop (or modify) this growth. However, algebraic growth is slow compared to an exponential growth. Therefore these perturbations are important when all the eigenmodes are damped. This new type of perturbations redefines the concept of flow stability. Indeed, as soon as these perturbations are able to reach a nonlinear regime they modify the basic flow and represent a true instability of it.

6.6.5 Non-Normal Operators

This non-trivial property of disturbances originates from the nature of the operators that govern their time evolution. Such operators like those of Orr–Sommerfeld–

Table 6.1 The energy gain of a few classical shear flows with the characteristics of the associated optimal disturbances given by the streamwise k_x and spanwise k_y wavenumbers (data are from Schmid and Henningson 2001)

Flow	Gain (10^{-3})	$t_{\max}$	k_x	k_y
Plane Poiseuille	$0.20\,\mathrm{Re}^2$	$0.076\,\mathrm{Re}$	0	2.04
Plane Couette	$1.18\,\mathrm{Re}^2$	$0.117\,\mathrm{Re}$	$36/\mathrm{Re}$	1.6
Cylindrical Poiseuille	$0.07\,\mathrm{Re}^2$	$0.048\,\mathrm{Re}$	0	1
Blasius	$1.51\,\mathrm{Re}^2$	$0.778\,\mathrm{Re}$	0	0.65

Fig. 6.12 Time evolution of
the solution (6.70) when
$\varepsilon = 0.01$, $x_0 = 1$ and
$y_0 = 0$. Note that the scale in
x is strongly dilated
compared to that of y

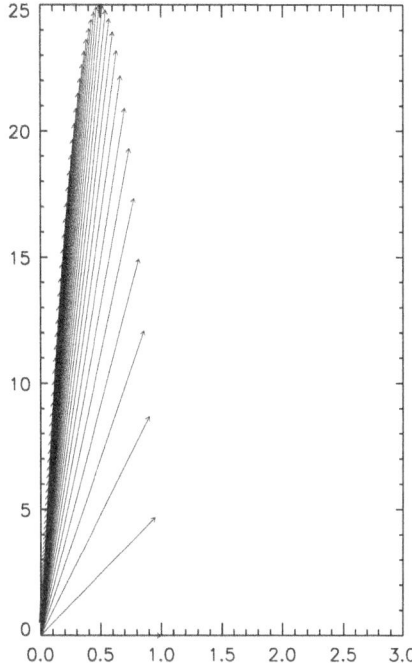

Squire (6.68) are said to be *non-normal*: their eigenfunctions do not make an
orthogonal basis (or not even a basis) of summable functions.

Let us consider again our simplified model and compute the eigenvectors
associated with the two eigenvalues $-\varepsilon$ and -2ε of the operator. We easily find
that these two vectors read:

$$\mathbf{X}_1 = \begin{pmatrix} \varepsilon \\ 1 \end{pmatrix}, \qquad \mathbf{X}_0 = \begin{pmatrix} 0 \\ 1 \end{pmatrix}$$

We note that $\mathbf{X}_1 \cdot \mathbf{X}_2 = 1$. The two eigenvectors are never orthogonal, whatever the
value of ε. In addition, when $\varepsilon \to 0$, the two vectors are no longer independent and
the matrix can no longer be diagonalized.

In fact it is precisely because the two eigenvectors are never orthogonal that short
time growth is possible.

To see this property, we briefly examine the case where the system (6.69) is
replaced by

$$\frac{d}{dt} \begin{pmatrix} x \\ y \end{pmatrix} = \begin{pmatrix} -\varepsilon & 0 \\ 0 & -2\varepsilon \end{pmatrix} \begin{pmatrix} x \\ y \end{pmatrix}$$

that is to say when the coupling between components is suppressed. Solutions are
then

$$\begin{cases} x(t) = x_0 e^{-\varepsilon t} \\ y(t) = y_0 e^{-2\varepsilon t} \end{cases} \tag{6.72}$$

The energy of these solutions is $E(t) = x_0^2 e^{-2\varepsilon t} + y_0^2 e^{-4\varepsilon t}$ and its temporal derivative is

$$\dot{E}(t) = -2\varepsilon x_0^2 e^{-2\varepsilon t} - 4\varepsilon y_0^2 e^{-4\varepsilon t},$$

which is strictly negative. These solutions are therefore strictly decreasing. Associated eigenvectors are obviously orthogonal.

This remark shows that the coupling term is absolutely essential for the transient growth of $y(t)$ to exist.

6.6.6 Spectra, Pseudo-Spectra and the Resolvent of an Operator

6.6.6.1 Some Definitions

In order to better understand the nature of non-normal operators, it is necessary to get acquainted with some properties of differential operators.

A first important characteristic of a differential operator is its *spectrum*. The spectrum $\sigma(\mathcal{L})$ of the linear operator $\mathcal{L}$ is the set of complex numbers λ such that

$$\lambda\,\mathrm{Id} - \mathcal{L}$$

is not invertible (Id is the identity operator). Its complementary set in $\mathbb{C}$ is called the *resolvent set* $\rho(\mathcal{L})$. It is the set of numbers where the operator

$$R_\lambda = (\lambda\,\mathrm{Id} - \mathcal{L})^{-1},$$

called the *resolvent* of $\mathcal{L}$ is defined.

The spectrum is divided in three parts: the *point spectrum* $\sigma_p(\mathcal{L})$ or *the eigenvalue spectrum*, the *continuous spectrum* $\sigma_c(\mathcal{L})$ and the *residual spectrum*. The residual spectrum $\sigma_r(\mathcal{L})$ is what remains of the spectrum when the point spectrum and the continuous spectrum have been removed.

The point spectrum is the usual set of eigenvalues. It is defined as the set of complex numbers such that

$$\lambda\,\mathrm{Id} - \mathcal{L}$$

is not an injection, namely a function in the image set of the operator may have more than one antecedent by this operator. If we consider the null function, we retrieve the usual property

$$\mathcal{L}(f) = \lambda f$$

of an eigenfunction associated with an eigenvalue λ. The continuous spectrum is the set of complex numbers where $\lambda Id - \mathcal{L}$ is injective but not surjective (the operator is not invertible when λ belongs to the spectrum). The continuous spectrum is not an eigenvalue spectrum and should not be confused with a continuous spectrum of eigenvalues like that of the Rayleigh operator (6.20), which is a set of continuous eigenvalues (i.e. belonging to the point spectrum).

Besides the spectrum, another useful concept is that of the norm of an operator. It is based on the norm of the functions at hands. In Fluid Mechanics, interesting functions are square-integrable functions, namely such that

$$\int_a^b f(x)^2 dx$$

exists. $[a, b]$ is the interval of definition of the function. Such an integral is usually related to the kinetic energy of the system. We thus introduce the norm

$$\|f\| = \sqrt{\int_a^b |f(x)|^2 dx}$$

of a function f. The norm of an operator is defined as

$$\|\mathcal{L}\| = \max_f \left(\frac{\|\mathcal{L}(f)\|}{\|f\|} \right)$$

Mathematics show the following property: for complex numbers z not belonging to the spectrum of $\mathcal{L}$

$$\|(z - \mathcal{L})^{-1}\| \geq \frac{1}{\text{dist}(z, \sigma(\mathcal{L}))} \tag{6.73}$$

Namely, the norm of the resolvent is larger than the inverse of the distance to the spectrum.

We can now introduce the *pseudo-spectrum* $\sigma_\varepsilon(\mathcal{L})$ of the operator $\mathcal{L}$, or rather the ε-pseudospectrum, which is the set of complex numbers z such that

$$\|(z - \mathcal{L})^{-1}\| \geq \varepsilon^{-1} \tag{6.74}$$

6.6.6.2 Physical Interpretation of the Pseudospectrum

Let us consider an operator whose spectrum is only composed of eigenvalues. The norm of its resolvent R_z goes to infinity when z approaches an eigenvalue. z enters the ε-pseudospectrum when the resolvent norm gets over ε^{-1}. The ε-pseudospectrum of an operator $\mathcal{L}$ is therefore a part of the complex plane limited by a contour defined by ε and which surrounds the eigenvalues (see Fig. 6.13).

Normal operators have a pseudospectrum that is in a neighbourhood of the eigenvalues while non-normal operators have a pseudospectrum that extends far away from the eigenvalues.

To give a picture, we may say that non-normal operators have an ill-defined spectrum in the sense that high values of the resolvent occupy large parts of the complex plane. On the contrary a normal operator has a pseudo-spectrum that remains in the neighbourhood of the eigenvalues.

Let us now examine the relation between the non-normality of an operator and the amplification of some disturbances. In order to do so, we consider the following problem:

$$\frac{\partial f(x,t)}{\partial t} = \mathcal{L}(f) \qquad \text{and} \qquad f = f_0(x) \quad \text{at} \quad t = 0$$

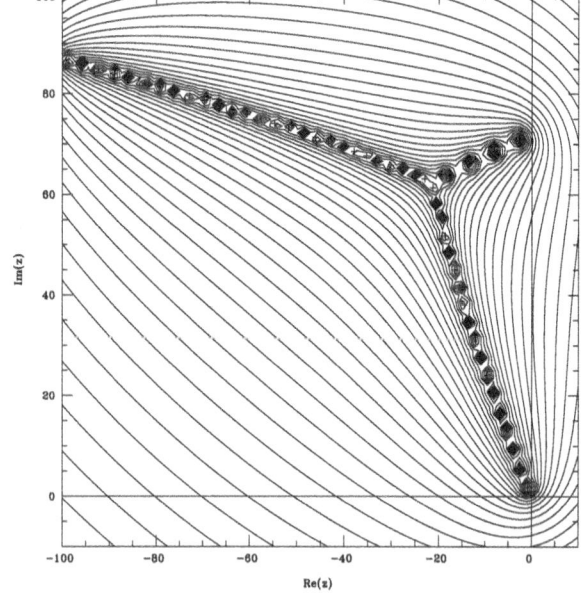

Fig. 6.13 Isocontours of the distance to the eigenvalue spectrum of the Davies operator $\frac{d^2}{dx^2} + (ax^2 - bx^4)$, with $a = 3 + 3i$ and $b = 1/16$. This is a Schrödinger equation with a complex potential. As indicated by (6.73), the ε-pseudospectrum is inside a contour associated with the value $1/\varepsilon$

We shall use the Laplace transform on time, $\tilde{f}$, of function f:

$$\tilde{f}(x, p) = \int_0^\infty f(x, t)e^{-pt}dt$$

Applying Laplace transform to the equation determining f, we get

$$\int_0^\infty \frac{\partial f(x, t)}{\partial t}e^{-pt}dt = \int_0^\infty \mathcal{L}(f)e^{-pt}dt$$

so that

$$\tilde{f} = (p - \mathcal{L})^{-1}f_0 \tag{6.75}$$

The solution for f is then derived from the inverse Mellin–Fourier transform, namely

$$f(x, t) = \frac{1}{2i\pi}\int_{c-i\infty}^{c+i\infty} \tilde{f}(x, p)e^{pt}dp$$

where c is real and larger than the largest real part of the eigenvalues of $\mathcal{L}$.

The foregoing formula shows that the transient response is controlled by the resolvent of $\mathcal{L}$ applied to the initial conditions f_0. If the long time evolution of f is a damping, that is if the eigenvalues of $\mathcal{L}$ are all in the half-plane $Re(z) < 0$, the transient response can nevertheless be large if the operator is non-normal and the initial conditions chosen properly. Indeed, (6.75) shows that the non-normality of the operator is not sufficient. Adapted f_0 are also needed, meaning an optimal choice.

Before ending this section it is interesting to consider another property of the pseudo-spectrum in relation with the stability of flows. Indeed, the pseudospectrum might also be viewed as the union of the spectra of all the operators $\mathcal{L} + \mathcal{E}$ where $\|\mathcal{E}\| \leq \varepsilon$. In other words, if we consider all the possible perturbations of the operator $\mathcal{L}$ by any operator of norm less than ε, the union of all the spectra of these operators defines a part of the complex plane that is identical to the ε-pseudospectrum of $\mathcal{L}$ (Trefethen and Embree 2005). It may well be that $\mathcal{L} + \mathcal{E}$ has unstable modes, namely that perturbing the operator generates exponentially growing modes. This is to say that non-normal operators are sensitive operators: a small change may strongly modify their spectrum.

We here touch finite amplitude perturbations: the small change of the operator (of order ε for its norm) may be viewed as a finite-amplitude disturbance that slightly modifies the background flow. If the operator is normal, nothing happens, but if it is non-normal the new flow may be prone to some exponentially growing modes.

The concept of pseudo-spectrum has many other implications, especially in the numerical calculation of the eigenvalues of matrices where it is associated with the influence of round-off errors (e.g. Valdettaro et al. 2007). We shall stop here the discussion of this subject which would turn into pure mathematics and refer the reader to specialized literature (e.g. Trefethen and Embree 2005).

6.6.7 Examples of Optimal Perturbations in Flows

After this mathematical digression, it is time to reconsider fluid flows. One may wonder if these perturbations actually exist and if they have been observed. As may be guessed from Table 6.1, plane-parallel shear flows are the best candidates for showing such perturbations. The most remarkable example is certainly that of streaks that appear in a boundary layer flow of Blasius type. Figure 6.14 shows the formation of these structures. We note that the flow varies rapidly in the spanwise direction y and slowly in the streamwise direction x. This is just the condition $k_x/k_y \ll 1$.

We may understand the appearance of streaks if we go back to Orr–Sommerfeld and Squire (6.66 and 6.67). If we set $\nu = 0$ and $k_x = 0$ then we get

$$\partial_t u_x = U'(z) u_z \tag{6.76}$$

$$(D^2 - k_y^2) \partial_t u_z = 0 \tag{6.77}$$

the solution of which are of the form

$$\partial_t u_z = A e^{-k_y z} + B e^{k_y z}$$

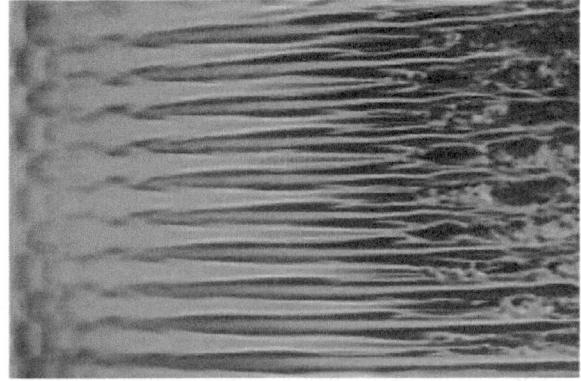

Fig. 6.14 Streaks as a consequence of the lift-up effect in a Blasius boundary layer. The flow is from *left to right*. Note that the disturbance generating the streaks is characterized by a spatial periodicity in the spanwise direction and that the streaks are themselves unstable at some downstream position (Photo by Elofsson and Matsubara, in Elofsson 1998)

but boundary conditions at infinity and at $z = 0$ (namely on the bounding plane) imply that $u_z = 0$ at these two places so that $\partial_t u_z = 0$, or that $u_z = \text{Cst} = u_z(t = 0) = u_z^0$. The first equation (6.76) gives us the time evolution of u_x:

$$u_x = U'(z)u_z^0 e^{ik_y y} \, t \tag{6.78}$$

Hence a disturbance of the vertical velocity, characterized by the wavenumber k_y, generates a local increase of the velocity of the flow in the streamwise direction. This effect is now known as the *lift-up effect*. We indeed observe that u_x may be written

$$u_x = U'(z) \, \Delta z \, e^{ik_y y}$$

where $\Delta z = u_z^0 t$ is the displacement of matter in the z direction induced by the initial perturbation $u_z^0 e^{ik_y y}$. $U'(z)\Delta z$ is just the first variation in z of the background flow:

$$U(z + \Delta z) = U(z) + U'(z)\Delta z$$

The initial perturbation has therefore lifted by Δz the background flow and yielded in $z + \Delta z$ the flow field $U(z) + u_x$ with $u_x = U'(z)u_z^0 \, t \, e^{ik_y y}$. This disturbance thus generates streaks of high and small speed whose wavelength is determined by the condition of optimal growth. If the initial conditions are that of a flow disturbed by some white noise, these perturbations emerge in the end.

The lift-up effect has been first described by the work of Ellingsen and Palm (1975). This is the first mechanism that has been recognized as being associated with optimal perturbations. However, there exist other mechanisms like Orr mechanism where a vorticity disturbance controls the dynamics (see Farrell and Ioannou 1993).

6.7 Exercises

1. *The interstellar cloud:* We consider a sphere of radius R filled with an ideal gas of constant density and constant temperature. Establish the condition on the radius which governs the stability of the sphere according to Jeans criterion. Propose a physical interpretation of this criterion. Make a numerical application for an interstellar cloud composed of molecular hydrogen, with a mass of 100 $M_\odot$ and a temperature of 50 K. What is the stability of this cloud if its diameter is 1 or 10 light-years?

2. Let us consider the flow of an inviscid and incompressible fluid such that

$$\mathbf{v} = s\Omega(s)\mathbf{e}_\varphi$$

a) Recall the condition, on $\Omega(s)$, of the stability of this flow with respect to axisymmetric disturbances.

b) We now study the stability of the following flow:

$$
\begin{cases}
\Omega = 0 & s \leq \eta \\
\Omega = A + B/s^2 & \eta \leq s \leq 1 \\
\Omega = \Omega_0/s^2 & s \geq 1
\end{cases}
\tag{6.79}
$$

where the constants A and B are such that $\Omega(s)$ is continuous in the whole domain occupied by the fluid.

We are interested in the non-axisymmetric two-dimensional disturbances. The pressure and the velocity perturbations are of the form

$$
f(s)e^{im\varphi+\lambda t}
$$

while $v_z = 0$.

Give the linearized equations controlling the evolution of disturbances.

c) What boundary conditions are met by the disturbances at the interfaces at $s = \eta$ and $s = 1$?

d) Show that the radial velocity u of the perturbations verify the same differential equation in the three regions and that it can be written

$$
\frac{d}{ds}\left(s\frac{d(su)}{ds}\right) = m^2 u
\tag{6.80}
$$

Note that in each domain, $\frac{\partial(s^2\Omega)}{\partial s} = 2as$ where a is either zero or equal to A.

e) Give the expression of $u(s)$ in each subdomain (one should look for solutions of the type s^α).

f) Determine the form of the pressure perturbations in each domain.

g) Show that the eigenmodes verify the following dispersion relation

$$
\left(\lambda + \frac{im\Omega_0}{2}\right)^2 = \Omega_0^2\left[\frac{\eta^{2m}-1}{(1-\eta^2)^2} + \frac{m}{1-\eta^2} - \frac{m^2}{4}\right]
\tag{6.81}
$$

h) Show that the modes $m = 1$ and $m = 2$ are always stable.

3. *Fjørtoft Theorem.* Extract the real part of (6.21) and show, using (6.22), that equation

$$
\int_a^b \left[(|D\psi|^2 + k^2|\psi|^2) + \frac{k^2|\psi|^2 U''(U-A)}{|\lambda + ikU|^2}\right]dz = 0
\tag{6.82}
$$

must be verified for any A. Deduce Fjørtoft theorem.

Further Reading

There are two well-known monographs on flow stability. The one of Drazin and Reid (1981), *Hydrodynamic stability* and the one of Chandrasekhar (1961), *Hydrodynamic and hydromagnetic stability*. Drazin and Reid's one is more modern and pedagogical in its presentation. It also discusses the question of nonlinear stability. However, the one of Chandrasekhar is very complete, especially detailed in the derivation and makes a large use of variational principles. For a very recent introduction to instabilities, the reader may also consult *Hydrodynamic Instabilities* by Charru (2011).

On the applications of dynamical systems to Fluid Mechanics, we suggest *Order within chaos* by Bergé et al. (1984), and also *Instabilities, Chaos And Turbulence: An Introduction To Nonlinear Dynamics And Complex Systems*, by Manneville (2004). As far as optimal perturbations are concerned, the reader may deepen the subject with the monograph of Schmid and Henningson (2001) and the recent review of Schmid (2007).

References

Bergé, P., Pomeau, Y. & Vidal, C. (1984). *Order within chaos*. New York: Wiley.

Cairns, R. (1979). The role of negative energy waves in some instabilities of parallel flows. *The Journal of Fluid Mechanics, 92*, 1–14.

Chandrasekhar, S. (1961). *Hydrodynamic and hydromagnetic stability*. Oxford: Clarendon Press.

Charru, F. (2011) *Hydrodynamic Instabilities*. Cambridge: Cambridge University Press.

Drazin, P. & Reid, W. (1981). *Hydrodynamic stability*. Cambridge: Cambridge University Press.

Ellingsen, T. & Palm, E. (1975). Stability of linear flow. *Physics of Fluids, 18*, 487–488.

Elofsson, P. (1998). Experiments on oblique transition in wall-bounded shear flows. Ph.D. thesis, Royal institute of Technology, Stockholm.

Farrell, B. F. & Ioannou, P. J. (1993). Optimal excitation of three-dimensional perturbations in viscous constant shear flow. *Physics of Fluids, 5*, 1390–1400.

Joggerst, C. C., Almgren, A. & Woosley, S. E. (2010). Three-dimensional Simulations of Rayleigh–Taylor Mixing in Core-collapse Supernovae with Castro. *The Astrophysical Journal, 723*, 353–363.

Manneville, P. (2004). *Instabilities, chaos and turbulence: An introduction to nonlinear dynamics and complex systems*. London: Imperial College Press.

Landau, L., Lifchitz, E. (1971–1989). Mécanique des fluides. Mir.

Pearson, J. (1958). On convection cells induced by surface tension. *The Journal of Fluid Mechanics, 4*, 489–500.

Richardson, L. F. (1920). The supply of energy from and to atmospheric eddies. *Royal Society of London Proceedings Series A, 97*, 354–373.

Schmid, P. & Henningson, D. (2001). *Stability and transition in shear flows*. New York: Springer.

Schmid, P. J. (2007). Nonmodal stability theory. *The Annual Review of Fluid Mechanics, 39*, 129–162.

Trefethen, L. & Embree, M. (2005). *Spectra and pseudospectra*. Princeton: Princeton University Press.

Valdettaro, L., Rieutord, M., Braconnier, T. & Fraysse, V. (2007). Convergence and round-off errors in a two-dimensional eigenvalue problem using spectral methods and arnoldi-chebyshev algorithm. *Journal of Applied Mathematics and Computing, 205*, 382–393.

Chapter 7
Thermal Convection

7.1 Introduction

Thermal convection is the transport of internal energy by the motion of a fluid. Two types of convection are usually distinguished: free or natural convection and forced convection. Natural convection is a fluid flow whose origin is always a thermal imbalance: it disappears when the temperature gradients vanish. In forced convection, on the other hand, the flow persists even if the temperature gradients[1] are eliminated. In this chapter we shall concentrate on natural thermal convection, which we simply call thermal convection. To begin with let us give some examples.

The most familiar example is doubtless that of the motion of water in a container heated from below. Well before boiling (i.e. before the appearance of steam bubbles), one may notice upward and downward motions in the liquid. These motions are easily interpreted in a qualitative manner. The water heated at the bottom of the container is lighter and rises to the surface, where it cools, falls down, reheats and ascends again etc. In this cycle, the water carries the heat from the bottom to the top of the layer. This is the phenomenon of natural thermal convection. With this example, we understand that convection plays an important role in heat exchanges realized or experienced by the fluids, in particular because it turns out to be much more effective than thermal conduction.

Convection occurs at various scales, but it is mostly at the largest scales that it easily arises. We shall see that, in many cases, an imposed temperature gradient triggers fluid motion if the size of the fluid domain is large enough. In other words, fluid motion is more efficient to transport heat on the large scale than on the small scales where conduction dominates. This is why insulating materials that use air as the insulating component (because of its small conductivity), are made of fibers

---

[1]Present wording tends to replace the terminology "forced convection" by *advection*. In this case, temperature is more like a passive scalar and does not, or little, influence the fluid flow.

© Springer International Publishing Switzerland 2015   241
M. Rieutord, *Fluid Dynamics*, Graduate Texts in Physics,
DOI 10.1007/978-3-319-09351-2_7

(like glass wool for instance). Fibres reduce the scale of fluid flows and hamper heat transport.

On much larger scales, like in the Earth atmosphere, thermal convection is very common: clouds (like cumulus) keep their droplets of water because of rising flows of thermal convection. Thermals, so praised by sail plane flight amateurs are also an example of fluid flows generated by temperature gradients. At still larger scales, thermal convection is the main heat carrier in the core of massive stars or in the envelope of solar type stars. In giant planets, like Jupiter, the radial heat flow is mainly insured by convectively driven fluid flows.

In the following we first retrieve basic properties of hydrostatic equilibrium of a fluid submitted to vertical temperature gradient. Then, we introduce the Boussinesq and anelastic approximations that much simplify the analysis. We complete the case of equilibria by examining the so-called baroclinic situation which renders equilibrium impossible. We proceed with the heart of thermal convection, namely the Rayleigh–Bénard instability and its nonlinear development. As an illustration of large-scale instabilities, we present the case of fixed-flux convection (a section that may be skipped at first reading). Finally, the route to turbulent convection is briefly discussed.

7.2   The Conductive Equilibrium

7.2.1   Equilibrium of an Ideal Gas Between Two Horizontal Plates

In Chap. 2 we saw that a fluid in hydrostatic and thermal equilibrium verifies:

$$\begin{cases} -\nabla P + \rho \mathbf{g} = \mathbf{0} \\ \nabla \cdot (\chi \nabla T) + \mathcal{Q} = 0 \\ P \equiv P(\rho, T) \qquad \text{and} \qquad \chi \equiv \chi(\rho, T) \end{cases} \tag{7.1}$$

In order to simplify the derivation as much as possible, we shall consider an ideal gas with no heat source ($\mathcal{Q}=0$), and for which we can neglect the variations of thermal conductivity. Furthermore, we suppose that the fluid is contained between two horizontal plates at a distance d apart. The upper plate has a temperature T_u and the lower one T_l (see Fig. 7.1).

The temperature field verifies:

$$\Delta T = 0$$

Fig. 7.1 A schematic view of the system

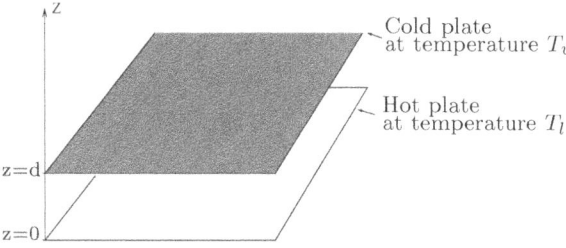

Assuming that the equilibrium configuration is independent of the horizontal coordinates x and y, we have

$$\frac{d^2T}{dz^2} = 0 \qquad \Longleftrightarrow \qquad T = az + b$$

The boundary conditions $T(0) = T_l$ and $T(d) = T_u$ determine the two constants a and b whence

$$T_{eq}(z) = T_l + (T_u - T_l)z/d \tag{7.2}$$

which can also be written:

$$T_{eq}(z) = T_l(1 - z/z_0) \qquad \text{where} \qquad z_0 = d/(1 - T_u/T_l) \tag{7.3}$$

z_0 is the temperature scale height. The hydrostatic (7.1) and the ideal gas equation of state $P = R_*\rho T$ lead to the expressions of P and ρ, namely

$$P_{eq}(z) = P_l(1 - z/z_0)^m \tag{7.4}$$

$$\rho_{eq}(z) = \rho_l(1 - z/z_0)^{m-1} \tag{7.5}$$

where the exponent m is given by:

$$m = \frac{gd}{R_*(T_l - T_u)}$$

7.2.2   The Adiabatic Gradient

The solution (7.2) shows that in adjusting the temperature of the plates, we can choose the temperature gradient in the fluid. Among all the possible gradients, there is one which plays a special role: namely that for which the fluid is isentropic. It may be derived from the thermodynamic relation:

$$dh = Tds + dP/\rho$$

where s and h are respectively the specific entropy and specific enthalpy. If the gas is isentropic, then $ds = 0$ and $dh = dP/\rho$ so that $\nabla h = \rho^{-1}\nabla P = \mathbf{g}$. For the ideal gas, $h = c_p T$ and thus

$$\nabla T = \frac{\mathbf{g}}{c_p} \equiv (\nabla T)_{ad} \tag{7.6}$$

which is called *the adiabatic gradient*. Its name is due to the fact that a fluid particle moved adiabatically of Δz experiences a temperature change of

$$\left(\frac{dT}{dz}\right)_{ad} \Delta z$$

We shall see below that the adiabatic gradient is tightly related to the threshold of the Rayleigh–Bénard instability.

7.2.3 The Potential Temperature

In atmospheric sciences the concept of potential temperature is often used. This temperature is defined as follows

$$T_{pot} = T \left(\frac{P_0}{P}\right)^{R_*/c_p} \tag{7.7}$$

We note that if the gas is isentropic, this new temperature is constant. In fact, for an ideal gas (see Sect. 1.7.1), we have

$$s = c_p \ln\left[\left(\frac{T}{T_0}\right)\left(\frac{P_0}{P}\right)^{R_*/c_p}\right] + s_0$$

whence

$$T_{pot} = T_0 e^{(s-s_0)/c_p} \tag{7.8}$$

The potential temperature thus only depends on the entropy of the fluid. This shows that for a perfect fluid

$$\frac{Ds}{Dt} = 0 \quad \Longrightarrow \quad \frac{DT_{pot}}{Dt} = 0$$

so that the potential temperature is conserved for each fluid particle when diffusion phenomena are neglected. A little manipulation of the foregoing expressions shows that

$$\nabla T_{pot} = \frac{T_{pot}}{T}(\nabla T - \nabla T_{ad}) \tag{7.9}$$

The potential temperature gradient measures the gap between the actual temperature gradient and the adiabatic gradient.

7.3 Two Approximations

The dynamics of a fluid whose motion is controlled by the temperature field is governed by three partial differential equations. Its analysis is therefore much more involved than the "simple" flow of an incompressible fluid. However, all the additional physics connected to the temperature field does not have the same importance in the dynamics of the system. Very often, notably when one deals with liquids, equations can be much simplified. These simplifications come out of approximations derived from asymptotic expansions. The most popular one is the Boussinesq approximation, which was introduced qualitatively by Joseph Boussinesq at the beginning of the twentieth century in a treatise, *Théorie analytique de la chaleur*, (1903). Another one, called the *anelastic approximation*, is slightly less restrictive and very useful in Astrophysics and Geophysics. Since we shall try to keep our analysis as simple as possible, these approximations will be extremely useful to us. We thus present them both.

7.3.1 The Boussinesq Approximation: A Qualitative Presentation

When studying thermal convection in liquids, one is tempted to neglect the variations of density. Liquids are indeed weakly compressible. However, such a simplification cannot be done blindly because density variations are important in the buoyancy force, which may drive the flow. Hence, J. Boussinesq suggested to neglect all the density variations except in the buoyancy term. Thus, according to this simplification, the perturbations of a hydrostatic and thermal equilibrium described by (7.1) should be controlled by

$$\begin{cases} \rho\dfrac{D\mathbf{v}}{Dt} = -\nabla\delta P + \delta\rho\mathbf{g} + \mu\Delta\mathbf{v} \\[2mm] \dfrac{D\delta T}{Dt} + \mathbf{v}\cdot\nabla T_{eq} = \kappa\Delta\delta T \\[2mm] \nabla\cdot\mathbf{v} = 0 \\[2mm] \dfrac{\delta\rho}{\rho} = -\alpha\delta T \end{cases} \qquad (7.10)$$

Using the thickness of the layer d as the length scale and the associated diffusion time for the time scale, i.e. d^2/κ, these equations are often written in the following dimensionless form

$$\begin{cases} \nabla\cdot\mathbf{u} = 0 \\[2mm] \dfrac{D\mathbf{u}}{D\tau} = -\nabla p + \mathcal{P}\mathrm{Ra}\theta\mathbf{e}_z + \mathcal{P}\Delta\mathbf{u} \\[2mm] \dfrac{D\theta}{D\tau} - u_z = \Delta\theta \end{cases} \qquad (7.11)$$

where θ and p are the dimensionless disturbances of temperature and pressure. $\mathcal{P}$ is the Prandtl number (see 1.36) and we introduced *the Rayleigh number*, defined by

$$\mathrm{Ra} = \frac{\alpha|T_u - T_l|gd^3}{\kappa\nu}$$

where α is the coefficient of thermal expansion. In fact, the Rayleigh number measures the temperature gradient imposed by the boundary conditions.

As proposed by Boussinesq, the density fluctuations appear only in the buoyancy term. Moreover, they only depend on the temperature fluctuations (not on the pressure ones). We also note that the diffusion coefficients are assumed to be constant.

If these simplifications seem to be reasonable for a liquid, we may wonder to what extent they can be applied to a gas. To answer this question, we need a more rigorous derivation of (7.10) starting from the complete equations.

7.3.2 The Asymptotic Expansions

Let us return to the equations of motion (1.13), (1.32) and (1.37), assuming that the transport coefficients, viscosity and conductivity are constant.[2]
We thus assume that the fluid is governed by the following equations:

$$
\begin{cases}
\frac{\partial \rho}{\partial t} + \nabla \cdot \rho \mathbf{v} = 0 \\[2mm]
\rho \frac{D\mathbf{v}}{Dt} = -\nabla P + \rho \mathbf{g} + \mu \left[\Delta \mathbf{v} + \frac{1}{3} \nabla (\nabla \cdot \mathbf{v}) \right] \\[2mm]
\rho \frac{De}{Dt} = \chi \Delta T - P \nabla \cdot \mathbf{v} + \mathcal{D} \\[2mm]
\rho T \frac{Ds}{Dt} = \rho \frac{De}{Dt} + P \nabla \cdot \mathbf{v}
\end{cases}
\tag{7.12}
$$

where $\mathcal{D}$ is the viscous dissipation. We shall also write this quantity $\frac{\mu}{2}(\nabla : \mathbf{v})^2$ to emphasize its quadratic nature and its dependence on velocity gradients. (7.12) is also completed by the equations of state of an ideal gas:

$$
\begin{cases}
P = R_* \rho T \\
e = c_v T
\end{cases}
\tag{7.13}
$$

First of all, we rewrite the equations (7.12) by subtracting the equilibrium solution (7.3), (7.4), (7.5) and by introducing the fluctuations of ρ, P, T associated with the fluid motion. The new equations are:

$$
\begin{cases}
\frac{\partial \delta \rho}{\partial t} + \nabla \cdot [(\rho_{eq} + \delta \rho)\mathbf{v}] = 0 \\[2mm]
(\rho_{eq} + \delta \rho)\frac{D\mathbf{v}}{Dt} = -\nabla \delta P + \delta \rho \mathbf{g} + \mu \left[\Delta \mathbf{v} + \frac{1}{3} \nabla (\nabla \cdot \mathbf{v}) \right] \\[2mm]
(\rho_{eq} + \delta \rho)c_v \left(\frac{D\delta T}{Dt} + \mathbf{v} \cdot \nabla T_{eq} \right) = \chi \Delta \delta T - (P_{eq} + \delta P)\nabla \cdot \mathbf{v} + \frac{\mu}{2}(\nabla : \mathbf{v})^2 \\[2mm]
\rho T \frac{Ds}{Dt} = \rho c_v \frac{DT}{Dt} + P \nabla \cdot \mathbf{v} \\[2mm]
\delta P = R_* (T_{eq}\delta \rho + \rho_{eq}\delta T + \delta \rho \delta T)
\end{cases}
\tag{7.14}
$$

As usual, we move to dimensionless variables. We thus introduce the following scales:

---

[2]Taking into account their variations with thermodynamic variables would not change the results or the method, but would make the whole derivation more obscure. For this very reason, we shall also neglect the second viscosity ζ.

Length scale . . . . d
Velocity scale V
Time scale . . . . d/V
Temperature scale . . . . T_*
Density scale . . . . ρ_*
Entropy scale . . . . c_p

We write

$$P_{eq} = R_* \rho_* T_* P_0, \quad \rho_{eq} = \rho_* \rho_0, \quad T_{eq} = T_* \theta_0$$

$$\mathbf{v} = V\mathbf{u}, \quad \delta P = \rho_* V^2 p_1, \quad \delta T = T_* \theta_1, \quad \delta\rho = \rho_* \rho_1$$

$$\mu = \rho_* \rho \nu, \quad \chi = \rho_* c_p \rho \kappa, \quad \rho = \rho_0 + \rho_1$$

and then obtain

$$
\begin{cases}
\frac{\partial \rho_1}{\partial \tau} + \nabla \cdot [(\rho_0 + \rho_1)\mathbf{u}] = 0 \\[2mm]
\frac{D\mathbf{u}}{D\tau} = -\frac{1}{\rho}\nabla p_1 - \frac{\rho_1}{\rho}\frac{gd}{V^2}\mathbf{e}_z + \frac{\nu}{Vd}(\Delta\mathbf{u} + \frac{1}{3}\nabla\nabla\cdot\mathbf{u}) \\[2mm]
\frac{D\theta_1}{D\tau} + \mathbf{u}\cdot\nabla\theta_0 = \gamma\frac{\kappa}{Vd}\Delta\theta_1 - \frac{(\gamma-1)P_0 + \gamma M'^2 p_1}{\rho}\nabla\cdot\mathbf{u} + \gamma\frac{M'^2}{2}\frac{\nu}{Vd}(\nabla:\mathbf{u})^2 \\[2mm]
\rho\theta\frac{Ds}{D\tau} = \frac{\rho}{\gamma}\frac{D\theta}{D\tau} + \frac{(\gamma-1)}{\gamma}(P_0 + \gamma M^2 p_1)\nabla\cdot\mathbf{u} \\[2mm]
\gamma M^2 p_1 = \rho_0\theta_1 + \theta_0\rho_1 + \rho_1\theta_1
\end{cases}
\qquad (7.15)
$$

where

$$M^2 = \frac{V^2}{c_*^2} = \frac{V^2}{\gamma R_* T_*} \qquad \text{and} \qquad M'^2 = \frac{V^2}{c_p T_*} = (\gamma - 1)M^2$$

In these expressions, c_* is the speed of sound at temperature T_* and M is therefore a Mach number. From these equations, we note that in order to recover (7.10), we need:

       ① $\rho_0 = \text{Cst}$,
       ② $\rho_1 \ll \rho_0$,
       ③ $M'^2 \ll 1$,
       ④ $\theta_1 = \mathcal{O}(1)$

We now deduce under which conditions the equations of fluid motion at the Boussinesq approximation can be derived.

Condition ① will be satisfied if the equilibrium configuration is such that the variations in temperature are equally small. Equation (7.2) leads to

$$\theta_0 = \theta_{00}\left(1 - \frac{|T_u - T_l|}{T_l}z\right), \qquad \text{with} \quad \theta_{00} = T_l/T_*$$

We therefore require that

$$\varepsilon = \frac{|T_u - T_l|}{T_l} \ll 1$$

Choosing $\rho_* = \rho_i$, we find from (7.5) that

$$\rho_0 = 1 + \mathcal{O}(\varepsilon)$$

Condition ④ requires the choice of the temperature scale T_* to be such that the temperature fluctuations δT are of order unity and therefore of the order of the imposed temperature difference $|T_u - T_l|$. We therefore choose

$$T_* = |T_u - T_l|$$

which immediately implies that

$$\theta_{00} = \mathcal{O}(1/\varepsilon)$$

but that

$$\nabla\theta_0 = \mathcal{O}(1)$$

In order that the fluctuations of density be controlled solely by temperature, (7.15-d) requires that

$$\rho_1\theta_0 = \mathcal{O}(\rho_0\theta_1) \qquad \Longrightarrow \qquad \rho_1 = \mathcal{O}(\varepsilon)$$

and that

$$M^2 \ll 1$$

In this way we recover conditions ② and ③ and we see that a second infinitesimal parameter, M^2, appears. In this case

$$\rho_1 = -\frac{\theta_1}{\theta_{00}} + \mathcal{O}(\varepsilon^2)$$

and the fluid velocity must be small compared to that of sound. We note that $1/\theta_{00} = \mathcal{O}(\varepsilon)$ is the coefficient of isobaric expansion of the fluid.[3]

Let us now rewrite (7.15) at leading order for each equation:

$$
\begin{cases}
\nabla \cdot \mathbf{u} = 0 \\[2ex]
\frac{D\mathbf{u}}{D\tau} = -\frac{1}{\rho}\nabla p_1 - \frac{\rho_1}{\rho}\frac{gd}{V^2}\mathbf{e}_z + \frac{\nu}{Vd}(\Delta\mathbf{u} + \frac{1}{3}\nabla\nabla\cdot\mathbf{u}) \\[2ex]
\frac{D\theta_1}{D\tau} + \mathbf{u}\cdot\nabla\theta_0 = \gamma\frac{\kappa}{Vd}\Delta\theta_1 \\[2ex]
\rho\theta\frac{Ds}{D\tau} = \frac{\rho}{\gamma}\frac{D\theta}{D\tau} \\[2ex]
0 = \rho_0\theta_1 + \theta_0\rho_1
\end{cases}
\tag{7.16}
$$

We note that the velocity scale is still arbitrary. If we choose, as in (7.11), $V = \kappa/d$, the buoyancy term is $\mathcal{O}(\varepsilon gd^3) = \mathcal{O}(\mathrm{Ra}\,\mathrm{Pr})$.

Finally, we observe that we did not pay attention to the diffusion terms $\frac{\nu}{Vd}\Delta\mathbf{u}$ and $\frac{\kappa}{Vd}\Delta\theta$. These terms are in fact retained in Boussinesq's approximation and their presence or absence can only result from new approximations. We can now rewrite (7.16) as

$$
\begin{cases}
\nabla \cdot \mathbf{u} = 0 \\[2ex]
\frac{D\mathbf{u}}{D\tau} = -\nabla p_1 + \mathcal{P}\mathrm{Ra}\theta_1\mathbf{e}_z + \mathcal{P}\Delta\mathbf{u} \\[2ex]
\frac{D\theta_1}{D\tau} - u_z = \gamma\Delta\theta_1 \\[2ex]
\theta_1 + \theta_{00}\rho_1 = 0
\end{cases}
\tag{7.17}
$$

which is identical to (7.11) if we recall that for liquids $\gamma \approx 1$.

The foregoing analysis shows that thermal convection of an ideal gas will be like that of a liquid if (*i*) the velocities stay small compared to that of sound and (*ii*) if the scale height of the equilibrium configuration is large compared to the vertical size of the volume occupied by the fluid.

Mathematically, the equations of motion at the Boussinesq approximation come from a series expansion using two small quantities: the square of the Mach number and the relative density variation across the layer, i.e. $\alpha|T_u - T_l|$.

---

[3] We can make the connection with the coefficient α introduced for the liquids: we have $\delta\rho/\rho = -\alpha\delta T$, let $\rho_1/\rho_0 = -\alpha T_*\theta_1 = -\theta_1/\theta_{00}$, therefore $\alpha = 1/\theta_{00}T_* = 1/T_l$.

7.3.3 Anelastic Approximation 🍄

When one deals with thermal convection in stars or in the atmosphere of a planet, the Boussinesq approximation is too restrictive because its second hypothesis is usually not verified. The scale height is not large compared to the vertical dimension of the system. However, the conditions of subsonic flows are still realized. The approximation, which consists in allowing only the condition of very subsonic motions, is called *the anelastic approximation*. In this approximation, as in Boussinesq's one, the sound waves are filtered out. The term anelastic means that the "elasticity" of the fluid, which allows the propagation of sound waves, has been neglected.

Working out the equations of fluid motion within this approximation is basically simpler than the foregoing one; we just need to expand the solutions into powers of the Mach number. Assuming $V = Mc_*$, we observe from (7.15) that the buoyancy term reads

$$-\frac{\rho_1}{\rho}\frac{gd}{M^2 c_*^2} = \frac{d}{z_* M^2} = \mathcal{O}(M^{-2})$$

where we noticed that the thickness and the scale height z_* of the layer are of the same order of magnitude. Since the buoyancy term should be of order unity, it turns out that

$$\frac{\rho_1}{\rho} = \mathcal{O}(M^2), \qquad p_1 = \mathcal{O}(1)$$

The equation of state results in

$$\theta_1 = \mathcal{O}(M^2)$$

At zeroth order, the heat equation leads to

$$\rho_0 \mathbf{u} \cdot \nabla \theta_0 = -(\gamma - 1) P_0 \nabla \cdot \mathbf{u} \qquad (7.18)$$

while the equations of continuity and momentum, at the same order, yield

$$\nabla \cdot [\rho_0 \mathbf{u}] = 0 \qquad (7.19)$$

$$\frac{D\mathbf{u}}{D\tau} = -\frac{1}{\rho_0}\nabla p_1 - \frac{\rho_1' gd}{\rho_0 c_*^2}\mathbf{e}_z + \frac{\nu}{Mc_* d}\left[\Delta \mathbf{u} + \frac{1}{3}\nabla(\nabla \cdot \mathbf{u})\right]$$

where we set $\rho_1 = M^2 \rho_1'$.

This system needs to be completed by the $\mathcal{O}(M^2)$ term, either of the energy equation

$$\rho_0 \frac{D\theta_1'}{D\tau} = \gamma \frac{\kappa \rho_0}{Vd}\Delta \theta_1' - \gamma(\gamma - 1)p_1 \nabla \cdot \mathbf{u} + \frac{\gamma(\gamma - 1)\rho_0}{2}\frac{\nu}{Vd}(\nabla : \mathbf{u})^2 \ .$$

or, of the entropy equation

$$\theta_0 \frac{Ds_1'}{D\tau} = \frac{\kappa}{Vd} \Delta \theta_1' + \frac{(\gamma - 1)\nu}{2Vd} (\nabla : \mathbf{u})^2$$

The equation of state is also necessary:

$$\gamma p_1 = \rho_0 \theta_1' + \theta_0 \rho_1'$$

where we set $\theta_1' = \theta_1/M^2$ and $s_1' = s_1/M^2$. Combining (7.18) and (7.19) leads to an interesting result, namely

$$\mathbf{u} \cdot \nabla s_0 = 0 \qquad \Longleftrightarrow \qquad \partial_z s_0 = 0 \qquad\qquad (7.20)$$

It means that the equilibrium solution must be quasi-isentropic, or that the temperature gradient of the equilibrium must be close to the adiabatic gradient. This result may be understood if we realize that our hypothesis (low Mach number) means that the velocity stays small with regard to that of sound, which means that the forcing of the flow is weak. Thus, when convection arises, the temperature gradient is close to the adiabatic one, according to Schwarzschild's criterion (see below in Sect. 7.5.1). The consequence of this result is that we must take the isentropic solution as the reference solution. This solution is always stable as we shall see in Sect. 7.5.1. When convection appears, the temperature gradient is superadiabatic and is imposed by the boundary conditions. In order to be able to make use of the anelastic approximation, we should include within the disturbances (ρ_1, θ_1), the difference between the (unstable) equilibrium solution and the isentropic solution (which is certainly a solution to static equations).

Finally, setting $\text{Re} = Vd/\nu$, the equations of the flow at the anelastic approximation are:

$$
\begin{cases}
\nabla \cdot (\rho_0 \mathbf{u}) = 0 \\[2mm]
\dfrac{D\mathbf{u}}{D\tau} = -\dfrac{1}{\rho_0} \nabla p_1 - \dfrac{\rho_1'}{\rho_0} \dfrac{gd}{c_*^2} \mathbf{e}_z + \dfrac{1}{\text{Re}} \Delta \mathbf{u} \\[3mm]
\theta_0 \dfrac{Ds_1'}{D\tau} = \dfrac{1}{\text{Pe}} \Delta \theta_1' + \dfrac{\gamma - 1}{2\text{Re}} (\nabla : \mathbf{u})^2 \\[3mm]
\gamma p_1 = \rho_0 \theta_1' + \theta_0 \rho_1'
\end{cases}
\qquad (7.21)
$$

where we introduced the *Péclet number* Pe, which is nothing but the Reynolds number where the kinematic viscosity is replaced by thermal diffusivity. This number is

$$\text{Pe} = \frac{Vd}{\kappa} = \mathcal{P}\text{Re}$$

(7.21) also show that the pressure fluctuation plays a part in the dynamics, as it influences the buoyancy.

The preceding result has another consequence. If a flow strongly disturbs a static solution which is far from the isentropic solution, then the fluid velocity is necessarily comparable to that of the sound. Indeed, when the mixing due to convection is important, the entropy distribution tends to be homogeneous. Hence, if the initial state is far form an isentropic state, variations of the thermodynamic variables may be of the order of the initial values, namely $\delta\rho \sim \rho_{eq}$, $\delta T \sim T_{eq}$ and $\delta P \sim P_{eq}$. Since pressure variations induced by the flow are $\sim \rho v^2$, it turns out that $v^2 \sim P_{eq}/\rho \sim R_* T_{eq} \sim c_s^2$. Hence, the fluids velocity is not small compared to that of the sound. This argument underlines the restriction of the anelastic approximation, namely that the variations of the thermodynamic variables must remain small compared to the values of the static solution. The use of the anelastic approximation is therefore not always possible: for instance in the surface layers of the Sun, the mean pressure and density drop to very small values leading to small values of the sound speed. In this case convective velocities get close to sonic values and reshape the mean density and pressure profiles.

In the case where the nonlinear effects are negligible, the anelastic approximation eliminates the sound waves thus assuming that the density fluctuations are not modified by the pressure ones.

7.4 Baroclinicity or the Impossibility of Static Equilibrium

We noted in Chap. 2 that the equilibrium of a fluid in a gravitational field can only be achieved if $P \equiv P(\rho)$, namely when the fluid is barotropic. Usually, $P \equiv P(\rho, T)$ but in some situations $T \equiv T(\rho)$ and thus $P \equiv P(\rho)$. The example of an ideal gas between two horizontal plates is typical of a non-barotropic fluid that is in a barotropic configuration. Using (7.4) and (7.5), we see that

$$P \propto \rho^{\frac{m}{m-1}} \qquad \text{and} \qquad T \propto \rho^{\frac{1}{m-1}} \ .$$

Equilibrium is therefore possible.

If this condition is not satisfied, a torque density appears and produces vorticity: the fluid cannot stay at rest. In order to illustrate this type of situation, called *baroclinic*, we now study an example where the static equilibrium does not exist if the temperature gradient is non-zero.

7.4.1 Thermal Convection Between Two Vertical Plates

Let us consider a system where the fluid is contained between two vertical plates with different temperatures. This situation, where the temperature gradient is perpendicular to the gravity, occurs in a double-paned window: the interior pane is warm and the exterior pane is cold (or vice-versa). We show below that the captive air between the two panes develops a flow that attempts to re-establish thermal equilibrium between the two panes by transferring the heat from the warm one to the cold one.

In order to ease the analysis of this system, we consider a set-up where the Boussinesq approximation can be applied. Furthermore, we assume the bounding panes to be infinite in size and of uniform temperature, T_c and T_w, respectively. These temperatures are taken only slightly different so that the flow is of small amplitude and may be described by linear equations. This will allow us to write the solution of the problem as a disturbance of the equilibrium state that exists when $T_c = T_w$. We further simplify by considering only the steady state.

Considering (7.10), eliminating the time derivatives and the nonlinear terms, we find

$$\begin{cases} -\nabla \delta P + \delta \rho \mathbf{g} + \mu \Delta \mathbf{v} = 0 \\[2mm] \mathbf{v} \cdot \nabla T_{eq} = \kappa \Delta \delta T \\[2mm] \nabla \cdot \mathbf{v} = 0 \\[2mm] \dfrac{\delta \rho}{\rho} = -\alpha \delta T \end{cases} \tag{7.22}$$

In our case $\nabla T_{eq} = \mathbf{0}$ since the temperature is constant at equilibrium. The temperature field δT is therefore a solution of Laplace's equation. Let x be the coordinate perpendicular to the plates, we then have $\delta T \equiv \delta T(x)$ and the solution of Laplace's equation immediately yields:

$$\delta T(x) = -(T_w - T_c)x/d$$

We have placed the warm plate at $x = -d/2$ and the cold one at $x = d/2$. The expression of the temperature gives the expression of the density perturbation:

$$\delta \rho = \frac{\alpha(T_w - T_c)\rho_0}{d} x$$

The velocity field therefore satisfies:

$$\mu \Delta \mathbf{v} - \nabla \delta P = \frac{\alpha(T_w - T_c)\rho_0 g}{d} x \mathbf{e}_z \qquad \text{and} \qquad \nabla \cdot \mathbf{v} = 0$$

which is just Stokes' equation with a forcing term. We look for a solution which depends only on x; in this case $\nabla \cdot \mathbf{v} = 0$ implies that $v_x = 0$. As a consequence

$\partial \delta P / \partial x = 0$ and $\partial \delta P / \partial z = G_p$, G_p being a constant. The z-component of the momentum equation gives v_z:

$$v_z(x) = Ax^3 + \frac{G_p}{2\mu}x^2 + Bx + C \qquad (7.23)$$

where B and C are two constants of integration and $A = \frac{\alpha(T_w - T_c)\rho_0 g}{6\mu d}$. The boundary conditions $v_z(\pm d/2) = 0$ imply that $B = -Ad^2/4$ but no constraint is imposed on the constants C and G_p. We can lift this degeneracy by imposing that the mass flux across a plane $z =$Cst be zero. This condition is, in fact, realized when the domain occupied by the fluid is finite; it expresses as

$$\int_{-d/2}^{d/2} v_z(x)dx = 0$$

which implies that $G_p = C = 0$. Finally, the velocity field has the following form:

$$v_z(x) = -\frac{\alpha(T_w - T_c)\rho_0 g}{24\mu d}x(d^2 - 4x^2) \qquad (7.24)$$

The form of this solution is simply that of a parallel shear flow which does not transfer heat since $\mathbf{v} \cdot \nabla \delta T = 0$ (see Fig. 7.2). In a realistic case, the streamlines are closed curves and a heat transfer exists. However, if the temperature difference is large enough the preceding flow is unstable and produces turbulence. In this case the convective heat transfer is quite significant. We see that the design of double-paned windows should achieve a compromise between a great thickness, d, which reduces the losses by conduction (by lowering the temperature gradient) and a small thickness, which inhibits the development of instabilities and losses by fluid motions.

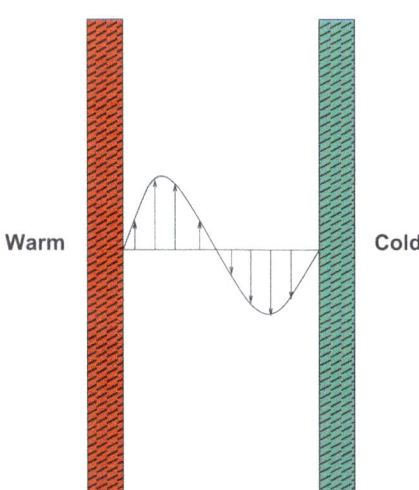

Warm   **Cold**

Fig. 7.2 The flow generated by a horizontal temperature gradient within the air of a double-paned window

7.5   Rayleigh–Bénard Instability

As we pointed it out while introducing the chapter, thermal convection is essentially
the result of an instability that develops thanks to an unstable temperature gradient.
Since this is an important source of fluid motion in Nature, we shall analyse it
in some details. However, before tackling the associated mathematics, we first
study a qualitative approach which leads to the Schwarzschild criterion, famous
in Astrophysics for locating the convection zones of stars.

7.5.1   Qualitative Analysis of Stability: Schwarzschild's Criterion

Like in Sect. 7.2.1, we shall first consider an ideal gas at rest between two horizontal
plates in a uniform gravity field. We further assume that

$$(T_u - T_l)/d > -g/c_p \tag{7.25}$$

namely, that the temperature gradient is *sub-adiabatic*. This means that the temperature decreases more slowly with altitude than in the case the gas were isentropic.

Let us now consider a fluid element located at an altitude z, and that we artificially
displace, by Δz, as it could result from a spontaneous fluctuation of the system.
Assuming that the displacement is sufficiently fast so that no heat exchange with
the surrounding medium takes place, the element undergoes an adiabatic expansion;
its temperature changes to

$$T_{el}(z_i + \Delta z) = T_{el}(z_i) + \Delta z \left(-\frac{g}{c_p} \right)$$

while that of the surrounding medium is

$$T_{env}(z_i + \Delta z) = T_{env}(z_i) + \Delta z \left(\frac{T_u - T_l}{d} \right)$$

Noting that $T_{el}(z_i) = T_{env}(z_i)$ and using (7.25), we find that

$$\Delta z > 0 \quad \Longrightarrow \quad T_{el}(z_i + \Delta z) < T_{env}(z_i + \Delta z)$$

$$\Delta z < 0 \quad \Longrightarrow \quad T_{el}(z_i + \Delta z) > T_{env}(z_i + \Delta z)$$

In other words, if a fluid element is artificially raised (alternatively, moved down),
it will be colder (alternatively, warmer), than its environment. During this motion
the fluid particle is always in pressure balance with its environment. Thus a colder

element is denser than the environment while a warmer one is less dense. The foregoing inequalities show that the buoyancy force pulls the fluid element back to its original equilibrium position. *The fluid's equilibrium is stable.*

Let us now consider the opposite case where

$$(T_u - T_l)/d < -g/c_p \tag{7.26}$$

The temperature gradient is now termed as *super-adiabatic*. The temperature decreases faster with altitude than if controlled by the adiabatic gradient. The difference in temperature between the fluid element and the surrounding medium is now reversed with respect to the preceding case:

$$\Delta z > 0 \quad \Longrightarrow \quad T_{\mathrm{el}}(z_i + \Delta z) > T_{\mathrm{env}}(z_i + \Delta z)$$

$$\Delta z < 0 \quad \Longrightarrow \quad T_{\mathrm{el}}(z_i + \Delta z) < T_{\mathrm{env}}(z_i + \Delta z)$$

This time an element which is displaced upwards will be warmer than the ambient medium and the buoyancy force will enhance this motion. Similarly, a displacement downwards is also amplified by the buoyancy force. Disturbances are thus amplified and the equilibrium of the fluid is now unstable (see Fig. 7.3).

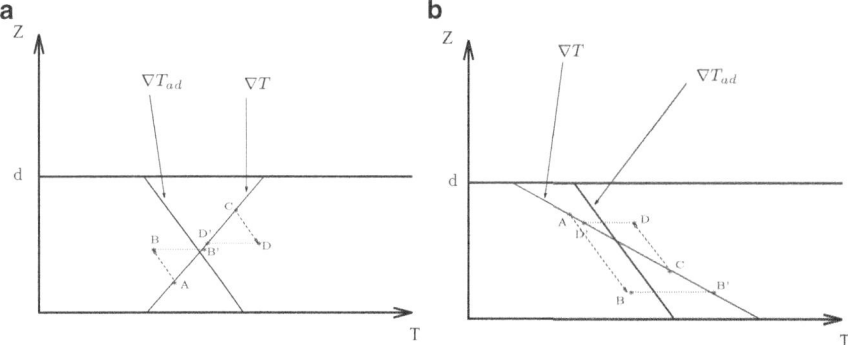

Fig. 7.3 An illustration of Schwarzschild's criterion: (**a**) the sub-adiabatic case: a fluid element moved from A to B (*upwards*) becomes colder than the surrounding fluid which is at temperature B' (according to the orientation of the temperature axis, *B* is below *B'*). It is brought back to its initial position by the buoyancy force. Similarly, a fluid element moved from C to D (thus downward) is hotter than the surrounding fluid and is pushed upwards by the buoyancy force. Thus, this temperature profile is stable. (**b**) The super-adiabatic case: a fluid element moved (*downwards*) from A to B becomes colder than its environment whose temperature is that of B'. The buoyancy now helps this motion and the fluid element continues downwards. Similarly, an initial rise from C to D let the fluid element warmer than its environment and the buoyancy force also helps this motion. The fluid equilibrium is now unstable

The foregoing analysis shows the crucial role played by the adiabatic gradient. When the temperature gradient is equal to the adiabatic gradient the equilibrium is neutral: disturbances are neither damped nor amplified. The criterion

$$\partial_z T - (\partial_z T)_{ad} \geq 0 \qquad \Longleftrightarrow \qquad \text{stability} \qquad\qquad (7.27)$$

is called *Schwarzschild's criterion*. Note that if we use the potential temperature (7.7), Schwarzschild's criterion reads

$$\partial_z T_{\text{pot}} \geq 0 \qquad \Longleftrightarrow \qquad \text{stability} \qquad\qquad (7.28)$$

or, since potential temperature only depends on entropy,

$$\partial_z s \geq 0 \qquad \Longleftrightarrow \qquad \text{stability} \qquad\qquad (7.29)$$

Schwarzschild's criterion determines the stability of the equilibrium of a perfect fluid. Indeed, in the preceding discussion, we neglected thermal conductivity and viscosity. The thermal conductivity, by reducing the differences in temperature, and the viscosity, by inhibiting the displacement, both contribute to the damping of perturbations. However, taking into account, quantitatively, these effects is not a simple game: a conventional stability analysis, as those of Chap. 6, is needed and this is our next step in the study of convection.

7.5.2 Evolution of Disturbances

Our first effort will be to derive the equations governing the evolution of perturbations under the influence of diffusion but in a simplified set-up where the Boussinesq approximation is valid with the further assumption (to be removed later) of a two-dimensional velocity field. We thus return to (7.11) and consider a 2D velocity like:

$$\mathbf{u} \begin{vmatrix} u_x(x, z, t) \\ 0 \\ u_z(x, z, t) \end{vmatrix}$$

This allows us to introduce the stream function ψ such that

$$u_x = -\frac{\partial \psi}{\partial z}, \qquad u_z = \frac{\partial \psi}{\partial x} \qquad\qquad (7.30)$$

observing that

$$\mathbf{u} = \nabla \times (\psi \mathbf{e}_y) = \nabla \psi \times \mathbf{e}_y, \qquad \nabla \times \mathbf{u} = -\Delta \psi \mathbf{e}_y$$

and taking the curl of the momentum equation, we obtain the system

$$\begin{cases} \dfrac{\partial \Delta \psi}{\partial \tau} - \mathcal{P} \mathrm{Ra} \dfrac{\partial \theta}{\partial x} - \mathcal{P} \Delta \Delta \psi = J[\Delta \psi, \psi] \\[4mm] \dfrac{\partial \theta}{\partial \tau} - \dfrac{\partial \psi}{\partial x} - \Delta \theta = J[\theta, \psi] \end{cases} \qquad (7.31)$$

where we note that the nonlinear terms appear in the form of a Jacobian, namely

$$J[f, g] = \frac{\partial f}{\partial x} \frac{\partial g}{\partial z} - \frac{\partial g}{\partial x} \frac{\partial f}{\partial z}$$

At first, we consider disturbances of infinitesimal amplitude; their evolution is governed by linear equations

$$\begin{cases} \dfrac{\partial \Delta \psi}{\partial \tau} - \mathcal{P} \mathrm{Ra} \dfrac{\partial \theta}{\partial x} - \mathcal{P} \Delta \Delta \psi = 0 \\[4mm] \dfrac{\partial \theta}{\partial \tau} - \dfrac{\partial \psi}{\partial x} - \Delta \theta = 0 \end{cases} \qquad (7.32)$$

To solve such a system we develop the solutions in Fourier modes as follows:

$$\psi = \psi_k(z)e^{ikx+\lambda t}, \qquad \theta = \theta_k(z)e^{ikx+\lambda t} \qquad (7.33)$$

If we observe that

$$\Delta \longrightarrow D^2 - k^2, \qquad D = \frac{\partial}{\partial z}$$

$$\frac{\partial}{\partial x} \longrightarrow ik, \qquad \frac{\partial}{\partial \tau} \longrightarrow \lambda$$

then (7.32) becomes

$$\begin{cases} (D^2 - k^2)(\mathcal{P}(D^2 - k^2) - \lambda)\psi_k + ik\mathcal{P}\mathrm{Ra}\theta_k = 0 \\[4mm] (D^2 - k^2 - \lambda)\theta_k + ik\psi_k = 0 \end{cases} \qquad (7.34)$$

7.5.3   Expression of the Solutions

The vertical profile of the perturbation, $\psi_k(z)$ and $\theta_k(z)$, is therefore given by the solutions of the sixth order linear differential system (7.34). These solutions are in the form:

$$\begin{pmatrix} \psi \\ \theta \end{pmatrix} = \sum_{n=1}^{6} A_n \begin{pmatrix} p_n \\ q_n \end{pmatrix} e^{\alpha_n z} \tag{7.35}$$

where $\{\alpha_n\}_{n=1,6}$ are the roots of the polynomial

$$(\alpha^2 - k^2 - \lambda)((\alpha^2 - k^2)\mathcal{P} - \lambda)(\alpha^2 - k^2) + k^2 \mathcal{P}\mathrm{Ra} = 0 \tag{7.36}$$

The solutions of this third degree equation in α^2 give the expressions of the roots as a function of λ. We then require that the solution satisfy the boundary conditions. Coefficients A_n, being non-zero, the determinant of the 6×6 system thus formed must be zero. This yields the dispersion equation $\lambda(k)$ of the modes of the system. By examining the dependency of λ as a function of k and Ra, we can find the condition of existence of unstable modes for which $Re(\lambda) > 0$. This method is the general one. It is quite arduous as we easily imagine.

We shall avoid momentarily these difficulties by considering a case where we can shortcut this general way. This is possible when the fluid meets stress-free boundary conditions on both bounding plates. This configuration is certainly not the most realistic but it is very educational.[4]

We recall that for such conditions, the two surfaces are fixed planes and disturbances satisfy:

$$\begin{cases} v_z = 0 \\ \\ \sigma_{xz} = \sigma_{yz} = 0 \end{cases} \quad \text{at} \quad z = 0, z = d \tag{7.37}$$

for the velocity. We further assume that the bounding plates are perfect conductors so that their temperature is fixed and no perturbation is allowed there (see Sect. 1.8.2). Thus

$$\theta = 0 \quad \text{at} \quad z = 0, z = d \tag{7.38}$$

for the temperature.

---

[4]We could, however, approach such a set-up by confining, for example, an oil layer between a layer of mercury in $z < 0$, and a layer of liquid sodium in $z > d$!

Noting that $\sigma_{xz} = 0$ at a fixed z implies that $\partial_x \sigma_{xz} = 0$ on the same plane, using mass conservation $\nabla \cdot \mathbf{v} = 0$, we find that $\partial_x^2 v_z - \partial_z^2 v_z = 0$ at the bounding plate. Since $v_z = 0$ there, the stress-free boundary conditions imply that

$$\frac{\partial^2 v_z}{\partial z^2} = 0 \quad \text{at} \quad z = 0, \ z = d \tag{7.39}$$

We can now derive the boundary conditions satisfied by ψ_k on the planes $z = 0$, $z = d$. According to (7.30) we find

$$\psi_k = D^2 \psi_k = 0 \quad \text{at} \quad z = 0, \ z = d \tag{7.40}$$

By using the differential equations (7.34), we find that all the even derivatives of ψ_k are zero on the $z = 0, z = d$ planes and that the same is true for θ_k:

$$D^{2m} \psi_k = D^{2m} \theta_k = 0 \quad \text{at} \quad z = 0, \ z = d \tag{7.41}$$

for all $m \in \mathbb{N}$. Recalling that ψ_k and θ_k are linear combinations of exponentials, we note that (7.41) imposes a severe constraint on the solution. Actually, the functions

$$\sin(n\pi z), \quad n \in \mathbb{N}$$

are the only linear combination of exponentials that verify these conditions. Therefore we have no choice, ψ_k and θ_k must be written like

$$\psi_k(z) = \sum_n A_{kn} \sin(n\pi z) \qquad \text{and} \qquad \theta_k(z) = \sum_n B_{kn} \sin(n\pi z)$$

where n is a (positive) non vanishing integer.

7.5.4 Criterion of Stability

The foregoing results give the value of α without solving (7.36). Actually, setting the value $\alpha = n\pi$ in (7.36) yields an equation for λ, the frequency of the eigenmodes. Setting

$$\ell^2 = \pi^2 n^2 + k^2$$

(7.36) is changed into

$$\lambda^2 + \ell^2(\mathcal{P} + 1)\lambda - \mathcal{P}\left(\frac{k^2 \mathrm{Ra}}{\ell^2} - \ell^4\right) = 0$$

whose solutions are

$$\lambda_{\pm} = \frac{\ell^2(\mathcal{P}+1)}{2}\left(-1 \pm \sqrt{1 + \frac{4\mathcal{P}}{\ell^4(\mathcal{P}+1)^2}\left(\frac{k^2 \mathrm{Ra}}{\ell^2} - \ell^4\right)}\right) \qquad (7.42)$$

We observe that the eigenmode associated with λ_- is always damped since $Re(\lambda_-) < 0$. On the other hand, the other mode can be amplified. If

$$\frac{k^2 \mathrm{Ra}}{\ell^2} - \ell^4 > 0$$

then, $Re(\lambda_+) > 0$. This inequality shows the existence of a critical Rayleigh number Ra_c which is such that if $\mathrm{Ra} > \mathrm{Ra}_c$ some perturbations grow. This critical Rayleigh number depends on the wavenumber of the perturbation and we easily get from the previous inequality that

$$\mathrm{Ra}_c(k,n) = \frac{\ell^6}{k^2}$$

In Fig. 7.4, we plot $\mathrm{Ra}_c(k,n)$ for the first values of n, thus showing a few critical curves. Among them, the $n = 1$ is the most important, since its minimum gives the true critical Rayleigh number below which every disturbance is damped out. We shall evaluate this minimum from the function

$$\mathrm{Ra}_c(k, n=1) = \frac{(\pi^2 + k^2)^3}{k^2}$$

A simple calculation shows that its minimum value occurs at the wavenumber

$$k_c = \frac{\pi}{\sqrt{2}} \qquad (7.43)$$

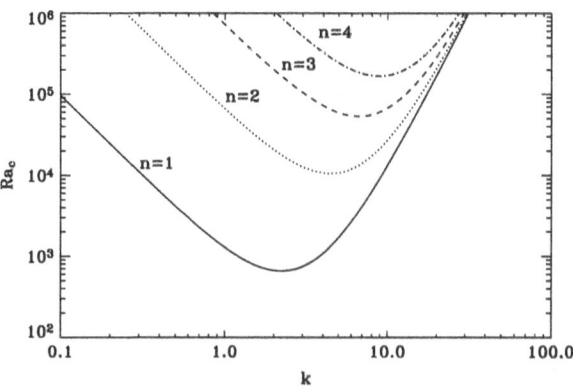

Fig. 7.4 Critical curves of the most unstable modes

which is the *critical wavenumber*. The value of the Rayleigh number is:

$$\text{Ra}_c = \frac{27\pi^4}{4} \simeq 657.51$$

Thus if $\text{Ra} < \text{Ra}_c$ all modes are damped and the fluid layer is in stable equilibrium, while if $\text{Ra} > \text{Ra}_c$, there is at least one growing mode starting thermal convection.

7.5.5 The Other Boundary Conditions ✿

What happens if we apply more realistic boundary conditions to the velocity field? For example those existing in a laboratory experiment where the fluid is confined between two solid plates on which the velocity vanishes? In this case (7.37) is replaced by

$$v_x = v_y = v_z = 0 \quad \text{at} \quad z = 0, \ z = d \tag{7.44}$$

These conditions lead, like the preceding ones, to two conditions on v_z:

$$v_z = \partial_z v_z = 0 \quad \text{at} \quad z = 0, \ z = d \tag{7.45}$$

We have seen before that v_z is a linear combination of exponentials; in the present case, there is no simple linear combination that satisfies these boundary conditions. We must, therefore return to (7.36) and its general solution. Equation (7.36) is a third degree polynomial equation in α^2, which solutions have no simple expressions in general. However, we just wish to determine the effects of the new boundary conditions on the critical Rayleigh number. For this, we set $\text{Ra} = \text{Ra}_c$ and $\lambda = 0$. It may be shown indeed that λ vanishes at the stability threshold. This property, which we demonstrate in the box below, is known as the *principle of the exchange of stability* (Chandrasekhar 1961). The three roots in α^2 of (7.36) are then easily found.

$$\alpha_1^2 = -k^2(R-1), \quad \alpha_{2,3}^2 = k^2(1 + R e^{\pm i\pi/3}) \tag{7.46}$$

where we have set $R = \text{Ra}^{1/3} k^{-4/3}$. Let us assume that R is greater than unity and therefore that all α are complex. Noting that α_2 is the complex conjugate of α_3, we write

$$\alpha_1 = \pm i a, \quad \alpha_2 = \pm \alpha, \quad \text{et} \quad \alpha_3 = \pm \alpha^* \tag{7.47}$$

The general solution of the problem is thus

$$\begin{pmatrix} \psi \\ \theta \end{pmatrix} = \begin{pmatrix} p \\ q \end{pmatrix} e^{iaz} + \begin{pmatrix} p \\ q \end{pmatrix}^* e^{-iaz} + \cdots$$

that we can also write as

$$\begin{pmatrix} \psi \\ \theta \end{pmatrix} = A_1 \cos az + A_2 \sin az + A_3 \operatorname{ch} \alpha z + A_3^* \operatorname{ch} \alpha^* z + A_4 \operatorname{sh} \alpha z + A_4^* \operatorname{sh} \alpha^* z$$

$$(7.48)$$

In this expression we grouped the complex conjugate terms since the solution is real.

We now require that this solution satisfy the boundary conditions on the planes $z = 0$ and $z = 1$. It is advantageous to shift the boundaries to $z = \pm 1/2$ so that we can benefit from the symmetries of the functions of expression (7.48). The cosine part of the solutions is symmetric with respect to $z = 0$ while the sine part is antisymmetric. If we write the boundary conditions in $z = \pm 1/2$, by adding and subtracting the equations, the symmetrical and antisymmetrical parts separate. For the symmetrical part, let us write

$$\begin{pmatrix} \psi \\ \theta \end{pmatrix} = p_1 \begin{pmatrix} 1 \\ q_1 \end{pmatrix} \cos az + p_2 \begin{pmatrix} 1 \\ q_2 \end{pmatrix} \operatorname{ch} \alpha z + p_3 \begin{pmatrix} 1 \\ q_3 \end{pmatrix} \operatorname{ch} \alpha^* z \ .$$

The boundary conditions then lead to the following equations:

$$\begin{cases} p_1 \cos a/2 + p_2 \operatorname{ch} \alpha/2 + p_3 \operatorname{ch} \alpha^*/2 = 0 \\ p_1 q_1 \cos a/2 + p_2 q_2 \operatorname{ch} \alpha/2 + p_3 q_3 \operatorname{ch} \alpha^*/2 = 0 \\ -p_1 a \sin a/2 + p_2 \alpha \operatorname{sh} \alpha/2 + p_3 \alpha^* \operatorname{sh} \alpha^*/2 = 0 \end{cases} \quad (7.49)$$

This system has a non-trivial solution if and only if its determinant is zero, namely

$$\begin{vmatrix} 1 & 1 & 1 \\ q_1 & q_2 & q_3 \\ -a \tan a/2 & \alpha \operatorname{th} \alpha/2 & \alpha^* \operatorname{th} \alpha^*/2 \end{vmatrix} = 0$$

after some simple rearrangements. We get the expression of the q_i with (7.34), namely

$$q_1 = \frac{ik}{\alpha_1^2 + k^2} = \frac{i}{kR}$$

$$q_2 = \frac{ik}{k^2 - \alpha_2^2} = -\frac{i}{kR} e^{-i\pi/3}$$

$$q_3 = \frac{ik}{k^2 - (\alpha_2^*)^2} = -\frac{i}{kR} e^{i\pi/3}$$

which allows us to simplify the determinant as

$$
\begin{vmatrix}
1 & 1 & 1 \\
1 & (i\sqrt{3}-1)/2 & -(i\sqrt{3}+1)/2 \\
-a\tan a/2 & \alpha \operatorname{th}\alpha/2 & \alpha^*\operatorname{th}\alpha^*/2
\end{vmatrix} = 0
\tag{7.50}
$$

Note that a and α are two functions of Ra_c and k. By using the preceding equation we can determine for each value of k the corresponding value of Ra_c. We can then draw the critical curve $\mathrm{Ra}_c(k)$ whose form is similar to that of Fig. 7.4. Of course the solution of (7.50) can only be found numerically.

The critical curve $\mathrm{Ra}_c(k)$ goes through a minimum when

$$
k = k_c = 3.11632355 \quad \text{and} \quad \mathrm{Ra}_c = 1707.76178
\tag{7.51}
$$

We note that the critical Rayleigh number is higher than in the preceding case: this agrees with intuition since the friction on the walls demands a greater forcing in order to start the flow. We also observe that the streamlines are quasi-circular since $\lambda/2 = \pi/k_c \approx 1$ (see Sect. 7.6).

In the foregoing calculation we focused on the modes which are symmetric with respect to the mid-layer plane. To be complete we should now examine the antisymmetric modes. The equation giving the critical Rayleigh number of these modes as a function of k is of the same form as (7.50) where it suffices to replace the tangents with cotangents. Its solution shows that the curves $\mathrm{Ra}_c(k)$ have an absolute minimum at $k_c = 5.365$ where $\mathrm{Ra}_c = 17610.39$. Clearly, antisymmetric modes are more difficult to destabilize and the true critical Rayleigh number is therefore $\mathrm{Ra}_c = 1707.76178$.

The calculation of antisymmetric modes is, however, not denied of interest because it allows us to obtain the critical Rayleigh number when one of the boundaries is no-slip and the other is stress-free. Indeed, we showed that on a bounding plane with stress-free conditions all the even derivative of ψ are zero. This is precisely the property of an antisymmetric solution at mid-layer, at $z = 0$. Thus the antisymmetric mode between $z = -1/2$ and $z = 0$ is the same as the solution meeting a no-slip boundary at the bottom and a stress-free one at the top. Conversely, if we know the solution meeting no-slip boundary conditions on the bottom and stress-free one on the top, we may obtain the antisymmetric solution by antisymmetrizing it. Thus the critical Rayleigh number associated with mixed-type boundary conditions will lead to the one associated with no-slip boundary conditions and anti-symmetric modes if we double the thickness of the layer and the temperature difference (so as to preserve the temperature gradient). Therefore

$$
\mathrm{Ra}_c(antisym.) = 17610 = \frac{\alpha(2\Delta T)g(2d)^3}{\kappa\nu} = 16\frac{\alpha\Delta T g d^3}{\kappa\nu}
$$

$$
= 16\,\mathrm{Ra}_c(\text{no}-\text{slip}/\text{stress}-\text{free})
$$

which gives Ra_c (no-slip/stress-free) $= 1100.65$, an intermediate value between the two preceding cases. The critical wavenumber is half of the antisymmetric modes. Indeed, the dimensional wavenumber is the same as the one of antisymmetric modes but the nondimensional one is scaled with a thickness which is half-size. Thus,

$$k_c(\text{no} - \text{slip/stress} - \text{free}) = \frac{k_c(\text{antisym.})/d}{1/(d/2)} = \frac{k_c(\text{antisym.})}{2} = 2.68$$

which is also intermediate between the no-slip and stress-free values.

The principle of the exchange of stabilities ♠

The idea of an exchange of stability at a critical value of a parameter, like the Rayleigh number, is due to Poincaré. It has been popularized by Chandrasekhar in his monograph on hydrodynamic stability. The principle is of course a theorem. It says that at the threshold of stability perturbations are stationary, i.e. non-oscillatory so that $\lambda(\text{Ra}_c) = 0$.

To show this result, we start from (7.34) and set $\chi = (D^2 - k^2)\psi$. It turns out that:

$$\begin{cases} (D^2 - k^2)\chi + ik\text{Ra}\theta = \frac{\lambda}{\mathcal{P}}\chi \\ \\ (D^2 - k^2)\theta + ik\psi = \lambda\theta \end{cases} \tag{7.52}$$

We complete this system by the following boundary conditions:

$$\psi = 0, \qquad D\psi = 0 \quad \text{or} \quad D^2\psi = 0, \qquad \theta = 0$$

From the definition of χ, we get

$$\int_0^1 \psi^* D^2 \chi dz = \int_0^1 |\chi|^2 dz + k^2 \int_0^1 \chi\psi^* dz$$

We then multiply the first equation of (7.52) by ψ^* and integrate over the thickness; using the preceding equality we now get:

$$\int_0^1 |\chi|^2 dz + ik\text{Ra} \int_0^1 \theta\psi^* dz + \frac{\lambda}{\mathcal{P}} \int_0^1 (|D\psi|^2 + k^2|\psi|^2) dz = 0$$

Similarly, using the equation of temperature, we get

$$\int_0^1 (|D\theta|^2 + k^2|\theta|^2) dz - ik \int_0^1 \psi\theta^* dz + \lambda \int_0^1 |\theta|^2 dz = 0$$

We now multiply this equation by Ra and add it to the foregoing one; we obtain:

$$\int_0^1 \left(\text{Ra}(|D\theta|^2 + k^2|\theta|^2) + |\chi|^2 \right) dz - 2k\text{Ra} \int_0^1 \text{Im}(\theta\psi^*) dz$$

$$+ \frac{\lambda}{\mathcal{P}} \int_0^1 (|D\psi|^2 + k^2|\chi|^2 + \text{Ra}\mathcal{P}|\theta|^2) dz = 0 \tag{7.53}$$

This last equality shows that when Ra is positive then $\text{Im}(\lambda)=0$ since every term of (7.53) is real except λ and the coefficient of λ cannot be zero. Thus if $\text{Ra} = \text{Ra}_c$ then $\lambda = 0$.

7.6 Convection Patterns

In the foregoing section, we presented the physical conditions which lead to the development of thermal convection. We may now wonder about the shape of the flows that replace the unstable hydrostatic equilibrium. Thus after discussing the eigenvalues of the problem (i.e. the growth rate of the instability), we shall now focus on the eigenmodes. Actually, since we are still dealing with linear quantities, we may easily, for a while, consider three-dimensional perturbations.

7.6.1 Three-Dimensional Disturbances

Let us assume that the disturbances have the form

$$\mathbf{u} = \mathbf{u}_0(z)e^{\lambda\tau + ik_x x + ik_y y}$$

Equation (7.11), when linearized, read

$$\begin{cases} \lambda u_x = -ik_x p + \mathcal{P}(D^2 - k^2)u_x \\ \lambda u_y = -ik_y p + \mathcal{P}(D^2 - k^2)u_y \\ \lambda u_z = -Dp + \mathrm{Ra}\mathcal{P}\theta + \mathcal{P}(D^2 - k^2)u_z \\ Du_z + ik_x u_x + ik_y u_y = 0 \\ (D^2 - k^2 - \lambda)\theta = -u_z \end{cases} \tag{7.54}$$

where we set $k = \sqrt{k_x^2 + k_y^2}$. If we use Squire's transformation (see Chap. 6), then we set

$$k\tilde{u} = k_x u_x + k_y u_y$$

and we easily show that the preceding system leads to

$$\begin{cases} \lambda\tilde{u} = -ikp + \mathcal{P}(D^2 - k^2)\tilde{u} \\ \lambda u_z = -Dp + \mathrm{Ra}\mathcal{P}\theta + \mathcal{P}(D^2 - k^2)u_z \\ Du_z + ik\tilde{u} = 0 \\ (D^2 - k^2 - \lambda)\theta = -u_z \end{cases} \tag{7.55}$$

By eliminating $\tilde{u}$, we find the two following equations

$$\begin{cases} [\mathcal{P}(D^2 - k^2) - \lambda][D^2 - k^2]u_z = k^2\mathcal{P}\mathrm{Ra}\theta \\ (D^2 - k^2 - \lambda)\theta = -u_z \end{cases} \tag{7.56}$$

which are strictly identical to (7.34) if we replace u_z by $ik\psi_k$.

The conclusion of these little calculations is that the three-dimensional disturbances have exactly the same critical Rayleigh number as their two-dimensional
counterpart. *This number does not depend on the orientation of the wavevector* **k**.

If then we increase the Rayleigh number beyond critical value, we destabilize
all the modes which have a wavevector of modulus k_c. The flow which appears,
depends on the wavevectors selected in the final solution. This selection depends on
the horizontal boundary conditions and on the stability of various possible solutions.
We shall not discuss this thorny and wide subject, and will restrict ourselves to
describing several solutions that are observed in Nature.

7.6.2 Convection Rolls

Convection rolls are nothing but the two-dimensional solutions. Let us write the
temperature perturbation as

$$\theta(x, z) = A \sin \pi z \cos k_c x$$

From this expression and the last equation of (7.56) we derive u_z and then u_x from
the equation of continuity. Hence, it follows

$$u_x = -\frac{3A\pi^2}{\sqrt{2}} \cos \pi z \sin k_c x \qquad \text{and} \qquad u_z = \frac{3A\pi^2}{2} \sin \pi z \cos k_c x$$

This solution is illustrated in Fig. 7.5.

7.6.3 Other Patterns of Convection

In order to get other patterns of convection, it is sufficient to combine in a linear
manner several wave vectors of different directions. Let us consider a temperature
disturbance of the form:

$$\delta T = A(\cos k_c x + \cos k_c y) \sin \pi z$$

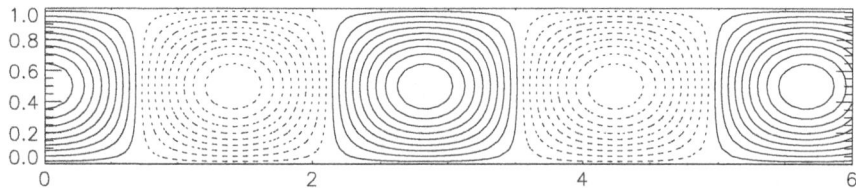

Fig. 7.5 Shape of isotherms for roll convection near the threshold of stability. Boundary conditions are stress-free for the velocity and fixed temperature

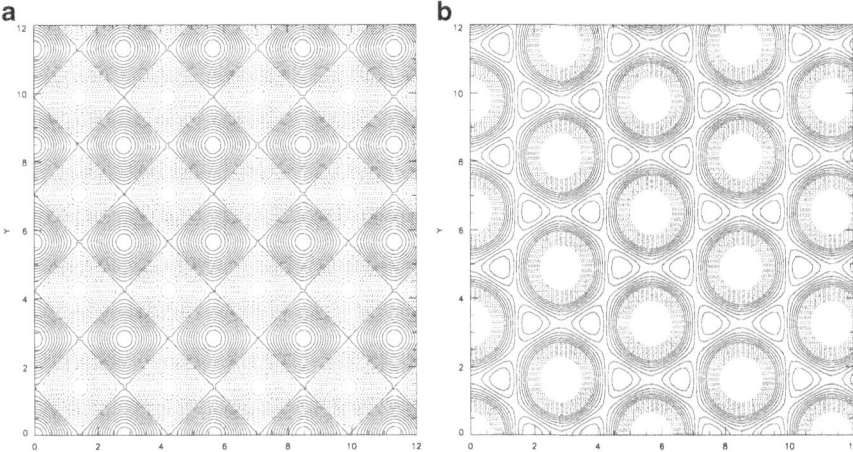

Fig. 7.6 Square (**a**) or hexagonal (**b**) convection cells viewed with their isotherms. *Dotted lines* are for $\delta T < 0$ and *solid lines* for $\delta T > 0$

This is the linear combination of two perpendicular waves of the same amplitude. If we look for the points of the maximum of δT, these are

$$x_{nm} = \frac{2\pi m}{k_c} \qquad y_{nm} = \frac{2\pi n}{k_c}$$

where m and n are integers. These maxima draw a network of squares. The convection flow thus appears as a set of square cells covering the horizontal plane (see Fig. 7.6a). Another pattern is however possible, if we use a symmetry frequently found in Nature, namely the invariance by rotations with a $2\pi/3$ angle. This symmetry indeed leads to hexagonal patterns. We retrieve a convection flow with this symmetry by superimposing three waves whose wave vectors make an angle of $2\pi/3$ between them:

$$\mathbf{k} = k_c \begin{vmatrix} 1 \\ 0 \end{vmatrix}, \qquad k_c \begin{vmatrix} \cos\frac{2\pi}{3} \\ \sin\frac{2\pi}{3} \end{vmatrix}, \qquad k_c \begin{vmatrix} \cos\frac{4\pi}{3} \\ \sin\frac{4\pi}{3} \end{vmatrix}.$$

This leads to the following field for the temperature fluctuations:

$$\delta T = A \sin \pi z \left[\cos k_c x + \cos(k_c(-x/2 + y\sqrt{3}/2)) + \cos(k_c(x/2 + y\sqrt{3}/2)) \right]$$
$$(7.57)$$

As expected, the isotherms of this solution, taken in a plane at constant z, display hexagonal cells as shown in Fig. 7.6b.

7.7 The Weakly Nonlinear Amplitude Range

When we impose a Rayleigh number greater than the critical value, all the disturbances belonging to the unstable band grow exponentially. But those of wavelength λ_c grow the fastest. Therefore, we may expect that these modes control the dynamics of the system. Thus if we wish to have a first view on the dynamics of thermal convection we may focus our attention on these modes and try to determine how their amplitude saturates at a finite value. For this we should take into account the nonlinear terms that we have overlooked until now.

7.7.1 Periodic Boundary Conditions

In order to simplify as much as possible the following analysis, we again restrict ourselves to the two-dimensional case, with which we are now well acquainted. In addition, we also "isolate" the mode of wavelength λ_c by imposing to our system periodic horizontal boundary conditions. Indeed, in this case all the functions satisfy $f(x) = f(x + L)$, where L is the length of our periodic "box". From (7.33) this implies that

$$e^{ikL} = 1 \qquad \text{or} \qquad k = \frac{2m\pi}{L}, \quad m \in \mathbb{N}$$

so that the possible horizontal wavenumbers k now form a discrete set. It is then easy to choose a box length and a Rayleigh number so that only one mode is unstable. For example, if we take $L = 2\sqrt{2}$ and $\mathrm{Ra} = 1.5(27\pi^4/4)$ the mode corresponding to $k = k_c = \pi/\sqrt{2}$ is the only unstable one.

These boundary conditions may seem rather artificial, but more realistic boundary conditions would not change the situation dramatically. The form of the horizontal base functions would no longer be e^{ikx}, but some other functions adapted to the horizontal boundary conditions, and also characterized by a typical wavenumber similar to k. The periodic boundary conditions are the most convenient for taking into account the finite horizontal size of a physical system.

Finally, we note that if the size of the system grows, the number of modes in the unstable spatial frequency band also grows. Thus, for a given Rayleigh number, the number of unstable modes grows with the size of the system (see Fig. 7.7).

7.7.2 Small Amplitudes

Solving the nonlinear equations is usually feasible only numerically. However, much can be learnt from the weakly nonlinear case which is accessible to analytic work. Here, we shall focus on this latter case and introduce two restrictions:

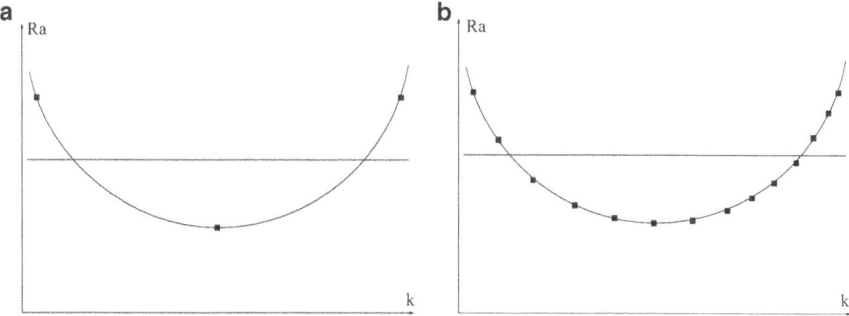

Fig. 7.7 (**a**) The case of a small-size system: for the imposed Rayleigh number, a single mode is unstable. (**b**) The case of a system of larger size. For the same imposed Rayleigh number now nine modes are unstable

- The Rayleigh number is slightly supercritical $Ra = (1 + \varepsilon)Ra_c$
- The amplitudes of the perturbations are small and $\mathcal{O}(\varepsilon')$.

At first, the small parameters ε and ε' seem to be independent, but we shall see below that the consistency of the solutions imposes a relation between them.

Recalling the results of the linear analysis (7.42), we find that the two eigenvalues have now the following form:

$$\lambda_+ = \varepsilon \frac{\ell^2 \mathcal{P}}{\mathcal{P} + 1} + \mathcal{O}(\varepsilon^2) \tag{7.58}$$

$$\lambda_- = -\ell^2(\mathcal{P} + 1) + \mathcal{O}(\varepsilon) \tag{7.59}$$

Thus, near the threshold, the growth rate of the instability is proportional to $Ra - Ra_c$, or

$$\lambda_+ = \frac{\ell^2 \mathcal{P}}{\mathcal{P} + 1} \frac{Ra - Ra_c}{Ra_c}$$

The first approximation implies that the instability grows on an $\mathcal{O}(\varepsilon^{-1})$ time scale, which is large compared to unity.

Let us write the instable mode as follows:

$$\begin{cases} \psi(x, z, t) = \psi_{11}(t) \sin \pi z \sin kx \\ \theta(x, z, t) = \theta_{11}(t) \sin \pi z \cos kx \end{cases} \tag{7.60}$$

We shall denote it also (θ_{11}, ψ_{11}). $k = k_c = 2\pi/L$ is its wavenumber. We aim at obtaining the differential equations verified by the amplitudes $\psi_{11}(t)$ and $\theta_{11}(t)$.

Let us start by computing the nonlinear terms with (7.60). We find

$$J[\theta, \psi] = J[\theta_{11}(t)\sin \pi z \cos kx, \; \psi_{11}(t)\sin \pi z \sin kx] = -\psi_{11}(t)\theta_{11}(t)k\pi/2 \sin 2\pi z$$

$$J[\psi, \Delta\psi] = J[\psi_{11}(t)\sin \pi z \sin kx, \; \Delta(\psi_{11}(t)\sin \pi z \sin kx)] = 0$$

We see that the first nonlinear role played by $(\psi_{11}(t), \theta_{11}(t))$ is to excite the harmonic characterized by $(k_x = 0, k_z = 2\pi)$, or, in our notations (ψ_{02}, θ_{02}). Using (7.31), we find that the evolution of this harmonic is governed by:

$$\begin{cases} \dfrac{d\psi_{02}}{d\tau} + 4\pi^2 \mathcal{P}\psi_{02} = 0 \\[2mm] \dfrac{d\theta_{02}}{d\tau} + 4\pi^2 \theta_{02} = -(k\pi/2)\psi_{11}\theta_{11} \end{cases} \qquad (7.61)$$

We should notice here that ψ_{02} is always damped very quickly and it would just be the same for θ_{02} if the nonlinear forcing were absent. If the forcing term evolves on a time scale large compared to unity, the derivative $d\theta_{02}/d\tau$ is always small compared to $4\pi^2\theta_{02}$; thus we may write

$$\theta_{02} = -\frac{k}{8\pi}\psi_{11}\theta_{11} \qquad (7.62)$$

In this case the mode (ψ_{02}, θ_{02}) is said to be a *slaved mode*. It closely follows the evolution imposed by the mode (ψ_{11}, θ_{11}) *as long as this evolution is slow*. We thus understand the reason why we chose a slightly supercritical Rayleigh number: the growth of the unstable mode last on a long time scale compared to unity. Finally, we also observe that θ_{02} is $\mathcal{O}(\varepsilon'^2)$. Moreover, since ψ_{02} is rapidly damped, we can set it to zero hereafter.

We may seek for other nonlinear effects. The most important comes from the interaction between (ψ_{11}, θ_{11}) and $(0, \theta_{02})$. This interaction modifies the evolution of θ_{11}. Indeed,

$$J[\theta_{02}\sin 2\pi z, \; \psi_{11}\sin \pi z \sin kx] = -\pi k\theta_{02}\psi_{11}(\sin 3\pi z - \sin \pi z)\cos kx \qquad (7.63)$$

thus

$$\frac{d\theta_{11}}{d\tau} - k\psi_{11} + (\pi^2 + k^2)\theta_{11} = \pi k\theta_{02}\psi_{11}$$

Moreover, (7.63) shows that the mode (θ_{13}, ψ_{13}) is excited as well, and thus interact with other modes, etc. Hence, we see that a whole chain of mode is excited. However, this chain can be truncated thanks to the small amplitude hypothesis. At this stage of the analysis, it is necessary to evaluate the order of magnitude of each term. Each nonlinear interaction increases the order of the term, and if we keep only terms up to the third order in ε', three equations are necessary:

$$\begin{cases} \dfrac{d\psi_{11}}{d\tau} - \mathcal{P}\mathrm{Rak}/\ell^2\theta_{11} + \mathcal{P}\ell^2\psi_{11} = 0 \\[2mm] \dfrac{d\theta_{11}}{d\tau} - k\psi_{11} + \ell^2\theta_{11} = \pi k\theta_{02}\psi_{11} \\[2mm] \dfrac{d\theta_{02}}{d\tau} + 4\pi^2\theta_{02} = -k\pi/2\psi_{11}\theta_{11} \end{cases} \tag{7.64}$$

7.7.3   Derivation of the Amplitude Equation

The three equations (7.64) are those which control the dynamics of the system just above the threshold of instability. They contain three modes: two are stable and one is unstable. The following manipulations aim at isolating the amplitude equation controlling the amplitude $A(\tau)$ of the unstable mode. The two other modes are slaved to the unstable one.

We first insert the solution (7.62) into the second equation of (7.64). The system changes into

$$\begin{cases} \dfrac{d\psi_{11}}{d\tau} - \mathcal{P}\mathrm{Rak}/\ell^2\theta_{11} + \mathcal{P}\ell^2\psi_{11} = 0 \\[2mm] \dfrac{d\theta_{11}}{d\tau} - k\psi_{11} + \ell^2\theta_{11} = -k^2\theta_{11}\psi_{11}^2/8 \end{cases} \tag{7.65}$$

We now write this new system in a compact form like:

$$\frac{d\mathbf{X}}{d\tau} = [L]\mathbf{X} + \mathbf{N} \tag{7.66}$$

where

$$\mathbf{X} = \begin{pmatrix} \psi_{11} \\ \theta_{11} \end{pmatrix}, \qquad [L] = \begin{pmatrix} -\mathcal{P}\ell^2 & \mathcal{P}\mathrm{Rak}/\ell^2 \\ k & -\ell^2 \end{pmatrix}, \qquad \mathbf{N} = -\frac{k^2}{8}\begin{pmatrix} 0 \\ \theta_{11}\psi_{11}^2 \end{pmatrix}$$

For the moment, we leave aside the nonlinear terms $\mathbf{N}$. The solution of the remaining linear system may be written as

$$\begin{pmatrix} \psi_{11} \\ \theta_{11} \end{pmatrix} = \begin{pmatrix} 1 \\ q_+ \end{pmatrix} A e^{\lambda_+\tau} + \begin{pmatrix} 1 \\ q_- \end{pmatrix} B e^{\lambda_-\tau} \tag{7.67}$$

where $\begin{pmatrix} 1 \\ q_- \end{pmatrix}$ and $\begin{pmatrix} 1 \\ q_+ \end{pmatrix}$ are two eigenvectors associated with the eigenvalues λ_- and λ_+ of the matrix $[L]$. Setting $A(\tau) = Ae^{\lambda_+\tau}$ and $B(\tau) = Be^{\lambda_-\tau}$, (7.67) looks like

$$\begin{pmatrix} \psi_{11} \\ \theta_{11} \end{pmatrix} = \begin{pmatrix} 1 & 1 \\ q_+ & q_- \end{pmatrix} \begin{pmatrix} A(\tau) \\ B(\tau) \end{pmatrix}, \qquad (7.68)$$

where

$$q_+ = \frac{k}{\lambda_+ + \ell^2} = \frac{k}{\ell^2} + \mathcal{O}(\varepsilon), \qquad q_- = \frac{k}{\lambda_- + \ell^2} = -\frac{k}{\mathcal{P}\ell^2} + \mathcal{O}(\varepsilon)$$

Equation (7.68) is just a change of the projection base. Let

$$[M] = \begin{pmatrix} 1 & 1 \\ q_+ & q_- \end{pmatrix} \qquad \Longleftrightarrow \qquad [M]^{-1} = \frac{1}{q_+ - q_-} \begin{pmatrix} -q_- & 1 \\ q_+ & -1 \end{pmatrix}$$

If we set $\mathbf{A} = \begin{pmatrix} A \\ B \end{pmatrix}$, (7.66) has the following shape

$$[M]\frac{d\mathbf{A}}{d\tau} = [L][M]\mathbf{A} + \mathbf{N}$$

Multiplying by the inverse of $[M]$, we transform $[L]$ into its diagonal form. Hence, we get

$$\frac{d}{d\tau}\begin{pmatrix} A \\ B \end{pmatrix} = \begin{pmatrix} \lambda_+ & 0 \\ 0 & \lambda_- \end{pmatrix}\begin{pmatrix} A \\ B \end{pmatrix} - \frac{k^2}{8}[M]^{-1}\begin{pmatrix} 0 \\ \theta_{11}\psi_{11}^2 \end{pmatrix}$$

which may be rewritten more explicitly as

$$\begin{cases} \dfrac{dA}{d\tau} = \lambda_+ A - \dfrac{\mathcal{P}\ell^2 k}{8(\mathcal{P}+1)}(A+B)^2(q_+A + q_-B) \\[4mm] \dfrac{dB}{d\tau} = \lambda_- B + \dfrac{\mathcal{P}\ell^2 k}{8(\mathcal{P}+1)}(A+B)^2(q_+A + q_-B) \end{cases} \qquad (7.69)$$

To solve this system, some remarks are in order. First, we have assumed a slightly supercritical Rayleigh number so that λ_+ is $\mathcal{O}(\varepsilon)$ and λ_- is $\mathcal{O}(1)$. If B were alone, it would be damped very quickly, but the slow growth of A let B "survive". Now, since A evolves slowly we may assume (and check later) that B does so. B changes on the same time scale as A, therefore $\frac{dB}{d\tau}$ is negligible compared to $\lambda_- B$. Just as

θ_{02} above, B is a slave mode. We are now left with a polynomial equation of cubic order in B, still difficult to solve. However, if we assume that the amplitude of B is very small compared to that of A, the solution is straightforward:

$$B = \frac{\mathcal{P}k^2}{8(\mathcal{P}+1)^2\ell^2}A^3 = \mathcal{O}(\varepsilon'^3)$$

Such a solution is consistent with our hypothesis $B \ll A$, which turns out to be the right one.[5] Finally, $A(\tau)$ verifies the differential equation

$$\frac{dA}{d\tau} = \lambda_+ A - \mathcal{L}A^3 . \tag{7.70}$$

This is *Landau equation* that we introduced in Chap. 6. The Landau constant can be computed explicitly, namely

$$\mathcal{L} = \frac{\mathcal{P}k^2}{8(1+\mathcal{P})} > 0$$

It is positive, showing that the bifurcation is supercritical. The diagram of this bifurcation is given in Fig. 7.8.

Let us now come back to the relation between ε and ε'. For consistency, each term in (7.70) needs to be of the same order of magnitude (otherwise one of them could

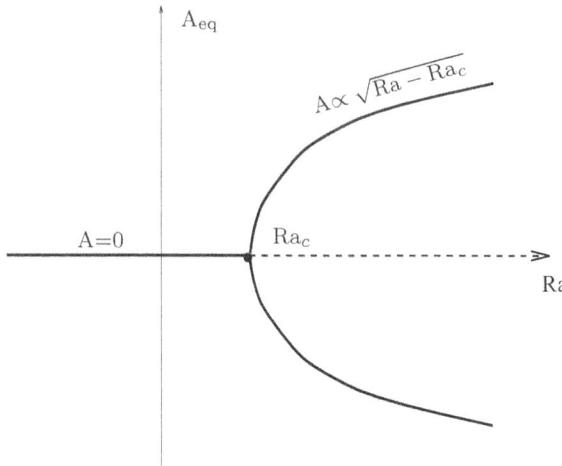

Fig. 7.8 The bifurcation diagram

---

[5]Other solutions may exist where B is not small compared to A, but they are uninteresting for us as we are focusing on the case where A grows first and B follows.

be suppressed); recalling that A is $\mathcal{O}(\varepsilon')$ and that it evolves on a time scale $\mathcal{O}(1/\varepsilon)$, $\frac{dA}{d\tau}$ and $\lambda_+ A$ are both $\mathcal{O}(\varepsilon\varepsilon')$ whereas $\mathcal{L}A^3$ is $\mathcal{O}(\varepsilon'^3)$. Hence, in order that all terms be of the same order (in the $\varepsilon, \varepsilon'$ expansion), it is necessary that $\varepsilon' = \mathcal{O}(\sqrt{\varepsilon})$. This relation enlights the link between the smallness of the growth rate and that of the amplitude.

Using the asymptotic amplitude, we have

$$A_{eq} = \sqrt{24\varepsilon} \tag{7.71}$$

From which we derive the amplitude of the slave mode θ_{02}

$$\theta_{02}(z) = -\frac{\varepsilon}{\pi} \sin 2\pi z$$

This mode gives the modification of the temperature profile when thermal convection sets in. This new profile has a steeper gradient near the boundaries and a weaker one in the middle of the layer. As shown in Fig. 7.9, thermal layers appear near the walls. We now understand the mechanism by which the instability saturates. On the one hand the flow reduces the temperature gradient in the central part of the layer, thus locally lowering the Rayleigh number, on the other hand, a stronger gradient near the wall arises but it is not destabilizing since it applies to a thinner layer (recall that Ra varies like d^3).

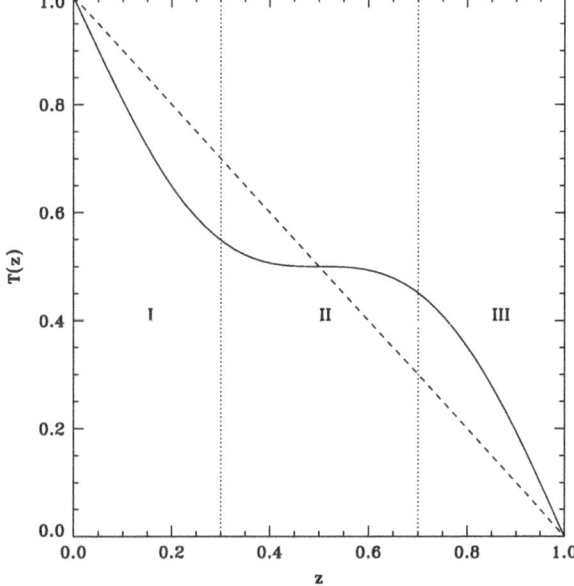

Fig. 7.9 The temperature profile above the threshold; the amplitude $\varepsilon = 0.5$ has been strongly exaggerated so as to clearly show the three layers appearing in the convecting fluid. The I and III zones show the thermal "boundary layers" while the middle layer II is a region with a quasi-adiabatic profile

7.7.4 Heat Transport: The Nusselt Number

We can now compute the heat flux between the two plates and compare it to the situation where the fluid is at rest. The ratio of these two fluxes is called the *Nusselt number*. Following this definition, we write

$$\text{Nu} = \frac{\text{Heat flux with convection}}{\text{Heat flux without convection}} \quad \Longleftrightarrow \quad \text{Nu} = \frac{\int_{(S)} (\rho c_p T \mathbf{v} - \chi \nabla T) \cdot d\mathbf{S}}{\int_{(S)} (-\chi \nabla T_{eq}) \cdot d\mathbf{S}}$$

Let us compute this number when the Rayleigh number is slightly supercritical. Since T_{eq} is solely a function of z, the Nusselt number depends only on horizontal means which we define as:

$$\langle f \rangle = \frac{\int_{(S)} f \, dS}{\int_{(S)} dS}$$

Using dimensionless quantities and noting that $d_z \theta_{eq} = -1$, then

$$\text{Nu} = \langle (\theta + \theta_{eq}) u_z - \partial_z (\theta + \theta_{eq}) \rangle \tag{7.72}$$

so that

$$\text{Nu} = 1 + \langle \theta u_z \rangle - \partial_z \theta_{02}$$

but

$$\langle \theta u_z \rangle = k_c \theta_{11} \psi_{11} (1 - \cos 2\pi z)/4 = \frac{k_c^2 A^2}{4\ell^2} (1 - \cos 2\pi z)$$

Finally,

$$\text{Nu} = 1 + 2\varepsilon \tag{7.73}$$

Showing that, near the threshold of instability, the Nusselt number increases linearly with ε or equivalently with the difference $\text{Ra} - \text{Ra}_c$.

7.8   Fixed Flux Convection ♣

7.8.1   *Introduction*

In all the foregoing matter, we considered that the plates limiting the fluid are
perfect heat conductors, so that their temperature remained constant (fixed by a
thermostat). Hence, we demanded that the temperature fluctuations vanished on the
boundaries.

In the case of a laboratory experiment the boundary conditions are not so
simple. As mentioned in Sect. 1.8.2 the only conditions that have to be satisfied
are

$$\chi_f \frac{dT_f}{dz} = \chi_s \frac{dT_s}{dz}, \qquad T_f = T_s$$

where the index f refers to the fluid and the index s to the solid. However, the
temperature field in the solid is not known and needs to be computed as well. The
general case is thus quite tedious and we refer the reader to the work of Hurle et al.
(1966) for a detailed study.

Here, we shall concentrate on the limit $\chi_s/\chi_f \to 0$ which is the case where the
solid is a very poor heat conductor compared to the fluid. This case corresponds to
the ideal insulator. Hence, after studying the ideal conductor case, we now explore
the other extreme. From a physical point of view, it means that the temperature
field in the solid is fixed (or evolve on a very long time scale compared to that of
the fluid). Thus, the temperature gradient, and therefore the energy flux, is fixed in
the solid and the temperature fluctuations at the interface do not propagate inside
the solid. Hence, one imposes that the temperature gradient does not fluctuate,
or that

$$\frac{\partial \theta}{\partial z} = 0 \qquad\qquad\qquad (7.74)$$

on $z = 0, 1$. We note that in such conditions the Nusselt number remains fixed to
unity.

The interesting point of this system is that the convective instability occurs
with a vanishing critical wavenumber. Hurle et al. (1966) indeed noticed that as
$\chi_s/\chi_f \to 0$ then $k_c \to 0$. Convection sets in at a scale all the larger that the
solid is less conductive. It is then possible to find out a weakly nonlinear solution
taking advantage of the fact that the horizontal scale is very large compared to
the height. The resolution of this problem is a typical example of *a multi-scale
analysis.*

7.8.2   *Formulation*

We start again from the equation of motion (7.31) and introduce the small parameter ε such that:

$$\psi = \varepsilon \phi, \quad \partial_x = \varepsilon \partial_X, \quad \partial_t = \varepsilon^4 \partial_\tau, \quad \mathrm{Ra} = \mathrm{Ra}_c + \mu^2 \varepsilon^2$$

where μ measures the rate of supercriticality. Thus doing, we rescaled the horizontal lengths, introducing the scaled variable $X = \varepsilon x$ of order unity. We also rescaled the time and introduced the new time variable $\tau = \varepsilon^4 t$. Hence, we can focus on very large horizontal scales and very long time scales. The choice of the ε^4 factor in the time scale is justified a posteriori by the consistency of the solutions. The two equations of (7.31) now read:

$$\varepsilon^6 \partial_\tau \partial_X^2 \phi + \varepsilon^4 \left(\partial_\tau D^2 \phi + \partial_X \phi \partial_X^2 D\phi - \partial_X^3 \phi D\phi \right) + \varepsilon^2 \left(\partial_X \phi D^3 \phi - \partial_x D^2 \phi D\phi \right) =$$
$$\mathcal{P} \left[(Ra_c + \mu^2 \varepsilon^2) \partial_X \theta + D^4 \phi + 2\varepsilon^2 \partial_X^2 D^2 \phi + \varepsilon^4 \partial_X^4 \phi \right]$$

$$\varepsilon^4 \partial_\tau \theta + \varepsilon^2 \left(\partial_X \phi D\theta - \partial_x \theta D\phi \right) = \varepsilon^2 \partial_X \phi + (D^2 + \varepsilon^2 \partial_X^2)\theta$$

Here, the functions depends on the three variables (τ, X, z). The boundary conditions at $z = \pm \frac{1}{2}$ are[6]

$$D\theta = 0$$

for the temperature and

$$u_z = \varepsilon^2 \partial_X \phi = 0 \quad \text{and} \quad \sigma_{xz} = 0 \iff D^2 \phi = 0$$

for the velocity. Note that we chose the stress-free boundary conditions; for no-slip conditions we would ask $u_x = 0$ or $D\phi = 0$.

7.8.3   *The Chapman–Proctor Equation*

We now develop the solution in powers of the small parameter up to the fourth order,

$$\theta = \theta_0 + \varepsilon^2 \theta_2 + \varepsilon^4 \theta_4 + \cdots, \qquad \phi = \phi_0 + \varepsilon^2 \phi_2 + \varepsilon^4 \phi_4 + \cdots$$

---

[6]We place the boundaries at $z = \pm \frac{1}{2}$ rather than at $z = 0, 1$ so as to be able to use the symmetry or the anti-symmetry of the functions with respect to the $z = 0$ plane.

Note that with the choice made on the amplitudes and the horizontal scales, the velocity field is $\mathcal{O}(\varepsilon)$, or

$$\mathbf{u} = -\frac{\partial \psi}{\partial z}\mathbf{e}_x + \frac{\partial \psi}{\partial x}\mathbf{e}_z = -\varepsilon D\phi\mathbf{e}_x + \varepsilon^2 \partial_X \phi\mathbf{e}_z$$

whereas the temperature field remains $\mathcal{O}(1)$.

At zeroth order, the equations of motion reduce to

$$D^2\theta_0 = 0 \quad\text{and}\quad \text{Ra}_c\partial_X\theta_0 + D^4\phi_0 = 0$$

which lead to the following type of solution

$$\theta_0 = f(X, \tau) \quad\text{and}\quad \phi_0 = \text{Ra}_c\, P(z)f'(X, \tau)$$

where $D^4 P(z) = -1$. $f(X, \tau)$ is an unknown function which needs to be determined; $f'(X, \tau)$ is its derivative with respect to X. We note that the boundary conditions on θ_0 are automatically satisfied whereas those on the velocity demand that $P(\pm\frac{1}{2}) = 0$ and $P'(\pm\frac{1}{2}) = 0$ for no-slip conditions or $P''(\pm\frac{1}{2}) = 0$ for stress-free ones. These last two conditions and the differential equation allow us to specify completely the function $P(z)$. In the no-slip case

$$P(z) = -\frac{1}{24}z^4 + \frac{z^2}{48} - \frac{1}{384} = -\frac{1}{24}\left(z^2 - \frac{1}{4}\right)^2$$

while in the stress-free one

$$P(z) = -\frac{1}{24}z^4 + \frac{z^2}{16} - \frac{5}{384}.$$

Let us now consider the ε^2-order of the temperature equation. We have

$$D^2\theta_2 = -\text{Ra}_c DPf'^2 - (\text{Ra}_c P + 1)f'' \tag{7.75}$$

This equation is interesting as it has a solution only if the right-hand side verifies a solvability condition. Indeed, if we integrate the equation on z, then the left-hand side is zero whereas the right-hand side implies:

$$\text{Ra}_c = -\left(\int_{-1/2}^{+1/2} P(z)dz\right)^{-1}$$

giving the value of the critical Rayleigh number. This expression leads to the numerical values $\text{Ra}_c = 720$ in the no-slip case and $\text{Ra}_c = 120$ in the free-slip one, values which were first derived by Hurle et al. (1966).

The equation (7.75) can now be solved. We find

$$\theta_2 = f_2(X, \tau) + W(z)\, f'^2 + Q(z)\, f''$$

where we introduced $W(z)$ and $Q(z)$ such that

$$W'' + \text{Ra}_c\, P' = 0 \qquad \text{and} \qquad Q'' + \text{Ra}_c\, P + 1 = 0$$

These new functions verify the boundary conditions $Q'(\pm\frac{1}{2}) = W'(\pm\frac{1}{2}) = 0$ since $D\theta = 0$ on the boundaries. We infer that $W' = -\text{Ra}_c\, P$.

The ε^2-order of the momentum equation leads to

$$D^4\phi_2 = -\mu^2 f' - \text{Ra}_c\left[f_2' + W(f'^2)' + (Q + 2P'')f''' \right]$$
$$+ \frac{\text{Ra}_c^2}{P}\left[PP''' - P'P'' \right] f'f''$$

which is solved in the same way as the equation for θ_2; we find

$$\phi_2 = \mu^2 P f' + \text{Ra}_c\, P f_2' + U f''' + S f' f''$$

with

$$D^4 U = -\text{Ra}_c(Q + 2P'') \qquad \text{and} \qquad D^4 S = -2\text{Ra}_c W + \frac{\text{Ra}_c^2}{P}\left(PP''' - P'P'' \right)$$

The boundary condition $u_z = 0$ imposes that

$$U(\pm 1/2) = S(\pm 1/2) = 0$$

The last step consists in writing the fourth order ε^4-term of the temperature equation. Integrating this equation on z between $\pm\frac{1}{2}$, we obtain

$$\partial_\tau f + A\mu^2 f'' + B f^{(4)} + C\left(f'^3 \right)' + D\left(f'f'' \right)' = 0 \tag{7.76}$$

which is the *Chapman–Proctor equation*. It controls the horizontal dynamics of small-amplitude convection at fixed flux (Chapman and Proctor 1980). The constants A, B, C, D are given by

$$A = \frac{1}{\text{Ra}_c}, \qquad B = -\int_{-1/2}^{1/2} (U + Q)\, dz,$$

$$C = -\text{Ra}_c^2 \int_{-1/2}^{1/2} P^2\, dz, \qquad D = \int_{-1/2}^{1/2} (\text{Ra}_c\, PQ' - S - 2W)\, dz$$

The evaluation of the foregoing integrals is not straightforward. Let us illustrate their derivation in the case of stress-free conditions on both boundaries. Because of the symmetry of the set-up $D = 0$. The calculation of B is a little tedious. We first remark that

$$-\int_{-1/2}^{1/2} U\,dz = \int_{-1/2}^{1/2} UD^4 P\,dz = \int_{-1/2}^{1/2} PD^4 U\,dz$$

Then, the differential equation verified by U implies that

$$B = \int_{-1/2}^{1/2} \left(2\text{Ra}_c (DP)^2 - (DQ)^2\right) dz$$

Noting that

$$DQ = z^5 - \frac{5}{2}z^3 + \frac{9}{16}z$$

we finally obtain

$$B = \frac{1091}{5544} \simeq 0.197$$

In the same way, one can derive that

$$C = -\frac{155}{126}$$

7.8.4 Properties of the Small-Amplitude Convection

Chapman–Proctor's equation gives a good description of the dynamics when the temperature gradient is slightly supercritical.

To start with, let us examine the linear case and search for a solution proportional to $e^{\lambda t}$; if $\mu = 0$ (i.e. $\text{Ra} = \text{Ra}_c$), then the growth rate of a disturbance is just $-Bk^4$ and the critical wavenumber is $k = 0$ as expected. If the Rayleigh number is now slightly supercritical, we may linearize (7.76) and find the dispersion relation

$$\lambda = k^2 A\mu^2 - Bk^4 \tag{7.77}$$

which shows that the wavenumber of the fastest growing mode is

$$k_m = \frac{\mu}{\sqrt{2BRa_c}} \tag{7.78}$$

This results shows that the fastest growing mode is not necessarily the one with the critical wavenumber. In the present case, the mode with the critical wavenumber ($k = 0$) has a zero growth rate!

We should also note that the transition from the hydrostatic state to the convective one is independent of the Prandtl number. The growth rate is real so that convection is steady (no oscillation).

Let us now examine the nonlinear régime. If the boundary conditions are identical on the top and bottom plates, the solution is symmetric with respect to the mid-layer $z = 0$ plane. The integrand defining D is antisymmetric and thus D should be zero in this case. The Chapman–Proctor equation therefore simplifies in this case and may be written

$$\partial_\tau g + A\mu^2 g'' + B g^{(4)} + C \left(g^3\right)'' = 0$$

where we took the derivative of the equation and set $g = f'$. Now, introducing the new variable $u = \sqrt{\frac{C}{A}\frac{g}{\mu}}$ and changing the time scale as well as the X-scale, we find the Cahn–Hilliard equation:

$$u_t = -u'' - \mathcal{B}u^{(4)} + \left(u^3\right)'' \tag{7.79}$$

where $\mathcal{B} = BC^{1/2}/A^{5/2}\mu^5$. This equation was uncovered by John Cahn and John Hilliard in 1958 when they studied the dynamics of the phase separation phenomena.[7]

We note that this equation, as the Chapman–Proctor one, is richer than Landau equation which allowed us to study the nonlinear evolution of disturbances leading to convection rolls. The Landau equation indeed controls the time evolution of the amplitude of perturbations (whose structure is fixed by the linear analysis), while the two foregoing equations control both the time evolution and the spatial structure of the solutions (being partial differential equations). They are much simpler than the original ones, but still contain a rich variety of solutions. For instance, one can solve the Cahn–Hilliard equation in a stationary case (the solution is expressed with elliptic integrals) and then study the stability of these nonlinear solutions. Chapman and Proctor have shown that, in a periodic box, the stable flow is made of very flattened contra-rotating rolls.

---

[7]Cahn and Hilliard (1958).

7.9   The Route to Turbulent Convection

7.9.1   The Lorenz Model

System (7.64) has been derived using the hypothesis of a slightly supercritical Rayleigh number. However, if we "forget" this restriction, we have at our disposal a set of three nonlinear equations which control a very rich variety of solutions, actually. They may give us interesting informations on the development of convection when we increase the Rayleigh number, if the modes that are dynamically active remain limited to these three ones.

In order to show the parameters which control this system, it is useful to make the following change of variables:

$$
\begin{cases}
\psi_{11} = \dfrac{\ell^2 \sqrt{2}}{\pi k} X, \qquad \theta_{11} = \dfrac{\ell^6 \sqrt{2}}{\pi k^2 \mathrm{Ra}} Y \\[3mm]
\theta_{02} = -\dfrac{\ell^6}{\pi k^2 \mathrm{Ra}} Z, \quad \tau = t/\ell^2
\end{cases}
\tag{7.80}
$$

We thus find the equation of *the Lorenz system:*

$$
\begin{cases}
\dfrac{dX}{dt} = \mathcal{P}(Y - X) \\[3mm]
\dfrac{dY}{dt} = rX - Y - XZ \\[3mm]
\dfrac{dZ}{dt} = -bZ + XY
\end{cases}
\tag{7.81}
$$

where we set

$$
r = \frac{\mathrm{Ra}}{\mathrm{Ra}_c} \qquad \text{and} \qquad b = \left(\frac{2\pi}{\ell}\right)^2
\tag{7.82}
$$

since $\mathrm{Ra}_c = \ell^6/k^2$. Three parameters control the Lorenz system: the Prandtl number $\mathcal{P}$, the reduced Rayleigh number r and b which measures the aspect ratio of the convection cells (the ratio between their height and their width).

As the Landau equation, this system has one or three fixed points (points of "equilibrium"), namely

$$
X = Y = Z = 0 \qquad \text{if} \qquad r < 1
$$

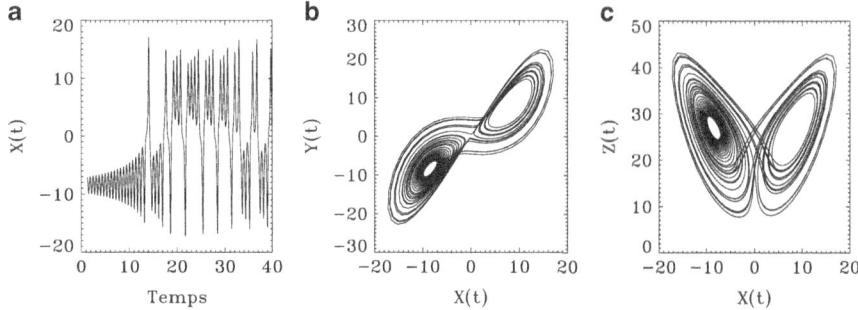

Fig. 7.10 Three representations of the time evolution of the Lorenz system in a chaotic régime: (**a**) X(t) shows the evolution of the X variable, especially the growth of the amplitude due to the Hopf bifurcation and the start of the chaotic régime. (**b, c**) give for the same time interval, the evolution of the system in the phase space showing evidence of a strange attractor

or

$$X = Y = Z = 0$$

or   if $r \geq 1$

$$X = Y = \pm\sqrt{b(r - 1)} \qquad \text{and} \qquad Z = r - 1$$

The first fixed point $(X = Y = Z = 0)$ turns unstable when the critical value $r = 1$ is overpassed. The new fixed point is linearly stable in the interval $[1, r_c[$ with $r_c = 24.74$. At $r = r_c$ there is a subcritical Hopf bifurcation and the system evolves towards a chaotic state where one finds the famous Lorenz attractor. This situation is illustrated in Fig. 7.10 where we clearly see the exponential growth and the beginning of a chaotic sequence. The subcritical nature of the bifurcation indicates that one may find a chaotic state[8] when $r < r_c$. A study of the stability of the branch $X = Y = \pm\sqrt{b(r - 1)}$ shows that the chaotic state disappears when $r < 13.926$ (if $b = 8/3$ and $\mathcal{P} = 10$).

7.9.2   The Domain of Very Large Rayleigh Numbers

In nature thermal convection usually appears with very large Rayleigh numbers because of the large size of the systems. For instance, at the Sun's surface, the convective cells (see Fig. 7.11) are controlled by a Rayleigh number larger than 10^{20}. Hence, many studies have explored the properties of thermal convection when $Ra \gg Ra_c$.

---

[8]This is a metastable chaos.

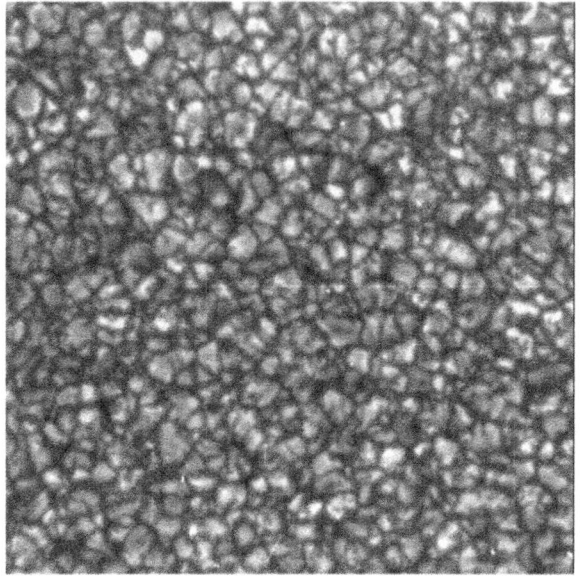

Fig. 7.11 Convection at the Sun's surface: the rising gas is hot and appears brighter than the cold downflowing gas. The temperature difference between the hot and "cold" gas is about 200 K around an average of 5,800 K. This gives the granular aspect to the surface. Solar convection cells are thus called granules. Their size is about 1,000 km and their lifetime less than 10 min (Credit T. Roudier, Lunette Jean Rösch - Observatoire Midi-Pyrénées)

When one progressively increases the Rayleigh number, for instance by increasing the temperature difference, the foregoing solutions are destabilized and after a few bifurcations a chaotic régime may set in. If the temperature difference is still increased, convection reaches a turbulent régime: a continuous spectrum of spatial and temporal scales appears.

Although the turbulent régime is very complicated (see Chap. 9), we may expect that some simple laws govern the mean quantities. For instance, the heat flux is a typical quantity of interest when one deals with turbulent convection. We may wonder whether there exists any asymptotic law governing this quantity when $\mathrm{Ra} \to \infty$. This question is still open, but some simple models may give us a first description of this asymptotic state. One of them (see Fig. 7.12) considers that the essential part of the temperature drop across the layer occurs in thin boundary layers attached to the bounding plates.

The thickness δ of the boundary layers is such that these layers are stable with respect to the convection, thus

$$\frac{\delta^3 \alpha g \Delta T}{\nu \kappa} \approx \mathrm{Ra}_c$$

but, by definition $\mathrm{Ra} = \frac{d^3 \alpha g \Delta T}{\nu \kappa}$ so

$$\frac{\delta}{d} \approx \left(\frac{\mathrm{Ra}_c}{\mathrm{Ra}} \right)^{1/3}$$

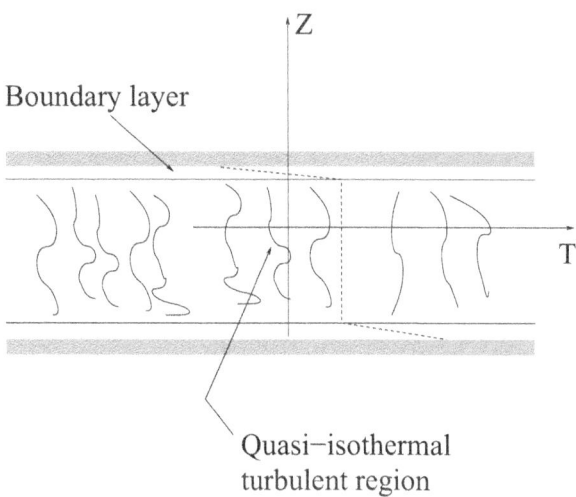

Fig. 7.12 Schematic view of turbulent convection between two plates

The Nusselt number is proportional to the ratio of the total flux with convection and the flux without convection. We use this number to measure the heat flux. We note that the flux without convection is proportional to $\Delta T/d$ whereas the flux with convection is proportional to $\Delta T/\delta$. Indeed, the boundary layers carry by conduction the whole heat flux. We thus write

$$\text{Nu} \propto \frac{d}{\delta} \sim \left(\frac{\text{Ra}}{\text{Ra}_c}\right)^{1/3}$$

which shows that the Nusselt number grows like the one-third power of the Rayleigh number.

Many experiments have attempted to find out the actual scaling law and eventually confirm the foregoing approach. For instance, Niemela et al. (2000) explored the relation between Nu and Ra using helium. While varying Ra between 10^6 and 10^{17}, they found that

$$\text{Nu} \simeq 0.124 \, \text{Ra}^{0.31}$$

in this range of Rayleigh numbers (see Fig. 7.13). But more recently Ahlers et al. (2012) using another gas (sulfur hexafluoride) found this law solely when $\text{Ra} \lesssim 10^{13}$, suggesting that beyond this value variations of the Prandtl number (see Fig. 7.13) influence the scaling law.

Fig. 7.13 Nusselt number versus Rayleigh number in the experiment of Niemela et al. (2000)

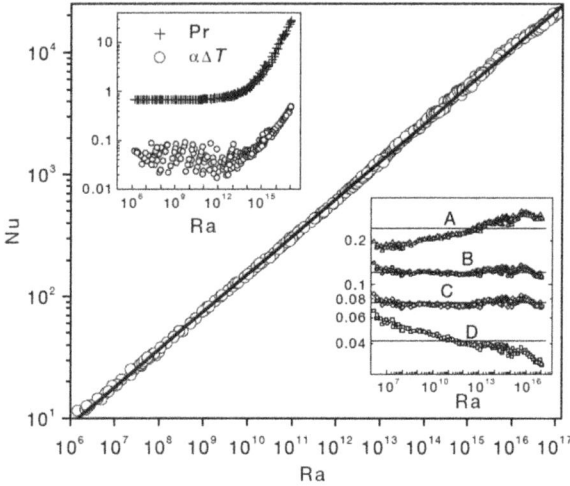

7.10 Exercises

1. a) Using the differential form of the enthalpy, show that for an ideal gas in equilibrium in a gravity field **g**,

$$\nabla T - (\nabla T)_{ad} = \frac{T}{c_p}\nabla s$$

   b) Derive (7.9).

2. From the values of the temperature gradient in the stratosphere what can we say about the convective stability of this layer? Use the values of the standard atmosphere given in table 2.1.

3. In the case of convection between two vertical plates described in Sect. 7.4.1, compute the Reynolds number of this flow. Give the numerical values, taking $T_c - T_f = 20$ K, d $= 1$ cm, $(T_c + T_f)/2 = 283$ K, $\nu = 10^{-5}$ m²/s (case of air) and g $= 9.8$ m/s². What do you conclude?

4. Compute the maximum Reynolds number of thermal convection near the threshold when using stress-free boundary conditions.

5. For the Lorenz system, show that the solution $X = Y = Z = 0$ is unstable when $r > 1$.

6. Here is a practical exercise: In a pan with, preferably, a white bottom, dispose a thin layer of oil (sunflower for instance) of 2 or 3 mm thick. Add a small amount of cocoa powder (less than half a tea spoon), an mix vigorously so as to obtain an homogeneous mixture. Put the pan on a cold electric heater and wait for the fluid be at complete rest. Then turn on the heater at minimum power, after a few minutes, a network of (nearly) hexagonal cells appear.

Further Reading

For all the questions around the problems of linear convective instability in various configuration (with rotation, magnetic field or in spherical geometry), one should consult *Hydrodynamic and hydromagnetic stability* by S. Chandrasekhar. To a lesser extent, this problem is also discussed in *Hydrodynamic stability* by Drazin and Reid (1981). More about the Lorenz attractor may be found in *Order within chaos* by Bergé et al. (1984). Another side of thermal convection not discussed in this book, namely heat transfer associated with fluid flows, may be found in *Convection Heat Transfer* by Bejan (1995).

References

Ahlers, G., He, X., Funfschilling, D. & Bodenschatz, E. (2012). Heat transport by turbulent Rayleigh–Bénard convection for $Pr \simeq 0.8$ and $3 \times 10^{12} \lesssim Ra \lesssim 10^{15}$: Aspect ratio $\Gamma = 0.50$. *New Journal of Physics, 14*(10), 103012.

Bejan, A. (1995). *Convection heat transfer*. New York: Wiley.

Bergé, P., Pomeau, Y. & Vidal, C. (1984). *Order within chaos*. New York: Wiley.

Cahn, J., & Hilliard, J. (1958). Free energy of a non-uniform system I. Interfacial free energy. *The Journal of Chemical Physics, 28*, 258–267.

Chandrasekhar, S. (1961). *Hydrodynamic and hydromagnetic stability*. Oxford: Clarendon Press.

Chapman, C.J., Proctor, M.R.E. (1980). Nonlinear Rayleigh-Benard convection between poorly conducting boundaries. *Journal of Fluid Mechanics, 101*, 759–782.

Drazin, P. & Reid, W. (1981). *Hydrodynamic stability*. Cambridge: Cambridge University Press.

Hurle, D., Jakeman, E. & Pike, E. (1966). On the solution of the bénard problem with boundaries of finite conductivity. *Proceedings of the Royal Society of London A, 225*, 469–475.

Niemela, J. J., Skrbek, L., Sreenivasan, K. R. & Donnelly, R. J. (2000). Turbulent convection at very high Rayleigh numbers. *Nature, 404*, 837–840.

Chapter 8
Rotating Fluids

8.1 Introduction

The most spectacular effect of rotation on a fluid flow is certainly the huge hurricanes surging up in the Earth's atmosphere when the waters of the ocean are warm enough. These huge flows, so typical in pictures of the Earth, would not exist if the Earth were not rotating. They owe their existence to the Coriolis acceleration.

In this chapter we wish to introduce the reader to the fundamentals of fluid dynamics in a rotating frame. Rotating fluids are indeed those fluids whose motion is essentially a solid body rotation supplemented by a small velocity field. Thus, even if hurricanes generate terrific winds, let say with speeds of 60 m/s, this is still small compared to the Earth rotation velocity (460 m/s). Such a velocity field is thus conveniently analysed in a rotating frame. As we shall see, all the novelties come from the Coriolis force, which deeply modifies the dynamics, imposing the quasi-bidimensionality of steady flows, generating new sorts of waves, new boundary layers, etc.

8.1.1 The Equation of Motion

The basic change in the equations governing a fluid flow in a rotating frame comes from the existence of inertial forces associated with the Coriolis and centrifugal accelerations. Thus, the equation of momentum is the only one to be modified. Its expression is easily derived from Newton equation, which controls the motion of a point mass particle. Let $\mathbf{r}$ be the position of the particle; it evolves according to

$$\rho \frac{d^2 \mathbf{r}}{dt^2} = \mathbf{f}$$

© Springer International Publishing Switzerland 2015
M. Rieutord, *Fluid Dynamics*, Graduate Texts in Physics,
DOI 10.1007/978-3-319-09351-2_8

in a galilean frame. When the frame rotates at an angular velocity $\boldsymbol{\Omega}$ the same equation reads

$$\rho \left(\frac{d^2 \mathbf{r}}{dt^2} + 2\boldsymbol{\Omega} \times \frac{d\mathbf{r}}{dt} + \boldsymbol{\Omega} \times (\boldsymbol{\Omega} \times \mathbf{r}) \right) = \mathbf{f}$$

This equation also gives the trajectory of a fluid particle according to the Lagrangian approach. Going back to the eulerian formalism, the preceding equation is translated into

$$\rho \left(\frac{D\mathbf{v}}{Dt} + 2\boldsymbol{\Omega} \times \mathbf{v} + \boldsymbol{\Omega} \times (\boldsymbol{\Omega} \times \mathbf{r}) \right) = \mathbf{f} \tag{8.1}$$

which gives the evolution of the velocity field. This expression could have been derived directly from the one we met in the first chapter; however, this derivation is lengthy and left to the reader as an exercise.

8.1.2   New Numbers

The importance of rotation may be appreciated if we use the right non-dimensional numbers. For this, we first introduce a length scale L, a velocity scale V and a time scale that we relate to rotation. This time scale is $(2\Omega)^{-1}$. 2Ω is known as *the Coriolis frequency*. In order to concentrate on the effects of rotation, we shall consider a simple fluid like the incompressible viscous fluid.

The momentum and continuity equation read:

$$\rho \left(\frac{D\mathbf{v}}{Dt} + 2\boldsymbol{\Omega} \times \mathbf{v} + \boldsymbol{\Omega} \times (\boldsymbol{\Omega} \times \mathbf{r}) \right) = -\nabla P + \mu \Delta \mathbf{v} \tag{8.2}$$

$$\nabla \cdot \mathbf{v} = 0 \tag{8.3}$$

where we left aside an eventual gravity force. If we observe that

$$\boldsymbol{\Omega} \times (\boldsymbol{\Omega} \times \mathbf{r}) = -\nabla[\frac{1}{2}(\boldsymbol{\Omega} \times \mathbf{r})^2]$$

namely, the fact that the centrifugal acceleration may be derived from a potential, then, we can rewrite the momentum equation as:

$$\frac{D\mathbf{v}}{Dt} + 2\boldsymbol{\Omega} \times \mathbf{v} = -\nabla \Pi + \nu \Delta \mathbf{v} \tag{8.4}$$

where $\Pi = P/\rho - \frac{1}{2}(\mathbf{\Omega} \times \mathbf{r})^2$ is called the *reduced pressure*. We are now in a position to use non-dimensional quantities and we find:

$$\frac{\partial \mathbf{u}}{\partial \tau} + \mathbf{e}_z \times \mathbf{u} + \mathrm{Ro}\,\mathbf{u} \cdot \nabla \mathbf{u} = -\nabla p + \mathrm{E}\Delta \mathbf{u} \tag{8.5}$$

where we set $\mathbf{\Omega} = \Omega \mathbf{e}_z$ and $p = \Pi/(2\Omega L V)$. Two numbers appeared:

$$\mathrm{E} = \frac{\nu}{2\Omega L^2} \quad \text{and} \quad \mathrm{Ro} = \frac{V}{2\Omega L} \tag{8.6}$$

which are respectively the *Ekman number* and the *Rossby number*. We note that the Ekman number measures the ratio of the viscous force to the Coriolis one, while the Rossby number shows the importance of the nonlinear advection terms with respect to the Coriolis acceleration.

When a fluid flow, in some inertial frame, is essentially a solid body rotation, we should write $\mathbf{V} = \mathbf{\Omega} \times \mathbf{r} + \mathbf{v}$ where $||\mathbf{v}|| \ll ||\mathbf{\Omega} \times \mathbf{r}||$. Since $||\mathbf{v}||$ is just the magnitude of the flow in the rotating frame, we see that flows dominated by rotation are such that their Rossby number is very small compared to unity.

We may observe that the Rossby and Ekman numbers decrease when the scale of the flow increases. Rotation is therefore expected to be important in the large scales. Let us consider two examples: a wind of 20 m/s in the Earth atmosphere is dominated by the Earth rotation when it affects a scale larger than 140 km. For these scales, the Rossby number is less than unity. An ocean current, like the Gulf Stream, is even more affected by rotation since its speed is much lower, typically 1 m/s. For this value, rotation is important for all scales larger than 7 km. This shows that an oceanic current, spanning thousands of kilometers, is very much dominated by the effects of rotation.

Now, if we turn to the Ekman number, it is usually extremely small. For instance, a water flow with a scale of 7 km, has an Ekman number around 10^{-10}. This implies, as we shall see, the existence of very thin boundary layers.

8.2   The Geostrophic Flow

8.2.1   Definition

The geostrophic flow is a steady flow where the viscous force and the nonlinear terms play a negligible part. The momentum equation is therefore reduced to

$$\rho 2\mathbf{\Omega} \times \mathbf{v} = -\nabla P \tag{8.7}$$

This is called *the geostrophic balance*. The pressure gradient balances the Coriolis force.

8.2.2 The Taylor–Proudman Theorem

The geostrophic flow has one remarkable property: it is independent of the
coordinate parallel to the rotation axis. Indeed, let us take the curl of (8.7); we find

$$\nabla \times (2\mathbf{\Omega} \times \mathbf{v}) = 0 \quad \Longleftrightarrow \quad (\mathbf{\Omega} \cdot \nabla)\mathbf{v} = 0 \quad \Longleftrightarrow \quad \frac{\partial \mathbf{v}}{\partial z} = 0 \qquad (8.8)$$

where we used (12.41). The velocity field therefore only depends on the coordinates
in the plane orthogonal to $\mathbf{\Omega}$. This result is known as *Taylor–Proudman Theorem*.

8.2.3 The Expression of the Geostrophic Flow

The geostrophic (8.7) can easily be solved. One finds:

$$\mathbf{v} = \frac{1}{2\Omega\rho}\mathbf{e}_z \times \nabla P + F(x, y)\mathbf{e}_z \qquad (8.9)$$

In this expression $F(x, y)$ is an arbitrary function to be determined with the
boundary conditions. This solution shows that the pressure also depends solely on
the plane coordinates. The pressure plays the role of a stream function since isobars
are also streamlines.

To further illustrate the properties of geostrophic flows, let us consider the case
where the rotating fluid is bounded by a surface defined by:

$$\begin{cases} z - f(x, y) = 0 & \text{if} \quad z \geq 0 \\ z + g(x, y) = 0 & \text{if} \quad z \leq 0 \end{cases} \qquad (8.10)$$

The outgoing (unnormalized) normal vector is

$$\begin{cases} \mathbf{n} = \mathbf{n}_{sup} = \nabla(z - f(x, y)) = \mathbf{e}_z - \nabla f \\ \mathbf{n} = \mathbf{n}_{inf} = -\nabla(z + g(x, y)) = -\mathbf{e}_z - \nabla g \end{cases} \qquad (8.11)$$

from which we derive the equality:

$$\mathbf{n}_{sup} - \mathbf{n}_{inf} + \nabla(f - g) = 2\mathbf{e}_z$$

However, on the bounding surface $\mathbf{n}_{sup} \cdot \mathbf{v} = 0$ or $\mathbf{n}_{inf} \cdot \mathbf{v} = 0$, but since $\mathbf{v}$ does not
depend on z, the foregoing equality may be used everywhere. Thus, taking the scalar
product with $\mathbf{v}$, we find

$$2v_z = 2F(x, y) = \mathbf{v} \cdot \nabla(f - g)$$

which we report in (8.9). This yields

$$\mathbf{v} = \frac{1}{4\Omega\rho}\left[\mathbf{n}_{sup} - \mathbf{n}_{inf} + \nabla(f - g)\right] \times \nabla P + \frac{1}{2}\mathbf{v} \cdot \nabla(f - g)\mathbf{e}_z$$

This new expression may be simplified if we note that

$$\mathbf{v} \cdot \nabla(f - g) = (\mathbf{e}_z \times \nabla P) \cdot \nabla(f - g)/(2\Omega\rho);$$

since $\nabla(f - g) \times \nabla P$ is parallel to $\mathbf{e}_z$. It turns out that

$$\mathbf{v} = \frac{1}{4\Omega\rho}(\mathbf{n}_{sup} - \mathbf{n}_{inf}) \times \nabla P \tag{8.12}$$

This expression may further be arranged as follows. If we take the scalar product of (8.12) with $\mathbf{n}_{inf}$, we find

$$(\mathbf{n}_{sup} \times \mathbf{n}_{inf}) \cdot \nabla P = 0$$

but $\mathbf{n}_{sup} \times \mathbf{n}_{inf} = \nabla(f + g) \times \mathbf{e}_z + \nabla f \times \nabla g$, so that

$$(\nabla h \times \mathbf{e}_z) \cdot \nabla P = 0 \quad \Longleftrightarrow \quad \nabla P \times \nabla h = \mathbf{0}$$

We introduced $h = f + g$ and observed that $\nabla f \times \nabla g$ and $\nabla h \times \nabla P$ are along $\mathbf{e}_z$. One should note that $h(x, y)$ is just the height of the container at (x, y). The foregoing relation shows that the pressure only depends on h. Noting that $\mathbf{n}_{sup} + \mathbf{n}_{inf} = -\nabla h$, (8.12) may be rewritten in its final form:

$$\mathbf{v} = \frac{1}{2\Omega\rho}\left(\frac{dP}{dh}\right)\mathbf{n}_{inf} \times \mathbf{n}_{sup} \tag{8.13}$$

This solution is valid only if the normal vectors are continuous in the x, y-plane. It may be observed that $\mathbf{v}$ is parallel to the curves of constant height since $\mathbf{v} \cdot \nabla h = 0$ because $\mathbf{n}_{inf} \times \mathbf{n}_{sup} = \mathbf{e}_z \times \nabla h$. These curves are also called *geostrophic contours*. Since they are streamlines they must be closed.

Another property of the geostrophic flow is that it possesses circulation around the rotation axis. Indeed, along a geostrophic contour

$$\oint_{(C)} \mathbf{v} \cdot \mathbf{dl} = \frac{1}{2\Omega\rho}\left(\frac{dP}{dh}\right)\oint_{(C)} ||\mathbf{n}_{inf} \times \mathbf{n}_{sup}||dl \neq 0 \tag{8.14}$$

Thus, in general, the geostrophic flow owns angular momentum.

8.2.4   Examples

8.2.4.1   The Geostrophic Flow in a Sphere

To give a simple illustration of the foregoing results, we now take the case of a geostrophic flow in a spherical container, which is typical of planetary or stellar situations. In this case, geostrophic contours are just circles of constant latitude and the velocity is constant on cylinders centered on the rotation axis.

Expliciting the results of the previous section, we note that for a sphere of radius R, the equations of the boundary are such that

$$f = g = \sqrt{R^2 - x^2 - y^2}$$

Letting $s^2 = x^2 + y^2$, the direction of the velocity is

$$\mathbf{n}_{inf} \times \mathbf{n}_{sup} = -\frac{2s}{\sqrt{R^2 - s^2}}\,\mathbf{e}_\varphi$$

which confirms that the velocity is purely azimuthal. If we now observe that $h = 2\sqrt{R^2 - s^2}$, the solution (8.13) gives

$$\mathbf{v} = \frac{1}{2\Omega\rho}\frac{\partial P}{\partial s}\mathbf{e}_\varphi \tag{8.15}$$

This relation could have been derived directly from (8.7), of course.

Solution (8.13) is more interesting when one deals with a more complicated geometry, like a spheroid for instance. One just needs to derive h and normal vectors from the shape of the surface boundary.

Let us note that if the sphere is truncated, like in Fig. 8.1b, some geostrophic contours are no longer closed. This ruins the existence of the geostrophic solution which disappears. As shown by Greenspan (1969), no steady state is possible, and the geostrophic flow is replaced by a set of Rossby waves, which form a subset of inertial modes (see below for their detailed presentation).

8.2.4.2   The Vortex of an Emptying Reservoir

When a reservoir like a bath tube is emptied, a strong vortex is often observed above the exit. The question of whether the rotation of this vortex is controlled by the Coriolis force due to the Earth rotation is often raised. Should the vortex rotate in opposite directions when one makes the experiment in the northern or southern hemisphere?

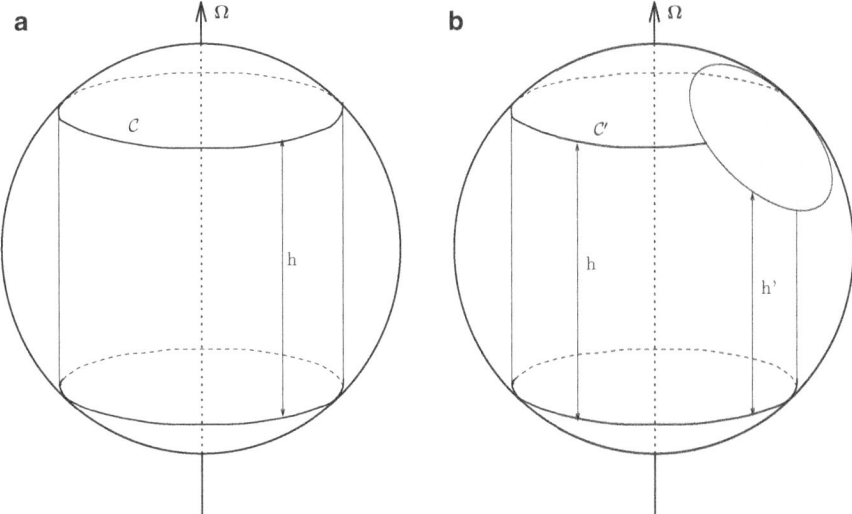

Fig. 8.1 (**a**) Geostrophic contours on a sphere. (**b**) A truncated sphere: some geostrophic contours are not closed (like $\mathcal{C}'$)

The answer is negative because the Coriolis force imposed by the Earth rotation is much too weak to be effective compared to other forces. To make things clear, it is useful to appreciate the orders of magnitude associated with such a flow. First, we may observe that the scale at which the Rossby number passes below unity for a flow whose typical velocity is 10 cm/s, is 700 m. Thus, unless the bath tube is of the size of a lake, the Rossby number will be very large, letting the $(\mathbf{v} \cdot \nabla)\mathbf{v}$ term a thousand time greater than the Coriolis acceleration.

Another way of understanding this question is to suppose that the flow is geostrophic (it is indeed almost steady, and viscous effects are small). In such a case, the amplitude of the fluid velocity would be $V \sim |\nabla P|/2\Omega$. We may estimate the pressure gradient by noting that on the bottom of the bath tube the pressure varies between $\rho g h + P_{atm}$ and P_{atm}, far from the exit and at the exit. Taking $h=10$ cm, we find a fluid velocity of 20,000 km/s which is absurd.

So what's going on in reality? The key point is to be found in the initial conditions. In general, the fluid is not strictly at rest when one empties a bath tube. With respect to the exit, the water owns some residual angular momentum. When the emptying is started, conservation of this momentum implies an amplification of the rotation near the exit. Actually, the convergence of the streamlines on exit strongly amplifies the vorticity. Thus, a flow which was not perceptible to the eye before the reservoir is emptying, shows up neatly when the exit is open. The low pressure at the vortex centre makes this structure clearly visible.

8.3   Waves in Rotating Fluids

We continue our exploration of the properties of rotating fluids by focusing on the waves that are specific to them.

8.3.1   Inertial Waves

Inertial waves owe their existence to the Coriolis force which is their restoring force. We recall that the existence of the Coriolis force is the consequence of the conservation of angular momentum. If this force were absent, the free motion in a rotating frame would not conserve the angular momentum.

To fully appreciate its effects, it is useful to consider the motion of a particle which is solely driven by this force. Its velocity verifies

$$\frac{d\mathbf{v}}{dt} + 2\mathbf{\Omega} \times \mathbf{v} = 0$$

This equation is easily solved and yields:

$$v_x = v_0 \cos(2\Omega t) \qquad \text{and} \qquad v_y = v_0 \sin(2\Omega t)$$

if we choose that $v_x = v_0$ and $v_y = 0$ at $t = 0$. A further integration gives the trajectory:

$$x = x_0 + \frac{v_0}{2\Omega} \sin(2\Omega t) \qquad \text{and} \qquad y = y_0 - \frac{v_0}{2\Omega} \cos(2\Omega t)$$

This shows that particles have a circular motion. The Coriolis force brings the particles back to their initial position after making a circular trajectory with a radius $v_0/2\Omega$.

Let us now focus on the dispersion relation of these waves. We take (8.5) and set $E = Ro = 0$. As needed, we assume that the pressure and velocity perturbations are plane waves, namely:

$$(p, \mathbf{v}) = (p, \mathbf{v})_0 e^{i(\omega t - \mathbf{k} \cdot \mathbf{x})}$$

Incompressibility implies that

$$\mathbf{k} \cdot \mathbf{v} = 0 \tag{8.16}$$

which shows that these waves are transversal. The equation of momentum, $i\omega \mathbf{v} + 2\Omega \mathbf{e}_z \times \mathbf{u} = i\mathbf{k}P$, leads to

$$\begin{cases} 2\Omega \mathbf{e}_z \cdot (\mathbf{u} \times \mathbf{k}) = ik^2 P \\ i\omega v_z = ik_z P \\ i\omega \mathbf{k} \times \mathbf{v} = 2\Omega k_z \mathbf{v} \end{cases} \tag{8.17}$$

from which we derive the following dispersion relation:

$$\omega^2 = (2\Omega)^2 \frac{k_z^2}{k^2} \tag{8.18}$$

This relation shows that the frequency of inertial waves is bounded by 2Ω, since $k_z \leq k$. Inertial waves are thus long period waves whose shortest period is half the rotation period.

The dispersion relation also shows that these waves propagate in a very anisotropic way. This is shown by the phase velocity:

$$\mathbf{v}_\phi = \frac{\omega}{k} \mathbf{e}_k = 2\Omega \frac{k_z}{k^3} \mathbf{k} \tag{8.19}$$

This expression shows that no propagation is possible if it is restricted to a plane perpendicular to the rotation axis rotation. Propagation preferentially occurs along the rotation axis.

Now let us consider the group velocity. We find

$$\mathbf{v}_g = \nabla_k \omega(\mathbf{k}) = 2\Omega \frac{\mathbf{k} \times (\mathbf{e}_z \times \mathbf{k})}{k^3} \tag{8.20}$$

This expression shows that $\mathbf{v}_g \cdot \mathbf{k} = 0$: like for internal gravity waves, energy propagates perpendicularly to the phase!

8.3.2   Inertial Modes

If the fluid domain is bounded, the equations of motion need to be completed by the boundary conditions $\mathbf{u} \cdot \mathbf{n} = 0$. The inertial modes are the oscillation modes of a rotating inviscid fluid contained in a reservoir. Setting ω as the mode frequency, the associated flow verifies

$$\begin{cases} i\omega \mathbf{u} + \mathbf{e}_z \times \mathbf{u} = -\nabla P \\ \nabla \cdot \mathbf{u} = 0 \\ \mathbf{u} \cdot \mathbf{n} = 0 \quad \text{on } S \end{cases} \tag{8.21}$$

which we wrote with non-dimensional variables following (8.5). Now, let ω_n and ω_m be two distinct eigenfrequencies, then

$$\int_{(V)} \mathbf{u}_n \cdot \mathbf{u}_m^* \, dV = 0 \qquad (8.22)$$

i.e. inertial modes are orthogonal with respect to this scalar product. This property is a consequence of (8.21) when it is written for two different modes of frequency ω_n and ω_m. Indeed,

$$\begin{cases} i\omega_n \mathbf{u}_n + \mathbf{e}_z \times \mathbf{u}_n = -\nabla P_n \\ -i\omega_m \mathbf{u}_m^* + \mathbf{e}_z \times \mathbf{u}_m^* = -\nabla P_m^* \end{cases} \qquad (8.23)$$

Taking the scalar product of the first equation with the complex conjugate $\mathbf{u}_m^*$ and the second equation with $\mathbf{u}_n$, adding the results and integrating over the whole fluid volume leads to

$$i(\omega_n - \omega_m) \int_{(V)} \mathbf{u}_n \cdot \mathbf{u}_m^* dV + \int_{(V)} \left[\mathbf{u}_m^* \cdot (\mathbf{e}_z \times \mathbf{u}_n) + \mathbf{u}_n \cdot (\mathbf{e}_z \times \mathbf{u}_m^*) \right] dV = 0$$

where we used the boundary conditions to eliminate the pressure term. The last two terms are of opposite sign so that we are left with (8.22) since $\omega_m \neq \omega_n$.

Another important property of inertial modes is that, like their wavy counterpart, their frequency is less that 2Ω or, for the scaled ω, less than unity. This result comes from the momentum equation when projected on $\mathbf{u}^*$ and integrated over the fluid's volume. It turns out that

$$\omega = \frac{\displaystyle\int_{(V)} \mathrm{Im}[(\mathbf{u}^* \times \mathbf{u}) \cdot \mathbf{e}_z] dV}{\displaystyle\int_{(V)} |\mathbf{u}|^2 dV}$$

Schwarz inequality, $|\mathrm{Im}[(\mathbf{u}^* \times \mathbf{u}) \cdot \mathbf{e}_z]| \leq |\mathbf{u}^* \times \mathbf{u}| \leq |\mathbf{u}|^2$, leads to

$$|\omega| \leq \frac{\displaystyle\int_{(V)} |\mathrm{Im}[(\mathbf{u}^* \times \mathbf{u}) \cdot \mathbf{e}_z]| dV}{\displaystyle\int_{(V)} |\mathbf{u}|^2 dV} \leq 1 \qquad (8.24)$$

For some simple containers, the spectrum, namely all the possible values of ω, can be computed. In this case, the eigenvalues are dense in the interval $[0,1]$. This means that for any real number in this interval, we may find a frequency ω as close to this number as we wish (see the box "The inertial modes in the sphere" for a detailed example).

8.3.3 The Poincaré Equation

If we now take the divergence of (8.21), the system can be reduced to a single equation on the pressure, namely

$$\Delta P - \frac{1}{\omega^2}\frac{\partial^2 P}{\partial z^2} = 0 \tag{8.25}$$

This equation is known as *Poincaré's equation* since the work of Élie Cartan (1922). This equation is completed by the boundary condition $\mathbf{u}\cdot\mathbf{n} = 0$, which can be reexpressed with the pressure as:

$$-\omega^2\mathbf{n}\cdot\nabla P + (\mathbf{n}\cdot\mathbf{e}_z)(\mathbf{e}_z\cdot\nabla P) + i\omega(\mathbf{e}_z\times\mathbf{n})\cdot\nabla P = 0 \tag{8.26}$$

The Poincaré equation completed with the foregoing boundary condition is peculiar as it constitutes a mathematically ill-posed problem. Indeed, since $\omega < 1$, this equation is of hyperbolic type, like the wave equation. However, unlike the wave equation, boundary conditions are imposed to the solutions of the Poincaré equation. This makes an ill-posed problem in the sense of Hadamard. In the general case, solutions own many singularities which endow inertial modes with very unusual properties as illustrated in Fig. 8.2.

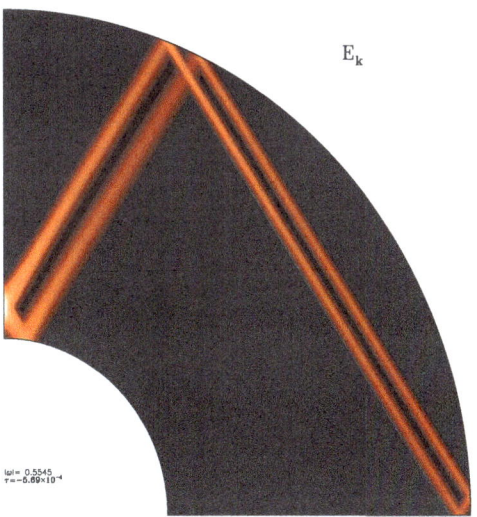

E_k

$|\omega| = 0.5545$
$\tau = -5.89\times10^{-4}$

Nr=560 L=2000 M=0 E=8.0×10⁻¹⁰ η=0.350 CL=ff

Fig. 8.2 A singular inertial mode: this figure shows the kinetic energy of an inertial mode inside a spherical shell. This meridian cut shows that the mode is concentrated along a periodic path of the characteristics of the Poincaré equation (this periodic path is called an attractor). When viscosity decreases (here the Ekman number is 8×10^{-10}), the mode gets more focused around the attractor and becomes singular at a vanishing viscosity (see Rieutord et al. 2001, for more details)

The inertial modes in the sphere

Let us write the Poincaré equation (8.25) as

$$\frac{\partial^2 P}{\partial x^2} + \frac{\partial^2 P}{\partial y^2} - \left(\frac{1-\omega^2}{\omega^2}\right)\frac{\partial^2 P}{\partial z^2} = 0 \tag{8.27}$$

The boundary condition $\mathbf{v}\cdot\mathbf{n} = 0$ yields

$$s\frac{\partial P}{\partial s} + \frac{1}{i\omega}\frac{\partial P}{\partial \varphi} + \left(1 - \frac{1}{\omega^2}\right)z\frac{\partial P}{\partial z} = 0 \tag{8.28}$$

when cylindrical coordinates are used. It is derived by using

$$\mathbf{n} = \frac{1}{\sqrt{r^2+z^2}}(r\mathbf{e}_r + z\mathbf{e}_z)$$

$$v_s = -\frac{1}{1-\omega^2}\left(i\omega\frac{\partial P}{\partial s} + \frac{1}{s}\frac{\partial P}{\partial \varphi}\right) \tag{8.29}$$

$$v_z = -\frac{1}{i\omega}\frac{\partial P}{\partial z} \tag{8.30}$$

The dispersion relation of inertial modes in a full sphere has been first obtained by Bryan (1889), who proposed to change the z-coordinate into

$$z' = -\frac{i\omega}{\sqrt{1-\omega^2}}z$$

so that Poincaré equation turns into Laplace's one. With this new system of coordinates, the bounding sphere of radius R becomes:

$$\frac{x^2+y^2}{R^2} - \frac{z'^2}{B^2} = 1 \tag{8.31}$$

with $B^2 = \frac{\omega^2}{1-\omega^2}R^2$. This is the equation of a one sheet axisymmetric hyperboloid. To solve Laplace equation, we need to use a coordinate system which is appropriate to this new geometry. These coordinates are those of the oblate ellipsoid Angot (1949,1972). This coordinate system uses the following surfaces:

$$\frac{x^2+y^2}{a^2\cos^2\chi} - \frac{z'^2}{a^2\sin^2\chi} = 1$$

$$\frac{x^2+y^2}{a^2\cosh^2\xi} + \frac{z'^2}{a^2\sinh^2\xi} = 1$$

where we identify

$$a^2 = \frac{R^2}{1-\omega^2}, \qquad \sin^2\chi = \omega^2$$

The ellipsoidal coordinates ξ, χ, φ are related to the cartesian ones by

$$\begin{cases} x = a\cosh\xi\cos\chi\cos\varphi \\ y = a\cosh\xi\cos\chi\sin\varphi \\ z' = a\sinh\xi\sin\chi \end{cases} \tag{8.32}$$

and the solutions of the Laplace equation are of the form:

$$P(\xi,\chi,\varphi) = \sum_{l,m} A_{l,m}P_l^m(\sin\chi)$$

$$\tilde{P}_l^m(i\sinh\xi)e^{im\varphi}$$

where the P_l^m are the Legendre polynomials. Noting that

$$z = \frac{\sqrt{1-\omega^2}}{\omega}a\,i\sinh\xi\sin\chi$$

we can set $\mu = i\sinh\xi$ and $\eta = a\sin\chi$; then

$$\begin{cases} s = \sqrt{x^2+y^2} = \sqrt{(a^2-\eta^2)(1-\mu^2)} \\ z = \frac{\mu\eta}{\sqrt{a^2-1}} \end{cases} \tag{8.33}$$

and $a = 1/(1-\omega^2)$ if we set $R = 1$. The solution is therefore:

$$P = \sum_{l,m} A_{l,m}P_l^m\left(\frac{\eta}{a}\right)P_l^m(\mu)e^{im\varphi}$$

$$\iff P = \sum_{l,m} A_{l,m}P_l^m(\eta\sqrt{1-\omega^2})P_l^m(\mu)e^{im\varphi}$$

The inertial modes in the sphere

The boundary conditions (8.28) need to be rewritten with the variables (η, μ). We find

$$\begin{cases} \frac{\partial \mu}{\partial s} = \frac{s\mu}{\Delta}, & \frac{\partial \mu}{\partial z} = \frac{(a^2-1)(1-\mu^2)z}{\mu\Delta} \\ \frac{\partial \eta}{\partial s} = -\frac{s\eta}{\Delta}, & \frac{\partial \eta}{\partial z} = \frac{\mu(\eta^2-a^2)\sqrt{a^2-1}}{\Delta} \end{cases}$$

(8.34)

with $\Delta = \eta^2 - a^2\mu^2$. Using these relations on the sphere, at $\eta = \frac{\omega}{\sqrt{1-\omega^2}}$, we finally obtain the dispersion relation:

$$(1 - \omega^2)\frac{dP_l^m}{d\omega} = mP_l^m \qquad (8.35)$$

which permits the computation of the frequencies of inertial modes in the sphere. If we consider the case of axisymmetric modes, $m = 0$, this relation is now simply $\frac{dP_l^m}{d\omega} = 0$ or $\omega = \pm 1$. The eigenfrequencies are thus the roots of the Legendre polynomial $P_l^1(\omega) = \sqrt{1 - \omega^2}\frac{dP_l^0}{d\omega}$. All these roots are between -1 and 1 (which meets the constraint (8.24)) and when $\ell \longrightarrow +\infty$, these root form a dense set in this interval.

8.3.4 Rossby Waves

Rossby waves constitute a wave category which is very important in planetary atmospheres. They are often called planetary waves.[1]

Let us restart the derivation of the dispersion relation for inertial waves but with the assumption that the fluid is contained in a very thin layer like the Earth atmosphere. We set the z-axis along the vertical, the x-axis towards East (parallel to the Earth rotation) and the y-axis to the North. We look for a purely two-dimensional wave solution, where vertical motions are negligible compared to the horizontal ones. The dispersion relation of such waves cannot be derived from the one of inertial waves since we now impose the condition $v_z = 0$. The simplification by v_z, which is needed to derive (8.18) is no longer possible. Thus, the derivation needs to be started *ab initio*. The equations of the flow are:

$$\begin{cases} i\omega\mathbf{v} + 2\mathbf{\Omega}(y) \times \mathbf{v} = -\nabla P \\ \nabla \cdot \mathbf{v} = 0 \end{cases} \qquad (8.36)$$

When writing this system, we explicitly mention the dependence $\mathbf{\Omega} \equiv \mathbf{\Omega}(y)$ since the local rotation vector depends on the latitude. We underline that since we restrict the motions to the horizontal ones, the horizontal part of $\mathbf{\Omega}$ does not play any role; we just need to consider the component of $\mathbf{\Omega}$ along the z-axis. We thus write:

$$i\omega\mathbf{v} + 2\Omega(y)\mathbf{e}_z \times \mathbf{v} = -\nabla P$$

---

[1] See Longuet-Higgins (1964).

where $\Omega(y) = \Omega \sin \lambda(y)$, λ being the latitude. We have

$$
\begin{cases}
i\omega v_x - 2\Omega(y)v_y = -\dfrac{\partial P}{\partial x} \\[3mm]
i\omega v_y + 2\Omega(y)v_x = -\dfrac{\partial P}{\partial y} \\[3mm]
\dfrac{\partial v_x}{\partial x} + \dfrac{\partial v_y}{\partial y} = 0
\end{cases}
\tag{8.37}
$$

Eliminating the pressure by taking the curl, we find the vertical component ζ of the vorticity:

$$
i\omega\zeta = 2v_y \frac{d\Omega}{dy}
\tag{8.38}
$$

This equation shows that the relation between Ω and the latitude is essential. Now a standard approximation, called *the β-plane approximation*, consists in setting $\frac{d\Omega}{dy}$ to a constant value. In atmospheric sciences, β is called the gradient of the planetary vorticity ($2d\Omega/dy$). With this assumption, we easily find the dispersion relation of Rossby waves:

$$
\omega = -\frac{2k_x}{k_x^2 + k_y^2}\left(\frac{d\Omega}{dy}\right)
\tag{8.39}
$$

This relation shows that $\omega k_x < 0$ since $\frac{d\Omega}{dy} > 0$. Rossby waves thus propagate to the $x < 0$, that is to say to the West, opposite to the Earth rotation. They are *retrograde* waves. The group velocity

$$
\mathbf{v}_g = 2\frac{d\Omega}{dy}\left((k_y^2 - k_x^2)\mathbf{e}_x + 2k_x k_y \mathbf{e}_y\right)/k^4
$$

shows that energy has no preferred direction.

The form of the dispersion relation of Rossby waves shows why we could not have derived it from the one of inertial waves: the variation of Ω is crucial. In particular, we note that if the velocity field of the perturbation is a plane wave, this is not the case for the pressure fluctuation because $\frac{\partial P}{\partial x} \neq ik_x P$. In fact, Rossby waves are rather a class of inertial modes which meet some constraints like bidimensionality. This is why, when a container does not admit closed geostrophic contours, the geostrophic flow is replaced by an infinite sum of Rossby waves, namely by inertial modes which are quasi 2D and of very low frequency.[2]

---

[2]A detailed discussion of this question may be found in Greenspan (1969).

8.3.4.1 Planetary Modes

Let us now generalize the foregoing results by considering the Rossby waves over the whole sphere. We still consider that they propagate in a very thin fluid layer covering a sphere, but we abandon the β-plane approximation. Such modes are usually called *planetary modes*.

Because the velocity field is two-dimensional, we may use a stream function to describe the flow. We thus introduce $\chi(\theta, \varphi)$, which is such that

$$\mathbf{v} = \nabla \times (\chi \mathbf{e}_r)$$

We derive the equation for χ by applying the $\mathbf{e}_r \cdot \nabla \times$ operator to (8.21). We get

$$i\omega \mathbf{e}_r \cdot \nabla \times \nabla \times (\chi \mathbf{e}_r) + \mathbf{e}_r \cdot \nabla \times (\mathbf{e}_z \times \mathbf{u}) = 0$$

which leads to

$$i\omega \Delta \chi + \frac{\partial \chi}{\partial \varphi} = 0$$

on a sphere of unit radius. It is natural to decompose the stream function χ on the set of spherical harmonics, which is a complete base for the functions defined on the sphere. Thus, setting

$$\chi = \sum_{\ell,m} \chi_m^\ell Y_\ell^m$$

we find that an eigenmode is represented by a single spherical harmonic to which corresponds the eigenfrequency

$$\omega_{\ell m} = \frac{m}{\ell(\ell + 1)} \tag{8.40}$$

We derived this dispersion relation using the spherical harmonics differential equation $\Delta Y_\ell^m = -\ell(\ell + 1) Y_\ell^m$ (see 12.31).

The expression of $\omega_{\ell m}$ shows that the (angular) phase velocity $-\omega/m = -1/\ell(\ell + 1)$ is always negative.[3] Thus, like Rossby waves, planetary modes propagate to the West. Figure 8.3 illustrates the wind pattern generated by these waves in the Earth atmosphere.

---

[3] We recall that we set χ proportional to $e^{i(\omega t + m\varphi)}$.

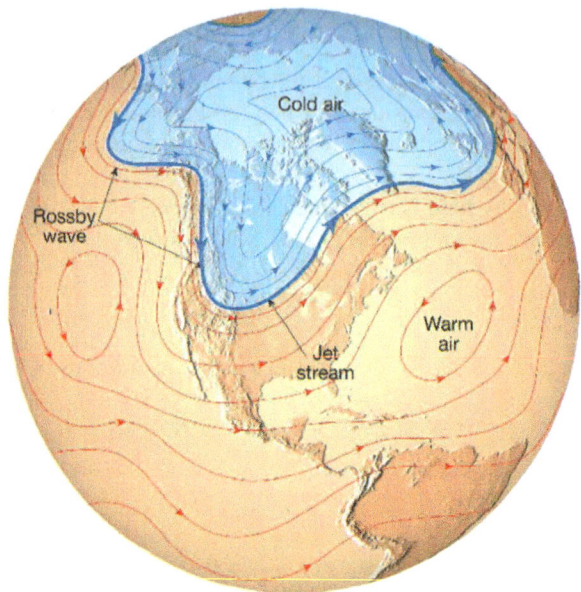

Fig. 8.3 Rossby waves in the Earth atmosphere: in shaping the interface between cold air and warm air, Rossby waves have a determining influence on the weather at mid-latitudes. Credit: City University of New York

8.4   The Effects of Viscosity

Until now we neglected the viscosity. We showed, while introducing the rotation time scale, that viscous terms are controlled by the Ekman number, which value is usually very small. Hence, the effects of viscosity are important only in places where the gradient of velocity is strong, namely in boundary (or shear) layers.

As above, we shall consider the limit of vanishing Rossby numbers so as to (again) neglect the nonlinear terms. Then, boundary layers usually result from a balance between viscous terms and the Coriolis term. They are called *Ekman layers*. The boundary layer flow can be formally solved, as we shall see below, because the equations governing the flow are linear, unlike the general boundary layers that we studied in Chap. 4 (like the Blasius flow for instance).

8.4.1   The Method

In order to simplify the discussion, we shall concentrate on the steady case only, namely on the geostrophic flow (an example of the unsteady case can be found in Rieutord 2001). The idea of the method is to divide the solution into small subsolutions which are easy to derive. For this, we first expand the solution into powers of the small parameter $\sqrt{E}$, which is the thickness of the layer (just like $1/\sqrt{Re}$ in Chap. 4). We thus write

$$\mathbf{u} = \mathbf{u}_0 + \sqrt{E}\,\mathbf{u}_1 + E\,\mathbf{u}_2 + \dots, \qquad p = p_0 + \sqrt{E}\,p_1 + \dots \qquad (8.41)$$

Then, each order $\mathbf{u}_n$ is split into two parts: the boundary layer part $\tilde{u}_n$ and the interior part $\bar{u}_n$. The derivation of each of these terms is much simpler than the full solution. Summing them together allows us to obtain a solution valid up to the chosen order (usually one or two).

8.4.2 The Boundary Layer Solution

The boundary layer solution is simpler than the general one because the flow is along the boundary and the velocity variations are dominated by the gradients along the normal to the wall (see Chap. 4).

Let $\mathbf{n}$ be the outer normal of the wall, and let us rewrite (8.5) with $Ro = \frac{\partial}{\partial \tau} = 0$. We find

$$\mathbf{e}_z \times \mathbf{u} = -\nabla p + E\Delta\mathbf{u} \qquad (8.42)$$

We now make the decomposition

$$\mathbf{u} = \bar{u}_0 + \tilde{u}_0, \qquad p = \bar{p}_0 + \tilde{p}_0$$

There, $\bar{u}_0$ is nothing but the geostrophic solution. $\tilde{u}_0$ and $\tilde{p}_0$ are the corrections to add to the geostrophic solution so that the boundary conditions are met. Since $\mathbf{e}_z \times \bar{u}_0 = -\nabla\bar{p}_0$ then

$$\mathbf{e}_z \times \tilde{u}_0 = -\nabla\tilde{p}_0 + E\Delta\tilde{u}_0 \qquad (8.43)$$

where we neglected $E\Delta\bar{u}_0$ since it is $\mathcal{O}(E)$ while other terms are of order unity.

Let ξ be the coordinate along the normal of the wall directed towards the container's interior. Projected along $\mathbf{n}$ (8.43) yields

$$\frac{\partial\tilde{p}_0}{\partial\xi} = \mathbf{n} \cdot (\mathbf{e}_z \times \tilde{u}_0)$$

Since $\tilde{p}_0$ is a boundary layer quantity, its variation along ξ is very fast. If $\sqrt{E}$ is the thickness of the layer as shown below, then $\partial\tilde{p}_0/\partial\xi$ is $\mathcal{O}(1/\sqrt{E})$ but $\mathbf{n} \cdot (\mathbf{e}_z \times \tilde{u}_0)$ is $\mathcal{O}(1)$. This implies that the normal derivative of $\tilde{p}_0$ is zero. Hence, $\tilde{p}_0$ is a constant across the boundary layer. We find here again the result derived from the Prandtl equations, which control a general boundary layer (see 4.37b). Since the value of $\bar{p}_0$ is zero outside the boundary layer, $\tilde{p}_0$ is vanishing everywhere. The pressure correction in the boundary layer is therefore of the next order, that is $\mathcal{O}(\sqrt{E})$. We thus have to write

$$p = \overline{p}_0 + \sqrt{E}\,(\overline{p}_1 + \tilde{p}_1)$$

Keeping only the $\mathcal{O}(1)$ terms in (8.43), we have

$$\mathbf{e}_z \times \tilde{u}_0 = \frac{\partial \tilde{p}_1}{\partial \zeta}\mathbf{n} + \frac{\partial^2 \tilde{u}_0}{\partial \zeta^2} \tag{8.44}$$

where we introduced the stretched coordinate ζ such that $\xi = \sqrt{E}\,\zeta$. Taking the cross product of this equation with $\mathbf{n}$ and observing that $\tilde{u}_0 \cdot \mathbf{n} = 0$, we find that

$$-(\mathbf{n} \cdot \mathbf{e}_z)\tilde{u}_0 = \frac{\partial^2 \mathbf{n} \times \tilde{u}_0}{\partial \zeta^2} \tag{8.45}$$

On the other hand $\tilde{u}_0 = (\tilde{u}_0 \cdot \mathbf{n})\mathbf{n} + (\mathbf{n} \times \tilde{u}_0) \times \mathbf{n} = (\mathbf{n} \times \tilde{u}_0) \times \mathbf{n}$ since we are dealing with boundary layer quantities. (8.44) may be rewritten as

$$(\mathbf{n} \cdot \mathbf{e}_z)(\mathbf{n} \times \tilde{u}_0) - \mathbf{e}_z \cdot (\mathbf{n} \times \tilde{u}_0)\mathbf{n} = \frac{\partial \tilde{p}_1}{\partial \zeta}\mathbf{n} + \frac{\partial^2 \tilde{u}_0}{\partial \zeta^2}$$

$$\implies \quad (\mathbf{n} \cdot \mathbf{e}_z)(\mathbf{n} \times \tilde{u}_0) = \frac{\partial^2 \tilde{u}_0}{\partial \zeta^2} \tag{8.46}$$

where we identified the vectors belonging to the tangent plane. Multiplying (8.46) by i and adding it to (8.45), we deduce that

$$\frac{\partial^2}{\partial \zeta^2}(\mathbf{n} \times \tilde{u}_0 + i\,\tilde{u}_0) = i\,(\mathbf{n} \cdot \mathbf{e}_z)(\mathbf{n} \times \tilde{u}_0 + i\,\tilde{u}_0) \tag{8.47}$$

This equation is easily solved. We find

$$(\mathbf{n} \times \tilde{u}_0 + i\,\tilde{u}_0) = (\mathbf{n} \times \tilde{u}_0 + i\,\tilde{u}_0)_{\zeta=0} \exp\left(-\zeta\sqrt{i\,(\mathbf{n} \cdot \mathbf{e}_z)}\right) \tag{8.48}$$

The integration constant $(\mathbf{n} \times \tilde{u} + i\,\tilde{u})_{\zeta=0}$ is given by the flow outside the boundary layer. For instance, if the boundary conditions are $\mathbf{u} = \mathbf{0}$ on the wall, then, the solution must be such that $\tilde{u}_0 + \overline{u}_0 = \mathbf{0}$ on the wall. Hence $(\mathbf{n} \times \tilde{u}_0 + i\,\tilde{u}_0)_{\zeta=0} = -(\mathbf{n} \times \overline{u}_0 + i\,\overline{u}_0)_{\text{wall}}$, so that

$$(\mathbf{n} \times \tilde{u}_0 + i\,\tilde{u}_0) = -(\mathbf{n} \times \overline{u}_0 + i\,\overline{u}_0)_{\text{wall}} \exp\left(-\zeta\sqrt{i\,(\mathbf{n} \cdot \mathbf{e}_z)}\right) \tag{8.49}$$

The solution (8.48) calls for some comments: First let us note that the velocity has a changing orientation within the boundary layer. Indeed, let us assume, for instance, that $\mathbf{n} = \mathbf{e}_z$ and $\bar{u}_0 = U\mathbf{e}_x$ on the wall. Then, (8.48) changes into

$$\begin{cases} \tilde{u}_x = -Ue^{-\zeta'}\cos\zeta' \\ \tilde{u}_y = Ue^{-\zeta'}\sin\zeta' \end{cases} \tag{8.50}$$

where we set $\zeta' = \zeta/\sqrt{2}$. The "complete" solution reads

$$\mathbf{u} = \bar{u}_0 + \tilde{u}_0 = U\left((1 - \cos\zeta'e^{-\zeta'})\mathbf{e}_x + \sin\zeta'e^{-\zeta'}\,\mathbf{e}_y\right) \tag{8.51}$$

To illustrate the shape of the velocity field, we draw the velocity vector as a function of the "depth" ζ'. This yields a spiral known as the *Ekman spiral* (see Fig. 8.4).

A second comment about (8.48) concerns the thickness of the Ekman layer which is

$$e = L\sqrt{\frac{2E}{|\mathbf{n}\cdot\mathbf{e}_z|}}$$

where L is the length scale. This expression shows that if the wall is parallel to the rotation axis, the thickness of the Ekman layer is infinite. In fact, in this very case, the analysis that led to (8.48) is no longer valid. This difficulty arises for instance when one deals with the geostrophic flow inside a sphere. At the equator, the boundary layer is singular: this is *the equatorial singularity*. It may be shown that for latitudes within an equatorial band of latitudinal extension $\mathcal{O}(E^{1/5})$, the thickness of the layer is $\mathcal{O}(E^{2/5})$. Thus, for a development in powers of $E^{1/2}$, the new thickness of the layer, scaling like $E^{2/5}$ appears to be of infinite size since $\lim_{E\to 0} E^{2/5-1/2} = \infty$. More details may be found in the original paper of Roberts and Stewartson (1963).

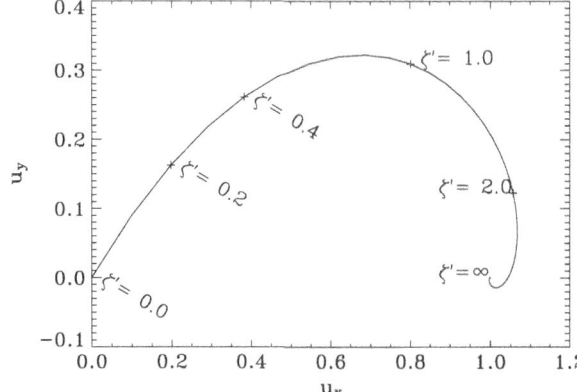

Fig. 8.4 The Ekman spiral:
on the boundary
$\mathbf{u}(\zeta' = 0) = \mathbf{0}$, while outside
the boundary layer
$\mathbf{u}(\zeta' \to \infty) = \mathbf{e}_x$

8.4.3   *Ekman Pumping and Ekman Circulation*

When we derived (8.48), we just solved the momentum equation, leaving aside mass conservation. It may well be that $\nabla \cdot \tilde{u}_0 \neq 0$. Fortunately, we need not throw away our solution (8.48). Let us be more explicit. If x and y are the coordinates in the tangent plane, the projection of (8.48) on this basis gives

$$\begin{cases} \tilde{u}_x^0 = -(\overline{u}_x^0 \cos \zeta' + \overline{u}_y^0 \sin \zeta')e^{-\zeta'} \\ \tilde{u}_y^0 = (\overline{u}_x^0 \sin \zeta' - \overline{u}_y^0 \cos \zeta')e^{-\zeta'} \end{cases} \tag{8.52}$$

where $\overline{u}_x^0$ and $\overline{u}_y^0$ are the x and y-components of the geostrophic flow taken on the wall. We now compute the divergence of $\tilde{u}_0$ noting that $\nabla \cdot \overline{u}_0 = 0$. We have:

$$\frac{\partial \tilde{u}_x^0}{\partial x} + \frac{\partial \tilde{u}_y^0}{\partial y} = \left(\frac{\partial \overline{u}_x^0}{\partial y} - \frac{\partial \overline{u}_y^0}{\partial x} \right) \sin \zeta' e^{-\zeta'} = -(\mathbf{n} \cdot \nabla \times \overline{u}_0) \sin \zeta' e^{-\zeta'}$$

This derivation is purely formal because we did not take into account the curvilinear nature of the coordinates; however, it keeps the dominant terms. This expression shows that the divergence is actually proportional to the normal component of the vorticity of the geostrophic flow.

This divergence is generally non-zero and is compensated by a flow along $\mathbf{n}$. Let us denote this flow $\tilde{u}'$. It verifies

$$\frac{\partial \tilde{u}_x^0}{\partial x} + \frac{\partial \tilde{u}_y^0}{\partial y} + \frac{\partial \tilde{u}'}{\partial \xi} = 0$$

Setting $R(x, y) = \mathbf{n} \cdot \nabla \times \mathbf{u}_{geo}$, then

$$\frac{\partial \tilde{u}'}{\partial \xi} = R(x, y) \sin \zeta' e^{-\zeta'}$$

which is easily integrated, remembering that $\xi = \sqrt{2E}\,\zeta'$; it turns out

$$\tilde{u}' = -\sqrt{E}\, R(x, y)\, e^{-\zeta'} \cos \left(\zeta' - \frac{\pi}{4} \right) \tag{8.53}$$

The important point shown by this expression is the fact that this new component of the boundary layer flow is of a higher order in powers of $\sqrt{E}$, so that the foregoing results are still valid, fortunately! We thus write

$$\tilde{u}' = \sqrt{E}\, \tilde{u}_1$$

This new component of the boundary layer flow is very important for the large-scale dynamics. We indeed observe that it is non-zero on the boundary at $\zeta = 0$. This means that in order for the boundary conditions to be verified at first order, the boundary layer flow needs to be completed by an interior one of the same order. Let us call this new interior flow $\overline{u}_1$. As $\overline{u}_0$ it verifies the geostrophic (8.7) but it meets a different boundary condition. Indeed, we now demand that

$$(\overline{u}_1 + \tilde{u}_1) \cdot \mathbf{n} = 0 \tag{8.54}$$

on the boundary.

The new component of the boundary layer flow, $\tilde{u}_1$, is called the *Ekman pumping*. This "pumping" is similar to the one we met with the Blasius flow. It may just be either positive of negative, meaning that the layer either pumps in or out the matter, depending on the sign of the local vorticity. This pumping forces the component $\overline{u}_1$ of the interior flow. This new component is known as *the Ekman circulation*. We shall see below that despite its small amplitude, Ekman circulation is crucial to the large-scale dynamics.

We have now all the pieces to write down the steady solution complete at first order. With obvious notations, we may write it:

$$\mathbf{u} = \mathbf{u}_{\text{geo}} + \tilde{u}_{\text{geo}} + \sqrt{E}(\tilde{u}_{\text{pump}} + \mathbf{u}_{\text{circ}}) + \mathcal{O}(E) \tag{8.55}$$

8.4.4   An Example: The Spin-Up Flow

The spin-up flow is the large-scale flow that arises within a rotating fluid when an exterior stress increases the angular velocity. For instance, when a liquid in some container rotates as a solid body, like the container, at an angular velocity $\mathbf{\Omega}$, a sudden change of the angular velocity of the container, by $\Delta\mathbf{\Omega}$, will generate a fluid flow, that will lead to the new solid body rotation at $\mathbf{\Omega} + \Delta\mathbf{\Omega}$. This transient flow may be split in several steps one of which is quasi-steady and called the spin-up (or spin-down) flow.

8.4.4.1   Spin-Up Driven by a Solid Plane

The simplest set-up to study a spin-up flow is to consider a viscous incompressible fluid in the neighbourhood of a solid plane staying at $z = 0$. The plane rotates uniformly at $\mathbf{\Omega} = \Omega\mathbf{e}_z$. The viscous fluid is in the half-space $z > 0$. The rotation of the plane is increased instantaneously by $\Delta\Omega\mathbf{e}_z$. After a transient of a few rotation periods, Ekman layers have formed and a quasi-steady flow takes place.

To study this flow we use a frame attached to the bounding plane. Far from this plane the fluid rotates at the angular velocity $-\Delta\Omega\mathbf{e}_z$. In this region, viewed from the plane, there is a geostrophic flow $\mathbf{v}_{geo} = -\Delta\Omega\mathbf{e}_z \times \mathbf{r}$. This is our basic geostrophic solution that needs to be completed by Ekman layers. Inserting this solution into (8.49), we get the needed boundary layer corrections so that $\mathbf{v} = \mathbf{0}$ at $z = 0$. Using cylindrical coordinates (s, φ, z), we get

$$\begin{cases} u_s = \Delta\Omega s \sin \zeta' e^{-\zeta'} \\ u_\varphi = \Delta\Omega s(\cos \zeta' e^{-\zeta'} - 1) \end{cases} \tag{8.56}$$

This solution shows that the spin-up flow is diverging in the boundary layer ($u_s > 0$), which shows that this boundary layer "sucks" the outer fluid. Since the boundary is plane we can use (8.53). Noting that $\mathbf{n} = \mathbf{e}_z$ and

$$R(x, y) = \mathbf{e}_z \cdot \nabla \times (-2\Delta\Omega\mathbf{e}_z \times \mathbf{r}) = -2\Delta\Omega$$

we deduce that

$$\tilde{u}_z = 2\Delta\Omega \sqrt{E} \cos \left(\zeta' - \pi/4\right) e^{-\zeta'}$$

This component of the boundary layer flow induces a pumping of the outer fluid into the boundary layer because $\tilde{u}_z(0) \neq 0$. Thus, in order that the boundary condition $u_z = 0$ be verified, the outer solution needs to be completed by an $\mathcal{O}(\sqrt{E})$ solution of the inviscid equations such that

$$\overline{u}(\zeta' = 0) = -\tilde{u}_z(\zeta' = 0)$$

In Fig. 8.5, we show schematically the radial and vertical components of the boundary layer flow. A solid body rotation should be added in thought.

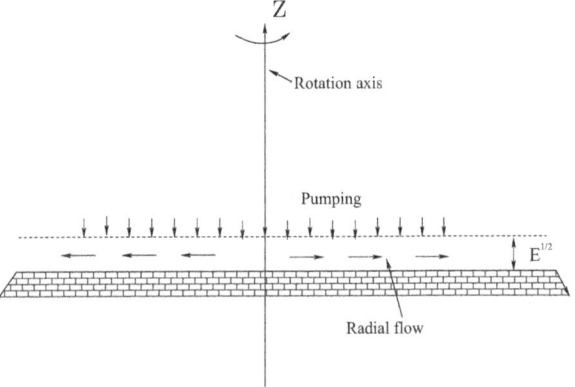

Fig. 8.5 Meridian view of a spin-up flow in the neighbourhood of a rotating plane. The radial component of the flow only exists in the boundary layer. To insure mass conservation the boundary layer absorb some mass from the interior. In a spin-dow flow all the flows would be reversed

The radial flow in the boundary layer is easily observable in a glass of water when we try to dissolve some sugar by stirring the water with a tea-spoon. When we stop stirring, we observe that the sugar gathers on the rotation axis of the water on the bottom of the glass. This is the signature of the radial component of the boundary layer flow, which is converging in the spin-down case, gathering the sugar at the centre.

8.4.4.2 Spin-Up Within a Sphere

As a second example, we now consider a viscous incompressible fluid inside a sphere whose angular velocity increases very slowly with time. In this ideal case the spin-up flow is steady. Let $\dot{\Omega}$ be the acceleration of the rotation, the natural scaling of the velocity field is

$$\mathbf{v} = \frac{\dot{\Omega} R}{2\Omega}\mathbf{u}$$

If we choose $(2\Omega)^{-1}$ as the time scale and the radius of the sphere R as the length scale, the momentum equation written in a frame corotating with the sphere reads

$$\frac{\partial \mathbf{u}}{\partial \tau} + \mathrm{Ro}(\mathbf{u} \cdot \nabla)\mathbf{u} + \mathbf{e}_z \times \mathbf{u} + \mathbf{e}_z \times \mathbf{r} = -\nabla p + E \Delta \mathbf{u}$$

The acceleration term $\dot{\boldsymbol{\Omega}} \times \mathbf{r}$ that yields the term $\mathbf{e}_z \times \mathbf{r}$ is sometimes called *the Euler force*. The Rossby number is assumed to be vanishingly small since we focus on very small accelerations. The nonlinear terms are therefore neglected and since we look for steady solutions, we'll have to solve

$$\begin{cases} \mathbf{e}_z \times \mathbf{u} + \mathbf{e}_z \times \mathbf{r} = -\nabla p + E \Delta \mathbf{u} \\ \nabla \cdot \mathbf{u} = 0 \\ \mathbf{u} = \mathbf{0} \qquad \text{on} \qquad r = 1 \end{cases} \qquad (8.57)$$

To solve this system, it is convenient to split the solution in the following way:

$$\mathbf{u} = 2z\mathbf{e}_z - s\mathbf{e}_s + \mathbf{u}_{\text{geo}}(s) + \tilde{u}$$

where we used the cylindrical coordinates (s, φ, z). The $2z\mathbf{e}_z - s\mathbf{e}_s$ terms represent a particular solution of vanishing divergence, that cancels the forcing term $\mathbf{e}_z \times \mathbf{r}$. But this particular solution does not meet the boundary condition $\mathbf{n} \cdot \mathbf{u} = 0$. Unfortunately the geostrophic solution which is parallel to $\mathbf{e}_\varphi$ cannot help. The mass flux of this particular solution on the bounding sphere needs thus to be compensated by the boundary layer mass flux. The particular solution therefore represents the

Ekman circulation part of the solution. Since this circulation is $E^{1/2}$ times smaller than the geostrophic part, we conclude that $\mathbf{u}_{\text{geo}}$ is $\mathcal{O}(E^{-1/2})$. It means that that $\mathbf{u}_{\text{geo}}$ diverges at zero viscosity, but this is not surprising since the sphere cannot entrains an inviscid fluid!

Let us come back to the resolution of our problem. We note that the Ekman pumping on the wall is such that

$$\tilde{u}_r + \mathbf{e}_r \cdot (2z\mathbf{e}_z - s\mathbf{e}_s) = 0 \quad \Longrightarrow \quad \tilde{u}_r = 3\sin^2\theta - 2$$

where θ is the polar angle of the spherical coordinates. On the other hand (8.15) shows that $\mathbf{u}_{\text{geo}}(s) = U(s)\mathbf{e}_\varphi$ and induces a boundary flow given by (8.49)

$$\begin{cases} \tilde{u}_\theta = -U(\sin\theta)\sin\alpha\, e^{-\alpha} \\ \tilde{u}_\varphi = -U(\sin\theta)\cos\alpha\, e^{-\alpha} \end{cases} \tag{8.58}$$

where

$$\alpha = \zeta\sqrt{\frac{\cos\theta}{2}}$$

Here, we'll assume that $\cos\theta > 0$ thus restricting our discussion to the Northern hemisphere. We note that on the bounding sphere $r = 1$ and $s = \sin\theta$. Finally, the geostrophic flow with its boundary layer correction reads

$$\begin{cases} u_\theta = -U(\sin\theta)\sin\alpha\, e^{-\alpha} \\ u_\varphi = U(s) - U(\sin\theta)\cos\alpha\, e^{-\alpha} \end{cases} \tag{8.59}$$

Mass conservation gives the relation between pumping and the foregoing flow. At the leading order we have

$$\frac{\partial \tilde{u}}{\partial r} + \frac{1}{\sin\theta}\frac{\partial}{\partial\theta}(\sin\theta\,\tilde{u}_\theta) = 0 \qquad \text{at} \quad r = 1$$

Using the boundary layer variable $\zeta = (1 - r)/\sqrt{E}$ and the previous expression of $\tilde{u}_\theta$, we get

$$\frac{\partial \tilde{u}}{\partial\zeta} = -\frac{\sqrt{E}}{\sin\theta}\frac{\partial}{\partial\theta}(\sin\theta\,U(\sin\theta)\sin\alpha\, e^{-\alpha})$$

This equation is integrated between 0 and $+\infty$ and leads to:

$$\tilde{u}_r(\zeta = 0) = \frac{\sqrt{E}}{\sin\theta}\frac{\partial}{\partial\theta}\left(\frac{\sin\theta\,U(\sin\theta)}{\sqrt{2\cos\theta}}\right)$$

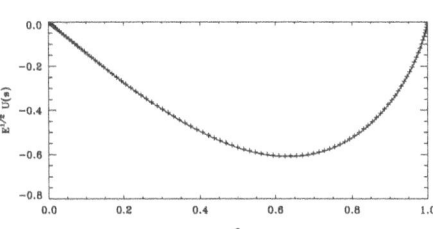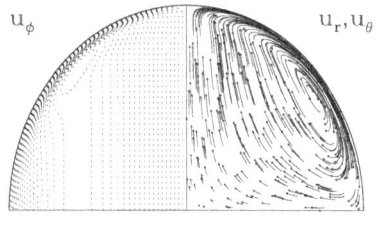

Fig. 8.6 *Left* the analytic solution (8.60) of the geostrophic flow associated with the spin-up flow in a sphere (*solid line*) and the numerical solution (pluses). The Ekman number is 10^{-7}. The difference between the two curves is less than a percent outside the Ekman layer. *Right* a meridian view of the velocity field. The Ekman number is $E = 4 \times 10^{-4}$ large enough to make the boundary layer flow clearly visible

Since we know the expression of $\tilde{u}_r (\zeta = 0)$, we get the differential equation for $U(\sin \theta)$. We finally get

$$U(\sin \theta) = -\sqrt{\frac{2}{E}} \sin \theta \left(1 - \sin^2 \theta\right)^{3/4}$$

valid on the sphere $r = 1$. The geostrophic flow therefore reads

$$\mathbf{u}_{\text{geo}}(s) = -\sqrt{\frac{2}{E}} s \left(1 - s^2\right)^{3/4} \mathbf{e}_\varphi \tag{8.60}$$

using the cylindrical coordinate $s = r \sin \theta$.

We have plotted this solution together with the full numerical solution of (8.57) in Fig. 8.6 (left) when $E = 10^{-7}$. The difference between the analytic and numerical solution is not noticeable. That would not be the case if we used $E = 4 \times 10^{-4}$ as in Fig. 8.6 (right) to better show the meridian flow. In fact, at $E = 4 \times 10^{-4}$ the boundary layer theory is not performing well (although E is small).

8.4.4.3   Conclusion: The Spin-Up Time

We have summarized in Fig. 8.7 all the components of a spin-up flow in a spherical container, including the boundary layer singularity.

One remarkable property of the spin-up flow is that the Ekman circulation controls the time scale of the spin-up. Indeed, this circulation insure the transport of angular momentum from the walls to the interior. We may thus evaluate the time scale of the process of synchronization between the fluid and the container. This is typically the turnover time scale of the Ekman circulation. If L is the characteristic size of the container, the amplitude of this circulation is $\Omega L \sqrt{E}$. It leads to the spin-up time scale:

Fig. 8.7 Schematic view of a spin-up flow within a sphere. $\bar{u}_0$ is the geostrophic azimuthal flow; $\tilde{u}_0$ is the meridional flow within the Ekman layer and $\tilde{u}_1$ is the Ekman pumping forcing the Ekman circulation $\bar{u}_1$

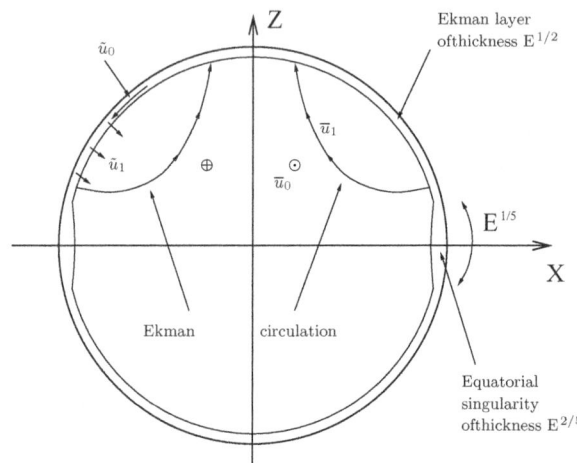

$$t_{\text{spin-up}} = \frac{L}{\Omega L \sqrt{E}} = \frac{\Omega^{-1}}{\sqrt{E}} = \frac{L}{\sqrt{\Omega \nu}} \qquad (8.61)$$

This expression shows that the spin-up of the fluid is realized on a time scale much shorter than the one imposed by viscous diffusion. Indeed,

$$T_{visc} = \frac{L^2}{\nu} = \frac{\Omega^{-1}}{E} \gg \frac{\Omega^{-1}}{\sqrt{E}}$$

since $E \ll 1$.

This new time-scale may be revealed by a simple experiment using a glass of water. If we make rotating the water within the glass, we can measure the time by which the water has ceased to rotate after our forcing has been stopped. Using a glass of water of 7 cm in diameter, rotating the water at one round per second, we find that the fluid flow has almost vanished 2.5 min after. Computing the diffusion time scale, we find $T_{visc} = 0.035^2/10^{-6} \approx 20$ min, which is much larger. This spin-down flow has a time scale, which we evaluate from (8.61), of 20 s, which is much closer to our observation. Within such an experiment, nonlinear effects are quite strong since the rotation ends at zero; however, orders of magnitude are correct, especially if we take a mean rotation of half a round per second.

8.5 Hurricanes

8.5.1 A Qualitative Presentation

In the introduction to this chapter, we mentioned one of the most violent phenomena in the terrestrial atmosphere, namely the hurricanes. We are now ready to explore their dynamics in some details.

First, let us observe that a low pressure region in the Earth atmosphere cannot be filled up by the geostrophic flow that it triggers: the winds are orthogonal to the pressure gradient. Thus, only non-geostrophic effects may fill up a low pressure region. Because of their weakness, the lifetime of such a pressure field is quite long.

In the case of a hurricane, the low pressure field has an especially long life time due to the existence of an energy source: the tropical ocean.

The dynamics of a hurricane can be understood with a simple model. One can then derive the value of the central depression as a function of the temperature of the ocean and of the upper atmosphere. However, before getting into these details, we shall first give a qualitative description of the hurricanes.

Let us consider the air near the surface of a tropical ocean: the percentage of water vapour may be quite high there, due to the important evaporation. Such a mixture is unstable to convection (see Chap. 7): a rising fluid element will face an adiabatic expansion which triggers the condensation of water vapour, releasing latent heat, which amplifies the rise. This process is at the origin of cloud formation and is called *wet convection.*

The low pressure created by the rising elements forces a geostrophic wind, which contributes to make the sea more rough. The fraction of water vapour within the air thus increases. The boundary layer has a radial drift which tries to fill up the depression. Within the centre of the depression air is forced to rise, releasing more and more latent heat and thus making the pressure even lower. Thus, the phenomenon amplifies. However, we may observe that there is a maximum value to the fraction of water vapour in the air: this is the saturation.

Thus we understand why hurricanes appear only in the tropics: the hotter the air, the larger the mass fraction of water vapour. At lower temperatures, the water vapour content is not enough to maintain the winds. It is also clear that above a continent a hurricane dies. Finally, computing the resulting Coriolis force on the ascending air column, we find that the depression should drift to the West as usually observed. This tendency is sometimes counteracted by an anticyclonic air mass.

8.5.2 The Steady State: A Carnot Engine

A hurricane is actually a true Carnot engine, running with a heat source, the ocean, and a cold source, the upper atmosphere. The thermal energy of the ocean is partly converted into mechanical energy: the wind.

In Fig. 8.8 we show the Carnot cycle followed by a fluid element. From A to B the fluid is heated at constant temperature: its water vapour mass fraction increases. From B to C, it follows an adiabatic expansion but during the rise of a fluid parcel water droplets form: it rains! From C to D, the fluid radiates its heat into space and cools down. Next, from D to A the model assumes that the fluid supports an adiabatic compression which is never realized actually. There is no streamline between A and D but this is no problem if the fluid elements have the same entropy at these two points. This is usually assumed.

Fig. 8.8 A meridional view
of a hurricane

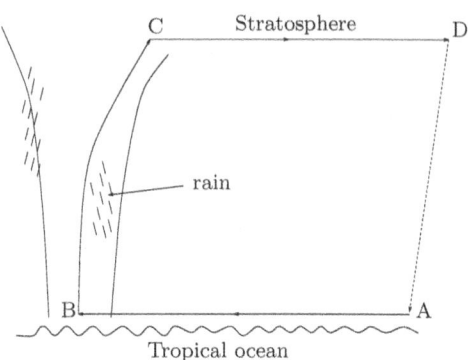

Let us now be a little more quantitative. In a steady regime, along a streamline, we have

$$d(\frac{1}{2}v^2 + gz) + \frac{dP}{\rho} - \mathbf{f} \cdot d\mathbf{l} = 0 \qquad (8.62)$$

where $\mathbf{f}$ represents the forces due to viscosity. We now integrate this relation along the cycle that we just described. It turns out that

$$\oint \mathbf{f} \cdot d\mathbf{l} = \oint \frac{dP}{\rho}$$

For a mixture of air and water vapour, assumed to be an ideal gas, we have

$$dh = c_p dT = Tds + \frac{dP}{\rho} - d(L_v q) \qquad (8.63)$$

where L_v is the latent heat of vapourization and q is the mass fraction of water. Therefore along the cycle $\oint dP/\rho = -\oint Tds$, and

$$\oint \mathbf{f} \cdot d\mathbf{l} = -\oint Tds$$

which shows that the entropy production is due to the friction (viscous dissipation). Now, if T_{sc} and T_{sf} are the temperature of the hot and cold sources respectively, s_A and s_B the entropy at A and B, then

$$-\oint \mathbf{f} \cdot d\mathbf{l} = \oint Tds = (T_{sc} - T_{sf})(s_B - s_A) \qquad (8.64)$$

since $s_C = s_B$ and $s_D = s_A$. Besides, from (8.63) and $\rho = P/(R_*T_{sc})$, we have

$$s_B - s_A = R_* \ln(P_A/P_B) + \frac{L_v}{T_{sc}}(q_B - q_A) \qquad (8.65)$$

Now we have to evaluate the power dissipated by friction. The main contribution to $\oint \mathbf{f} \cdot d\mathbf{l}$ comes from the leg AB which is in the atmospheric boundary layer. The flow there follows a spiral, namely the azimuthal geostrophic flow combined with a radial drift of the (turbulent) Ekman layer. Using (8.62) between A and B, we see that the work of friction forces comes from the pressure (just like for the Poiseuille flow). It turns out that

$$\oint_{AB} \mathbf{f} \cdot d\mathbf{l} = -\oint_{AB} \frac{dP}{\rho} = R_*T_{sc} \ln(P_A/P_B) \qquad (8.66)$$

since the temperature is constant on AB and $P = R_*\rho T$. Finally, combining (8.64), (8.65) and (8.66), we get

$$R_*T_{sf} \ln(P_A/P_B) = \varepsilon L_V(q_B - q_A) \qquad (8.67)$$

where we introduced $\varepsilon = (T_{sc} - T_{sf})/T_{sc}$ which is nothing but the efficiency of the Carnot cycle. Equation (8.67) shows that the depression of the hurricane will be all the stronger that q_B be the larger. However, the highest quantity of water vapour in the air is reached when the air is saturated. Setting q_B to this maximum value, we obtain the minimum central pressure, that is the strongest hurricane. If T_{sc} is expressed in Celsius degrees, a very good approximation of q at saturation is given by

$$q_{sat} = \frac{380.2}{P} \exp\left[\frac{17.67\, T_{sc}}{243.5 + T_{sc}}\right]$$

where P is expressed in Pascals.

Let us take the case of a hurricane blowing in the Northern Atlantic ocean. A typical temperature in the Caribbean sea is $28\,°C$, while that of the stratosphere is $T_{sf} = -60\,°C$; using $L_V = 2.3\ 10^6$ J/kg and assuming that outside the hurricane the partial pressure of water vapour is 75 %, we find that the ratio P_A/P_B verifies

$$\ln(P_A/P_B) = 0.256(P_A/P_B - 0.75)$$

which solution is $P_A/P_B \simeq 1.09$. The strongest hurricane has a central pressure about 930 hPa. For comparison, Emily (Fig. 8.9) had a central pressure at 960 hPa. However, some hurricanes in the Eastern Pacific ocean have reached pressures as low as 870 hPa.

Fig. 8.9 The hurricane
Emily near the coast of North
Carolina (USA) in September
1993

8.5.3   The Birth of Hurricanes

The foregoing very simplified model allows us to understand the way hurricanes work. However, it does not teach us how such vortices arise. Indeed, the conditions for their existence are realized most of the time in tropical oceans. Thus, we would expect that they would be always present. However, this is by far not the case: Hurricanes are rather rare features in the atmosphere. According to recent studies, this scarcity seems to be the consequence of the finite-amplitude nature of the instability that leads to a hurricane. At the origin of the phenomenon, we mentioned the wet convection. This convection is usually giving birth to gentle clouds like cumulus, which extend over a fraction of the troposphere. When a hurricane sets in, wet convection is able to connect the ocean (the heat source) with the stratosphere (the cold source), otherwise the Carnot engine does not work. This is like if a cumulus extends over the whole troposphere. Only, a small fraction of tropical storms reach such an amplitude and turn into a hurricane.

In fact, many sides of the hurricanes dynamics remain obscure because of their complexity. For instance, only very few hurricanes reach the strongest state. Likely, the storm sweeping the ocean, generate an upwelling of cold water, which cools the surface water and decreases the water vapour content of the air near the surface. Hence, a good model needs to take into account the dynamic coupling between the ocean and the atmosphere, and this is not an easy matter.

8.6 Exercises

1. Show by a direct transformation (8.1). This demonstration may be done in three steps: (i) splitting the velocity field into a solid rotation and a remaining flow, (ii) noting that this flow should be expressed using the unit vectors of the rotating frame and (iii) observing that the time dependence of the velocity changes.
2. Show that even with viscosity, the frequency of inertial is such that $|\omega| \leq 1$.
3. Show that the stream function of axisymmetric inertial modes obeys a hyperbolic equation similar to the Poincaré one.
4. Explain why hurricanes do not appear on the equator.

Further Reading

There is only one monograph dealing entirely with rotating fluids: *The theory of rotating fluids* by Greenspan (1969), unfortunately out of print. However, some insights may be found in *Geophysical fluid dynamics* by Pedlosky (1979) but in the context of the ocean and atmosphere dynamics. A recent review of spin-up flows is given by Duck and Foster (2001). A more detailed presentation of hurricanes may be found in Emanuel (1991).

References

Angot, A. (1949, 1972). *Compléments de mathématiques*. Paris: Masson.
Bryan, G. (1889). The waves on a rotating liquid spheroid of finite ellipticity. *Philosophical Transactions of the Royal Society London, 180*, 187–219.
Duck, P. & Foster, M. (2001). Spin-Up of Homogeneous and Stratified Fluids. *Annual Review of Fluid Mechanics, 33*, 231–263.
Emanuel, K. (1991). The theory of hurricanes. *The Annual Review of Fluid Mechanics, 23*, 179.
Greenspan, H. P. (1969). *The theory of rotating fluids*. Cambridge: Cambridge University Press.
Longuet-Higgins, M. S. (1964). Planetary waves on a rotating sphere. *Proceedings of the Royal Society of London, A279*, 446–473.
Pedlosky, J. (1979). *Geophysical fluid dynamics*. New York: Springer.
Rieutord, M. (2001). Ekman layers and the damping of inertial r-modes in a spherical shell: application to neutron stars. *The Astrophysical Journal, 550*, 443–447.
Rieutord, M., Georgeot, B. & Valdettaro, L. (2001). Inertial waves in a rotating spherical shell: Attractors and asymptotic spectrum. *The Journal of Fluid Mechanics, 435*, 103–144.
Roberts, P. & Stewartson, K. 1963 On the stability of a Maclaurin spheroid of small viscosity. *The Astrophysical Journal, 137*, 777–790.

Chapter 9
Turbulence

As we pointed it out in the first pages of this book, the understanding of turbulence remains one of the challenges of nowadays physics. The goal of this chapter is to introduce the reader to the main approaches that are used to deal with this difficult problem.

9.1 The Fundamental Problem of Turbulent Flows

9.1.1 How Can We Define Turbulence?

Defining a turbulent flow is not an easy matter because, to be precise, we need some notions that we shall develop below. However, if we are satisfied with general ideas, we can make it. First, let us observe that turbulent flows are quite disordered: a lot of vortices seem to be constantly appearing and disappearing in an essentially random way; this seems to be their main feature. To characterize this disorder, correlations are very useful. Let us introduce this notion in a simple way. If, in a turbulent flow, we record one component of the velocity at two distinct points A and B. The result is like the curves plotted in Fig. 9.1. These curves show an evolution of the physical quantities that looks like impredictable. To characterize the nature of this signal, one uses the autocorrelation function, which is the average over time of the product $V_A(t)V_A(t+T)$. The autocorrelation characterizes in some way the similarity of the function with itself at a different point. If the function changes nearly randomly, $V_A(t+T)$ is statistically independent of $V_A(t)$. Of course T cannot be too small, for the functions are continuous. The random nature appears when T is much larger than a specific time interval T_c which is called the *correlation time*. When this time is finite, the flow has a chaotic evolution. This is the case for a turbulent flow, but such a flow owns an "additional" chaos, namely a "spatial" chaos. Indeed, if we consider two points A and B at some large distance from one

© Springer International Publishing Switzerland 2015
M. Rieutord, *Fluid Dynamics*, Graduate Texts in Physics,
DOI 10.1007/978-3-319-09351-2_9

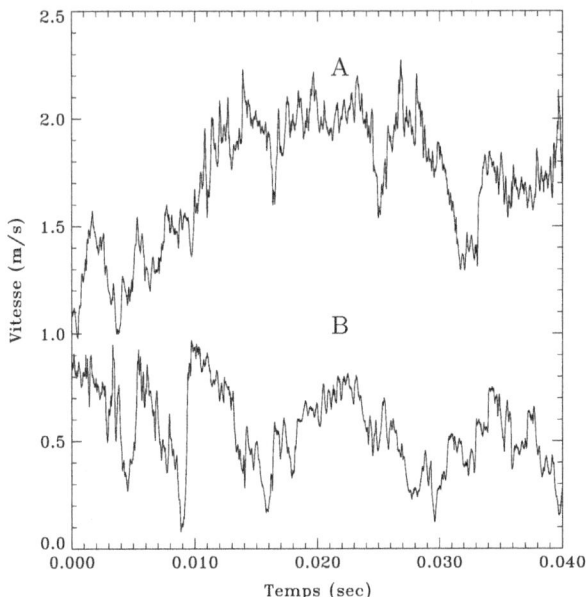

Fig. 9.1 Time evolution of the velocity at two distinct points A and B in a turbulent flow. No similarities between the two curves can be detected, except their random nature

another, the velocities at these points appear uncorrelated. Only when the distance is short enough, less than *the correlation length*, do the velocities show correlated values, so that $\langle V_A V_B \rangle \neq \langle V_A \rangle \langle V_B \rangle$, $\langle \, \rangle$ being a statistical average.

With these ideas in mind, it is easy to define turbulent flows: they are these flows where the correlation length is shorter that the size of the fluid domain and whose correlation time is shorter than the time scale we focus on. Turbulence thus appears as a fluid motion endowed with a "spatio-temporal chaos" or a "spatio-temporal decorrelation".

9.1.2 The Closure Problem of the Averaged Equations

The apparent random nature of a turbulent flow suggests that these flows should be studied using the tools of Statistics. We indeed suspect that only some mean quantities are really important for the understanding of the properties of the flow. Therefore, the effort should be concentrated on the derivation of the equations governing these mean quantities. This derivation raises a very difficult problem that has still not been circumvented. In order to present it in a simple way, let us consider the turbulent flow of some incompressible fluid. We write the equations of motion in a symbolic way:

$$\begin{cases} \partial_t v + \partial(vv) = -\nabla P + \Delta v \\ \nabla \cdot v = 0 \end{cases} \qquad (9.1)$$

Let us now write the averaged equation. We use the averaging operator $\langle\ \rangle$ which we assume to commute with all the derivative operators. Applying this to the foregoing system, we get:

$$\begin{cases} \partial_t \langle v \rangle + \partial \langle vv \rangle = -\nabla \langle P \rangle + \Delta \langle v \rangle \\ \nabla \cdot \langle v \rangle = 0 \end{cases} \qquad (9.2)$$

which governs the evolution of the mean velocity $\langle v \rangle$. However, this system is not closed since the quantity $\langle vv \rangle$ is unknown. It is usually different from $\langle v \rangle^2$. Thus, we need a new equation to constrain this quantity. The way to get this new equation is through (9.1) which is first multiplied by the velocity and then averaged. It turns out that we get a new equation like:

$$\frac{\partial \langle vv \rangle}{\partial t} + \partial \langle vvv \rangle = -\nabla \langle Pv \rangle + \langle v \Delta v \rangle$$

The important feature of this new equation is that it contains a term using $\langle vvv \rangle$, which is a triple correlation. Hence, the evolution of double correlations, like $\langle vv \rangle$, is controlled by the triple correlations and we easily guess that the evolution of the triple correlations depends on the one of fourth order correlations $\langle vvvv \rangle$ etc.

Hence, we always have a set of equations which contains a larger set of unknowns: this is the famous problem of the *closure* of the mean-field equations, for which the general solution is still missing.

9.2   The Tools

In order to continue this study, we now need to introduce the good tools which will allow us to deal with the random nature of turbulent flows. Very naturally, we shall borrow these tools to Statistical Physics.

9.2.1   Ensemble Averages

An ensemble average $\langle X \rangle$ of a quantity X is the average derived from N independent experiments measuring this quantity when N goes to infinity. Thus

$$\langle X \rangle = \lim_{N \to \infty} \frac{1}{N} \sum_{n=1}^{N} X_n$$

For the velocity field we have:

$$\langle \mathbf{v}(\mathbf{r}, t) \rangle = \lim_{N \to \infty} \frac{1}{N} \sum_{n=1}^{N} \mathbf{v}_n(\mathbf{r}, t)$$

Note that the mean field remains a function of space and time. We easily verify that taking the ensemble average of a quantity is an operation that commutes with all types of differentiation. Hence

$$\left\langle \frac{\partial \mathbf{v}}{\partial r} \right\rangle = \frac{\partial \langle \mathbf{v} \rangle}{\partial r}.$$

9.2.2 Probability Distributions

In the following of the chapter, the notion of probability distribution or that of probability density function will be used very often. We thus need to define the meaning of these tools.

Let us assume that we are analysing the pressure fluctuations and are interested in the probability of finding this quantity in the interval $]-\infty, x]$. We denote this probability $F_P(x)$, where the index P means that P is the random variable. The function F_P is called a *probability distribution function*. From this definition it turns out that

- $F_P(x)$ is not decreasing
- $F_P(x)$ is continuous on right
- $F_P(-\infty) = 0$ and $F_P(+\infty) = 1$.

If this function is differentiable, then $F_P'(x)$ is called the *probability density function*, often called the "pdf". Physically, it means that $F_P'(x)dx$ is the probability of finding P in the interval $]x, x + dx]$.

9.2.3 Moments and Cumulants

The moments of a probability distribution are the averages of the powers of the random variable:

$$M_n = \langle P^n \rangle = \int_{-\infty}^{+\infty} x^n F_P'(x)dx \tag{9.3}$$

The first order moment is just *the average or mean* also called *mathematical expectation or expected value*.

The moments of $P - \langle P \rangle$ are said to be *centered*. The variance is the centered moment of order two:

$$\langle (P - \langle P \rangle)^2 \rangle$$

and the root mean square is the square root of the variance.

The cumulant of order n of a random variable is the difference between the n-th order moment of this variable and the same moment of the gaussian distribution that has the same variance and the same mean as the given random variable.

9.2.4 Correlations and Structure Functions

In addition to the correlations that we already introduced at the beginning of this chapter, we shall also need n-points correlations which are the averages of a given function taken at n different points.

$$\langle f(x_1) f(x_2) \ldots f(x_n) \rangle$$

As far as vectorial quantities are concerned, like the velocity, we may mix the components like in

$$\langle v_i(\mathbf{r}_1) v_j(\mathbf{r}_2) \cdots v_k(\mathbf{r}_n) \rangle$$

which leads to the definition of new tensors.

Structure functions are quantities like

$$S_n = \langle ((f(\mathbf{r}_1) - f(\mathbf{r}_2))^n \rangle$$

which are easily related to two-points correlations.

9.2.5 Symmetries

The study of turbulent flows is much simplified when some symmetries are verified. Unfortunately, real flows usually own little symmetries or very approximately. Nevertheless, their use is very handy to reduce a given problem to its essential features.

Five symmetries may actually be verified by a turbulent flow:

- *Homogeneity*: This is the invariance of the properties of turbulence (all the moments for instance) with respect to space translations. This is a very strong hypothesis that is verified in only small regions. However, it generates important simplifications that are very useful to understand the problems generated by turbulence.

- *steadyness/stationarity*: This is invariance with respect to time translations. This is a less constraining hypothesis, but it is often difficult to really know when the flow is really steady on average.
- *Isotropy*: This is the invariance of the turbulence with respect to space rotation $\mathcal{O}(3)$. It is very useful in homogeneous turbulence in order to simplify the calculations (of course, it cannot exist in non-homogenous turbulence).
- *Parity*: This is the symmetry with respect to a point or a plane. One may see it as an invariance with respect to a change of sign of all the vectors. It cannot be reduced to a combination of rotations. It allows the elimination of quantities like helicity, which change sign when this symmetry is imposed.
- *Scale invariance*: This is the invariance with respect to transformations like $\mathbf{v}(\mathbf{r}, t) \rightarrow \lambda^h \mathbf{v}(\lambda \mathbf{r}, \lambda^{1-h} t)$. The solutions of Navier–Stokes equation satisfy this symmetry only if $h = -1$. This is just the similarity of flows at the same Reynolds number that we studied in Chap. 4. Now, if there is no viscosity, then h is not constrained (as long as $\lambda > 0$). In turbulent flows, which are usually at high Reynolds numbers, this symmetry may be verified in some range of scales.

In many textbooks, isotropic turbulence refers to a case which includes both isotropy and parity invariance. Here, we shall be more restrictive and always mention the use of parity invariance.

9.3   Two-Points Correlations

After this short presentation of the problem of turbulence and the basic tools that are used to deal with it, we shall now focus on the two-point correlations because they play an important role in the analysis of turbulent flows. To make this study easier we shall restrict ourselves to the case of *homogeneous turbulence* because this case owns all the universal properties of turbulence. Although more general, the non-homogeneous case is very dependent on this problem and can be discussed in a second step.

9.3.1   The Reynolds Stress

Let us come back to the mean-field equations using as in the introduction the case of an incompressible fluid. We now decompose the velocity field as:

$$\mathbf{v} = \langle \mathbf{v} \rangle + \mathbf{v}' \tag{9.4}$$

where $\langle \cdots \rangle$ is an ensemble average. The fluctuation $\mathbf{v}'$ is necessarily such that $\langle \mathbf{v}' \rangle = 0$. This decomposition is known as *Reynolds decomposition*. Inserting it in Navier–Stokes equation, we get

$$\rho \left(\frac{\partial \langle \mathbf{v} \rangle}{\partial t} + \frac{\partial \mathbf{v}'}{\partial t} + \langle \mathbf{v} \rangle \cdot \nabla \mathbf{v}' + \mathbf{v}' \cdot \nabla \langle \mathbf{v} \rangle + \langle \mathbf{v} \rangle \cdot \nabla \langle \mathbf{v} \rangle + \mathbf{v}' \cdot \nabla \mathbf{v}' \right) =$$
$$-\nabla \langle P \rangle - \nabla P' + \mu \Delta \langle \mathbf{v} \rangle + \mu \Delta \mathbf{v}' \quad (9.5)$$

$$\nabla \cdot \langle \mathbf{v} \rangle + \nabla \cdot \mathbf{v}' = 0 \quad (9.6)$$

Averaging these equations, we get:

$$\rho \left(\frac{\partial \langle \mathbf{v} \rangle}{\partial t} + \langle \mathbf{v} \rangle \cdot \nabla \langle \mathbf{v} \rangle + \langle \mathbf{v}' \cdot \nabla \mathbf{v}' \rangle \right) = -\nabla \langle P \rangle + \mu \Delta \langle \mathbf{v} \rangle \quad (9.7)$$

$$\nabla \cdot \langle \mathbf{v} \rangle = 0 \quad (9.8)$$

These new equations contain a new quantity, namely $\langle \rho \mathbf{v}' \cdot \nabla \mathbf{v}' \rangle$ which is related to the *Reynolds stress tensor*:

$$[R] = \rho \langle \mathbf{v}' \otimes \mathbf{v}' \rangle$$

where $\otimes$ denote the tensorial product. The components of $[R]$ are

$$R_{ij} = \rho \langle v'_i v'_j \rangle$$

Now, (9.6) and (9.8) imply

$$\nabla \cdot \mathbf{v}' = 0 \quad (9.9)$$

so that

$$\rho \langle \mathbf{v}' \cdot \nabla \mathbf{v}' \rangle_i = \rho \partial_k \langle v'_i v'_k \rangle = \partial_k R_{ik}$$

if ρ is constant.[1]

---

[1]Let us note here that the true stress induced by the correlation $\langle \mathbf{v}' \otimes \mathbf{v}' \rangle$ is rather $-R_{ij}$ since the momentum equation (9.7) may also be written

$$\rho \frac{D \langle v_i \rangle}{Dt} = \partial_j \sigma_{ij}$$

with $\sigma_{ij} = -\langle P \rangle + \mu (\partial_i \langle v_j \rangle + \partial_j \langle v_i \rangle) - R_{ij}$. Note also that the Reynolds tensor is often defined as $\langle \mathbf{v}' \otimes \mathbf{v}' \rangle$.

We observed in the introduction of this chapter that the turbulence problem comes from our ignorance of how to express R_{ik} as a function of the mean-fields $\langle \mathbf{v} \rangle$ and $\langle P \rangle$. Modeling turbulence is therefore equivalent to finding a way of relating these quantities. As we may suspect it, no universal way is known yet. Understanding the present ones, which are all ad hoc at some level, requires a good knowledge of the properties of the Reynolds tensor. This demands a study of the velocity two-point correlations.

9.3.2   The Velocity Two-Point Correlations

The first step in studying the properties of the Reynolds tensor needs the investigation of a slightly more general quantity, namely

$$Q_{ij}(\mathbf{r}_A, \mathbf{r}_B) = \left\langle v_i'(\mathbf{r}_A)v_j'(\mathbf{r}_B)\right\rangle \tag{9.10}$$

which is the *second order tensor of the velocity (fluctuations) correlations* taken at two points A and B. In order to simplify the study, we assume that there is no mean flow, namely $\langle \mathbf{v} \rangle = \mathbf{0}$. Thus, the velocity and its fluctuations are identical.

In full generality, this tensor is a function of six space variables and two time variables. Again, we simplify the matter by assuming that the velocities are taken at the same time. Moreover we drop the time variable since we work on the spatial properties of the tensor. We also set $\mathbf{r}_A = \mathbf{x}$ and $\mathbf{r}_B = \mathbf{x}' = \mathbf{x} + \mathbf{r}$. Thus

$$Q_{ij}(\mathbf{x}, \mathbf{x}') = \left\langle v_i(\mathbf{x})v_j(\mathbf{x} + \mathbf{r})\right\rangle \tag{9.11}$$

The first (elementary) property of $[Q]$ is that

$$Q_{ij}(\mathbf{x}, \mathbf{x}') = Q_{ji}(\mathbf{x}', \mathbf{x}) \tag{9.12}$$

In addition, since we are working with an incompressible fluid, (9.9) is verified and

$$\partial_i Q_{ij} = \partial_j' Q_{ij} = 0 \tag{9.13}$$

where $\partial_j' = \partial/\partial x_j'$.

Let us now assume that the turbulence is homogeneous. Its properties are thus independent of the point that is considered. Consequently, Q_{ij} depends only on the difference between the two vectors $\mathbf{x}$ and $\mathbf{x}'$, so that $Q_{ij} \equiv Q_{ij}(\mathbf{r})$. We further note that

$$Q_{ij}(\mathbf{0}) = R_{ij}$$

and that $R_{ii} = Tr([Q])(0)$ is just twice the turbulent kinetic energy per unit mass.

The homogeneity of turbulence allows us to reduce the number of independent components of $[Q]$. Actually, as we shall see, they are only three. First of all, (9.12) implies that

$$Q_{ij}(\mathbf{r}) = Q_{ji}(-\mathbf{r}) \qquad (9.14)$$

So that we are left with only six independent components. However, we have to build up a second order tensor that only depends on the vector $\mathbf{r}$. Since second order tensors are built up from tensor products using tensors of lower orders, the only possibility is

$$Q_{ij} = A(\mathbf{r})\delta_{ij} + B(\mathbf{r})\frac{r_i r_j}{r^2} + H(\mathbf{r})\epsilon_{ijk}\frac{r_k}{r} \qquad (9.15)$$

The three functions $A(\mathbf{r})$, $B(\mathbf{r})$ and $H(\mathbf{r})$ are the three independent components. They are unknown, but if we use (9.14), we find that they verify

$$A(-\mathbf{r}) = A(\mathbf{r}), \quad B(-\mathbf{r}) = B(\mathbf{r}), \quad H(-\mathbf{r}) = H(\mathbf{r})$$

Furthermore, H is a pseudo-scalar.[2] One consequence is: if the turbulence is parity-invariant then H is zero. We shall see below that this quantity is related to helicity.

Now we may further constraint the unknown components of Q_{ij} using the relation of incompressibility. Equation (9.13) leads to

$$\nabla A + \mathbf{e}_r \nabla \cdot (B\mathbf{e}_r) + \nabla H \times \mathbf{e}_r = 0$$

Let us further restrict our discussion to that of *isotropic turbulence*. In this case, the functions no longer depend on the direction of $\mathbf{r}$, but only on $r = \|\mathbf{r}\|$. Incompressibility then relates A and B

$$\frac{\partial A}{\partial r} + \frac{1}{r^2}\frac{\partial r^2 B}{\partial r} = 0$$

We now introduce the longitudinal and transversal velocity correlations. These quantities may indeed be measured experimentally, and they are usually used instead of A and B. The longitudinal component of the velocity is the one which is parallel to $\mathbf{r}$. We call it v_ℓ, and thus $v_\ell = \mathbf{v} \cdot \mathbf{r}/r$.

---

[2]A pseudo-scalar is a scalar quantity the sign of which depends on the orientation of the vector basis. For instance, the determinant of three vectors (in three dimensions) is a pseudo-scalar. In our case, if $\mathbf{X}$ et $\mathbf{Y}$ are two vectors, from the definition of $[Q]$, $X_i Y_j Q_{ij}$ is a true scalar. Thus

$$A(\mathbf{X} \cdot \mathbf{Y})^2 + (\mathbf{r} \cdot \mathbf{X})(\mathbf{r} \cdot \mathbf{Y})B/r^2 + H\epsilon_{ijk}X_i Y_j r_k/r$$

is a true scalar. In this expression we see that the last term is the determinant of three vectors times H. Thus H is a pseudo-scalar.

The transversal component is just the remaining vector $\mathbf{v}^t = \mathbf{v} - v_\ell \mathbf{r}/r$. The longitudinal and transversal correlations are usually denoted f and g. From these definitions and using the expression (9.15) of Q_{ij}, we easily find that

$$f = Q_{ij} r_i r_j / r^2 \quad \text{and} \quad g = Q_{ij} (\mathbf{v}^t)_i (\mathbf{v}^t)_j / \|\mathbf{v}^t\|^2$$

so that

$$f = A + B \quad \text{and} \quad g = A$$

Incompressibility allows us to relate g to f, namely

$$g = \frac{1}{2r} \frac{\partial r^2 f}{\partial r} \tag{9.16}$$

Finally, we get the expression of Q_{ij} for the homogeneous isotropic turbulence of an incompressible fluid:

$$Q_{ij} = \frac{1}{2r} \frac{\partial r^2 f}{\partial r} \delta_{ij} - \frac{r}{2} \frac{\partial f}{\partial r} \frac{r_i r_j}{r^2} + H(\mathbf{r}) \epsilon_{ijk} \frac{r_k}{r} \tag{9.17}$$

We end this discussion with a final remark on the behaviour of $Q(\mathbf{r})$ when the distance between the point grows without bound. We said in the introduction that turbulence was also characterized by a finite correlation length L_c. Hence, if $r \gg L_c$ then the velocity correlations should be negligible. It is therefore legitimate to assume:

$$\lim_{r \to \infty} Q_{ij}(r) = 0 \tag{9.18}$$

9.3.3 Vorticity and Helicity Correlations

We shall need later another tensor, namely the one of vorticity correlations at two points:

$$\Omega_{ij} = \langle \omega_i(\mathbf{x}) \omega_j(\mathbf{x}') \rangle \tag{9.19}$$

Just like the velocity correlation tensor, this tensor depends on six independent variables $\mathbf{x}$ and $\mathbf{x}'$. However, $\omega_i = \epsilon_{ikl} \partial_k v_l$, but $\mathbf{x}$ and $\mathbf{x}'$ are independent; we may thus write:

$$\Omega_{ij} = \epsilon_{ikl} \epsilon_{jmn} \frac{\partial^2}{\partial x_k \partial x'_m} \langle v_l(\mathbf{x}) v_n(\mathbf{x}') \rangle = -\epsilon_{ikl} \epsilon_{jmn} \frac{\partial^2}{\partial r_k \partial r_m} Q_{ln}(\mathbf{r})$$

where we used the homogeneity of turbulence and the relations

$$\frac{\partial Q}{\partial x_j} = -\frac{\partial Q}{\partial r_j}, \qquad \frac{\partial Q}{\partial x'_j} = \frac{\partial Q}{\partial r_j}$$

since $\mathbf{r} = \mathbf{x}' - \mathbf{x}$. Now, using (12.1) together with incompressibility (9.13), the foregoing relation leads to the following expression:

$$\Omega_{ij} = \frac{\partial^2 Q_{kk}}{\partial r_i \partial r_j} - \Delta Q_{ji} - \delta_{ij} \Delta Q_{kk} \qquad (9.20)$$

where Δ is the Laplacian.

Finally, we also need the cross-correlation vorticity-velocity, which is called helicity correlation or just mean helicity. This is

$$\langle \omega_i(\mathbf{x}) v_i(\mathbf{x}') \rangle = \epsilon_{ijk} \frac{\partial v_k}{\partial x_j} v_i(\mathbf{x}') = \epsilon_{ijk} \frac{\partial}{\partial x_j} \langle v_k(\mathbf{x}) v_i(\mathbf{x}') \rangle$$

$$= \epsilon_{ijk} \frac{\partial Q_{ki}}{\partial x_j} = \epsilon_{ijk} \frac{\partial Q_{ik}}{\partial r_j} = -\frac{\partial}{\partial r_j} \left(\frac{2H(\mathbf{r}) r_j}{r} \right) \qquad (9.21)$$

This definition does not depend on the order of points. Indeed, if we exchange $\mathbf{x}$ and $\mathbf{x}'$, we find

$$\langle \omega_i(\mathbf{x}') v_i(\mathbf{x}) \rangle = \epsilon_{ijk} \frac{\partial Q_{ik}(\mathbf{x}, \mathbf{x}')}{\partial x'_j} = \epsilon_{ijk} \frac{\partial Q_{ik}(\mathbf{r})}{\partial r_j}$$

since $Q_{ik}(\mathbf{x}, \mathbf{x}') = Q_{ik}(\mathbf{x}' - \mathbf{x}) = Q_{ik}(\mathbf{r})$.

9.3.4 The Associated Spectral Correlations

When we analysed instabilities, we found it convenient to decompose the unknowns on a basis of orthogonal functions. When we are dealing with turbulent flows such a decomposition is also useful. In the simple case of homogeneous turbulence, the Fourier basis is appropriate. The Fourier transform of the tensors describing the correlations own interesting properties which give another view of turbulence, in particular on its energy side.

As a first step we introduce the Fourier transform that are used to obtain the spectral quantities. Thus let us define $\hat{f}(\mathbf{k})$ the Fourier transform of a square integrable function $f(\mathbf{r})$ and the inverse transform. We have

$$\hat{f}(\mathbf{k}) = (2\pi)^{-3} \int f(\mathbf{r}) e^{-i\mathbf{k}\cdot\mathbf{r}} d^3\mathbf{r} \qquad \text{and} \qquad f(\mathbf{r}) = \int \hat{f}(\mathbf{k}) e^{i\mathbf{k}\cdot\mathbf{r}} d^3\mathbf{k}$$

Let us now introduce the Fourier transform of the two point velocity correlation tensor, namely:

$$\phi_{ij}(\mathbf{k}) = (2\pi)^{-3} \int Q_{ij}(\mathbf{r})e^{-i\mathbf{k}\cdot\mathbf{r}}d^3\mathbf{r}$$

From (9.14), we easily show that

$$\phi_{ji}(\mathbf{k}) = \phi_{ij}(-\mathbf{k}) = \phi_{ij}^*(\mathbf{k}) \tag{9.22}$$

while incompressibility implies that

$$k_i\phi_{ij}(\mathbf{k}) = 0 \tag{9.23}$$

The relation (9.22) implies that the symmetric part of $[\phi]$ is real while the antisymmetric part is purely imaginary. Indeed, if we set:

$$\phi_{ij} = \frac{1}{2}(\phi_{ij} - \phi_{ji}) + \frac{1}{2}(\phi_{ij} + \phi_{ji}) = \hat{A}_{ij} + \hat{S}_{ij}$$

then (9.22) gives

$$\hat{A}_{ij}^* = -\hat{A}_{ij} \qquad \hat{S}_{ij}^* = \hat{S}_{ij} \tag{9.24}$$

Thus quite generally, we may write

$$\hat{A}_{ij} = i\,\epsilon_{ijn}a_n$$

where $\mathbf{a}$ is an unspecified real vector. However, incompressibility implies that $k_i\hat{A}_{ij} = 0$ and thus $\mathbf{a} \times \mathbf{k} = \mathbf{0}$. Hence, we can set $\mathbf{a} = h(\mathbf{k})\mathbf{k}$ and

$$\hat{A}_{ij} = i\,\epsilon_{ijn}k_n h(\mathbf{k}) \tag{9.25}$$

The function $h(\mathbf{k})$ is a pseudo-scalar related, as we may guess, to the helicity of turbulence. Let us indeed take the Fourier transform of $H(\mathbf{r})$. One can show (see exercises) that (9.21) yields

$$\hat{H} = i\,\epsilon_{ijn}k_j\phi_{in}$$

Using (9.25), we also get $\hat{H} = 2k^2h(\mathbf{k})$. Thus the antisymmetric part of ϕ_{ij} is just proportional to the Fourier transform of helicity correlations. We thus write

$$\hat{A}_{ij} = \frac{i\hat{H}(\mathbf{k})}{2k^2}\epsilon_{ijn}k_n$$

We can treat in a similar way the symmetric part of ϕ_{ij}, since $\hat{S}_{ij}$ is a symmetric tensor that depends only on $\mathbf{k}$. Its most general form is therefore

$$\hat{S}_{ij}(\mathbf{k}) = \hat{e}(\mathbf{k})\delta_{ij} + \hat{g}(\mathbf{k})k_i k_j$$

Again, incompressibility can be used to simplify the expression and we obtain

$$\hat{S}_{ij} = \hat{e}(\mathbf{k}) \left(\delta_{ij} - \frac{k_i k_j}{k^2} \right) \tag{9.26}$$

We shall see below that the function $\hat{e}(\mathbf{k})$ is related to the kinetic energy spectrum of the turbulence. Finally, we may write the general form of ϕ_{ij} as:

$$\phi_{ij} = \hat{e}(\mathbf{k}) P_{ij} + \frac{i\hat{H}(\mathbf{k})}{2k^2}\epsilon_{ijn}k_n \tag{9.27}$$

In this latter expression we introduced $P_{ij} = \delta_{ij} - k_i k_j/k^2$ also called the *projection tensor*. Indeed, if $\mathbf{a}$ is some vector, $P_{ij}a_j$ is a vector which lies in a plane perpendicular to $\mathbf{k}$ since $k_i P_{ij}a_j = 0$. This tensor often appears when one deals with incompressible fluids, since the continuity equation implies that the Fourier transform of the velocity belongs to this plane.

In the same way we dealt with velocity correlations, we can consider vorticity correlations. Let $Z_{ij}(\mathbf{k})$ be the Fourier transform of Ω_{ij}. Using (9.20), $Z_{ij}(\mathbf{k})$ can be expressed as a function of $\phi_{ij}(\mathbf{k})$, namely

$$Z_{ij} = P_{ij}k^2\phi_{nn} - k^2\phi_{ji} \tag{9.28}$$

9.3.5 Spectra

We alluded above to the relation between the kinetic energy density and the Reynolds stress tensor which trace is just twice this quantity. Now, we may focus on the spectral energy density per unit mass, that is on the kinetic energy which is contained in the wavenumber interval $[k, k + dk]$. This quantity is $E(k)$ and it is defined by

$$E_{turb} = \frac{1}{2}\langle v^2 \rangle = \int_0^{+\infty} E(k)dk \tag{9.29}$$

We shall now relate this quantity to ϕ_{ij}. Indeed,

$$E_{turb} = \frac{1}{2}Q_{ii}(0) = \frac{1}{2}\int \phi_{ii}(\mathbf{k})d^3\mathbf{k} = \frac{1}{2}\int_0^{+\infty} k^2 dk \int_{(4\pi)} \phi_{ii}(\mathbf{k})d\Omega_k$$

$$\Longrightarrow \quad E(k) = \frac{1}{2}k^2 \int_{(4\pi)} \phi_{ii}(\mathbf{k})d\Omega_k \tag{9.30}$$

where $d\Omega_k$ is the elementary solid angle in the Fourier space.

In a similar way, we can define the enstrophy spectrum $Z(k)$ writing

$$Z = \frac{1}{2}\langle \omega^2 \rangle = \int_0^{+\infty} Z(k)dk \,;$$

where we see that enstrophy is analogous to turbulent kinetic energy, but using vorticity instead of the velocity field. Similarly as for kinetic energy, we write

$$Z(k) = \frac{1}{2}k^2 \int_{(4\pi)} Z_{ii}(\mathbf{k})d\Omega_k$$

Using (9.28) we deduce that $Z_{ii} = k^2 \phi_{ii}$. This allows us to relate the enstrophy and kinetic energy spectra by

$$Z(k) = k^2 E(k) \tag{9.31}$$

Finally, we may also introduce the helicity spectrum $H(k)$ such that

$$\langle \boldsymbol{\omega} \cdot \mathbf{v} \rangle = \int_0^{+\infty} H(k)dk \tag{9.32}$$

9.3.6 The Isotropic Case

We shall specialize a little more our discussion by focusing on the important case of isotropic turbulence.

In this case the tensor $[Q]$ depends only on the distance r between the two points. We introduce the function $R(r) = \frac{1}{2}Q_{ii}(r)$ which is just half the trace of $[Q]$. Let us note that the value of R at $r = 0$ is just the local mean kinetic energy per unit mass, $R(0) = E_{turb}$.

If the turbulence is isotropic, then the Fourier transform of Q_{ij} is independent of the direction of the wavevector $\mathbf{k}$, thus

$$E(k) = 2\pi k^2 \phi_{ii}(k) = 4\pi k^2 \hat{e}(k) \tag{9.33}$$

following (9.30). If we observe that $H(k) = 4\pi k^2 \hat{H}(k)$, then

$$\phi_{ij} = \frac{E(k)}{4\pi k^2} P_{ij} + \frac{iH(k)}{8\pi k^4} \epsilon_{ijn}k_n \tag{9.34}$$

The expression of $R(r)$ with respect to $E(k)$ may be derived using the expression of ϕ_{ii}. Indeed,

$$R(r) = \frac{1}{2} \int \phi_{ii}(k) e^{i\mathbf{k} \cdot \mathbf{r}} d^3\mathbf{k} = \frac{1}{2} \int \phi_{ii}(k) e^{ikr\cos\theta} k^2 dk \sin\theta \, d\theta \, d\varphi$$

from which it turns out that, after integration on the angular variables and use of (9.33),

$$R(r) = \int_0^\infty E(k) \frac{\sin kr}{kr} dk \tag{9.35}$$

Another property of R is its symmetry with respect to the origin $R(-r) = R(r)$ (see (9.14)). The values of its (even) derivatives at the origin are also related to $E(k)$. Namely

$$\left(\frac{\partial^{2n} R}{\partial r^{2n}}\right)_{r=0} = \int_0^\infty E(k) \frac{\partial^{2n}}{\partial r^{2n}} \left(\frac{\sin kr}{kr}\right)_{r=0} dk = \frac{(-1)^n}{2n+1} \int_0^\infty k^{2n} E(k) dk$$

where we used that

$$\left(\frac{d^{2n}}{dx^{2n}} \frac{\sin x}{x}\right)_{x=0} = \frac{(-1)^n}{2n+1}$$

In particular, the second order derivative verifies

$$\left(\frac{\partial^2 R}{\partial r^2}\right)_{r=0} = -\frac{1}{3} \int_0^\infty k^2 E(k) dk = -\frac{Z}{3} \tag{9.36}$$

which shows that it is related to the local enstrophy Z. It also emphasizes the fact that velocity correlations are, as expected, maximum at $r = 0$ since the derivative is zero and the second derivative is negative.

Relation (9.35) can be inverted (cf exercises) and yields the following relation:

$$E(k) = \frac{2}{\pi} \int_0^\infty kr \sin kr \, R(r) dr \tag{9.37}$$

which shows that if the wavelength of the Fourier mode is much larger than the correlation length, then

$$E(k) = \frac{2k^2}{\pi} \int_0^\infty r^2 R(r) dr$$

since for all the values where $R(r)$ is non-zero, $kr \ll 1$ and thus $\sin kr \sim kr$.

This result shows that whatever the dynamics, the infrared behaviour of a three-dimensional kinetic energy spectrum follows a k^2-law. As may be guessed, the exponent depends on the dimension of space (see Sect. 9.9).

9.3.7 Triple Correlations

The next step in our investigation of turbulence leads us to now examine the triple correlations since we know that they control the evolution of double correlations. As for these correlations, we consider the triple correlations in two points. A priori, we expect two types of triple correlations, namely

$$S_{ijk} = \langle v_i(\mathbf{x})v_j(\mathbf{x})v_k(\mathbf{x}+\mathbf{r})\rangle \quad \text{and} \quad S'_{ijk} = \langle v_i(\mathbf{x})v_j(\mathbf{x}+\mathbf{r})v_k(\mathbf{x}+\mathbf{r})\rangle \tag{9.38}$$

However, when the turbulence is homogeneous, these two quantities are related by

$$S'_{ijk}(\mathbf{r}) = S_{jki}(-\mathbf{r})$$

Thus, only one type of triple correlations exists for homogenous turbulence.

The tensor $[S]$ has some interesting properties that deserve some discussion. It is obviously symmetric with respect to the first two indices. If the turbulence is parity invariant, then it should be invariant if we inverse all the axis of coordinates.

$$S_{ijk}(\mathbf{r}) = \langle v_i(\mathbf{x})v_j(\mathbf{x})v_k(\mathbf{x}+\mathbf{r})\rangle$$
$$= \langle (-v_i(-\mathbf{x}))(-v_j(-\mathbf{x}))(-v_k(-\mathbf{x}-\mathbf{r}))\rangle = -S_{ijk}(-\mathbf{r}) \tag{9.39}$$

The third equality comes from the homogeneity of turbulence. Thus, the S_{ijk} are anti-symmetric with respect to the origin, and

$$S_{ijk}(\mathbf{0}) = 0 \tag{9.40}$$

As expected the one-point triple correlations are zero in a homogeneous and parity-invariant turbulence. It would not be the case with helical turbulence.

As for the double correlations, we shall reduce the expression of $[S]$ to a single scalar function: the longitudinal triple correlation:

$$k(\mathbf{r}) = \langle v_\ell(\mathbf{x})^2 v_\ell(\mathbf{x}+\mathbf{r})\rangle \tag{9.41}$$

We first express $S_{ijk}(\mathbf{r})$ with the tensors δ_{ij} and r_i, taking into account the symmetry with respect to the first two indices. Thus

$$S_{ijk}(\mathbf{r}) = A(r)r_i r_j r_k + B(r)(r_i\delta_{jk} + r_j\delta_{ik}) + C(r)\delta_{ij}r_k$$

The three functions A, B and C can be expressed with $k(r)$ if we assume the isotropy of turbulence and the fluid's incompressibility. Isotropy implies:

$$S_{iik} = 0 \quad \Longleftrightarrow \quad r^2 A + 2B + 3C = 0 \tag{9.42}$$

since S_{iik} is the average of a vector which therefore cannot indicate any privileged direction. Thus it is zero. Incompressibility implies:

$$\partial_k S_{ijk} = 0 \tag{9.43}$$

A short manipulation leads to

$$(rA' + 5A + \frac{2B'}{r})r_i r_j + (2B + rC' + 3C)\delta_{ij} = 0.$$

Thus, with (9.42), we get three relations:

$$\begin{cases} rA' + 5A + \frac{2B'}{r} = 0 \\[2mm] 2B + rC' + 3C = 0 \\[2mm] r^2 A + 2B + 3C = 0 \end{cases} \tag{9.44}$$

According to its definition,

$$k(r) = r^3 A + r(2B + C)$$

This equation combined with (9.42) leads to $C = -k/2r$, while the last two equations of (9.44) give $A = C'/r$. Thus, it turns out that $A = (k - rk')/2r^3$ and $B = (2k + rk')/4r$. Finally,

$$S_{ijk}(r) = \left(\frac{k - rk'}{2r^3}\right) r_i r_j r_k + \left(\frac{2k + rk'}{4r}\right)(r_i \delta_{jk} + r_j \delta_{ik}) - \frac{k}{2r}\delta_{ij} r_k \tag{9.45}$$

We easily show from (9.39) that the function k is antisymmetric: $k(r) = -k(-r)$ and thus $k(0) = 0$; in addition $k'(0) = 0$. This result may be shown as follows:

$$\left.\frac{\partial k}{\partial r}\right)_{r=0} = \left\langle v_\ell^2(\mathbf{x})\left(\frac{\partial}{\partial r}v_\ell(\mathbf{x} + \mathbf{r})\right)\right\rangle_{r=0} = \left\langle\left(v_\ell^2(\mathbf{x} + \mathbf{r})\frac{\partial}{\partial r}v_\ell(\mathbf{x} + \mathbf{r})\right)\right\rangle_{r=0}$$

$$= \frac{1}{3}\left\langle\left(\frac{\partial}{\partial r}v_\ell^3(\mathbf{x} + \mathbf{r})\right)\right\rangle_{r=0} = \frac{1}{3}\left\langle\left(\frac{\partial}{\partial x}v_\ell^3(\mathbf{x} + \mathbf{r})\right)\right\rangle_{r=0} = \frac{1}{3}\frac{\partial}{\partial x}\langle v_\ell^3\rangle = 0$$

since $\langle v_\ell^3\rangle$ is independent of $\mathbf{x}$ because of the homogeneity of turbulence.

9.4   Length Scales in Turbulent Flows

In order to characterize more completely turbulent flows, we need to precise the various spatial scales which control their dynamics.

9.4.1   Taylor and Integral Scales

The largest scale is the correlation length or *the integral scale*. It is defined by

$$\ell_0 = \int_0^\infty R(r)dr/R(0) \tag{9.46}$$

Using (9.35) and the fact that $\int_0^\infty \frac{\sin x}{x}dx = \pi/2$, we find the other following expression:

$$\ell_0 = \frac{\pi}{2} \frac{\int_0^\infty k^{-1} E(k)dk}{\int_0^\infty E(k)dk} . \tag{9.47}$$

It shows that the integral scale is the mean wavelength weighted by the spectral density of kinetic energy. This scales therefore points to the most energetic structures of a turbulent flow. We shall come back to it in the next section.

   Another scale, also very useful to characterize turbulent flows, is the *Taylor scale*. It is defined by:

$$\ell_T = \sqrt{\frac{E}{Z}} = \sqrt{\frac{\langle v^2 \rangle}{\langle \omega^2 \rangle}} \tag{9.48}$$

or

$$\ell_T = \left(\frac{\int_0^\infty E(k)dk}{\int_0^\infty k^2 E(k)dk} \right)^{1/2} \tag{9.49}$$

   Using (9.36), we see that this scale is related to the second derivative of the velocity autocorrelation since

$$\ell_T = \sqrt{-\frac{R(0)}{3R''(0)}}$$

The definition of this scale shows that it characterizes the velocity gradients: broadly speaking, this scale shows the mean size of vortices.

9.4.2 The Dissipation Scale

Let us first consider a chunk of fluid of unit mass, within a turbulent flow. Without any forcing, the turbulence would decay thanks to viscous dissipation. After some time, it would disappear altogether, the kinetic energy (of turbulence) being transformed into internal energy. In a steady state, turbulence is stationary because some energy is injected and compensates the losses by viscous dissipation. We shall denote by $\langle \varepsilon \rangle$ the power injected per unit mass into the turbulence (i.e. into the random like fluctuations of the flow). In a homogeneous and stationary turbulence, this quantity is a constant and because of the conservation of energy, this is also the power dissipated by unit mass.

If we observe that in the spectral space, the viscous force is proportional to $k^2 \hat{v}_k$, we easily guess that the small scales (large k) are the scales where most of the dissipation occurs. Let us now assume that this dissipation comes from a single wavenumber k_D. Conservation of energy implies that

$$\langle \varepsilon \rangle \sim \nu k_D^2 \hat{v}^2 (k_D) \tag{9.50}$$

for orders of magnitude. However, if $\hat{v}(k_D)$ is in the dissipative range, its associated Reynolds number is of order unity, thus $\hat{v}(k_D) \sim \nu k_D$. We easily derive from (9.50) that

$$k_D \sim \left(\frac{\langle \varepsilon \rangle}{\nu^3} \right)^{1/4} \tag{9.51}$$

With the wavenumber k_D, one usually associates the scale $\ell_D = 1/k_D$ called the *dissipation scale* or the *Kolmogorov scale*. This scale separates the spectrum into two domains: the one where viscosity dominates $\ell \ll \ell_D$ and the one where this force may be neglected $\ell \gg \ell_D$.

9.5 Universal Turbulence

After the long foregoing presentation of some kinematic aspects of turbulence, we shall now get closer to the real difficulties associated with turbulent flows, namely their dynamics. To get a first idea of it, we follow the work of Andrei Kolmogorov which was published in 1941. This pioneering work suggested for the first time the idea of a universal turbulence, which would be independent of the instabilities that maintain it. This ideal state has been investigated by Kolmogorov and his theory is often referred to as "K41", an acronym that we shall also use below.

9.5.1 Kolmogorov Theory

9.5.1.1 The Hypothesis

The basic idea of Kolmogorov is that there exist a universal state of turbulence that may be observed when we consider the flow in a box much smaller than the scales where the instabilities are working. In other words, a box following the mean flow and much smaller than the integral scale. Within this box, turbulence is characterized only by $\langle \varepsilon \rangle$ according to Kolmogorov who introduced this quantity. Kolmogorov also suggested that in this box the turbulence is (almost) homogeneous and isotropic and so that it should meet two hypothesis:

- **H1: First similarity hypothesis.** The structure functions for the velocity within an isotropic homogeneous turbulence just depend on $\langle \varepsilon \rangle$ and ν.
- **H2: Second similarity hypothesis.** If the distance between the points is large compared to the dissipation scale, then the structure functions just depend on $\langle \varepsilon \rangle$.

Kolmogorov concentrated on the structure functions because of the a priori idea that they are less sensitive to a large-scale flow hence to some non homogeneity.

A first consequence of these hypothesis is the existence of scaling laws when the distances are large compared to the dissipation scale. We shall come back to this when discussing the idea of intermittency.

9.5.1.2 The Kolmogorov Spectrum

We shall now derive the kinetic energy spectrum $E(k)$ under Kolmogorov's second hypothesis, namely when the effects of viscosity are negligible.

According to H2, $E(k)$ depends only on k and $\langle \varepsilon \rangle$. We thus need to build a quantity of the same dimension as $E(k)$ using k and $\langle \varepsilon \rangle$ only. Let us first observe that $\langle \varepsilon \rangle$ is a specific kinetic energy per unit time. Thus, if we use the velocity v_k typical of the scale $1/k$, dimensionally speaking

$$\langle \varepsilon \rangle \sim k v_k^3$$

However, still dimensionally, $v_k^2 \sim k E(k)$. Combining these two expressions, we find that

$$E(k) = \langle \varepsilon \rangle^{2/3} k^{-5/3} f(\langle \varepsilon \rangle, k)$$

where f is dimensionless. Moreover, the argument of f must be dimensionless too. But there isn't any combination of k and $\langle \varepsilon \rangle$ that is dimensionless. Therefore f is a constant C_K which is called the *Kolmogorov constant*. Scientists have tried very hard to determine the value of this constant, but this is a difficult task. Its value[3] seems to be close to 1.6.

Finally, the spectrum reads:

$$E(k) = C_K \langle \varepsilon \rangle^{2/3} k^{-5/3} \tag{9.52}$$

This is commonly called the *Kolmogorov spectrum*, even if its expression has first been given by Obukhov. The foregoing analysis is valid only for some range of scales: this range is known as the *inertial range*. These scales are much smaller than the integral scale but much larger than the dissipation one. The name "inertial range" comes from the idea that only inertial terms, like $(\mathbf{v} \cdot \nabla)\mathbf{v}$, and pressure terms are important in the dynamics of these scales.

The foregoing approach leads to a new view of turbulence through the evolution of the energy. Let us first observe that within the inertial range the kinetic energy is conserved: indeed, if we imagine a volume of fluid which would not exchange any mass with its surroundings and whose motion would be due to Fourier modes within the inertial range, for these modes viscous action is negligible and therefore the kinetic energy remains constant. On the contrary, the energy of modes with $k > k_D$ is rapidly transformed into heat by viscous friction. In a steady state, such a loss must be compensated. The energy is provided by the scales of the inertial range, however there is no energy source there. Thus, we must consider still larger scales, namely those where the forces driving the turbulence are working. These scales are generally in the domain $k \leq k_0$, also called the *injection range*.

Hence, the picture of the *turbulent cascade* is emerging: Energy injected in the large scales by the instabilities leading to the turbulent flow, is progressively carried through smaller and smaller scales until it reaches the dissipative range where it is transformed into heat. The physical picture behind this spectral dynamics is the repeated breaking of vortices. When the size of these structures is small enough, they are removed by viscosity. In Fig. 9.2 we show the kinetic energy spectrum in an ideal view and using real data. In Fig. 9.3 we illustrate the cascade process in the real space.

We should however be careful not to take this picture as an exact view of the reality. This is, unfortunately, only a partial view of it as we shall see below.

Let us now come back to the various scales that we introduced in 9.4 and let us plot them in Fig. 9.2a. Calculating the integral scale from (9.47) with the functions used to plot Fig. 9.2a, gives $k_0 = 1/\ell_0 \sim 0.6$ which is very close to the maximum value that we fixed at $k = 1$. As expected this scale is the one of the most energetic structures.

---

[3]First experimental values as those given by Monin and Yaglom (1975) are around 1.5. Recent measurements in the atmospheric boundary layer by Cheng et al. (2010) give 1.56. Numerical experiments have long given values around 2 (e.g. Vincent and Meneguzzi 1991), but recently it has been understood that the numerical resolution was an important issue. The latest results obtained with the very high resolution numerical simulations are getting closer to experimental values (Kaneda et al. 2003).

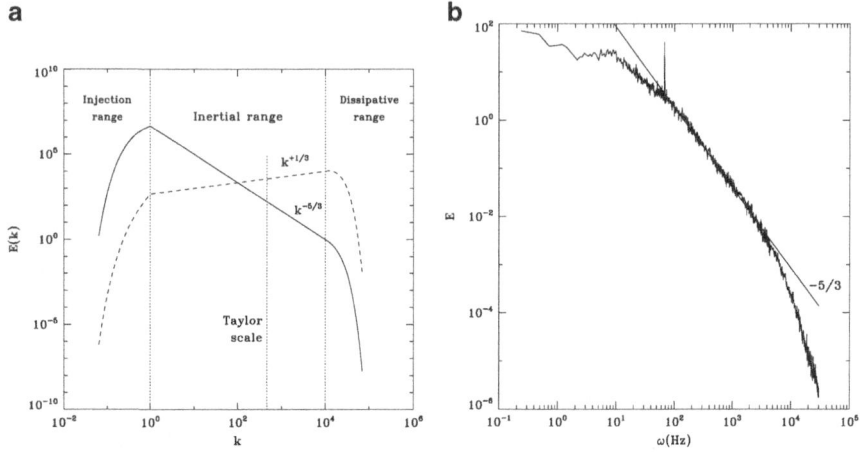

Fig. 9.2 (a) Idealized view of the different regions of the kinetic energy spectrum of a turbulent flow. The dashed line shows the enstrophy spectrum. (b) A kinetic energy spectrum derived from an experiment on turbulence with Helium. Note that the law $\omega^{-5/3}$ is visible on almost two decades (courtesy H. Willaime)

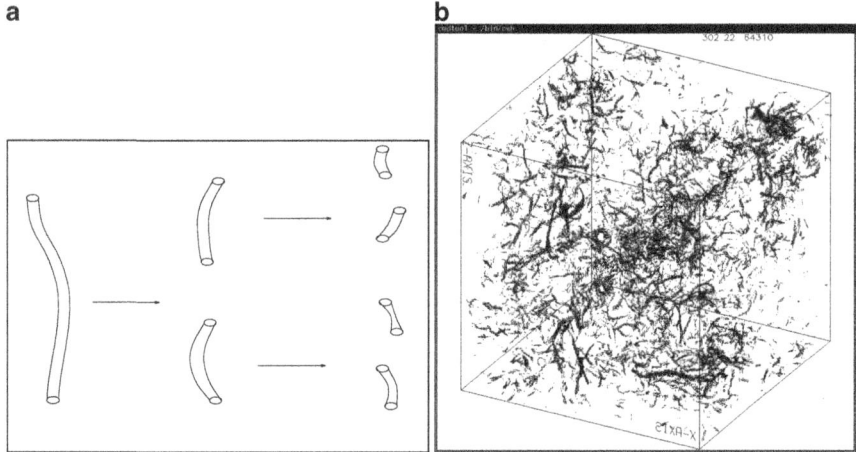

Fig. 9.3 (a) An illustration of the cascade of energy towards the small scales: vortices split into smaller and smaller pieces. (b) The vorticity field as computed by a direct numerical simulation of isotropic turbulence at Re~ 1000 from Vincent and Meneguzzi (1991)

Let us now estimate the Taylor scale. We derive its value from (9.49) assuming that the dissipation scale is much smaller than the integral one, namely $k_D \gg k_0$. We may show (see the exercises), that the order of magnitude of this scale is

$$\ell_T \sim \ell_0^{1/3} \ell_D^{2/3} \qquad (9.53)$$

The Taylor scale is therefore a kind of geometric mean between the integral and dissipation scales with more weight to the latter. The wavenumber $k_T = 1/\ell_T$ is in the middle of the inertial range as shown in Fig. 9.2. This scale therefore characterizes more specifically the "Universal Turbulence". The Reynolds number associated with this scale is usually taken as

$$\mathrm{Re}_\lambda = \frac{v_0 \ell_T}{\nu} \tag{9.54}$$

Using (9.53) and $\langle \varepsilon \rangle \sim v_0^3/\ell_0$, we see that $\mathrm{Re}_\lambda = \sqrt{\mathrm{Re}_0}$ where Re_0 is the Reynolds number associated with the integral scale.

9.5.2 Dynamics in the Spectral Space

The foregoing discussion is essentially qualitative and we may wonder what kind of constraints are given by the equations of motion as far as the spectral quantities are concerned. To investigate this point let us write the Navier–Stokes equation and that of mass conservation in the spectral space. We have

$$\begin{cases} \partial_t \hat{v}_i + i k_j \widehat{v_i v_j} = -i k_i \hat{p} - \nu k^2 \hat{v}_i \\ k_i \hat{v}_i = 0 \end{cases} \tag{9.55}$$

where the hat is for the Fourier transform. We then project this equation on the plane perpendicular to $\mathbf{k}$ thanks to the projector tensor P_{ij} (see 9.27). Thus

$$(\partial_t + \nu k^2)\hat{v}_i = -i P_{ij} k_n \widehat{v_j v_n} = -\frac{i}{2} P_{ijn} \widehat{v_j v_n} \tag{9.56}$$

where we set $P_{ijn} = P_{ij}k_n + P_{in}k_j$.

These expressions show that the evolution of the Fourier component $\hat{v}_i(\mathbf{k})$ comes from the damping by viscosity on a time scale $1/(\nu k^2)$ and a forcing from all the components verifying:

$$\mathbf{p} + \mathbf{q} = \mathbf{k}, \quad \text{since} \quad \widehat{v_j v_n} = \int \hat{v}_j(\mathbf{p}) \hat{v}_n(\mathbf{k} - \mathbf{p}) d^3 p$$

$$= \int \hat{v}_j(\mathbf{p}) \hat{v}_n(\mathbf{q}) \delta(\mathbf{p} + \mathbf{q} - \mathbf{k}) d^3 p \, d^3 q$$

These terms reflect a local interaction of Fourier modes when

$$\|\mathbf{p}\| \sim \|\mathbf{q}\| \sim \|\mathbf{k}\|$$

and a non-local interaction when

$$\|\mathbf{k}\| \ll \|\mathbf{p}\| \sim \|\mathbf{q}\|$$

Fig. 9.4 Interactions
between Fourier modes

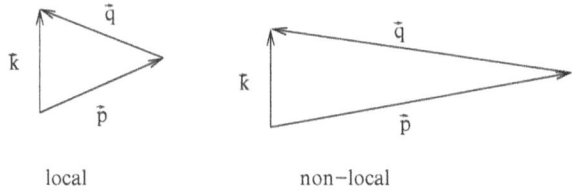

local   non−local

Figure 9.4 illustrates these two types of interactions.

Let us now focus on the evolution of the turbulence spectrum. We start from
(9.56) and use the result of exercise 1. We easily show that

$$(\partial_t + 2\nu k^2)\phi_{ij}(\mathbf{k})\delta(\mathbf{k} - \mathbf{k}') = \frac{i}{2}\left\langle P_{ilm}\widehat{v_l v_m}^*\hat{v}'_j - P'_{jlm}\widehat{v_l v_m}'\hat{v}^*_i\right\rangle$$

Noting that the right-hand side is also proportional to $\delta(\mathbf{k} - \mathbf{k}')$, we find

$$(\partial_t + 2\nu k^2)\phi_{ii}(\mathbf{k}) = -\text{Im}(P_{ilm}\langle\widehat{v_l v_m}^*\hat{v}_i\rangle)$$

which can be rewritten as

$$(\partial_t + 2\nu k^2)\phi_{ii}(\mathbf{k}) = -\text{Im}\left[P_{ilm}\int \langle\hat{v}_i(\mathbf{k})\hat{v}_l(\mathbf{p})\hat{v}_m(\mathbf{q})\rangle\,\delta(\mathbf{k} + \mathbf{p} + \mathbf{q})dpdq\right]$$
$$(9.57)$$

The evolution of the spectral density of kinetic energy in the isotropic case is now
driven by:

$$(\partial_t + 2\nu k^2)E(k,t) = T(k,t) \tag{9.58}$$

where $T(k,t)$ is called the transfer term. It comes from the nonlinear terms of
the Navier–Stokes equation: it explicits the energy exchange between three Fourier
modes when the wavenumbers of the triad are compatible ($\mathbf{k} + \mathbf{p} + \mathbf{q} = \mathbf{0}$).

We may wonder about the physical meaning of the transfer terms. Of course
this is the spectral translation of the nonlinear interactions: more specifically, it
expresses the mechanisms that allow one feature at a given scale to pump energy
from structures at other scales. This mechanism is essentially non-local because it
is intrinsically the result of instabilities. The mechanisms behind $T(k,t)$ are very
complex: many instabilities, like the ones we analysed in Chap. 6, pump the energy
from the large scales to the small ones. However, in turbulent flows the other way
round is also possible: some large-scale flows may grow using the energy available
in the small scales: this is a large-scale instability. When the turbulence is in a steady
state, the transfer between scales is in both directions: towards the small scales
and towards the large scales. Of course, the transfer to the small scales slightly

dominates, so that the kinetic energy cascades on average from the large scales to the small scales with a rate equal to $\langle \varepsilon \rangle$.

9.5.3 The Dynamics in Real Space

9.5.3.1 Kármán–Howarth Equation

Some years before Kolmogorov proposed his new approach of turbulence, von Kármán and Howarth (1938) derived the first equation controlling the dynamics of homogeneous isotropic and symmetric turbulence. This equation relates the double and triple longitudinal correlation f and k. We now derive this equation and for that purpose we write the Navier–Stokes equation at two independent points:

$$\frac{\partial v_i}{\partial t} + \partial_k (v_k v_i) = -\partial_i \, p + \nu \Delta v_i$$

$$\frac{\partial v'_j}{\partial t} + \partial'_k (v'_k v'_j) = -\partial'_j \, p' + \nu \Delta' v'_j$$

where we simplified the expressions setting $\mathbf{v}' = \mathbf{v}(\mathbf{x}')$ and $\partial' = \partial/\partial x'$. We then multiply the first equation by v'_j and the second by v_i. We also note that $\partial_j v'_i = \partial'_j v_k = 0$. Summing the results and taking the average we finally get:

$$\frac{\partial Q_{ij}}{\partial t} + \partial_k \left(\left\langle v_i v'_j v'_k \right\rangle - \left\langle v_i v_k v'_j \right\rangle \right) = \partial_i \left\langle v'_j \, p \right\rangle - \partial_j \left\langle v_i \, p' \right\rangle + 2\nu \Delta Q_{ij}$$

where the space derivatives are taken with respect to $\mathbf{r}$. With the definition of S_{ijk} we have

$$\frac{\partial Q_{ij}}{\partial t} - \partial_k [S_{ikj}(\mathbf{r}) - S_{jki}(-\mathbf{r})] = 2\nu \Delta Q_{ij} + \partial_i \langle p(\mathbf{x}) v_j (\mathbf{x} + \mathbf{r}) \rangle - \partial_j \langle p(\mathbf{x}) v_i (\mathbf{x} - \mathbf{r}) \rangle$$

Using the antisymmetry of S_{ijk} and taking the trace of the equation we get:

$$\frac{\partial R}{\partial t} - \partial_k (S_{iki}) = 2\nu \Delta R \tag{9.59}$$

because $R = Q_{ii}/2$ and because pressure-velocity correlations disappear thanks to isotropy. However, from (9.45)

$$S_{iki} = \frac{1}{2r^4} \frac{\partial (r^4 k)}{\partial r} \, r_k$$

so that

$$\frac{\partial R}{\partial t} - \frac{1}{2r^2}\frac{\partial}{\partial r}\left(\frac{1}{r}\frac{\partial(r^4 k)}{\partial r}\right) = 2v\frac{1}{r^2}\frac{\partial}{\partial r}\left(r^2\frac{\partial R}{\partial r}\right) \tag{9.60}$$

If we use longitudinal correlation rather than R, we can make the substitution

$$R = \frac{1}{2r^2}\frac{\partial r^3 f}{\partial r}\ , \tag{9.61}$$

Then the equation may be integrated and finally we get:

$$\frac{\partial f}{\partial t} - \frac{1}{r^4}\frac{\partial r^4 k}{\partial r} = 2v\left(\frac{\partial^2 f}{\partial r^2} + \frac{4}{r}\frac{\partial f}{\partial r}\right) \tag{9.62}$$

which is *the equation of Kármán–Howarth*.

When it is used at $r = 0$ this equation yields some additional informations. Indeed, we know that $f(0) = \langle v_l^2\rangle = \frac{1}{3}\langle v^2\rangle = \frac{2}{3}E_c$. However, at $r = 0$ (9.62) becomes

$$\frac{df(0)}{dt} = 2v\left(\frac{1}{r^4}\frac{\partial}{\partial r}r^4\frac{\partial f}{\partial r}\right)_{r=0} \tag{9.63}$$

which leads to

$$\frac{dE_c}{dt} = 15v f''(0)$$

The right-hand side represents the energy dissipation by viscosity, thus $-\langle\varepsilon\rangle$ by definition. Hence, we find that $\langle\varepsilon\rangle = -15v f''(0)$, but also that, using (9.36),

$$\langle\varepsilon\rangle = 2vZ = v\langle\omega^2\rangle \tag{9.64}$$

meaning that $\langle\varepsilon\rangle$ is directly proportional to enstrophy.

This latter equation gives a new interpretation of the Taylor scale. Indeed, we have:

$$\frac{dE_c}{dt} = -\langle\varepsilon\rangle = -\frac{v}{\ell_T^2}E_c$$

namely that the kinetic energy decreases on a time scale $\tau = \ell_T^2/v$. Thus, it is just like if turbulence was damped by viscosity but on an effective length-scale equal to the Taylor scale.

9.5.3.2 The Kolmogorov Equation

Kolmogorov manipulated furthermore Kármán–Howarth equation using the structure functions with longitudinal velocities, namely

$$S_2 = \langle (v_\ell(\mathbf{x} + \mathbf{r}) - v_\ell(\mathbf{x}))^2 \rangle \quad \text{and} \quad S_3 = \langle (v_\ell(\mathbf{x} + \mathbf{r}) - v_\ell(\mathbf{x}))^3 \rangle$$

These structure functions are easily expressed with f and k; indeed,

$$S_2 = 2(f(0) - f(\mathbf{r})) \quad \text{and} \quad S_3 = 6k(\mathbf{r})$$

If the turbulence is in a steady state, the viscous dissipation must be compensated by an energy source, the power of which is $\langle \varepsilon \rangle$. In freely decaying turbulence,

$$\frac{dE_c}{dt} = -\langle \varepsilon \rangle, \quad \text{while} \quad f(0) = \frac{2}{3} E_c$$

so that $df(0)/dt = -\frac{2}{3}\langle \varepsilon \rangle$. In a steady state, this loss is compensated by the same term with opposite sign. Hence,

$$\frac{df(0)}{dt} = 2v \left(\frac{1}{r^4} \frac{\partial}{\partial r} r^4 \frac{\partial f}{\partial r} \right)_{r=0} + \frac{2}{3} \langle \varepsilon \rangle = 0$$

Since $\langle \varepsilon \rangle$ is a constant, we should find this term also in (9.62). The steady state version of the Karman–Howarth equation (9.62), is therefore:

$$-\frac{1}{r^4} \frac{\partial r^4 k}{\partial r} = 2v \left(\frac{\partial^2 f}{\partial r^2} + \frac{4}{r} \frac{\partial f}{\partial r} \right) + \frac{2}{3} \langle \varepsilon \rangle$$

After simple integrations and use of the structure functions instead of f and k, we find

$$4\langle \varepsilon \rangle + \frac{1}{2r^4} \frac{\partial (r^4 S_3)}{\partial r} = 6v \left(\frac{\partial^2 S_2}{\partial r^2} + \frac{4}{r} \frac{\partial S_2}{\partial r} \right)$$

which can be integrated after multiplication by r^4. We obtain

$$\frac{4}{5} \langle \varepsilon \rangle r + S_3 = 6v \frac{\partial S_2}{\partial r} \tag{9.65}$$

which is called the *Kolmogorov equation* This new equation is quite interesting since it shows that if $r \gg \ell_D$, namely if we are considering a length scale in the inertial range, then S_3 verifies the scaling law

$$S_3 = -\frac{4}{5} \langle \varepsilon \rangle r \tag{9.66}$$

called the "four-fifth law". This is a remarkable result as it is non-trivial and exact for universal turbulence (but see Frisch 1995, for a more thorough discussion).

The Log-Normal theory of Obukhov-Kolmogorov 1962

When we wrote the scaling laws (9.70) we only used two quantities: the scale r and the mean dissipation $\langle \varepsilon \rangle$. This lead us to the law $S_p \sim r^{p/3} \langle \varepsilon \rangle^{p/3}$. However, with the same dimensional arguments we could have written $S_p \sim r^{p/3} \langle \varepsilon^{p/3} \rangle$. But the two quantities $\langle \varepsilon \rangle^{p/3}$ and $\langle \varepsilon^{p/3} \rangle$ are not identical (except for $p = 3$) because dissipation is a fluctuating quantity. Landau was the first to point out this problem with the K41 theory. Hence, some years later, Obukhov and Kolmogorov (1962) proposed a modification of the K41 approach. This new model is now known as the *Log-normal theory*. This theory may be presented as follows.

First, OK62 define a dissipation ε_r, averaged over a ball of size r, namely

$$\varepsilon_r = \frac{1}{V} \int_{(V_r)} \varepsilon(\mathbf{x}) dV$$

Obviously, $\langle \varepsilon_r \rangle^p$ is all the more different from $\langle \varepsilon_r^p \rangle$ that the fluctuations of ε_r are strong. However, these fluctuations increase when the volume decreases. Indeed, let us consider a flow with a very high Reynolds number. The velocity gradients may be very strong implying in some places very high values of viscous dissipation. Actually, we expect from the Kolmogorov spectrum that the fluctuations of dissipation are not bounded when the Reynolds number goes to infinity. OK62 thus proposed that the variance of the logarithm of ε_r is not bounded when L/r goes to infinity (L is a given large scale). They also assumed that this quantity obeys to a normal statistics (the probability density function is a gaussian). One may wonder why they considered the logarithm of ε_r ? This is because ε_r is a positive quantity which cannot follow a

normal law, while the logarithm symmetrizes the points 0 and $+\infty$ by moving zero to $-\infty$. The normal law is symmetric with respect to the mean value, hence we may expect that it applies more precisely to the logarithm.[a] OK62 thus proposed this formulation of the variance, which completely defines, with the mean, a normal distribution,

$$\sigma_r^2 = A(x, t) + \mu \log(L/r) \qquad (9.67)$$

where μ is a supposed universal constant. Now we may wonder why a logarithm dependence has been chosen for the variance. Essentially, because power laws are expected for spectra or moments.[b]

For a log-normal law, one has

$$\langle \varepsilon_r^{p/3} \rangle = e^{p/3 \langle \log \varepsilon_l \rangle + \frac{p^2 \sigma_r^2}{18}} \qquad (9.68)$$

if we set $p = 3$ in this formula and if we use (9.67), we find $\langle \varepsilon_r \rangle = (L/r)^{\mu/2} e^{\langle \log \varepsilon_l \rangle + A/2}$, so that

$$\frac{\langle \varepsilon_r^{p/3} \rangle}{\langle \varepsilon_r \rangle^{p/3}} = C_p(x) \left(\frac{L}{r} \right)^{\mu p(p-3)/18} .$$

We can thus derive a new expression for the exponent ζ_p of the structure functions of order p:

$$\zeta_p = \frac{p}{3} - \mu p(p - 3)/18 \qquad (9.69)$$

The first experimental measurements of ζ_p, obtained for small p's, gave $\mu \simeq 0.2$. Some years later, Arneodo et al. (1998) have shown that experimental data (obtained for $-10 \leq p \leq +10$) are well represented by a quadratic normal law with $\zeta_p = mp - \sigma^2 p^2/2$ with $m = 0.32$ and $\sigma^2 = 0.03$.

---

[a]However, this assumption is still approximate because there is no good reason that fluctuations towards small values are as probable as those towards high values.

[b]We should keep in mind that in 1962, the Kolmogorov spectrum had already been observed experimentally, and thus any new theory should reproduce this result.

9.5.4   Some Conclusions on Kolmogorov Theory

The foregoing discussion revealed to us some of the important properties of turbulence which we shall now summarize.

1. We first noticed that scales are important in a turbulent flow: the properties depend on the scale we are considering. Investigating the spectral side of turbulence, we could discriminate three different ranges of length scales: the integral, inertial and dissipative ranges.
2. We then understood that if a turbulent flow is very dependent of the instabilities at the integral scale, it may well be that within the inertial and dissipative ranges, turbulence reaches a universal state where (9.66) is certainly one of the first laws.
3. However, Kolmogorov's approach assumes rather strong hypothesis: homogeneity, isotropy, parity. Among the three assumptions, homogeneity is the strongest. If it is relaxed, there is some mean flow whose evolution is dictated by the transport properties of turbulence.
   These transport properties appear in the closure of averaged equations. We may have noticed that the Kármán–Howarth equation is not closed.
4. The Kolmogorov scenario "forgets" about the possible role of fluctuations of ε, which points to another side of turbulence, namely intermittency, to be discussed below.

9.6   Intermittency

9.6.1   Presentation

The intermittency of turbulence, which is sometimes called internal intermittency, is one of the ill-known sides of turbulence. We shall first present this phenomenon as it appears in the experiments.

Figure 9.5 shows a random function whose distribution function is gaussian, and its derivative. It also shows a plot of a record of the velocity of a turbulent flow as well as its derivative. The difference between these two random functions is quite clear: while we note that the gaussian random function and its derivative are rather similar, we see that the velocity and its derivative are quite different. In particular, the derivative of the velocity shows large amplitude fluctuations. Now, if we plot the probability density function of the velocity and the random function (Fig. 9.6), the difference between the gaussian random function and the velocity is even neater. The large amplitude events in a turbulent velocity field are more likely than if they were the results of a sum of random, uncorrelated events (which would lead to the gaussian distribution). We pinpoint here one of the true problem of turbulence: the phenomenon is really random in nature, but this chance is guided by the Navier–Stokes equation in a still obscure way.

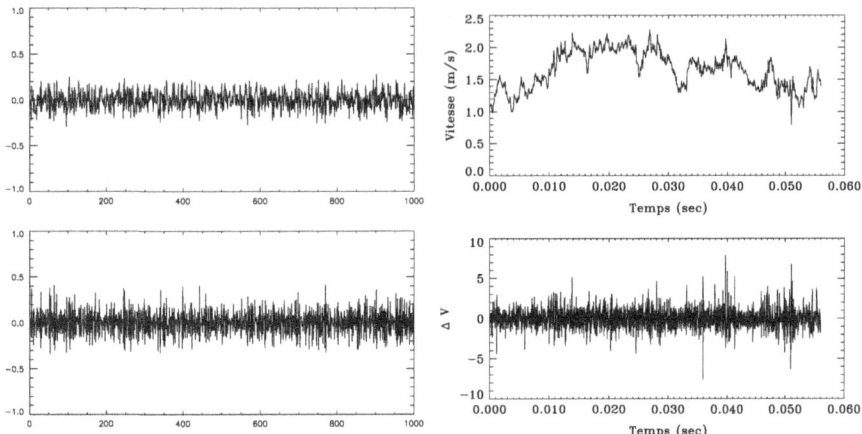

Fig. 9.5 On *left* (*top*) a random function with a gaussian distribution and its derivative (*bottom*). On *right* (*top*) the record of a turbulent velocity field and its derivative (*bottom*)

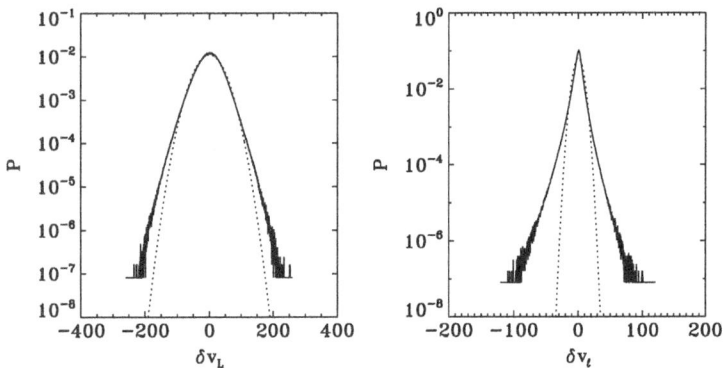

Fig. 9.6 The probability density function for the velocity difference between two points separated by a large scale L or by a small scale ℓ; these distributions have been derived using the data shown in Fig. 9.5. The normal, gaussian, distribution is shown as *dashed lines*

It is interesting to compare the random world of turbulence and the one of atoms and molecules within a gas. Indeed, the distribution of velocities of atoms or molecules of a gas in usual conditions is gaussian. This is a consequence of the fact that the velocity of a molecule at a given time results from the huge number of collisions that are statistically independent. Indeed, the central limit theorem states that the probability density function of a random variable that is the sum of an infinite number of independent random variable is a gaussian. Hence, the distribution of molecule velocities follows a normal statistics. We see that this statistical result is independent of the equation of motion of the molecules. In a turbulent flow the velocity at a given point is the combination of the influence of many vortices operating at various scales. In this respect many random processes contribute to the build up of the velocity field, but these processes are not independent: we know that long vortices tend to split thanks to instabilities and therefore correlations between scales are important.

However, intermittency does not only appear in the probability distributions; it also influences the scaling laws of structure functions. This is one aspect of the universal turbulence that we shall discuss now.

9.6.2 The Scaling Laws of Structure Functions

We already met the structure functions. These are important functions for many reasons: First they measure the relative velocity of two points of the flow: if the turbulence has a universal regime, such quantities will show it. Experimentally, it is difficult to create homogeneous and isotropic turbulence. The best approximation to this ideal situation is certainly grid turbulence,[4] which, in a frame comoving with the mean flow, is quasi-homogeneous and isotropic but is decaying with time. The structure functions eliminate the mean flow and are measurable quantities.

Using dimensional arguments, the structure function of order p may be written

$$S_p = \langle (v_\ell(\mathbf{x} + \mathbf{r}) - v_\ell(\mathbf{x}))^p \rangle = C_p(\langle \varepsilon \rangle r)^{p/3} \tag{9.70}$$

since the velocity scale is $(\langle \varepsilon \rangle r)^{1/3}$. The C_p's are constants which likely depend on the flow, except C_3 since

$$C_3 = -\frac{4}{5}$$

from (9.66). We should also note that C_2 is related to the Kolmogorov constant, and one may show, as an exercise, that when S_2 is proportional to $r^{2/3}$ then $E(k)$ is proportional to $k^{-5/3}$.

Let us now focus on variations of S_p with r. Setting

$$S_p \propto r^{\zeta_p}, \tag{9.71}$$

we see that the Kolmogorov theory implies that

$$\zeta_p = \frac{p}{3} \tag{9.72}$$

As shown by Fig. 9.7, experiments show a clear deviation to this suite of exponents with respect to the Kolmogorov one. This deviation is all the more marked that p is

---

[4]This is the turbulence which appears in the wake of a grid. It is homogeneous in the directions parallel to the grid

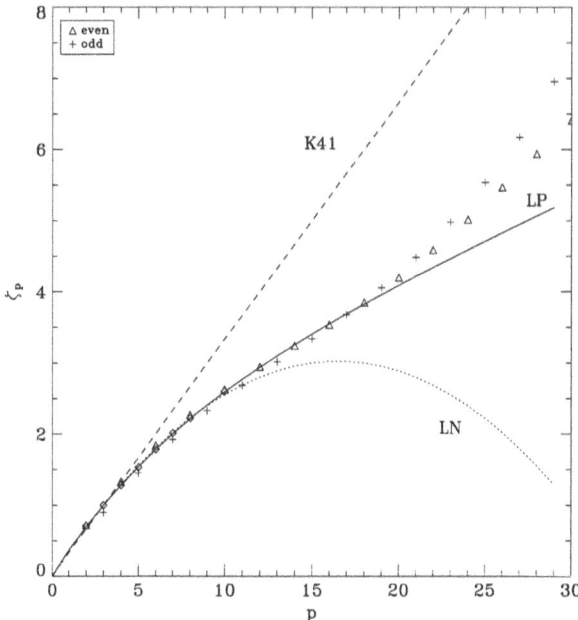

Fig. 9.7 The exponents ζ_p from various theories compared to some experimental data. Pluses and triangles are from a numerical simulation of Vincent and Meneguzzi (1991) and diamonds are from Benzi et al. (1993). The good fit of the Log-Poisson law (LP) is now understood as an effect of the small (not large!) value of the Reynolds number at Taylor scale used by the numerical simulation of Vincent and Meneguzzi (1991). Convergence to the Log-normal law would appear at very much higher Reynolds numbers. The straight line K41, is from the Kolmogorov theory and the LN curve shows the log-normal law with $\mu = 0.2$

high. However, large orders are sensitive to the wings of the probability distribution, that is to rare events. They are thus sensitive to the large amplitude events typical of the intermittency. The absence of intermittency in the K41 theory was soon noticed by Landau. Kolmogorov and Obukhov then proposed a modification of this theory, which is now known as the *Log-Normal Theory* (see the box). Unfortunately, this theory raised new questions and new theories have been developed (see the Log-Poisson box).

9.6.2.1 Two Properties of the Exponents

The exponents suite ζ_p verifies two general conditions:

$$\frac{d^2\zeta_p}{dp^2} \leq 0 \qquad \text{and} \qquad \zeta_{2p+2} \geq \zeta_{2p} \tag{9.73}$$

The first one comes from a Schwartz inequality verified by random variables. If A and B are two random variables, then

$$\langle AB \rangle \leq \langle A^2 \rangle^{1/2} \langle B^2 \rangle^{1/2}$$

taking $A = (v_\ell(\mathbf{x} + \mathbf{r}) - v_\ell(\mathbf{x}))^p$ and $B = (v_\ell(\mathbf{x} + \mathbf{r}) - v_\ell(\mathbf{x}))^q$, we get

$$S_{p+q} \leq \sqrt{S_{2p} S_{2q}}$$

If $S_p = A_p r^{\zeta_p}$, then

$$A_{p+q} r^{\zeta_{p+q}} \leq \sqrt{A_{2p} A_{2q}}\, r^{(\zeta_{2p} + \zeta_{2q})/2} \qquad \forall r \in \text{Inertial range} \tag{9.74}$$

The Log-Poisson Theory

Making more precise determinations of the exponents of structure functions has shown that neither the Kolmogorov theory, nor its Log-normal improvement could explain the variations of the ζ's with the order p. In 1994, She & Lévêque, She & Waymire and Dubrulle proposed a new approach which seemed to square much better with the experimental results available at the time (see Fig. 9.7).

This new approach was based on three hypothesis:

  i. The structure function of order p verifies the scaling law:

$$S_p \sim r^{p/3} \langle \varepsilon_r^{p/3} \rangle$$

  ii. The moments of the pdf of the energy dissipation obey the induction relation:

$$\frac{\langle \varepsilon_r^{p+1} \rangle}{\varepsilon_r^\infty \langle \varepsilon_r^p \rangle} = A_p \left(\frac{\langle \varepsilon_r^p \rangle}{\varepsilon_r^\infty \langle \varepsilon_r^{p-1} \rangle} \right)^\beta, \qquad \text{and} \qquad 0 < \beta < 1 \tag{9.75}$$

where A_p are constants and $\varepsilon_r^\infty = \lim_{p \to \infty} \langle \varepsilon_r^{p+1} \rangle / \langle \varepsilon_r^p \rangle$. We shall see that ε_r^∞ is a quantity specific to the most intermittent structures. The relation (9.75) could be a hidden symmetry of the Navier–Stokes equation.

  iii. When $r \longrightarrow 0$, $\varepsilon_r^\infty \sim r^{-2/3}$.

If we assume that $\langle \varepsilon_r^p \rangle$ verifies the scaling law $\langle \varepsilon_r^p \rangle \sim r^{\tau_p}$, then, it turns out from the definitions that $\zeta_p = p/3 + \tau_{p/3}$. the exponent τ_p measures the distance to the Kolmogorov law. Using its definition and the second hypothesis, we find the new relation

$$\tau_{p+2} - (1 + \beta)\tau_{p+1} + \beta\tau_p + \frac{2}{3}(1 - \beta) = 0$$

It is then convenient to set $\tau_p = -\frac{2}{3}p + 2 + f_p$. Finally,

$$f_{p+2} - f_{p+1} = \beta(f_{p+1} - f_p)$$

which is easily solved as

$$f_p = f_0 + A \left(\frac{1 - \beta^p}{1 - \beta} \right)$$

The initial conditions of this suite are given by the initial conditions on τ_p. By construction $\tau_1 = 0$ and we assume that $\tau_0 = 0$. This latter conditions is equivalent to the assumption that the volume of dissipative structures remains finite as viscosity tends to zero. We finally find

$$\zeta_p = \frac{p}{3} + \frac{2}{3}\left[\frac{1 - \beta^{p/3}}{1 - \beta} - \frac{p}{3}\right] \qquad (9.76)$$

The curve which fits the experimental data so well is obtained for $\beta = 2/3$.[a] What is the meaning of this new exponent? Obviously, it characterizes the degree of intermittency of viscous dissipation. If $\beta \to 1$ then $\zeta \to p/3$: we find K41 again.

Dubrulle (1994) has shown that the second hypothesis (ii) could be inferred if the pdf of $\varepsilon_r/\varepsilon_r^\infty$ was assumed to be the convolution of a Log-Poisson law with another undetermined law. The hidden symmetry underlying (9.75) is therefore not very stringent. The following work of Arneodo et al. (1998) has shown that experimental data at very high Reynolds numbers ($Re_\lambda \gtrsim 2000$) contradicted the Log-Poisson theory in favor of the Log-Normal approach. It seems that the Log-Poisson theory is more appropriate for $Re_\lambda \lesssim 800$, a range of Reynolds numbers where the Log-Normal theory does not give very good results. Today (2013), it is believed that the Log-Normal theory applies when $Re_\lambda \to \infty$, namely only asymptotically.

[a]We saw that the exponent ζ_2 was related to exponent of the energy density spectrum. The change implied by this new theory compared to the Kolmogorov one is very small: this exponent is now: $-\frac{5}{3} - 0.03$.

This inequality implies that

- if $r \gg 1$ then (9.74) is true only if $\zeta_{p+q} \leq (\zeta_{2p} + \zeta_{2q})/2$, that is to say if the function $\zeta(p)$ is concave.[5]
- if $r \ll 1$ then (9.74) is true only if $\zeta_{p+q} \geq (\zeta_{2p} + \zeta_{2q})/2$ that is to say if the function $\zeta(p)$ is convex.

Experiments readily show that the suite $\zeta(p)$ is convex and therefore the second case is the right one. The relevant scale in the inertial domain is such that everywhere $r < 1$. We should thus take the integral scale as the unity.

The second condition, which demands that the suite of exponents with the same parity is non-decreasing was obtained by Frisch (1991). It comes from the fact that the velocity is bounded. Indeed, if V_{max} is that bounding value, then

$$S_{2p+2} = \left\langle (v_\ell(\mathbf{x} + \mathbf{r}) - v_\ell(\mathbf{x}))^{2p+2} \right\rangle \leq \left\langle (v_\ell(\mathbf{x} + \mathbf{r}) - v_\ell(\mathbf{x}))^{2p} \right\rangle 4 V_{max}^2 = 4 V_{max}^2 S_{2p}$$

$$\implies A_{2p+2} r^{\zeta_{2p+2}} \leq 4 V_{max}^2 A_{2p} r^{\zeta_{2p}}$$

[5]A function f is concave, if the following inequality $f[(x+y)/2] \leq (f(x) + f(y))/2$ is verified. For a continuous and derivable function, this inequality is equivalent to $f''(x) \geq 0$.

Since this inequality is valid for all $r = \|\mathbf{x} - \mathbf{x}'\|/\ell_0 \leq 1$, exponents naturally verify

$$\zeta_{2p+2} \geq \zeta_{2p}. \tag{9.77}$$

9.7 Theories for the Closure of Spectral Equations

Until now, all the dynamical equations of the mean fields have been left "open". Those written in the spectral space like (9.57) or those written in the real space like (9.62). Closing these equations is equivalent to expressing the third order moments as a function of those of lower order.

Several theoretical approaches have been devised to close the equations in the spectral space. Here, we shall present the main ideas of these approaches and refer the reader to the specialized textbooks (Lesieur 1990; Leslie 1973; McComb 1990) for more details.

9.7.1 The EDQNM Theory

EDQNM means "Eddy-Damped Quasi-Normal Markovian" which means that the statistics are assumed to be quasi-normal (close to the gaussian laws), Markovian (there is no memory effect), and Damped (some terms are purposely damped). This is probably one of the most popular closure in the spectral space. It was elaborated in the sixties and one of its conceptors, Steven Orszag, has written a masterful synthesis in the Les Houches Lectures of 1973.

The fundamentals of this approach are the followings: we need to get a closure of (9.57), which means that we have to relate the third order moments to the second order ones. However, we know that the evolution of the third order moments depends on those of the fourth order. At this point, the first hypothesis of quasi-normality interrupts the chain of equations. It is assumed indeed, that the statistics of the Fourier components is quasi-normal and hence obey to the Gaussian law. This law has the property that fourth order moments can be expressed with the second order ones. Thus, no hypothesis is made on the third order moments.

The quasi-normality hypothesis is simple: we just neglect the cumulants of fourth order. Unfortunately, this simplification has a disastrous consequence: the kinetic energy spectrum may become negative! The cure of that is to avoid the complete neglect of fourth order cumulant and to replace them by a damping term; hence the Eddy-Damped. This improved very much the theory, but still did not guarantee the positiveness of the energy. The Markovian constraint was then added, inferring that the turbulence has no memory effects.[6] One can then show that if the energy spectrum is positive at one time, then it is positive at all later times.

The EDQNM theory therefore simplifies turbulence on two crucial aspects:

- One assumes that the fourth order cumulants are damping terms for the third order correlations.
- There is no memory effect in the evolution of the spectral quantities.

This approach is interesting since it allows us to compute the evolution of the various spectral quantities. Hence, the way the Kolmogorov spectrum forms from some given initial conditions can be studied (with no intermittency of course!), and the relative simplicity of the method allows some generalization to more complex situation like helical turbulence, or turbulence with a background rotation.

9.7.2 The DIA

DIA means "Direct Interaction Approximation". This is another way of attacking turbulence theory which was developed by Kraichnan at the beginning of the sixties. It relies on a rather severe simplification of reality, which is a drawback, but it is self-consistent. Nevertheless, it allowed the scientists who investigated its consequences, to understand some important points for the theory of turbulence: for instance, the fact that the Kolmogorov spectrum is related to the invariance of the theory in random galilean transform. The book of Leslie (1973) gives a detailed description of this theory.

9.7.3 The Renormalization Group Approach

To end this short review of the closure theories, we should mention that of the renormalization group, which was inspired by the technics of statistical physics in the study of critical phenomena.

Let us assume that we can represent a turbulent flow by a discrete set of Fourier modes k-bounded from above by k_0 which is in the dissipative range ($k_0 \sim k_D$).

Now we cut the spectral domain in two parts by introducing a wavenumber k_1 slightly smaller than k_0 and we consider the fluid motions associated with the spectral domain $k_1 < k < k_0$. Thus we are considering fluid motions at a scale slightly larger than $1/k_0$. Since this range is in the dissipative range, we may linearize Navier–Stokes equation and solve for the evolution of these modes as a function of those in $0 < k \leq k_1$. Now, the evolution of the modes in $0 < k \leq k_1$

---

[6]Markovian processes are such that the probability of an event does not depend on the history of the process.

is also a function (but nonlinear) of those of the band $k_1 < k < k_0$; using the formal expression of the modes $k_1 < k < k_0$ as a function of the modes $0 < k \leq k_1$, we can derive an equation where only the mode of the band $0 < k \leq k_1$ intervene. Then the process can be iterated by replacing k_1 by a slightly smaller k_2. Progressively, the spectral band of the small scales is eliminated; at each iteration the viscosity is "renormalized", since the elimination of a range increases the dissipation of the remaining range.

The method's principle is quite simple, but its setting out is extremely difficult. The reader is referred to the textbook of McComb (1990) for a more thorough presentation of this approach.

9.8 Inhomogeneous Turbulence

In the foregoing sections we focused on the homogeneous turbulence case. This allowed us to be more familiar with the numerous concepts and problems that arise when studying a turbulent flow. Of course, turbulence in real flows is far from homogeneous and it is time now to make the jump in this new jungle...

To fix ideas, we consider the turbulent flow of an incompressible fluid that is in a statistically steady state. We rewrite the equations of the mean quantities (9.7) and (9.8):

$$\rho \partial_j \langle v_i \rangle \langle v_j \rangle + \partial_j R_{ij} = -\partial_i \langle p \rangle + \mu \Delta \langle v_i \rangle \qquad \text{and} \qquad \partial_i \langle v_i \rangle = 0$$

Contrary to the homogeneous case, the Reynolds stress tensor $R_{ij} = \langle \rho v_i' v_j' \rangle$ is no longer constant. We need to find a way to relate it to the mean flow $\langle \mathbf{v} \rangle$. The methods are called closure models on the Reynolds tensor. These models are said to be at zero, one or two equations, according to the number of equations that are solved simultaneously with the evolution of the mean velocity. They might for instance prescribe the evolution of the turbulent kinetic energy or the turbulent dissipation. We shall also say a word about models using a closure on the second order moments, where the evolution of all the components of the Reynolds tensor is computed.

9.8.1 A Short Review of the Closure Models

9.8.1.1 Models with Algebraic Prescriptions: Turbulent Viscosity

Facing the problem of the expression of the Reynolds tensor as a function of $\langle \mathbf{v} \rangle$, we may try to adopt the same reasoning that we used to determine the expression of the viscous stress, assuming that the role of small-scale turbulence is similar to that of the molecules of a Newtonian gas. Small scale turbulence is therefore assumed to diffuse momentum, heat, etc. Thus we write

$$- R_{ij} = -p_{turb}\delta_{ij} + \rho \nu_{turb}(\partial_i \langle v_j \rangle + \partial_j \langle v_i \rangle) \tag{9.78}$$

This simple closure is due to Joseph Boussinesq who introduced the idea of a turbulent viscosity as soon as 1877. In the foregoing expression, the turbulent pressure, p_{turb} can be determined after the velocity field when the fluid is incompressible. Thus, in this case, the crucial part of the model is the "viscous" shape of the tensor and the expression of the viscosity. We shall present two methods which are rather popular for the determination of the turbulent viscosity: the mixing-length theory which was devised by Prandtl, and the Smagorinsky approach devised in 1963.

Prandtl proposed (Prandtl 1925) that turbulent viscosity be the result of momentum transport by fluid elements with a velocity typical of the turbulent fluctuations. Namely, the fluid motion of velocity $\sqrt{\langle v'^2 \rangle}$ at a scale ℓ_M, the mixing length, are the engine of the turbulent diffusion. When the fluid elements have run this length, they vanish, releasing the quantities they carry. Of course this mixing-length is unknown and should be evaluated for every problem.

For a plane-parallel shear flow, Prandtl proposed that $\sqrt{\langle v'^2 \rangle} = \ell_M \left| \frac{d\langle v_x \rangle}{dy} \right|$ so that $\nu_T = \ell_M^2 \left| \frac{d\langle v_x \rangle}{dy} \right|$. This kind of approximation is unfortunately not general since the turbulent viscosity vanishes where the mean velocity gradient vanishes. This is obviously not the case for a turbulent jet: on its axis turbulent diffusion is certainly not vanishing. However, this assumption leads to very acceptable results as far as wall-turbulence is concerned.

Using the same concept, Smagorinsky (1963) proposed to model turbulent viscosity by a formula like

$$\nu_T = \Delta^2 \sqrt{\langle c_{ij} \rangle \langle c_{ij} \rangle}$$

where Δ is a length scale to be precised. We may note the similarity with the Prandtl approach. However, the idea of Smagorinsky is less ambitious: This expression is not meant to be used to determined a mean flow, but just to represent subgrid motions in a numerical simulation. Indeed, in numerical simulations of a turbulent flow, the small scales are generally not computed because of the implied computational cost. At the grid scale the Reynolds number is still large, and the effects of the dismissed scales have to be modeled, namely replaced by something. This is the role of subgrid models. The Smagorinsky model is one such models. Therefore, the length scale Δ is taken as the smallest resolved length (usually the mesh size).[7]

---

[7]Let us mention that usually subgrid scale models are not categorized in models of turbulence since they give a local prescription that can be used only in numerical simulation. However, their similarity with the mean-field approach is strong enough that we discuss them here.

9.8.1.2 The K-ε Model: A Model with Two Equations

We shall leave aside the models using just one additional equation (like the ones of turbulent kinetic energy), which are no longer used, and focus on one using two equations like the celebrated K-ε model.

The K-ε model was proposed by Launder and Spalding (1972). The assumption is that the Reynolds tensor is a function of both the large scale shear $\langle c_{ij}\rangle = \partial_i \langle v_j\rangle + \partial_j \langle v_i\rangle$ and the local strength of turbulence characterized by the turbulent kinetic energy K and the viscous dissipation ε (both taken per unit mass). This dependence is similar as (9.78), namely

$$-\langle v_i' v_j'\rangle = -\frac{2}{3} K \delta_{ij} + v_T \left(\partial_i \langle v_j\rangle + \partial_j \langle v_i\rangle\right) \tag{9.79}$$

where $v_T = c_v K^2/\varepsilon$. This expression reveals the assumptions of this model: first the turbulent pressure depends only on the turbulent kinetic energy and is equal to $\frac{2}{3} K$ while the turbulent viscosity is determined by both the kinetic energy and thus the viscous dissipation. c_v is a dimensionless coefficient which is calibrated with experiments (one usually takes $c_v \sim 0.09$).

In such a model, the turbulent kinetic energy and the turbulent dissipation play a crucial role but need to be determined. The K-ε model proposes to compute them using equations that model their evolution. Hence, we write:

$$\begin{cases} \frac{\partial K}{\partial t} + \langle \mathbf{v}\rangle \cdot \nabla K = -\varepsilon + \frac{v_T}{2} \langle c_{ij}\rangle \langle c_{ij}\rangle + \nabla \cdot (v_T \nabla K) \\[2mm] \frac{\partial \varepsilon}{\partial t} + \langle \mathbf{v}\rangle \cdot \nabla \varepsilon = -c_2 \varepsilon^2/K + \frac{c_1 K}{2} \langle c_{ij}\rangle \langle c_{ij}\rangle + \nabla \cdot (v_\varepsilon \nabla \varepsilon) \end{cases} \tag{9.80}$$

These equations come from the ones verified by the velocity fluctuations. Third order correlations or velocity-pressure correlations are then approximated to close the system.

The equation of velocity fluctuations is derived by combining (9.5) and (9.7):

$$\frac{\partial v_i'}{\partial t} + \langle v_j\rangle \partial_j v_i' + v_j' \partial_j \langle v_i\rangle + v_j' \partial_j v_i' - \langle v_j' \partial_j v_i'\rangle = -\frac{1}{\rho}\partial_i P' + v \partial_j \partial_j v_i' \tag{9.81}$$

Taking the dot product with v_i' and averaging, we find the equation governing the evolution of K:

$$\frac{\partial K}{\partial t} + \langle v_j\rangle \partial_j K + r_{ij}\partial_j \langle v_i\rangle + \overbrace{\langle v_i' v_j' \partial_j v_i'\rangle}^{\text{I}} = -\partial_i \overbrace{\langle P' v_i'\rangle}^{\text{II}} + v \overbrace{\langle v_i' \Delta v_i'\rangle}^{\text{III}} \tag{9.82}$$

where $r_{ij} = R_{ij}/\rho$. In this equation, the three terms (I), (II) and (III) need to be modeled. The first of them can be rewritten $\langle v_j' \partial_j v'^2/2\rangle$: this is the advection of kinetic energy by the fluctuations of the velocity. Following an analogy with a purely diffusive process, the K-ε model assumes that this term can be represented by a turbulent diffusion, namely:

$$\left\langle v'_j \partial_j v'^2/2 \right\rangle = -\nabla \cdot (v_T \nabla K)$$

where the turbulent viscosity v_T remains to be determined. If the turbulence is locally isotropic, we may neglect the correlations with pressure: $\left\langle P' v'_i \right\rangle \sim 0$. Similarly, we can also rewrite $v \left\langle v'_i \Delta v'_i \right\rangle = \partial_j \left\langle v'_i \sigma'_{ij} \right\rangle - \varepsilon$ and, assuming again local isotropy, $\left\langle v'_i \sigma'_{ij} \right\rangle \sim 0$, we find that the power of the viscous force is just the opposite of the viscous dissipation, as expected.[8] Finally, K verifies the following equation:

$$\frac{\partial K}{\partial t} + \left\langle v_j \right\rangle \partial_j K + r_{ij} \partial_j \left\langle v_i \right\rangle = \nabla \cdot (v_T \nabla K) - \varepsilon \qquad (9.83)$$

Using (9.79), we find again the first equation of (9.80).

The equation governing the evolution of ε is much more difficult to derive and we leave the details of the derivation in an appendix of this chapter. It leads to the following expression:

$$\frac{\partial \varepsilon}{\partial t} + \left\langle v_k \right\rangle \partial_k \varepsilon + \left\langle c_{jk} \right\rangle \overbrace{\left\langle v c'_{ij} c'_{ik} \right\rangle}^{\text{I}} + \left\langle \Omega_{ik} \right\rangle \overbrace{\left\langle v c'_{ij} \Omega'_{kj} \right\rangle}^{\text{II}} + \overbrace{\left\langle v c'_{ij} v'_k \right\rangle}^{\text{III}} \partial_k \left\langle c_{ij} \right\rangle$$

$$+ \overbrace{\left\langle v'_k \partial_k v c'^2_{ij}/2 \right\rangle}^{\text{IV}} + 2v \overbrace{\left\langle c'_{ij} (\partial_j v'_k)(\partial_k v'_i) \right\rangle}^{\text{V}} = -2v \overbrace{\left\langle c'_{ij} \partial_i \partial_j p' \right\rangle}^{\text{VI}} + v^2 \overbrace{\left\langle c'_{ij} \Delta c'_{ij} \right\rangle}^{\text{VII}}$$

The seven numbered terms need a model. A first simplification is to assume that the turbulence is locally homogeneous, isotropic and parity invariant. This latter property with homogeneity eliminates (III) while isotropy zeroes terms (II) and (VI).[9] Hence, terms (I), (IV), (V) and (VII) need a more detailed model.

Term (I) is a second order tensor. It may be related to the large-scale shear $\left\langle c_{jk} \right\rangle$; making a Taylor expansion for weak shears (just like we did when dealing with Newtonian fluids), we get

$$\left\langle c_{jk} \right\rangle \left\langle v c'_{ij} c'_{ik} \right\rangle \sim -c_1 K \left\langle c_{jk} \right\rangle^2$$

where c_1 is a dimensionless constant. The turbulent kinetic energy is the quantity which characterizes turbulence.[10]

---

[8] We noted that $\sigma'_{ij} = v(\partial_j v'_i + \partial_i v'_j)$.

[9] See appendix for the demonstration.

[10] Indeed, the local properties of turbulence can only be, with this model, characterized by the two scalars K and ε. In the present case K is the only one dimensionly correct.

The fourth term (IV) is modeled is a similar way as the first term (I) of the K-equation, namely with a turbulent diffusion. We thus write

$$\left\langle v'_k \partial_k \nu c'^2_{ij}/2 \right\rangle \sim -\nabla \cdot (\nu_\varepsilon \nabla \varepsilon)$$

where ν_ε is a new turbulent diffusion, but for dissipation.

Finally, we are left with terms (V) and (VII). These two terms are slightly special since they are the only ones to remain in an isotropic homogeneous steady turbulence. In this latter case they compensate exactly. Hence, it is not necessary to separate their modelling, since their difference is the only important quantity. In the K-ε model these two terms are proportional to $\langle \varepsilon \rangle^2 / K$, namely

$$V - VII \sim c_2 \frac{\varepsilon^2}{K}$$

Finally, these equations need to be completed by the expression of the turbulent diffusions ν_T and ν_ε. Their expression is obtained from the only scalar that has the same dimension as a diffusivity. This leads to the expressions

$$\nu_T \sim c_\nu \frac{K^2}{\varepsilon} \quad \text{and} \quad \nu_\varepsilon \sim c_\varepsilon \frac{K^2}{\varepsilon}$$

As before, the non-dimensional coefficients are calibrated with experiments and usually the following values are adopted:

$$c_\nu = 0.09, \quad c_\varepsilon = 0.07, \quad c_1 = 0.126, \quad c_2 = 1.92$$

The K-ε model tries to establish a relation between inhomogeneous turbulence and universal turbulence by assuming that, locally, the fluctuations of the mean flow are universal. Moreover, this local turbulence is assumed to behave like a Newtonian fluid with a variable viscosity.

These hypothesis are obviously very strong and one may wonder whether the turbulent fluid can behave as a Newtonian fluid even with a variable viscosity. This would mean a separation of scales that is not observed, even approximately. The local homogeneity also implicitly assumes a separation of the spatial scales.

In addition to these questions about the physical validity of the model, some other problems on the internal consistency of the model arise. For instance, the model allows a computation of the kinetic energy and viscous dissipation. These two quantities are positive and their evolution should preserve this positivity. Presently, there is no general proof that this is indeed the case. A few demonstrations, applying to some restricted cases and showing that this is true, may reassure us.

9.8.1.3   Second Order Closure Models

As we mentioned it previously, a two-equations model like the K-ε one implies a very restrictive form to the Reynolds tensor. It is assumed to be like that of a Newtonian fluid even if its viscosity is not locally determined. However, the turbulent fluid has no reason to be isotropic, and anisotropy is likely not a local function as well. It may result from the past time evolution of a fluid element (memory effect) or from distant interactions like those coming through the pressure. It therefore seems simpler, if we wish to get a more realistic description, to directly compute the evolution of R_{ij}, through equations like:

$$\frac{DR_{ij}}{Dt} = \cdots$$

These equations of course introduce the third order correlations, which need a new modeling. Even if this modelling is coarse, it is hoped that it will give realistic values of the R_{ij}, just like the turbulent viscosity model is able to give realistic mean flows in some cases. Hence, if the R_{ij} are better, the mean flow may be much better. Comparison of the results of these models with experiments seem to comfort this hope.

9.8.2   Examples: Turbulent Jets and Turbulent Plumes

We end this section with the analysis of two very common turbulent flows: those of jets and plumes. As shown in Fig. 9.8, these flows have a conical shape outside of

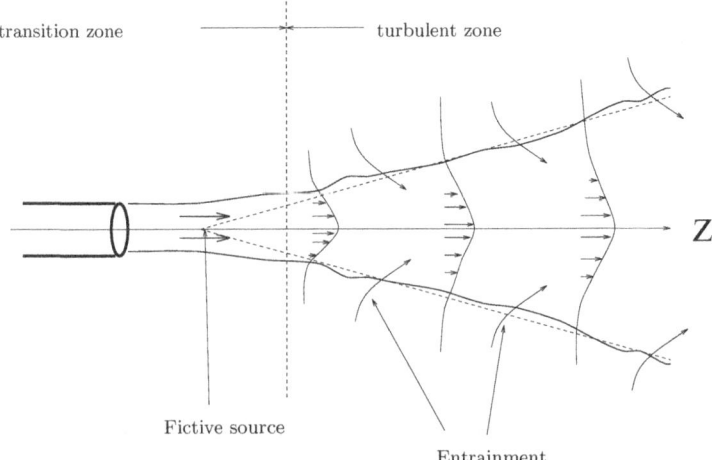

Fig. 9.8 A schematic view of a turbulent jet. The transition zone is the region where the shear instabilities give birth to the turbulence

which the turbulence is very low or absent. We shall see that this property, comes from the self-similarity of the solutions. Self-similarity is likely deeply rooted in turbulent flows.

Jets, plumes and wakes are often called free shear flows. Indeed, turbulence results from an imposed strong shear, actually a shear layer, which is, as we saw in Chap. 6, very unstable. The development of turbulence entrains the outer fluid inside the jet and the jet broadens as it progresses.

Let us assume that the fluid flow is self-similar, so that we may write the velocity field as:

$$\begin{cases} \langle v_r \rangle = V(z)g(r/b(z)) \\ \langle v_z \rangle = V(z)f(r/b(z)) \end{cases} \tag{9.84}$$

where we additionally assumed the jet axial symmetry. The dependence of the solutions with respect to $\xi = r/b(z)$, insures the similarity of the velocity profiles for all z. This profile has always the same shape, given by $f(\xi)$ for v_z, but its real width varies with the distance to the source z. Let us now assume incompressibility so that mass conservation implies:

$$\frac{\partial \langle v_r \rangle}{\partial r} + \frac{\langle v_r \rangle}{r} + \frac{\partial \langle v_z \rangle}{\partial z} = 0 \quad \Longleftrightarrow \quad \frac{1}{\xi}\frac{d(\xi g)}{d\xi} - b'(z)\xi f'(\xi) + \frac{V'(z)b(z)}{V(z)}f(\xi) = 0$$

where the prime indicates a derivative. The existence of solutions like (9.84) implies that the variables can be separated. This implies that $b'(z)$ and $B = V'(z)b(z)/V(z)$ are constant. This means that the width of the jet grows linearly. Then, noting that $V'(z)/V(z) \propto 1/z$ (taking the origin of z where b is zero), then $V(z)$ varies like z^α.

We need the equation of dynamics to infer α. Neglecting viscous effects, the steady mean flow verifies:

$$\frac{1}{r}\frac{\partial r \langle v_r \rangle \langle v_z \rangle}{\partial r} + \frac{\partial \langle v_z \rangle^2}{\partial z} = -\frac{1}{r}\frac{\partial r \langle v_r' v_z' \rangle}{\partial r} - \overbrace{\frac{\partial \langle v_z'^2 \rangle + \langle P \rangle}{\partial z}}^{I}$$

The term (I) is usually neglected in this type of flow. It is indeed a pressure gradient which is usually small: this comes from the fact that the ambient pressure where the jet develops is constant. Indeed, the pressure is almost constant in a section of the jet, because the mean streamlines are straight lines (recall the results of Chap. 3). Then this equation may be rewritten to show explicitly the flux of momentum; integrating over r, we find

$$\frac{d}{dz}\int_0^\infty b(z)^2 V(z)^2 f(\xi)^2 \xi d\xi + [r\langle v_r \rangle \langle v_z \rangle + r\langle v_r' v_z' \rangle]_0^\infty = 0$$

The second term is zero since v_z vanishes at infinity and rv_r is finite (see below). The correlation term is zero outside the turbulent zone. Finally, this equation shows that

$b(z)^2 V(z)^2 = P^2$ where P^2 is a constant (the momentum flux), which characterizes the jet. Consequently, it turns out that the jet mean velocity decreases as $1/z$ and that $B = -b'$.

Now, we apply the same treatment to the mass conservation equation. We find that

$$\lim_{r \to \infty} r \langle v_r \rangle = -\left(\int_0^\infty f(\xi)\xi d\xi \right) \frac{d[b^2 V(z)]}{dz} = -V(z)b(z) \underbrace{b' \int_0^\infty f(\xi)\xi d\xi}_{\alpha_j}$$

This expression shows that the radial velocity is proportional to the axial velocity $V(z)$ and to α_j which is called the *entrainment constant*.

The expression of α_j shows that it depends on the velocity profile $f(\xi)$ and on the aperture angle of the cone b'. In fact these two quantities are themselves dependent on the closure relations or the transport properties of turbulence. Many models have been proposed to explain the velocity profile of a turbulent jet, but none is completely satisfactory. Experiments show that in general the profile is a Gaussian. Thus, taking $f(\xi) = \exp\{-\xi^2\}$ is a good approximation. Measurements then give $\alpha_j = 0.054$ for the entrainment constant of jets. As an exercise, we may show that the entrainment constant and the aperture cone angle are functions of the velocity profile.

The self-similar turbulent jet with a gaussian profile thus obeys two simple equations:

$$\begin{cases} \dfrac{d[b^2 V(z)]}{dz} = 2\alpha_j b(z) V(z) \\[3mm] \dfrac{d[b(z)^2 V(z)^2]}{dz} = 0 \end{cases} \tag{9.85}$$

which translate respectively the conservation of mass and momentum. Their solution is obviously given by the proceeding laws: $b(z) = 2\alpha z$ and $V(z) = P/2\alpha z$. We note that the initial mass flux does not play any role in the solution. Actually, a short analysis of the solutions shows that the initial conditions are rapidly forgotten by the solution which quite quickly reaches the self-similar regime. In the final steady state the initial mass flux is a very small part of the actual mass flux, which has grown through entrainment of the surrounding fluid. The jet thus appears as generated by a pure source of momentum.

Let us now examine the case of a *turbulent plume*. The most common example is the smoke plume that raises over a chimney. The hot fluid raises in the atmosphere thanks to buoyancy which smoke particles render visible as a turbulent mixed flow.

The behaviour of the plume is very similar to that of the jet, but in this case, this is the initial enthalpy flux which controls the dynamics of the plume. Indeed, similarly as above, the initial mass flux and the initial momentum flux are forgotten by the flow. The initial mass flux disappears because of entrainment as in the turbulent jet, and the initial momentum flux also disappears because of the work of buoyancy that add new momentum to the flow. Thus the plume is made by a pure source of enthalpy.

Experiments show that the velocity profiles in plumes are close to those of jets. In addition to the mean velocity field, the plume is characterized by an "enthalpy jump" field δh, which measures the difference of enthalpy within the plume and outside the plume. As for the jets, self-similar solutions also exist and verify:

$$
\begin{cases}
\dfrac{d\,[b^2 V]}{dz} = 2\alpha_p b V \\[2mm]
\dfrac{d\,[b^2 V^2]}{dz} = 2g b^2 \Delta\rho \\[2mm]
\dfrac{d\,[b^2 \Delta\rho V]}{dz} = 0
\end{cases}
\tag{9.86}
$$

We derived these equations using the Boussinesq approximation and orienting the z-axis along the gravity field $\mathbf{g}$. We also note that another entrainment constant has been used because experiments say that $\alpha_p = 0.083$, which is different from the jet (why?).

The solutions of (9.86) are naturally power laws in z as prescribed by self-similarity. We easily find that:

$$
b(z) = \frac{6\alpha_p}{5} z, \quad V(z) = \left(\frac{25 g F_b}{24\alpha_p^2} \right)^{1/3} z^{-1/3}
$$

$$
\Delta\rho = \frac{5}{6} \left(\frac{5 F_b^2}{9\alpha_p^4 g} \right)^{1/3} z^{-5/3}
\tag{9.87}
$$

where $b^2 \Delta\rho V = F_b$ is the buoyancy flux. We also note that the aperture angle of the plume is quite similar to that of the jet, namely ~ 0.1 while the entrainment constant is somewhat different. This difference is certainly related to the transport of momentum and the active scalar δh.

9.9 Two-Dimensional Turbulence

Two-dimensional turbulence is quite different from its three-dimensional counterpart. Despite the strong approximation that is made (our world is three dimensional!), the turbulent flows in two dimensions deserve being studied because they enlight us on the dynamics of the Earth's atmosphere or oceans. Indeed, the fluid flows in these two thin layers of the Earth are almost two dimensional as soon as the scales considered largely exceeds the thickness of the layer (10 km for the atmosphere and 5 km for the oceans).

Two dimensionality implies new conservations law which strongly modify the dynamics, in particular when we consider the evolution of vorticity. Equation (3.41) indeed says that:

$$\frac{Df(\omega)}{Dt} = f'(\omega)\frac{D\omega}{Dt} = 0 \qquad \Longleftrightarrow \qquad \int_{(S)} f(\omega)dS = \text{Cst}$$

where S is a surface advected by the fluid. The main consequence of this peculiarity is that the picture of the turbulent cascade is completely modified.

9.9.1 Spectra and Second Order Correlations

As in three dimensions, it is interesting to examine the properties of the homogeneous and isotropic turbulence.

The tensors Q and ϕ have the same definition, but just four components. If we observe that in two dimensions there is no helicity (vorticity being orthogonal to velocity), then following the same steps as when deriving (9.17), we find:

$$Q_{ij} = (rf)'\delta_{ij} - f'(r)r_ir_j/r \tag{9.88}$$

where $f(r)$ is the longitudinal correlation. Similarly, as for (9.27), we find:

$$\phi_{ij} = e(k)P_{ij} = \frac{E(k)}{\pi k}P_{ij} \tag{9.89}$$

using expression (9.92) below. Indeed, we still have

$$E_{turb} = \frac{1}{2}\langle v^2 \rangle = \int_0^{+\infty} E(k)dk \tag{9.90}$$

which we relate to ϕ_{ij} by

$$E_{turb} = \frac{1}{2}Q_{ii}(0) = \frac{1}{2}\int \phi_{ii}(\mathbf{k})dS_k = \frac{1}{2}\int_0^{+\infty} kdk \int_{(2\pi)} \phi_{ii}(\mathbf{k})d\theta_k$$

$$\Longrightarrow \quad E(k) = \frac{1}{2}k\int_0^{2\pi} \phi_{ii}(\mathbf{k})d\theta_k \ . \tag{9.91}$$

In the isotropic case

$$E(k) = \pi k \phi_{ii}(k) \tag{9.92}$$

The function $R(r)$, defined as half the trace of $[Q]$, now reads,

$$R(r) = \frac{1}{2}Q_{ii}(r) = \frac{1}{2}\int \phi_{ii}(k)e^{i\mathbf{k}\cdot\mathbf{r}}dS_k = \frac{1}{2}\int \phi_{ii}(k)e^{ikr\cos\theta}kdkd\theta$$

The integration over the angular variable can easily be realized if we use the general expression of the zeroth order Bessel function, namely

$$J_0(z) = \frac{1}{\pi}\int_0^\pi \cos(z\cos\theta)d\theta \ .$$

Thus, we find

$$R(r) = \int_0^\infty E(k)J_0(kr)dk \tag{9.93}$$

Conversely

$$\phi_{ii}(\mathbf{k}) = \frac{1}{(2\pi)^2}\int Q_{ii}(\mathbf{r})e^{-i\mathbf{k}\cdot\mathbf{r}}d^2\mathbf{r} = \frac{1}{\pi}\int_0^{+\infty} rR(r)J_0(kr)dr$$

gives the expression of the spectrum

$$E(k) = \int_0^{+\infty} krJ_0(kr)R(r)dr \tag{9.94}$$

As in the three-dimensional case, this expression allows us to derive the behaviour of the spectrum at the very large scales. In this case

$$E(k) \sim k\int_0^{+\infty} rR(r)dr \quad \text{as} \quad k \to 0$$

showing that the spectral kinetic energy density grows like k.

9.9.2 Enstrophy Conservation and the Inverse Cascade

In order to understand the implication of enstrophy conservation on the spectral properties of two-dimensional turbulence, it is convenient to consider a set-up where there would be only three Fourier modes of wavenumbers $k_1 < k_2 < k_3$, and energy E_1, E_2, E_3. The modes are assumed to be in nonlinear interactions, in a one-dimensional flow. Hence, we set $k_3 = k_1 + k_2$. Neglecting furthermore the effects of viscosity, energy and enstrophy of this system are constant. Thus energy variations must meet:

$$\delta E_1 + \delta E_2 + \delta E_3 = 0$$

$$k_1^2 \delta E_1 + k_2^2 \delta E_2 + k_3^2 \delta E_3 = 0$$

from which we easily find

$$\delta E_1 = -\frac{k_3^2 - k_2^2}{k_3^2 - k_1^2} \delta E_2 \qquad \text{and} \qquad \delta E_3 = -\frac{k_2^2 - k_1^2}{k_3^2 - k_1^2} \delta E_2$$

Now let us suppose that the energy of the intermediate mode of wavenumber k_2 decreases. Then energy of the two others increases, namely $\delta E_2 < 0 \implies \delta E_1 > 0$ and $\delta E_3 > 0$. If the modes are now spectrally close, for instance if $k_2 = \lambda k_1$ and $\lambda < 1 + \sqrt{3}$, then $\delta E_1 > \delta E_3$. This means that in this case the energy of the second mode is preferentially transferred to the first mode. This illustrate the case of an inverse cascade of energy. Enstrophy conservation together with nonlinear interactions tends to transfer energy towards the larger scales. On the other hand we may observe that simultaneously, enstrophy would rather tend to cascade to the small scales for, generally, $\delta Z_3 = k_3^2 \delta E_3 > \delta Z_1 = k_1^2 \delta E_1$.

Let us now focus on the shape of the spectra. The evolution of kinetic energy is guided by

$$\frac{dE_{turb}}{dt} = \frac{d}{dt} \int_0^{+\infty} E(k,t) dk = -2\nu \int_0^{+\infty} k^2 E(k,t) dk$$

$$= -2\nu \int_0^{+\infty} Z(k,t) dk = -\varepsilon(t)$$

while enstrophy follows

$$\frac{dZ}{dt} = \frac{d}{dt} \int_0^{+\infty} Z(k,t) dk = -2\nu \int_0^{+\infty} k^2 Z(k,t) dk$$

$$= -2\nu \int_0^{+\infty} k^4 E(k,t) dk = -\zeta(t)$$

This equation shows that enstrophy can only decrease, and thus remain bounded from above by its initial value. However if $\int_0^{+\infty} Z(k,t) dk$ is bounded, this implies that $\varepsilon \to 0$ when $\nu \to 0$. This means that ε cannot be used to determine the kinetic energy spectrum in the inertial range in two dimensions. We are left with $\zeta(t)$, namely the dissipation rate of enstrophy. Assuming that it is the quantity which controls the two-dimensional turbulence, then, using similar arguments as in three-dimensions, we find that

$$E(k,t) \propto \zeta(t)^{2/3} k^{-3} \qquad (9.95)$$

The major consequence of this power law, is that the dissipation of kinetic energy $k^2 E \propto k^{-1}$ occurs in the large scales. Enstrophy dissipation, on the contrary, occurs in the small scales ($k^2 Z \propto k$).

Two-dimensional turbulence gives us a scenario that is quite different from its three-dimensional counterpart. Energy tends to accumulate in the large scale while enstrophy tends to be extracted by the small scales.

Numerical simulations have shown quite clearly what was going on in the physical space: vortices with similar vorticity (i.e. cyclonic or anti-cyclonic) tend to merge and form larger structures, signing the inverse cascade, while enstrophy, which is conserved by any fluid element, faces a filamentation producing smaller and smaller scales which are in the end erased by viscosity.

9.9.3 Turbulence with Rotation or Stratification

The shape of the container is not the only way to impose two-dimensionality to a flow. In Chap. 8, we saw that rotation, through the Coriolis acceleration could make a flow two-dimensional. In fact, two phenomena may also impose some two-dimensional dynamics to turbulence, i.e. a background rotation or a stable density stratification.

These two physical constraints can make a flow two-dimensional when the time-scale of the motion is much larger than those imposed either by rotation or stratification. Comparing time scales leads to the determination of the scale at which the effects of rotation or stratification start to be noticeable.

The turn-over time at a scale ℓ is $\tau_\ell = \ell / v_\ell$ but in Kolmogorov inertial range $v_\ell = (\ell \varepsilon)^{1/3}$, thus $\tau_\ell = \ell^{2/3} \varepsilon^{-1/3}$. This leads to the scale ℓ_t where the transition between three-dimensional and two-dimensional motions occurs:

$$\ell_t = \varepsilon^{1/2} \Omega^{-3/2} \qquad \text{or} \qquad \ell_t = \varepsilon^{1/2} N^{-3/2} \qquad (9.96)$$

for respectively the rotating and stratified cases. For stably stratified fluids, ℓ_t is also known as Ozmidov scale.

9.10 Some Conclusions on Turbulence

To conclude this rather long chapter, I would like to present to the reader some ideas in order to better appreciate the way we are from a solution to the problems that we crossed in the course of this chapter.

The models such as the K-ε one, try to make a parallel between the "turbulent fluid" and a real fluid. Like the real fluid, the turbulent fluid would have a "pressure", a viscosity, etc. However, if such a way of doing is relevant, the first step is to describe correctly the equilibrium state, namely the "thermodynamics" of such a material. Such a step is still not accomplished although attempts have flourished in the literature (see Castaing 1989, 1996; Chorin 1991, for instance). However, let us admit that we succeeded. The next step is to derive the transport coefficients of the turbulence. As we did for the Newtonian fluid, one should analyse the response of the turbulence to weak perturbations. However, this is a formidable task. Indeed, unlike a standard fluid, which own just a single (very small) scale (the mean-free path, see Chap. 11), the turbulent fluid owns a very large number of scales that strongly interact. Pushing this idea to its end, we see that the turbulent fluid should be compared to some non-Newtonian fluid with and extremely complex, non-local, rheological law.

Presently, we may hope that the situation be not so dramatic and that among all the scales which intervene, just a small number are truly important, the others following the firsts. This possibility is not so unrealistic since in many cases, turbulent flows tend to self-similar situations, emphasizing scale invariance.

9.11 Exercises

1. a) Show that

$$\langle \hat{v}_i^*(\mathbf{k})\hat{v}_j(\mathbf{k}') \rangle = \phi_{ij}(\mathbf{k})\delta(\mathbf{k} - \mathbf{k}')$$

   where $\hat{v}$ represents the Fourier transform of the velocity fluctuations and δ is Dirac distribution.

   b) Show the following equality:

$$Z_{ij} = (2\pi)^{-3} \int \langle \omega_i'(\mathbf{x})\omega_j'(\mathbf{x}') \rangle e^{-i\mathbf{k}\cdot\mathbf{r}} d^3\mathbf{r} = \epsilon_{ilm}\epsilon_{jnp}k_l k_n \phi_{mp}(\mathbf{k})$$

   and then, that this expression leads to (9.28).

   c) Show the reciprocal relation of (9.35), namely

$$E(k) = \frac{2}{\pi} \int_0^\infty kr \sin kr\, R(r)dr$$

2. Following a similar way as in Sect. 9.4.1, show that the correlation length of the vorticity may be written as

$$\ell_z = \frac{\pi}{2} \frac{\int_0^\infty k E(k) dk}{\int_0^\infty k^2 E(k) dk}$$

   Show that ℓ_z is also the dissipation scale ℓ_D.
3. Let us consider the relation linking the energy spectrum $E(k)$ and the scaling law of the structure function S_2.

   a) From (9.35) and (9.61) show that the two-point correlation of the longitudinal components of the velocity f verifies

   $$f(r) = 2 \int_0^\infty \frac{\sin kr - kr \cos kr}{(kr)^3} E(k) dk \qquad (9.97)$$

   b) Retrieve that $f(0) = \frac{2}{3} E_{turb}$ and derive that

   $$S_2(r) = 4 \int_0^\infty \left[\frac{1}{3} - \frac{\sin kr - kr \cos kr}{(kr)^3} \right] E(k) dk \qquad (9.98)$$

   c) Show that if $E(k) = C_K \langle \varepsilon \rangle^{2/3} k^{-5/3}$ then $S_2 = C_2 \langle \varepsilon \rangle^{2/3} r^{2/3}$ and that C_2 and C_K are proportional.
   d) Using (9.76), find the difference between the She & Leveque exponent and the Kolmogorov exponent of the energy spectrum.
4. We assume that the energy spectrum of some turbulence is such that:

   $$\begin{cases} E(k) = i(k) & k \le k_0 \\ E(k) = k^{-5/3} & k_0 \le k \le k_D \\ E(k) = d(k) & k \ge k_D \end{cases} \qquad (9.99)$$

   and that functions i and d verify:

   $$i(k) \le k_0^{-5/3}$$

   $$d(k) \le k_D^{-5/3} e^{-(k-k_D)/k_D}$$

   Show that in these conditions, if $k_D \gg k_0$, then

   $$\sqrt{\frac{8}{69}} (\ell_0 \ell_D^2)^{1/3} \lesssim \ell_T \lesssim \frac{2\sqrt{5}}{3} (\ell_0 \ell_D^2)^{1/3} \qquad (9.100)$$

   derive (9.53).

5. If the distribution of the logarithm of x obey to a normal law, show that

$$\langle x^p \rangle = e^{p\langle \ln x \rangle + p^2 \sigma^2 / 2}$$

where σ^2 is the variance of the distribution. The first step is to show that if y is a random variable with a normal distribution and zero mean, then

$$\frac{1}{\sigma\sqrt{2\pi}} \int_{-\infty}^{+\infty} y^{2p} e^{-y^2/2\sigma^2} dy = (2p-1)!!\,\sigma^{2s}$$

where $(2p-1)!! = 1 \times 3 \times 5 \times 7 \times \ldots \times (2p-1)$.

6. *Turbulent jet.*

   a) If we measure in Fig. 9.9, the half-aperture angle of the turbulent jet visualized by the water vapour (or rather the droplets of the condensing vapour), we find a value around 0.15 rd. What can we infer?

   b) A ping-pong ball is placed in a turbulent jet directed upwards. Although the thrust of the jet is maximum on its axis we observe that the ball remains in the

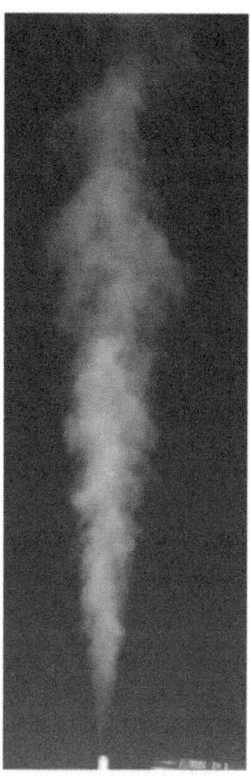

Fig. 9.9 Steam jet at the outlet of a pressure cooker. Note the conical shape of the jet steam

Fig. 9.10 Ping-pong ball
sustained by an inclined
turbulent air jet

jet, wandering around. We may even incline the jet about 30° without the ball
fall (see Fig. 9.10). Explain.

Appendix: Complements for the K-ε Model

Let us start from the equation of the velocity fluctuations (9.81),

$$\frac{\partial v_i'}{\partial t} + \langle v_k \rangle \, \partial_k v_i' + v_k' \partial_k \langle v_i \rangle + v_k' \partial_k v_i' - \langle v_k' \partial_k v_i' \rangle = -\partial_i P' + \nu \Delta v_i'$$

We rewrite it for v_j', and write down the one for the fluctuations of the shear, namely
$c_{ij}' = \partial_j v_i' + \partial_i v_j'$. We get

$$
\begin{aligned}
\frac{\partial c_{ij}'}{\partial t} &+ \langle v_k \rangle \, \partial_k c_{ij}' + \partial_j \langle v_k \rangle \, \partial_k v_i' + \partial_i \langle v_k \rangle \, \partial_k v_j' \\
&+ v_k' \partial_k \langle c_{ij} \rangle + \partial_j v_k' \partial_k \langle v_i \rangle + \partial_i v_k' \partial_k \langle v_j \rangle \\
&+ v_k' \partial_k c_{ij}' + \partial_j v_k' \partial_k v_i' + \partial_i v_k' \partial_k v_j' \\
&- \partial_j \partial_k R_{ik} - \partial_i \partial_k R_{jk} = -2 \partial_i \partial_j p' + \nu \Delta c_{ij}'
\end{aligned}
\tag{9.101}
$$

Since $\varepsilon = \frac{\nu}{2} \langle c_{ij}' c_{ij}' \rangle$, we now contract the foregoing equation with $\nu c_{ij}'$ and take the
average. The first two terms can be rewritten as

$$\frac{\partial \varepsilon}{\partial t} + \langle v_k \rangle \, \partial_k \varepsilon$$

Then the four terms $\partial_j \langle v_k \rangle \partial_k v_i' + \partial_i \langle v_k \rangle \partial_k v_j' + \partial_j v_k' \partial_k \langle v_i \rangle + \partial_i v_k' \partial_k \langle v_j \rangle$ give

$$2 \left\langle c_{ij}' \left(\partial_j \langle v_k \rangle \partial_k v_i' + \partial_i \langle v_k \rangle \partial_k v_j' \right) \right\rangle$$

because c_{ij}' is symmetric. They are further transformed into

$$\left\langle c_{ij}' \Omega_{kj}' \right\rangle \langle \Omega_{ik} \rangle + \left\langle c_{ij}' c_{jk}' \right\rangle \langle c_{ik} \rangle$$

where we set $\Omega_{ij} = \partial_i v_j - \partial_j v_i$. Noting that

$$c_{ij}' v_k' \partial_k c_{ij}' = v_k' \partial_k c_{ij}'^2 / 2 \qquad \text{and} \qquad c_{ij}' \left(\partial_j v_k' \partial_k v_i' + \partial_i v_k' \partial_k v_j' \right) = 2 c_{ij}' \partial_j v_k' \partial_k v_i'$$

Then we get the equation that we were looking for, namely

$$\frac{\partial \varepsilon}{\partial t} + \langle v_k \rangle \partial_k \varepsilon + \overbrace{\left\langle v c_{ij}' \Omega_{kj}' \right\rangle}^{(I)} \langle \Omega_{ik} \rangle + \left\langle v c_{ij}' c_{jk}' \right\rangle \langle c_{ik} \rangle + \overbrace{\left\langle v c_{ij}' v_k' \right\rangle}^{(II)} \partial_k \langle c_{ij} \rangle$$

$$+ \left\langle v v_k' \partial_k c_{ij}'^2 / 2 \right\rangle + 2v \left\langle c_{ij}' \partial_j v_k' \partial_k v_i' \right\rangle = -2v \underbrace{\left\langle c_{ij}' \partial_i \partial_j p' \right\rangle}_{(III)} + v^2 \left\langle c_{ij}' \Delta c_{ij}' \right\rangle$$

Let us show now that local isotropy removes terms (I) and (III). If ω is the vorticity, then

$$\Omega_{ij} = \epsilon_{ijk} \omega_k$$

thus (I) also reads

$$\epsilon_{kjl} \epsilon_{ikn} \left\langle v c_{ij}' \omega_l' \right\rangle \langle \omega_n \rangle = \left\langle c_{ij}' \omega_i' \right\rangle \langle \omega_j \rangle - \left\langle c_{ii}' \omega_l' \right\rangle \langle \omega_l \rangle$$

incompressibility implies that $c_{ii}' = 0$ and isotropy that $\left\langle c_{ij}' \omega_i' \right\rangle = 0$.

Term (III) is reshuffled as:

$$\left\langle c_{ij}' \partial_i \partial_j p' \right\rangle = \partial_i \left\langle c_{ij}' \partial_j p' \right\rangle - \left\langle \partial_i c_{ij}' \partial_j p' \right\rangle = \partial_i \left\langle c_{ij}' \partial_j p' \right\rangle - \partial_j \left\langle \partial_i c_{ij}' p' \right\rangle + \left\langle (\partial_i \partial_j c_{ij}') p' \right\rangle$$

Isotropy makes the first two terms zero, while the last one disappears because $\nabla \cdot \mathbf{v}' = 0$.

Term (II) can be also eliminated if the turbulence is locally homogeneous and parity-invariant. In this case, we introduce $C_{ijk}(\mathbf{r}) = \langle c_{ij}'(\mathbf{x}) v_k'(\mathbf{x} + \mathbf{r}) \rangle$ and we show that like for S_{ijk} (see 9.39) we have $C_{ijk}(\mathbf{r}) = -C_{ijk}(-\mathbf{r})$ so that $C_{ijk}(\mathbf{0}) = 0$. Noting that $\langle c_{ij}' v_k' \rangle = C_{ijk}(\mathbf{0})$, the result follows.

Further Reading

There are numerous textbooks devoted to this very rich subject. A recent thorough review may be found in Davidson (2004). In a slightly more comprehensive style, we recommend the book of Frisch (1995) *Turbulence: the legacy of A. N. Kolmogorov* where the case of intermittency is well discussed. *Turbulence in fluids* by Lesieur (1990) presents at length the spectral side of turbulence, while *The physics of fluid turbulence* de McComb (1990) is a monograph focusing on the renormalization group approach (for more acquainted readers). The Les Houches lectures of Orszag (1973), is still a very good introduction to turbulence and to EDQNM in particular. Let us also mention some now classical work like the monograph of Leslie (1973) dealing with the DIA, the two volumes of Monin and Yaglom, reviewing the knowledge in 1975 or the *Turbulence* of Hinze (1959,1975) focused on the engineering approach of turbulence, like Tennekes and Lumley (1972) or Piquet (2001).

References

Arneodo, A., Manneville, S. & Muzy, J. F. (1998). Towards log-normal statistics in high Reynolds number turbulence. *European Physical Journal B, 1*, 129–140.

Benzi, R., Ciliberto, S., Tripiccione, R., Baudet, C., Massaioli, F. & Succi, S. (1993). Extended self-similarity in turbulent flows. *Physical Review E, 48*, R29.

Castaing, B. (1989). Conséquences d'un principe d'extremum en turbulence. *Journal of Physique, 50*, 147.

Castaing, B. (1996). The temperature of turbulent flows. *Journal of Physique 2, 6*, 105–114.

Chapman, C. J. & Proctor, M. R. E. (1980). Nonlinear Rayleigh–Bénard convection between poorly conducting boundaries. *Journal of Fluid Mechanics, 101*, 759–782.

Cheng, X.-L., Wang, B.-L. & Zhu, R. (2010). Kolmogorov Constants of Atmospheric Turbulence over a Homogeneous Surface. *Atmospheric and Oceanic Science Letters, 3*, 195–200.

Chorin, A. (1991). *Vorticity and turbulence.* New York: Springer.

Davidson, P. (2004). *Turbulence: An introduction for scientists and engineers.* Oxford: Oxford University Press.

Dubrulle, B. (1994). Intermittency in fully developed turbulence: Log-Poisson statistics and generalized scale covariance. *Physical Review Letters, 73*, 959.

Frisch, U. (1991). From global scaling, à la Kolmogorov, to local multifractal scaling in fully developed turbulence. *Proceedings of the Royal Society of London, 434*, 89.

Frisch, U. 1995 *Turbulence: The legacy of A. N. Kolmogorov.* Cambridge: Cambridge University Press.

Kaneda, Y., Ishihara, T., Yokokawa, M., Itakura, K. & Uno, A. (2003). Energy dissipation rate and energy spectrum in high resolution direct numerical simulations of turbulence in a periodic box. *Physics of Fluids, 15*, L21–L24.

von Kármán, T. & Howarth, L. (1938). On the statistical theory of isotropic turbulence. *Proceedings of the Royal Society of London, 164A*, 192.

Launder, B. E. & Spalding, D. B. (1972). *Lectures in mathematical models of turbulence.* London: Academic Press.

Lesieur, M. (1990). *Turbulence in fluids.* Dordrecht: Kluwer.

Leslie, D. (1973). *Developments in the theory of turbulence.* Oxford: Oxford Science Publishier.

McComb, W. D. (1990). *The physics of fluid turbulence*. Oxford: Oxford Science Publishier.

Monin, A. & Yaglom, A. (1975). *Statistical fluid mechanics* (Vol. 1). Cambridge: MIT Press.

Orszag, S. (1973). Fluids Dynamics. In Balian, R., & Peube, J.L. (Eds), *Lectures on the statistical theory of turbulence*. (pp. 237–374), Gordon and Breach, New York, 1977.

Piquet, J. (2001). *Turbulent flows: Models and physics*. New York: Springer.

Prandtl, L. (1925). Bericht über Untersuchungen zur ausgebildeten Turbulenz. *Zeit. Ang. Math. Mech., 5*, 136–139.

She, Z.-S. & Lévêque, E. (1994). Universal scaling laws in fully developed turbulence. *Physical Review Letters, 72*, 336.

She, Z.-S. & Waymire, E. (1995). Quantized energy cascade and Log-Poisson statistics in fully developed turbulence. *Physical Review Letters, 74*, 262.

Smagorinsky, J. (1963). General circulation experiments with the primitive equations. I. The basic experiment. *Monthly Weather Review, 91*, 99.

Tennekes, H. & Lumley, J. (1972). *A first course in turbulence*. Cambridge: MIT Press.

Vincent, A. & Meneguzzi, M. (1991). The spatial structure and statistical properties of homogeneous turbulence. *Journal of Fluid Mechanics, 225*, 1–20.

Chapter 10
Magnetohydrodynamics

Magnetohydrodynamics (MHD for the experts) is often impressive for its complexity. However, it is only the dynamics of electrically conducting fluids. It is indeed complicated because of a new vector field that enters the game, namely the magnetic field. The dynamics is different because of a new force: the Laplace force. Since conducting fluids support electric currents that may generate magnetic fields, we easily imagine that the evolution of both velocity and magnetic fields may be quite complex. In this chapter we wish to remain introductive and therefore we shall focus only on the very basis of magnetohydrodynamics.

10.1 Approximations Leading to Magnetohydrodynamics

Magnetohydrodynamics deals with the motions of a conducting fluid where there is no free charge. This implies that some approximations are met.

The first one is that the fluid motion is not relativistic, namely that the fluid velocity is much less than that of light c, i.e.

$$V/c \ll 1 \qquad (10.1)$$

Two other conditions come from the absence of free charges. If our medium is a fully ionized plasma made of electrons (of charge $-e$) and ions (of charge Ze), then a first condition which should be met by a flow characterized by a length scale L is

$$L \gg \lambda_D$$

© Springer International Publishing Switzerland 2015
M. Rieutord, *Fluid Dynamics*, Graduate Texts in Physics,
DOI 10.1007/978-3-319-09351-2_10

where λ_D is the Debye length. This latter length is the mean distance beyond which the charge of an ion is screened by electrons. It depends on the temperature T and the electron density n_e of the medium, namely

$$\lambda_D = \sqrt{\frac{\varepsilon_0 k T}{n_e e^2}}$$

where ε_0 is the vacuum permittivity and k Boltzmann constant. Using international units this length reads $\lambda_D = 2.8 \times 10^{-12} \sqrt{AT/\rho Z}$ metres, where A is the mass number of the ions and ρ is the mass density. This scale may invalidate the use of the MHD equations if the fluid is too hot or too dilute.

Charge separation may also occur if the time frequency of the fluid motion is too large, i.e. larger than the plasma frequency ω_P. Thus we should also demand that the time scales of the fluid flow $T = L/V$ verifies

$$T \gg \omega_P^{-1} = \sqrt{\frac{m_e \varepsilon_0}{n_e e^2}} = 7.2 \times 10^{-16} \sqrt{\frac{A}{\rho Z}} \text{ seconds}$$

The two foregoing constraints are important in fluids with very low densities. A typical example is the solar wind. In this flow the density is very low and a more refined approach from plasma physics is often needed.

Finally, we shall suppose that the electrical conductivity is isotropic, namely that Ohm's law reads

$$\mathbf{j} = \sigma \mathbf{E} \qquad\qquad (10.2)$$

where $\mathbf{j}$ is the current density, σ is the electrical conductivity and $\mathbf{E}$ is the electric field. A common source of anisotropy of the electrical conductivity is the magnetic field which induces cyclotronic motion of the electrons. Here again, dilute plasmas are more likely to be prone to such anisotropy. Conductivity is indeed much higher in the direction parallel to the magnetic field than orthogonally to it. In dense plasma, however, collision frequency is much higher than the cyclotron one ($\omega_{cyclo} = eB/m_e$). The mean charge motion is thus hardly influenced by the magnetic field and thus conductivity is a scalar.[1]

---

[1]Let us recall that electrons moving in a magnetic field, without shocks, follow helicoidal trajectories around field lines. Their rotation frequency is the cyclotron frequency.

10.2 The Flow Equations

As stated in the introduction the main peculiarity of the dynamics of a conducting fluid is the action of the Laplace force[2]

$$\mathbf{F}_L = \mathbf{j} \times \mathbf{B}$$

where $\mathbf{B}$ is the local magnetic field. The momentum equation is therefore

$$\rho \frac{D\mathbf{v}}{Dt} = -\nabla P + \mu \Delta \mathbf{v} + \mathbf{j} \times \mathbf{B} \tag{10.3}$$

for an incompressible fluid. Two other equations are needed to complete the momentum equation: one should give $\mathbf{j}$ and the other $\mathbf{B}$. They will be derived from Maxwell equations and Ohm's law.

10.2.1 $\mathbf{j}$ and $\mathbf{B}$ Equations

Let us first recall the Maxwell equations for a medium whose dielectric properties are similar to those of the vacuum. Hence

$$\begin{cases} \nabla \cdot \mathbf{E} = \nabla \cdot \mathbf{B} = 0 \\[2mm] \dfrac{\partial \mathbf{B}}{\partial t} = -\nabla \times \mathbf{E} \\[2mm] \nabla \times \mathbf{B} = \mu_0 \left(\mathbf{j} + \varepsilon_0 \dfrac{\partial \mathbf{E}}{\partial t} \right) \end{cases} \tag{10.4}$$

Following a fluid particle, Ohm's law reads

$$\mathbf{j}' = \sigma \mathbf{E}' \tag{10.5}$$

where $\mathbf{j}'$ and $\mathbf{E}'$ are the current and electric field measured in a frame attached to the fluid particle. The motion of this fluid element is supposed to be non-relativistic.

---

[2]This force is called the Lorentz (1853–1928) force in the Anglo-Saxon world while this is Laplace force in the French literature. Laplace (1749–1827) actually gave the first analytic expression of the force that Biot & Savart measured for the action of a magnetic field on a wire carrying an electric current. It is therefore close to the force that we encounter in MHD. Lorentz force was derived for the charged particles and leads of course to the same expression for the action of a magnetic fields on an electrically conducting fluid.

Thus, the electric field viewed by the particle, $\mathbf{E}'$, is related to the one measured in the laboratory by

$$\mathbf{E} = \mathbf{E}' - \mathbf{v} \times \mathbf{B} \tag{10.6}$$

Since the fluid does not contain free charges, $\mathbf{j} = \mathbf{j}'$ so that (10.5) now reads

$$\mathbf{j} = \sigma(\mathbf{E} + \mathbf{v} \times \mathbf{B}) \tag{10.7}$$

Finally, the fourth Maxwell equation can be simplified using (10.1). This inequality allows us to estimate the order of magnitude of the displacement term $\varepsilon_0 \mu_0 \partial \mathbf{E}/dt$ compared to other terms:

$$\varepsilon_0 \mu_0 \frac{\partial \mathbf{E}}{\partial t} = \frac{1}{c^2} \frac{\partial \mathbf{E}}{\partial t} \sim \frac{E}{c^2 T} \sim \frac{LB}{c^2 T^2} \sim \frac{V^2}{c^2} \nabla \times \mathbf{B} \ll \nabla \times \mathbf{B}$$

where we noted that $E \sim VB$. The displacement field is therefore very small and will be neglected. The magnetic field thus verifies:

$$\begin{cases} \nabla \cdot \mathbf{B} = 0 \\[2mm] \nabla \times \mathbf{B} = \mu_0 \mathbf{j} \\[2mm] \dfrac{\partial \mathbf{B}}{\partial t} = -\nabla \times \mathbf{E} \\[2mm] \mathbf{j} = \sigma(\mathbf{E} + \mathbf{v} \times \mathbf{B}) \end{cases} \tag{10.8}$$

The third equation is called the *induction equation*. It is usually written without the electric field, namely as:

$$\frac{\partial \mathbf{B}}{\partial t} = \nabla \times (\mathbf{v} \times \mathbf{B}) - \nabla \times (\eta \nabla \times \mathbf{B}) \tag{10.9}$$

where $\eta = 1/(\mu_0 \sigma)$ is the *magnetic diffusivity*. This quantity is expressed in m²/s just as the kinematic viscosity. Thus, a new number arises, namely *the magnetic Prandtl number*, which is defined as

$$\mathcal{P}_m = \frac{\nu}{\eta}$$

When η is a constant, the induction equation reads:

$$\frac{\partial \mathbf{B}}{\partial t} = \nabla \times (\mathbf{v} \times \mathbf{B}) + \eta \Delta \mathbf{B} \tag{10.10}$$

We observe that this is a linear equation for $\mathbf{B}$. Using non-dimensional variables, we rewrite this equation as

$$\frac{\partial \mathbf{B}}{\partial \tau} = \nabla \times (\mathbf{u} \times \mathbf{B}) + \frac{1}{\mathrm{Re}_m} \varDelta \mathbf{B} \qquad (10.11)$$

where $\mathrm{Re}_m = \frac{VL}{\eta}$ is the *magnetic Reynolds number*.

10.2.2   *Boundary Conditions on the Magnetic Field*

The partial differential equations that we derived for the magnetic field need to be completed by boundary conditions. Basically, the magnetic field must be continuous. However, in many situations it is desirable to idealize the medium which bounds the flow. This is often a way to avoid the computation of the magnetic field outside the fluid domain. Just like temperature, two ideal cases are used: the perfect conductor or the perfect insulator.

10.2.2.1   Boundary Conditions at an Electrical Insulator

If the fluid is bounded by an electrical insulator, the current outside the domain is vanishing, namely $\mathbf{j}$ is zero. Hence, outside the fluid domain

$$\nabla \times \mathbf{B} = 0 \qquad \Longleftrightarrow \qquad \mathbf{B} = \nabla \varPhi$$

Since $\mathbf{B}$ is continuous at the surface, $\mathbf{B}$ must match a potential field. On the other hand, no current crosses the boundary so

$$\mathbf{j} \cdot \mathbf{n} = 0 \qquad \Longleftrightarrow \qquad \mathbf{n} \cdot \nabla \times \mathbf{B} = 0$$

This equation shows that if the field is continuous, this is not the case for all its derivatives. The normal component of the curl is the only combination of the derivatives that is continuous.

10.2.2.2   Boundary Conditions at a Perfect Electrical Conductor

These boundary conditions are definitely more delicate to establish. To be as clear as possible, we shall consider the case of a fluid meeting a (solid) conductor whose conductivity will be increased up to infinity. We then focus on the field inside the wall.

Let us assume that the magnetic field includes a time-variation like $e^{i\omega t}$ (this might just be the Fourier component of a more complex time dependence). In the solid, we assume that $\mathbf{v} = \mathbf{0}$. Hence, $\mathbf{B}$ verifies

$$i\omega\mathbf{B} = \eta_s\Delta\mathbf{B}$$

where η_s is the diffusivity inside the solid. We shall let this quantity vanish. If we assume the surface separating the fluid and the solid to be the plane $z = 0$ (solid $z \geq 0$, fluid $z < 0$), then we might also assume that the variations of $\mathbf{B}$ along z are much faster that along the other directions. In other words, we assume the existence of a boundary layer. In this layer, $\mathbf{B}$ verifies

$$i\omega\mathbf{B} = \eta_s\frac{\partial^2\mathbf{B}}{\partial z^2}$$

whose solution is

$$\mathbf{B} = \mathbf{B}_0(x, y)e^{-(1+i)z/\delta} \tag{10.12}$$

Here, $\mathbf{B}_0(x, y)$ is the field at $z = 0$ and δ is the boundary layer thickness (the skin depth in electromagnetism). Let us underline the similarity of (10.12) with the Ekman layer: the field shows an oscillatory damping (but without changing direction). The thickness of the layer reads:

$$\delta = \sqrt{\frac{2\eta_s}{\omega}} = \sqrt{\frac{2}{\omega\sigma_s\mu_0}} \tag{10.13}$$

The foregoing expression shows that the magnetic field does not penetrate into the solid if the product $\omega\sigma_s$ goes to infinity.

Now, if we use the flux conservation in the solid, namely that $\nabla \cdot \mathbf{B} = 0$, the normal component of $\mathbf{B}$ may be expressed with the divergence of the tangent field, i.e.

$$B_{0z} = (1 + i)\delta\nabla_h \cdot \mathbf{B}_0(x, y)$$

Again, this equation is very similar to that of the Ekman pumping (8.53). It shows that when the thickness of the layer goes to zero, this component vanishes. We thus find that at the boundary of a perfect conductor

$$\mathbf{B} \cdot \mathbf{n} = 0 \tag{10.14}$$

Namely, that the field does not penetrate into a perfect conductor. We may also observe that in the solid the field is proportional to $e^{-z/\delta}$ with $\delta \to 0$. Thus, the field is zero inside the perfect conductor. However, the tangential component of the field may remain finite at the surface, thus having a jump at the surface (note that the normal component is continuous though). The gradient of the field, and thus the current density diverges near the surface. Indeed,

$$\mathbf{j} = \frac{1}{\mu_0} \nabla \times \mathbf{B} = \frac{(1+i)}{\mu_0 \delta} \begin{vmatrix} B_{0y} \\ -B_{0x} \\ 0 \end{vmatrix} e^{-(1+i)z/\delta}$$

Thus, there is always some current at the surface of the conductor. One usually introduces a *surface current density* $\mathbf{j}_S$ defined by

$$\mathbf{j}_S = \int_0^{+\infty} \mathbf{j} dz = \mathbf{n} \times \mathbf{B}/\mu_0$$

which is finite and measures the jump in the tangential components of the magnetic field when one crosses the bounding surface.

Now, let us focus on the electric field in the solid. We have

$$\mathbf{E} = \frac{(1+i)}{\sigma_s \mu_0 \delta} \mathbf{e}_z \times \mathbf{B}_0 e^{-(1+i)z/\delta} = \mathcal{O}(\sqrt{\frac{\omega}{\sigma_s}})$$

It shows that this field disappears as expected when the conductivity is infinite. Thus at the interface with a perfect conductor we should write

$$\mathbf{E} \times \mathbf{n} = \mathbf{0} \qquad\qquad (10.15)$$

namely, the tangential electric field vanishes. Using Ohm's law and the fact that $\mathbf{B}$ et $\mathbf{j}$ are both tangential to the bounding surface, this condition also reads

$$\mathbf{j} \times \mathbf{n} = \mathbf{0} \qquad \text{or} \qquad (\nabla \times \mathbf{B}) \times \mathbf{n} = \mathbf{0} \qquad (10.16)$$

10.2.3   The Energy Equation with a Magnetic Field

10.2.3.1   The Maxwell Tensor

We shall first note that the Laplace force $\mathbf{j} \times \mathbf{B}$ is also the divergence of a tensorial field. Indeed,

$$(\mathbf{j} \times \mathbf{B})_i = \partial_j \Sigma_{ij}$$

where

$$\Sigma_{ij} = \frac{1}{\mu_0} (B_i B_j - \frac{1}{2} B^2 \delta_{ij})$$

$[\Sigma]$ is the magnetic stress tensor or Maxwell tensor.

10.2.3.2 Joule Heating

To derive the equation governing the local evolution of internal energy we need to start with the energy balance that lead us to (1.29) and to introduce the magnetic terms.

For that, we consider some volume independent of time because the magnetic field is not attached to the fluid. The energy balance in this volume reads

$$\frac{d}{dt}\int_{(V)}\left[\rho(\frac{1}{2}v^2+e)+\frac{B^2}{2\mu_0}\right]dV = -\int_{(S)}\rho(\frac{1}{2}v^2+e)\mathbf{v}\cdot d\mathbf{S} - \int_{(S)}\mathbf{E}\times\mathbf{B}/\mu_0\cdot d\mathbf{S}$$

$$+\int_{(V)}\mathbf{f}\cdot\mathbf{v}dV + \int_{(S)}v_i\sigma_{ij}dS_j$$

$$-\int_{(S)}\mathbf{F}\cdot d\mathbf{S} + \int_{(V)}\mathcal{Q}dV$$

In this expression the new terms are the magnetic energy density that completes internal and kinetic energies, and the Poynting flux $\mathbf{E}\times\mathbf{B}/\mu_0$ that represents the surface flux of electromagnetic energy through the boundary. We may be surprised of the absence of the Laplace force. This is natural: the Laplace force does not modify the energy content of the volume. It just permits exchanges between the kinetic and magnetic energies reservoirs. To simplify the foregoing energy balance we need using the magnetic field (10.9). After a scalar product by $\mathbf{B}/\mu_0$ and using (12.40), we obtain

$$\frac{\partial}{\partial t}\left(\frac{B^2}{2\mu_0}\right) = \nabla\cdot[(\mathbf{v}\times\mathbf{B}-\eta\nabla\times\mathbf{B})\times\mathbf{B}]/\mu_0 - \mathbf{v}\cdot(\mathbf{j}\times\mathbf{B}) - \eta(\nabla\times\mathbf{B})^2/\mu_0 \quad (10.17)$$

This expression shows that the power of the Laplace force $\mathbf{v}\cdot(\mathbf{j}\times\mathbf{B})$ extracts energy from the magnetic reservoir (and so fills that of kinetic energy).

Now, combining the magnetic energy equation with that of kinetic energy (1.28) (completed with the Laplace force work), we finally get:

$$\rho\frac{De}{Dt} = \nabla\cdot(\chi\nabla T) - P\nabla\cdot\mathbf{v} + \frac{\mu}{2}(\nabla:\mathbf{v})^2 + \zeta(\nabla\cdot\mathbf{v})^2 + \eta(\nabla\times\mathbf{B})^2/\mu_0 \quad (10.18)$$

This equation shows that the magnetic field is at the origin of a new source of internal energy (and entropy) through the term $\eta(\nabla\times\mathbf{B})^2/\mu_0$ which represents the Joule heating (see exercises).

10.3 Some Properties of MHD Flows

10.3.1 The Frozen Field Theorem

When the diffusion time of the magnetic field L^2/η is large compared to the advection time L/V, the magnetic field is just like "frozen" in the fluid. The field lines are attached to the fluid particles. Then, we can show the following theorem:

When the magnetic Reynolds number increases to infinity, the magnetic field flux through a surface attached to fluid particles is constant.

To prove this theorem, we need to show that the integral $\int \mathbf{B} \cdot d\mathbf{S}$ is constant when S is attached to fluid particles. If we use the vector potential $\mathbf{A}$ of the magnetic field, the demonstration is quite similar to that of Kelvin's theorem that we studied in Chap. 3. Indeed, setting $\eta = 0$, we have

$$\frac{\partial \mathbf{B}}{\partial t} = \nabla \times (\mathbf{v} \times \mathbf{B}) \qquad \Longleftrightarrow \qquad \frac{\partial \mathbf{A}}{\partial t} = \mathbf{v} \times \nabla \times \mathbf{A} + \nabla Q \qquad (10.19)$$

where Q is an arbitrary function.

Let us call $\phi(t)$ the magnetic flux through the surface $S(t)$ which leans on the contour $C(t)$ carried by the fluid. We write

$$\phi(t) = \int_{S(t)} \mathbf{B} \cdot d\mathbf{S} = \oint_{C(t)} \mathbf{A} \cdot d\mathbf{l} \qquad (10.20)$$

Just like in Kelvin's theorem demonstration, we use relation (1.13) and find

$$\frac{d\phi}{dt} = \oint_{C(t)} \left(\frac{\partial A_i}{\partial t} - (\mathbf{v} \times \mathbf{B})_i \right) dl_i = \oint_{C(t)} \partial_i Q \, dl_i = 0$$

where we used (10.19). The flux of $\mathbf{B}$ through an open surface given by a contour attached to fluid particles is therefore a constant.

10.3.2 Magnetic Pressure and Magnetic Tension

The Laplace force may also be written as

$$\mathbf{j} \times \mathbf{B} = \frac{1}{\mu_0}(\nabla \times \mathbf{B}) \times \mathbf{B} = \frac{1}{\mu_0}(\mathbf{B} \cdot \nabla)\mathbf{B} - \nabla \left(\frac{B^2}{2\mu_0} \right)$$

thanks to (12.43). In this new expression we note that part of the force derives from a scalar potential which is known as the *magnetic pressure*. We set

$$P_m = \frac{B^2}{2\mu_0} \tag{10.21}$$

The magnetic pressure is therefore identical to the magnetic energy density.[3]

The remaining term, $(\mathbf{B} \cdot \nabla)\mathbf{B}$ can be split, just as $(\mathbf{v} \cdot \nabla)\mathbf{v}$ in Euler's equation (see Chap. 3, 3.12), as

$$(\mathbf{B} \cdot \nabla)\mathbf{B} = B\frac{\partial B}{\partial s}\mathbf{e}_s + \frac{B^2\mathbf{n}}{R_s} \tag{10.22}$$

The first term is called the *magnetic tension* since it is parallel to the field line while the second term is the *curvature force* since it grows as the radius of curvature R_s decreases. Note that the tension term is equal to the longitudinal component of the magnetic pressure gradient. This is a consequence of the fact that the Laplace force has no component along $\mathbf{B}$.

10.3.3   Force-Free Fields

When the current density is parallel to the magnetic field, the Laplace force vanishes. Such situations are thought to exist (approximately) in regions where the magnetic pressure is strong enough to control the distribution of matter, and therefore that of currents. The most famous example is the atmosphere of the Sun: there the magnetic field is dominating and shapes the distribution of matter. The most spectacular illustration of this situation is given by solar prominences (see Fig. 10.1). The structure of these magnetic features is often approximated using a force-free field. Below, we shall also use a force-free field to get a simple example of a dynamo.

If the Laplace force is vanishing, then the magnetic field verifies the following extra-equation

$$\nabla \times \mathbf{B} = K(\mathbf{r})\mathbf{B} \tag{10.23}$$

The function $K(\mathbf{r})$ is unknown, but since $\nabla \cdot \mathbf{B} = 0$, it must verify:

$$\mathbf{B} \cdot \nabla K = 0$$

which means that it is constant along the field lines.

---

[3]The reader may verify that an energy volumic density is dimensionally identical to a pressure.

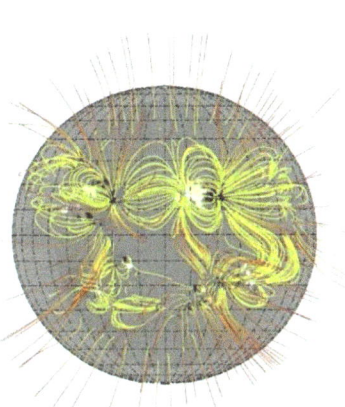

Fig. 10.1 *Left* a modelling of the solar coronal magnetic fields using force-free fields by Tadesse *et al.* (2013). *Right* extrapolation of the magnetic field of the star V374 Pegasi from spectropolarimetric observations. This star is a red dwarf of 0.28 solar mass with a radius about one third that of the Sun. Its magnetic field is quite strong (0.2 T) and generated by a turbulent dynamo triggered by the thermal convection that transport heat from the central regions to the surface (see Morin *et al.*, 2008). (Picture by M. Jardine & J.-F. Donati)

One example of a force-free field may be obtained if we assume K to be constant. Then, $-\Delta \mathbf{B} = K \nabla \times \mathbf{B} = K^2 \mathbf{B}$, which means that $\mathbf{B}$ verifies Helmholtz equation, namely:

$$(\Delta + K^2)\mathbf{B} = \mathbf{0}$$

Since we took the curl of (10.23), the solution of this equation are too general, but among them, there are some where the curl of $\mathbf{B}$ is parallel to $\mathbf{B}$. A simple example is given by

$$\mathbf{B} = B_0 \begin{vmatrix} \cos Kz \\ \sin Kz \\ 0 \end{vmatrix}$$

which is indeed a force-free field.

There are other solutions in cylindrical or spherical geometry, but they are more complex (see Moffatt 1978 for instance).

10.3.4   The Equipartition Solutions and Elsässer Variables

When the fluid is incompressible and when diffusive effects are neglected ($\nu = \eta = 0$), there exists a simple steady solution to the equations of MHD. The equations for a steady flow indeed read:

$$\rho \mathbf{v} \cdot \nabla \mathbf{v} = -\nabla(P + P_m) + \frac{1}{\mu_0} \mathbf{B} \cdot \nabla \mathbf{B}$$

$$\nabla \times (\mathbf{v} \times \mathbf{B}) = \mathbf{0}$$

$$\nabla \cdot \mathbf{v} = \nabla \cdot \mathbf{B} = 0$$

These equations are satisfied if

$$\mathbf{v} = \pm \frac{\mathbf{B}}{\sqrt{\rho \mu_0}} \qquad \text{and} \qquad P = \text{Cst} - P_m \qquad (10.24)$$

This is the *equipartition solution* because

$$\frac{1}{2} \rho \mathbf{v}^2 = \frac{\mathbf{B}^2}{2\mu_0}$$

The kinetic energy density equals the magnetic energy density. This solution shows that the quantity $B/\sqrt{\rho \mu_0}$ is a velocity. This is the *Alfvén speed*. The equipartition solution is a solution of the nonlinear equations. It slowly fades with time when diffusion is included (over a typical diffusion time $\min(L^2/\nu, L^2/\eta)$). Chandrasekhar (1961) has shown that this solution is linearly stable.

We shall come back later on the physical meaning of the Alfvén speed. We may however notice that it naturally introduces new variables, called the *Elsässer variables*, which are defined as:

$$\mathbf{z}^{\pm} = \mathbf{v} \pm \frac{\mathbf{B}}{\sqrt{\rho \mu_0}} \qquad (10.25)$$

Combining the momentum and induction equations (and taking into account diffusion terms), we can derive this other form of the MHD equations, namely

$$\begin{cases} \dfrac{\partial \mathbf{z}^+}{\partial t} + \mathbf{z}^- \cdot \nabla \mathbf{z}^+ = -\nabla \pi + \nu_+ \Delta \mathbf{z}^+ + \nu_- \Delta \mathbf{z}^- \\[4mm] \dfrac{\partial \mathbf{z}^-}{\partial t} + \mathbf{z}^+ \cdot \nabla \mathbf{z}^- = -\nabla \pi + \nu_- \Delta \mathbf{z}^+ + \nu_+ \Delta \mathbf{z}^- \end{cases} \qquad (10.26)$$

where we set $\pi = (P + P_m)/\rho$ and $\nu_{\pm} = \frac{1}{2}(\nu \pm \eta)$. The equipartition solutions are simply $\mathbf{z}^{\pm} = \mathbf{0}$.

10.4   The Waves

10.4.1   Alfvén Waves

Let us consider an incompressible fluid bathed by a uniform magnetic field **B**. The fluid is in equilibrium. Small amplitude perturbations are denoted **b** for the magnetic field, **v** for the velocity and p for the pressure. We neglect diffusion. These perturbations are governed by the following equations:

$$\begin{cases} \rho_0 \dfrac{\partial \mathbf{v}}{\partial t} = -\nabla p + \dfrac{1}{\mu_0}(\nabla \times \mathbf{b}) \times \mathbf{B} \\[2mm] \dfrac{\partial \mathbf{b}}{\partial t} = \nabla \times (\mathbf{v} \times \mathbf{B}) \\[2mm] \nabla \cdot \mathbf{v} = \nabla \cdot \mathbf{b} = 0 \end{cases} \tag{10.27}$$

We first derive the dispersion relation of the associated freely propagating waves. We set

$$\mathbf{v} \propto \exp(i\omega t + i\mathbf{k} \cdot \mathbf{x}), \qquad \mathbf{b} \propto \exp(i\omega t + i\mathbf{k} \cdot \mathbf{x})$$

The system (10.27) turns into

$$\begin{cases} \rho_0 \omega \mathbf{v} = -p\mathbf{k} + (\mathbf{k} \times \mathbf{b}) \times \mathbf{B}/\mu_0 \\[2mm] \omega \mathbf{b} = \mathbf{k} \times (\mathbf{v} \times \mathbf{B}) \\[2mm] \mathbf{v} \cdot \mathbf{k} = \mathbf{b} \cdot \mathbf{k} = 0 \end{cases} \tag{10.28}$$

which leads to

$$\begin{cases} \mathbf{b} = \dfrac{\mathbf{k} \cdot \mathbf{B}}{\omega} \mathbf{v} \\[2mm] \left(\omega^2 - \dfrac{(\mathbf{B} \cdot \mathbf{k})^2}{\rho_0 \mu_0} \right) \mathbf{v} \times \mathbf{k} = 0 \end{cases} \tag{10.29}$$

so that the dispersion relation is

$$\omega^2 = \frac{(\mathbf{B} \cdot \mathbf{k})^2}{\rho_0 \mu_0} \tag{10.30}$$

We now introduce the Alfvén speed $V_A = B/\sqrt{\rho_0 \mu_0}$ and θ the angle between the wave vector $\mathbf{k}$ and $\mathbf{B}$. We get

$$\frac{\omega}{k} = V_A \cos \theta \tag{10.31}$$

The Alfvén speed is therefore the maximum velocity of the waves. These waves are the *Alfvén waves*. Just like inertial waves or gravity waves, these waves do not propagate isotropically. They cannot propagate in a direction perpendicular to the field, and are the fastest in the direction parallel to the field lines. These waves may be thought as the ones propagating along a string, where the magnetic field lines play the role of the string. Their group velocity,

$$\mathbf{v}_g = \nabla_k \omega(\mathbf{k}) = \frac{\mathbf{B}}{\sqrt{\rho_0 \mu_0}} = V_A \frac{\mathbf{B}}{B}$$

shows that the energy only propagates along the field lines at the Alfvén speed. Unlike inertial and gravity waves, the phase and group velocities are not orthogonal.

10.4.2 *Magnetosonic Waves*

When the compressibility of the fluid cannot be neglected, Alfvén waves are coupled with acoustic waves and form the set of *magnetosonic waves* which we now study.

To analyse their properties, we start from (10.27) but now taking into account the density perturbations. Still considering infinitesimal amplitudes, mass conservation now implies:

$$\frac{\partial \rho}{\partial t} + \rho_0 \nabla \cdot \mathbf{v} = 0$$

Just like for acoustic waves, the density perturbation is related to the pressure one by $p = c_s^2 \rho$, where c_s is the sound speed (see 5.16). As usual, we decompose the disturbances on the plane waves and get the following relations between the amplitudes:

$$\begin{cases} \mathbf{b} \cdot \mathbf{k} = 0 \\ \omega \rho + \rho_0 \mathbf{v} \cdot \mathbf{k} = 0 \\ \omega \mathbf{b} = \mathbf{k} \times (\mathbf{v} \times \mathbf{B}) \\ \rho \omega \mathbf{v} = -p \mathbf{k} + (\mathbf{k} \times \mathbf{b}) \times \mathbf{B}/\mu_0 \\ p = c_s^2 \rho \end{cases} \tag{10.32}$$

Eliminating the pressure, the magnetic field and the density, we are left with an equation where there is only the velocity amplitude:

$$\rho_0 \mu_0 \omega^2 \mathbf{v} = (\rho_0 \mu_0 c_s^2 + B^2)(\mathbf{v} \cdot \mathbf{k})\mathbf{k} + (\mathbf{k} \cdot \mathbf{B})^2 \mathbf{v} - (\mathbf{v} \cdot \mathbf{B})(\mathbf{k} \cdot \mathbf{B})\mathbf{k} - (\mathbf{k} \cdot \mathbf{B})(\mathbf{k} \cdot \mathbf{v})\mathbf{B}$$

The dispersion relation is conveniently derived if we write this equation in a matrix form (like 5.8). Then, we set to zero the determinant. Very generally, we may fix the direction of some vectors. For instance, we may choose $\mathbf{B} = B\mathbf{e}_z$ and, using cylindrical coordinates, set $\mathbf{k} = k_s\mathbf{e}_s + k_z\mathbf{e}_z$. As before, θ is the angle between $\mathbf{B}$ and $\mathbf{k}$. We also set $B^2 = \rho_0\mu_0 V_A^2$. Thus, the foregoing equation reads

$$\Omega^2\mathbf{v} = [(c_s^2 + V_A^2)\mathbf{k}\cdot\mathbf{v} - V_A^2 k_z v_z]\mathbf{k} - V_A^2(\mathbf{k}\cdot\mathbf{v})k_z\mathbf{e}_z$$

where $\Omega^2 = \omega^2 - k_z^2 V_A^2$. More explicitly,

$$\begin{cases} \Omega^2 v_s = (V_B^2 k_s^2)v_s + (c_s^2 k_s k_z)v_z \\[2mm] \Omega^2 v_\varphi = (V_B^2 k_\varphi k_s)v_s + (c_s^2 k_\varphi k_z)v_z \\[2mm] \Omega^2 v_z = (c_s^2 k_s k_z)v_s + (c_s^2 - V_A^2)k_z^2 v_z \end{cases} \qquad (10.33)$$

where $V_B^2 = V_A^2 + c_s^2$. Zeroing the determinant of this system leads to the dispersion relation:

$$(\omega^4 - \omega^2 k^2(V_A^2 + c_s^2) + k^2 c_s^2 k_z^2 V_A^2)(\omega^2 - k_z^2 V_A^2) = 0 \qquad (10.34)$$

This new relation contains three types of waves that are specified by their phase velocity ω/k. The first one is the pure Alfvén wave:

$$\frac{\omega}{k} = V_A \cos\theta$$

For this wave, compressibility does not play any role. Indeed, taking the last two equations of (10.33), we find that $k_s v_s = k_z v_z = 0$ and v_φ undetermined. This is always a transverse wave, namely $\mathbf{k}\cdot\mathbf{v}=0$, and the density fluctuation is always zero. The velocity field is perpendicular to the plane formed by the wave vector and the magnetic field.

The two other waves have the following phase velocity:

$$\begin{cases} V_f = \sqrt{\dfrac{V_A^2 + c_s^2 + \sqrt{(V_A^2 + c_s^2)^2 - 4\cos^2\theta\, V_A^2 c_s^2}}{2}} \\[8mm] V_s = \sqrt{\dfrac{V_A^2 + c_s^2 - \sqrt{(V_A^2 + c_s^2)^2 - 4\cos^2\theta\, V_A^2 c_s^2}}{2}} \end{cases} \qquad (10.35)$$

They correspond, respectively, to *the fast magnetosonic wave* and *the slow magnetosonic wave*. In general these waves are neither transverse nor longitudinal.

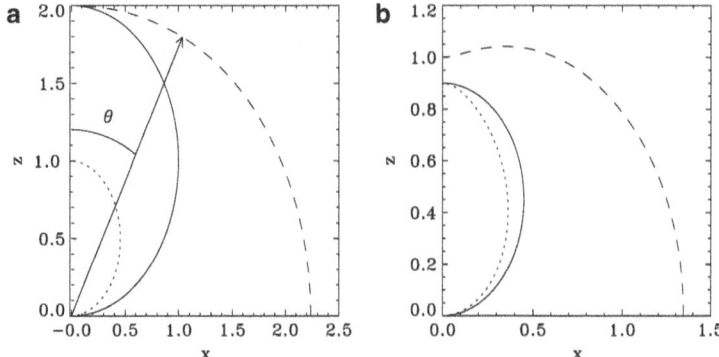

Fig. 10.2 Phase velocity of the three magnetosonic waves in polar coordinates. The *solid line* denotes the Alfvén wave, the *dotted line* is for the slow magnetosonic wave and the *dashed line* the fast magnetosonic wave. *Left* The case where the medium is such that $V_A > c_s$. *Right* when $V_A < c_s$ (here $V_A = 0.9c_s$)

Two cases deserve some attention: the case when the wave propagates either perpendicularly ($\theta = \pi/2$) to the magnetic field or along it ($\theta = 0$).

- If $\theta = 0$, then $V_f = c_s$ and $v_s = v_\varphi = 0$. The wave is longitudinal. This is just a plain acoustic wave. There is no magnetic field perturbation. The slow wave verifies $V_s = V_A$: this is a pure Alfvén wave.
- If $\theta = \pi/2$, we find only the fast magnetosonic wave, propagating with the phase velocity $V_f = \sqrt{V_A^2 + c_s^2}$. It is longitudinal. In fact, this is an acoustic wave which propagates in a fluid whose pressure is increased by the magnetic pressure.

We show in Fig. 10.2 the phase velocity of all these waves for every angle θ. This diagram is sometimes called *Friedrich diagram*.

10.5   The Dynamo Problem

One of the fascinating properties of conducting fluids is their ability to generate magnetic fields by the famous *dynamo effect*. Thanks to this effect, planets like the Earth or Jupiter or stars like the Sun own a magnetic field.

The dynamo problem is also one of the most complex in Fluid Mechanics, because no simple solution of this problem exists.

10.5.1 The Kinematic Dynamo

To make a first step into the dynamo problem, it is convenient to start with the one of the kinematic dynamo. A kinematic dynamo is a velocity field that amplifies the magnetic field without being perturbed by the Lorentz force. Such a velocity field is prescribed and whether the solution of the induction equations grows or not the velocity field is considered as a dynamo or not. Thus we give $\mathbf{v}$ and solve:

$$\begin{cases} \dfrac{\partial \mathbf{B}}{\partial t} = \nabla \times (\mathbf{v} \times \mathbf{B}) + \eta \Delta \mathbf{B} \\[2mm] \nabla \cdot \mathbf{B} = 0 \end{cases} \tag{10.36}$$

with boundary conditions. Note that this problem is linear for $\mathbf{B}$. We may set

$$\mathbf{B}(\mathbf{r}, t) = \mathbf{B}(\mathbf{r}) e^{\lambda t}$$

and the velocity field is a kinematic dynamo if and only if there exist a critical diffusivity η_{crit} such that if $\eta < \eta_{\text{crit}}$ then $\mathrm{Re}(\lambda) > 0$. Introducing a length scale L and a velocity scale V, we may associate with η_{crit} a critical magnetic Reynolds number beyond which the magnetic field is amplified.

The reader may have guessed that finding a kinematic dynamo is much more difficult than determining the stability of a flow. Indeed, in this new problem, a very large number of "parameters" control the stability of the magnetic field. These are all the values of the function $\mathbf{v}(\mathbf{r})$ in the fluid's domain. In fact, a kinematic dynamo is to be found in a function space. In addition, we shall see below that, when a dynamo exists, it cannot be a simple velocity field.

Two kinds of kinematic dynamos are usually distinguished: the *fast* dynamos and the *slow* dynamos. If the time scale which controls the growth of the magnetic field is the diffusive one, namely L^2/η then the dynamo is said to be slow. If, on the other hand, this time scale is the advective one, i.e. L/V, then the dynamo is said to be fast.

Presently, nobody knows a criterion on the velocity field that tells whether a dynamo is slow or fast. We only know that some characteristics of the flow are favourable for a fast dynamo. For instance, if the trajectory of the Lagrangian particles are chaotic or if the velocity field owns shear discontinuities, the amplification of the magnetic field may be fast.

10.5.2 The Amplification of the Magnetic Field

As a first step, we shall examine the ways magnetic fields can be amplified by a flow. This is actually the role of the term $\nabla \times (\mathbf{v} \times \mathbf{B})$ in the induction equation.

To show this, let us consider the case where the velocity field is zero and show that necessarily the magnetic field disappears. $\mathbf{B}$ verifies

$$\begin{cases} \dfrac{\partial \mathbf{B}}{\partial t} = \eta \Delta \mathbf{B} \\[2mm] \nabla \cdot \mathbf{B} = 0 \\[1mm] \text{Boundary conditions} \end{cases} \qquad (10.37)$$

We scalarly multiply the first equation by $\mathbf{B}$ and integrate it over the whole space. This gives us the evolution of the magnetic energy (up to the μ_0 factor):

$$\frac{d}{dt} \int \frac{1}{2} B^2 dV = \eta \int \mathbf{B} \cdot \Delta \mathbf{B} dV \qquad (10.38)$$

with (12.40) we find

$$\nabla \cdot (\mathbf{B} \times (\nabla \times \mathbf{B})) = (\nabla \times \mathbf{B})^2 - \mathbf{B} \cdot \nabla \times \nabla \times \mathbf{B}$$

thus

$$\int_{(V)} \mathbf{B} \cdot \Delta \mathbf{B} dV = -\int_{(V)} \mathbf{B} \cdot \nabla \times \nabla \times \mathbf{B} dV = \int_{(S)} (\mathbf{B} \times \nabla \times \mathbf{B}) \cdot d\mathbf{S} - \int_{(V)} (\nabla \times \mathbf{B})^2 dV$$

Now we assume that the conducting fluid does not fill the whole space, so that $\nabla \times \mathbf{B}$ is zero on some sufficiently large surface S. The magnetic energy verifies:

$$\frac{dE_M}{dt} = -\eta \int_{(V)} (\nabla \times \mathbf{B})^2 dV \qquad (10.39)$$

which shows that it decreases with time. If there is some amplification, it must come from the $\nabla \times (\mathbf{v} \times \mathbf{B})$-term, which we now study.

Using the equality (12.41), the induction equation may be rewritten as:

$$\frac{D\mathbf{B}}{Dt} = \overbrace{(\mathbf{B} \cdot \nabla)\mathbf{v}}^{(I)} - \overbrace{\mathbf{B}\nabla \cdot \mathbf{v}}^{(II)} + \eta \Delta \mathbf{B} \qquad (10.40)$$

Two terms are potentially able to amplify the magnetic field. The role of the second one (II) is easy to understand: when the (compressible) fluid flow is convergent ($\nabla \cdot \mathbf{v} < 0$) the field lines are gathered, the flux density (namely, the field) increases (see Fig. 10.3a). This term disappears when the compressibility of the fluid vanishes. It remains however the first term (I) which is also able to increase the magnetic field. We see that this phenomenon occurs when the velocity gradient is parallel to the magnetic field. In this case the component of the magnetic field along the velocity vector grows. This is illustrated in Fig. 10.3b. There, we see that if the magnetic field

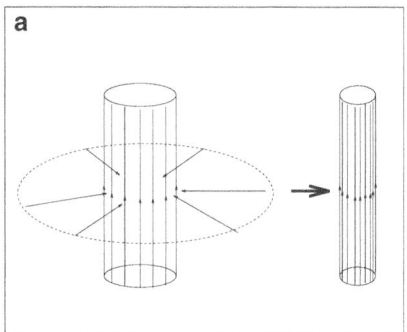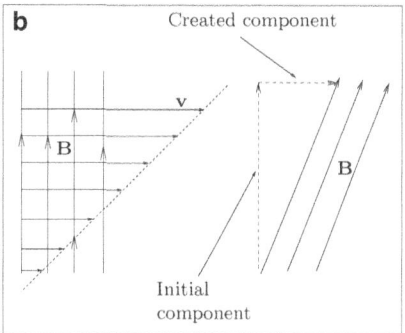

Fig. 10.3 Amplification of **B** by (**a**) a converging flow: the flow convergence gathers the *field lines* thus increasing **B** but does not change its direction. In (**b**) **B** is amplified through the raise of a new component

owns a single component and the velocity field is in the direction orthogonal to it, with some shear, then the magnetic field gets a new component parallel to **v** while the initial component is not affected. Hence, the magnetic energy increases locally. One may note the analogy with vorticity (see Chap. 3).

10.5.3   Some Anti-Dynamo Theorem

The dynamo problem is a difficult one because there is no simple velocity field that is able to amplify a magnetic field. Generally speaking, a dynamo has a low degree of symmetry. We show below that no purely axisymmetric dynamo exists. This result is the first antidynamo theorem. It was first demonstrated by Cowling (1933) and may be stated as follows:

An axisymmetric magnetic field cannot be sustained by an axisymmetric velocity field.

We demonstrate this theorem by showing that if both **B** and **v** are axisymmetric, then **B** necessarily decays. For this, we first write **B** in cylindrical coordinates:

$$\mathbf{B}(s, z, t) = \begin{vmatrix} \dfrac{1}{s}\dfrac{\partial A}{\partial z} \\[2mm] B \\[2mm] -\dfrac{1}{s}\dfrac{\partial A}{\partial s} \end{vmatrix} \qquad (10.41)$$

Here B is the toroidal component of the field while A is related to the toroidal component of the vector potential by $A = -sA_\varphi$. A controls the meridional field (also called the poloidal component). We now derive equations governing the evolution of A and B (details are given in appendix).

Using the components along $\mathbf{e}_s$ and $\mathbf{e}_z$ of the induction equation, we find the equation for A:

$$\frac{\partial A}{\partial t} + \mathbf{v} \cdot \nabla A = \eta \left(\Delta - \frac{2}{r} \frac{\partial}{\partial r} \right) A \tag{10.42}$$

This equation shows that A is simply advected and diffused. As time passes, it evolves towards a constant, which means that the meridional magnetic fields go to zero.

When A is a constant, then B verifies

$$\frac{\partial B}{\partial t} + s\mathbf{v} \cdot \nabla(B/s) = \eta(\Delta - 1/s^2)B \tag{10.43}$$

where we assumed here that $\nabla \cdot \mathbf{v} = 0$. This is a similar equation as the one for A and therefore B also converges to a constant, which is necessarily zero (why?).

Thus, no axisymmetric velocity field can sustain an axisymmetric magnetic field.

10.5.3.1 Other Cases

The foregoing theorem is one case among a larger set of theorems which state cases where a magnetic field cannot be (re)generated. Here are some examples:

1. No magnetic field independent of one space coordinate can be sustained by a velocity field of the same type.
2. A divergence-free velocity field without any radial component (namely always tangent to a sphere) cannot sustain a magnetic field.
3. A two-dimensional flow cannot sustain a magnetic field.
4. A purely radial flow cannot sustain a magnetic field.

10.5.3.2 Conclusions

All these theorems show that a two-dimensional velocity field cannot sustain a two-dimensional magnetic field, whatever the surface we work on (plane, cylinder or sphere). A magnetic field can be generated only in three dimensions.

10.5.4 An Example: The Ponomarenko Dynamo

According to the foregoing discussion, a simple example of a dynamo flow is not easy to find. The Ponomarenko dynamo is one of them and was found not so long ago. The flow has the following form:

$$\mathbf{v} = \omega s \mathbf{e}_\varphi + U \mathbf{e}_z \quad s < a$$

$$\mathbf{v} = \mathbf{0} \quad s > a$$

in a fluid that fills the whole space. This is an axisymmetric flow and therefore only non-axisymmetric magnetic fields can be amplified. We are thus looking for solutions of the form:

$$\mathbf{B} = \mathbf{B}_0(s)e^{i(m\varphi+kz)+\lambda t}$$

We leave the resolution of this problem to the reader as an exercise. The result is the following: if the product ωU is large enough, then there exist unstable magnetic modes, for which $Re(\lambda) > 0$ and $m \neq 0$ of course. This flow can thus amplify magnetic fields. Two ingredients are indeed very favourable to this property: first, it is a helical flow. Helicity

$$H = \mathbf{v} \cdot \nabla \times \mathbf{v} = 2\omega U$$

is non-zero and we note that a critical value of it determines the dynamo action. We shall see below that this is indeed an important quantity for dynamos. Second, the flow owns a very steep (actually infinite) velocity gradient at $r = a$. This is a very useful feature for a dynamo because we noticed that magnetic field amplification only depends on the velocity gradients. This infinite gradient implies that all the scales of magnetic field are amplified, and in fact lead to a fast dynamo (see Gilbert 1988).

10.5.5   The Turbulent Dynamo

We learnt that dynamos are necessarily flows of low symmetry. Hence, it is no surprise that turbulent flows are very good candidates to be dynamos. Actually, natural dynamos in stars or planets are all turbulent flows. Of course, using turbulence to make magnetic fields is not an obvious way since we have no general theory of turbulent flows as we saw in the previous chapter. Nevertheless, the analysis of equations reveals some general laws that are helpful to understand what we see on the Sun or planets (including the Earth) as far as magnetic field are concerned (see Fig. 10.4 for an illustration of the solar magnetic cycle).

To study the role of turbulence in the generation of magnetic field it is useful to split the fields into their (ensemble) average and fluctuating parts. Thus

$$\mathbf{B} = \langle \mathbf{B} \rangle + \mathbf{B}' \quad \text{and} \quad \mathbf{v} = \langle \mathbf{v} \rangle + \mathbf{v}'$$

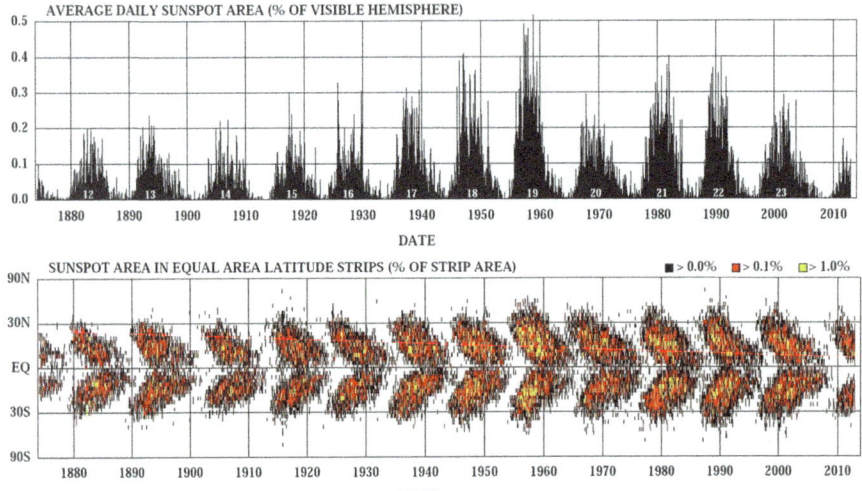

Fig. 10.4 The magnetic cycle of the Sun. *Up* The surface covered by sunspots as a function of time. *Bottom* The distribution of sunspots as a function of latitude and time (color indicates the importance of the spotted area). This diagram shows that spots appear at a latitude that decreases with time. When the equator is reached, a new cycle starts with the emergence of new spots around latitudes $\pm 30\,^\circ$ (source: Dr. David Hathaway, NASA)

Reporting this decomposition into the induction equation and taking the average, we get

$$\frac{\partial \langle \mathbf{B} \rangle}{\partial t} = \nabla \times (\langle \mathbf{v} \rangle \times \langle \mathbf{B} \rangle) + \nabla \times \langle \mathbf{v}' \times \mathbf{B}' \rangle + \eta \Delta \langle \mathbf{B} \rangle$$

Correlations between velocity and magnetic field fluctuations appear. They generate a mean electric field:

$$\mathcal{E}_i = \langle \mathbf{v}' \times \mathbf{B}' \rangle_i$$

To go forward, we need to model this correlation. When the mean magnetic field is not too strong this may be done rather precisely. Indeed, the magnetic field fluctuations verify:

$$\frac{\partial \mathbf{B}'}{\partial t} = \nabla \times \left(\langle \mathbf{v} \rangle \times \mathbf{B}' + \mathbf{v}' \times \langle \mathbf{B} \rangle \right) + \nabla \times (\mathbf{v}' \times \mathbf{B}' - \langle \mathbf{v}' \times \mathbf{B}' \rangle) + \eta \Delta \mathbf{B}'$$

If the velocity fluctuations are independent of the mean magnetic field, which is expected if this mean field is weak enough, then the magnetic field fluctuations depend linearly on $\langle \mathbf{B} \rangle$, as well as $\langle \mathbf{v}' \times \mathbf{B}' \rangle$. Thus, we may write:

$$\langle \mathbf{v}' \times \mathbf{B}' \rangle_i = a_{ij}\langle B_j \rangle + b_{ijk}\partial_j \langle B_k \rangle + \cdots \tag{10.44}$$

where the tensors $[a]$ and $[b]$ are functions of the turbulent velocity field (not perturbed by the mean magnetic field). We may observe that they are two pseudo-tensors that are not invariant by parity transformations. For an isotropic turbulence we can write

$$a_{ij} = \alpha \delta_{ij} \qquad \text{and} \qquad b_{ijk} = \beta \epsilon_{ijk} .$$

β is a true scalar but α is a pseudo-scalar which vanishes if turbulence is parity-invariant. β has the dimension of a diffusivity and is consequently interpreted as the turbulent diffusivity of the magnetic field. α has the dimension of a velocity and gives birth to the now famous *alpha effect*, which we now discuss.

10.5.6 The Alpha Effect

The alpha effect is important in natural dynamo because this is an efficient way to generate magnetic fields. To get a more precise idea of the way it works, we take the example of a mean force-free magnetic field with an alpha effect. We assume that the mean velocity field is zero. Thus, the mean magnetic field verifies:

$$\frac{\partial \mathbf{B}}{\partial t} = \alpha \nabla \times \mathbf{B} + \eta_{turb}\Delta \mathbf{B} \tag{10.45}$$

where we assumed that α and β are constants. We also defined $\eta_{turb} = \eta + \beta$. Let us now determine the condition under which the mean magnetic field is amplified by turbulence. We set

$$\mathbf{B} = \mathbf{B}_0(\mathbf{r})e^{\lambda t}$$

where $\nabla \times \mathbf{B}_0(\mathbf{r}) = K\mathbf{B}_0(r)$. Using (10.45), we find the following dispersion relation of the Fourier modes:

$$\lambda = \alpha K - \eta_{turb}K^2 \tag{10.46}$$

It shows that if $\alpha > \eta_{turb}K$, the field is amplified.

Another way to see the amplification effect of the new term $\alpha \nabla \times \mathbf{B}$ is to reconsider Cowling's antidynamo theorem. Equation (10.42) now reads:

$$\frac{\partial A}{\partial t} + \mathbf{v}\cdot\nabla A = \alpha s B + \eta_{turb}\left(\Delta - \frac{2}{s}\frac{\partial}{\partial s}\right)A \tag{10.47}$$

There we see that we no longer have the simple advection-diffusion of the potential A that leads to the disappearance of the field. The toroidal component of the field B comes into play and allows the regeneration of A, which in turn regenerates B.

10.6   Exercises

1. Show that the electric current that would result form the presence of free charges in the non-relativistic flow of a conducting fluid is always negligible compared to the induced current. Show that it implies that the force exerted on the fluid by the electrostatic forces is always very small compared to the Laplace force.
2. Show that the power dissipated by Joule effect in an electric circuit (RI^2 where R is the resistance of the circuit and I the intensity of the current that circulates) has the same origin as the magnetic dissipation that appears in (10.18).
3. Study the dispersion relation of a plane wave in an homogeneous isotropic turbulence where the alpha effect is present and where the mean velocity is zero.
4. *The magnetorotational instability*: This instability is much studied in Astrophysics since it is thought to be the main source of turbulence in accretion discs (see Chap. 6). Here we propose a simplified study of this instability.

    We start with the system made of a differentially rotating incompressible fluid contained between two infinitely long cylinders (see Sect. 6.2.1). The fluid is now bathed by a uniform magnetic field parallel to the rotation axis $\mathbf{e}_z$. Let $\mathbf{U} = U(s)\mathbf{e}_\varphi = s\Omega(s)\mathbf{e}_\varphi$ be the basic differential rotation and $\mathbf{u}$ the velocity perturbation, $B_0\mathbf{e}_z$ the imposed magnetic field and $\delta\mathbf{B} = B_0\mathbf{b}$ its perturbation. We assume that all perturbations are axisymmetric, of vanishing amplitude and proportional to $\exp(i\omega t)$.

    (a) Show that

    $$i\omega\mathbf{u} = 2\Omega u_\varphi\mathbf{e}_s - \frac{\kappa^2}{2\Omega}u_s\mathbf{e}_\varphi - \nabla p/\rho + v_a^2\nabla\times\mathbf{b}\times\mathbf{e}_z \qquad (10.48)$$

    $$i\omega\mathbf{b} = \nabla\times(\mathbf{u}\times\mathbf{e}_z) + (\mathbf{b}\cdot\nabla)\mathbf{U} - (\mathbf{U}\cdot\nabla)\mathbf{b} \qquad (10.49)$$

    where magnetic diffusion is neglected. κ is the epicyclic frequency as given by (6.10). What is the expression of v_a^2 ?

    (b) Show that

    $$(\mathbf{b}\cdot\nabla)\mathbf{U} - (\mathbf{U}\cdot\nabla)\mathbf{b} = s\frac{d\Omega}{ds}b_s\mathbf{e}_\varphi$$

    (c) We now assume that disturbances just depend on z and are proportional to $\exp(ikz)$. Show that

    $$\nabla\times(\mathbf{u}\times\mathbf{e}_z) = ik\mathbf{u} \qquad \text{and} \qquad u_z = b_z = p = 0$$

    (d) and deduce that

$$\begin{cases} i\omega u_s - 2\Omega u_\varphi = v_a^2 ikb_s \\ i\omega u_\varphi + \frac{\kappa^2}{2\Omega}u_s = v_a^2 ikb_\varphi \\ i\omega b_s = iku_s \\ i\omega b_\varphi = iku_\varphi + s\frac{d\Omega}{ds}b_s \end{cases} \quad (10.50)$$

(e) From the foregoing relations show that the dispersion relation reads

$$\omega^4 - (\kappa^2 + 2v_a^2 k^2)\omega^2 + v_a^2 k^2 \left(v_a^2 k^2 + s\frac{d\Omega^2}{ds} \right) = 0 \quad (10.51)$$

(f) Show that at least one root of this equation may lead to an instability. Derive the following condition for instability:

$$s\frac{d\Omega^2}{ds} < -v_a^2 k^2 \quad (10.52)$$

What is the general condition on the flow that can be deduced?

(g) We set $y = v_a^2 k^2$. Show that the growth rate of the instability is maximum when

$$y = -s\frac{d\Omega^2}{ds}\left(\frac{1}{4} + \frac{\kappa^2}{16\Omega^2} \right)$$

(h) Show that the maximal growth rate is given by

$$\omega_{max} = \frac{s}{2}\left| \frac{d\Omega}{ds} \right| \quad (10.53)$$

(i) Show that the keplerian flow $\Omega \propto s^{-3/2}$ of an accretion disc of thickness H can be unstable if the background magnetic field is less than a limiting value. Give the expression of this upper limit.

Appendix: Equations of the Axisymmetric Field

We start from (10.41) and write the diffusion term and the curl of the electric field $E = v \times B$:

$$\Delta B \begin{vmatrix} (\Delta - s^{-2})\left(\frac{1}{s}\frac{\partial A}{\partial z} \right) \\ (\Delta - s^{-2})B \\ -\Delta\left(\frac{1}{s}\frac{\partial A}{\partial s} \right) \end{vmatrix} \qquad \nabla \times E \begin{vmatrix} -\frac{\partial E_\varphi}{\partial z} \\ \frac{\partial E_s}{\partial z} - \frac{\partial E_z}{\partial s} \\ \frac{1}{s}\frac{\partial s E_\varphi}{\partial s} \end{vmatrix}$$

The $\mathbf{e}_s$ and $\mathbf{e}_z$-components of the induction (10.10) lead to:

$$\frac{\partial A}{\partial t} = -sE_\varphi + \eta s(\Delta - s^{-2})A/s + f(s)$$

$$-\frac{\partial}{\partial t}\frac{\partial A}{\partial s} = \frac{\partial s E_\varphi}{\partial s} - \eta s\Delta\left(\frac{1}{s}\frac{\partial A}{\partial s}\right)$$

We take the s-derivative of the first equation and add it to the second equation. We find:

$$0 = f'(s) + \eta\left(\frac{\partial}{\partial s}s(\Delta - s^{-2})A/s - s\Delta\left(\frac{1}{s}\frac{\partial A}{\partial s}\right)\right)$$

but

$$\frac{\partial}{\partial s}s(\Delta-s^{-2})A/s - s\Delta\left(\frac{1}{s}\frac{\partial A}{\partial s}\right) = \frac{\partial^2}{\partial s^2}\left(s\frac{\partial}{\partial s}\frac{A}{s}\right) - \frac{\partial}{\partial s}\left(\frac{A}{s^2}\right) - \frac{\partial}{\partial s}\left[s\frac{\partial}{\partial s}\left(\frac{1}{s}\frac{\partial A}{\partial s}\right)\right]$$

which shows that the terms in parenthesis cancel and that $f'(s) = 0$. Noting that

$$sE_\varphi = v_s\frac{\partial A}{\partial s} + v_z\frac{\partial A}{\partial z}$$

we find (10.42) up to a constant.

To derive the equation for B we take the φ-component of the induction equation; hence

$$\frac{\partial B}{\partial t} = \frac{\partial E_s}{\partial z} - \frac{\partial E_z}{\partial s} + \eta(\Delta - s^{-2})B$$

setting $v_\varphi = s\omega$, then

$$E_s = -Bv_z - \omega\frac{\partial A}{\partial s} \qquad\text{and}\qquad E_z = Bv_s - \omega\frac{\partial A}{\partial z}$$

Considering the case where $A \to$ Cst, the equation for B now reads

$$\frac{\partial B}{\partial t} + \mathbf{v}\cdot\nabla B = -B\left(\frac{\partial v_z}{\partial z} + \frac{\partial v_s}{\partial s}\right) + \eta(\Delta - s^{-2})B$$

using $\nabla\cdot\mathbf{v} = 0$, we rearrange the terms so that

$$\frac{\partial B}{\partial t} + sv_s\frac{\partial}{\partial s}\left(\frac{B}{s}\right) + sv_z\frac{\partial}{\partial z}\left(\frac{B}{s}\right) = \eta(\Delta - s^{-2})B$$

from which we find (10.43).

Further Reading

A classical reference to the subject of fluid dynamos is the book of K. Moffatt *Magnetic field generation in fluids* (1978) unfortunately out of print. The lectures of A. Pouquet and P. Roberts in Les Houches volume *Astrophysical Fluid Dynamics* (1992) give another introduction to MHD turbulence and dynamos, but see also *An Introduction to Magnetohydrodynamics* by Davidson (2001). One may also consult *Lectures on Solar and Planetary Dynamos* (Proctor & Gilbert Edts, 1994), or *Principles of Magnetohydrodynamics: With Applications to Laboratory and Astrophysical Plasmas* by Goedbloed & Poedts (2004).

References

Chandrasekhar, S. (1961). *Hydrodynamic and hydromagnetic stability*. Oxford: Clarendon Press.

Cowling, T. G. (1933). The magnetic field of sunspots. *Monthly Notices of the Royal Astronomical Society, 94*, 39–48.

Davidson, P. (2001). *An introduction to magnetohydrodynamics*. Cambridge: Cambridge University Press.

Gilbert, A. D. (1988) Fast dynamo action in the Ponomarenko dynamo. *Geophysical and Astrophysical Fluid Dynamics, 44*, 241.

Goedbloed, J. P., & Poedts, S. (2004). *Principles of magnetohydrodynamics: With applications to laboratory and astrophysical plasmas*. Cambridge: Cambridge University Press.

Moffatt, K. (1978). *Magnetic field generation in fluids*. Cambridge: Cambridge University Press.

Morin, J., Donati, J. -F., Forveille, T., Delfosse, X., Dobler, W., Petit, P., & et al. (2008). The stable magnetic field of the fully convective star V374 Peg. *Monthly Notices of the Royal Astronomical Society, 384*, 77–86.

Pouquet, A. (1992). Magnetohydrodynamic turbulence. In J. -P. Zahn & J. Zinn-Justin (Eds.), *Les Houches 1987: Astrophysical fluid dynamics* (pp. 139–227). North-Holland.

Proctor, M., & Gilbert, A. (1994). *Lecture on solar and planetary dynamos, Publications of the Newton Institute*. Cambridge: Cambridge University Press.

Roberts, P. (1992). Dynamo theory. In J. -P. Zahn & J. Zinn-Justin (Eds.), *Les Houches 1987: Astrophysical fluid dynamics* (pp. 229–323). North-Holland.

Tadesse, T., Wiegelmann, T., Inhester, B., MacNeice, P., Pevtsov, A., & Sun, X. (2013). Full-disk nonlinear force-free field extrapolation of SDO/HMI and SOLIS/VSM magnetograms. *Astronomy and Astrophysics, 550*, A14.

Chapter 11
Beyond Fluid Mechanics: An Introduction to the Statistical Foundations of Gas Dynamics

11.1 Introduction

While introducing Fluid Mechanics, we had to introduce also the idea of continuous media, which is the mathematical idealization of real fluids (or solids). In many circumstances, the limits of this approach arose: for instance the rheological laws, which relate strain and stress, are not given by Fluid Mechanics, they need another model. Fluid Mechanics considers these laws as given. In the first chapter we observed that in the limit of small perturbations of the basic thermodynamic equilibrium by the flow, we could derive the functional form of the rheological laws, namely that of Newtonian fluids, but the specificity of the fluid was then condensed in its viscosity or, more generally, in its transport coefficients.

Fluid Mechanics does not say anything about transport coefficients: it is in the same position as Thermodynamics which does not say anything either on thermoelastic coefficients of the various materials. Fluid Mechanics and Thermodynamics are two theories of the macroscopic world. They give the general laws that are followed by matter independently of its nature. But to be predictive, both of them need to be completed either by a more detailed approach that includes its microscopic nature, or by experimental measurements.

In the case of Fluid Mechanics, the derivation of the macroscopic rheological laws from the microscopic properties of fluids is really well developed in the case of gases. The case of liquids is much more complicated and thus less explored.[1]

The theory of the statistical properties of gases beyond equilibrium, which is based on their microscopic properties, is usually called the *kinetic theory of gases*. The present chapter offers an introduction to this difficult but fascinating subject. We shall discover for instance how the dependence of the viscosity of a gas with

---

[1] A taste of this approach may be found in the book of Guyon et al. (2001a).

M. Rieutord, *Fluid Dynamics*, Graduate Texts in Physics,
DOI 10.1007/978-3-319-09351-2_11

respect to temperature gives a constraint on the potential of interaction between its atoms (or molecules), thus opening another window on the microscopic world.

As the reader may guess, this is a wide subject which would require a whole book. We wish to remain introductive therefore shall restrict ourselves to one question: how can we derive Navier–Stokes equation and Fourier's law for a gas assumed to be a set of interacting particles? In other words, how can one move from matter described as a set of interacting particles to a continuous medium?

In the following section we try to give a qualitative answer guided by our intuition. Since this approach may not be fully satisfying to some readers, we pursue with a more rigorous path, leading us to the Navier–Stokes equation. Although more rigorous, this derivation still simplifies reality, but thus doing the reader will be acquainted with Boltzmann equation and will uncover how to derive the viscosity of a gas from its sole microscopic characteristics.

11.2   A Qualitative Approach

11.2.1   Back to the Continuous Medium

Let us imagine the change of the number of gas particles in a volume δV when this volume decreases from a macroscopic value to an infinitesimal one. Plotting the particle density $n = \delta N/\delta V$ (δN is the number of particles in δV), as a function of the scale $(\delta V)^{1/3}$, one obtains a curve like the one drawn in Fig. 11.1. Two values need to be underlined: L_M is a macroscopic scale beyond which density variations are noticeable because the gas is not at equilibrium and L_m a microscopic scale below which the particle density decreases until vanishing. Approximating the gas by a continuous medium is assuming that the plateau at $n = n_0$ continues until $\delta V = 0$.

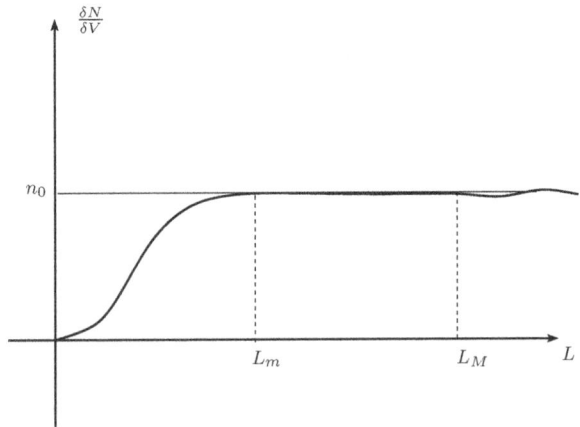

Fig. 11.1 A schematic representation of the number density as a function of scale

Thus, we can build a continuum that represents the average properties of the plateau appearing at scale L_M. For this new medium all the variables are continuous and derivable functions of space and time, except, eventually on some surface. Of course it may well be that this plateau does not exist, if $L_m \sim L_M$. In this case the continuous medium model cannot be used. This model relies on the assumption that there exist a separation of scale between the macroscopic and microscopic world, namely $L_m \ll L_M$. This is this separation of scale that misses when one tries to build a mean-field theory of turbulence.

We shall remember that the continuous medium is a model of matter where all the points of space are gifted of mean fields values, namely $\langle v \rangle (\mathbf{x}, t)$, $\langle n \rangle (\mathbf{x}, t)$, $\langle \frac{1}{2} mv^2 \rangle (\mathbf{x}, t)$ etc. We'll have to give the way to compute the averages, but this is not necessary for our qualitative discussion and we postpone the precise definitions for the next step. Presently, we just need to imagine that we take an average over a small fluid volume that contains enough particles.

11.2.2 Particles Interactions, Collisions and the Mean Free Path

To make progress we now need to review some important features of the microscopic nature of the fluid, in other words we need to describe our model of atoms or molecules that make the fluid. As mentioned previously, we restrict ourselves to gases. At the microscopic level, these are characterized by the fact that their particles do not permanently interact. Interactions represent a small fraction of their trajectories and therefore can be called *collisions*. The main part of a particle trajectory is a free fly at constant speed. This means that we assume an effective short-range potential for the interaction between particles. We easily conceive that this is valid only for dilute gases. To make this argument quantitative we introduce the van der Waals radius r_w of the particles: it can be thought as that of the sphere where the interaction potential energy is of the same order as the kinetic energy of the particles. Thus, a gas is dilute when

$$nr_w^3 \ll 1 \qquad (11.1)$$

where n is the numeric density of the particles. (11.1) means that the volume of the particles is negligible compared to the volume occupied by the gas or else that

$$r_w \ll n^{-1/3}$$

namely that the van der Waals radius is very small compared to the mean distance between particles.

For such a gas, the first step to understand its properties is to assume that the particles are like *elastic hard spheres* of radii r_w. For such a model, the interaction potential is very simple: it is either zero or infinite. More realistic potentials will be considered in the second step.

Fig. 11.2 Collision between two elastic hard sphere of diameter d. The exclusion sphere has a radius d

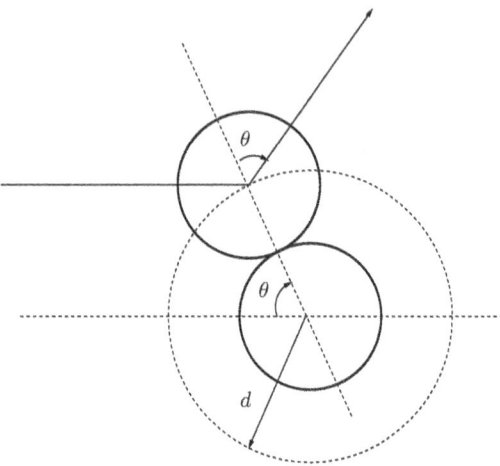

The hard elastic spheres model will serve us as a guide along this chapter. As a start, it will allow us to define the mean free path, a crucial concept to understand the microscopic world.

Let us first note that the distance between two balls is equal to their diameter d when they collide (see Fig. 11.2). Thus, during its motion, a particle sweeps a cylindrical tube of section πd^2. Collisions occur when another particle is in this cylinder when the particle passes.

Let v be the velocity of the particles, the volume swept per unit time is just $\pi d^2 v$. If the number density is n, we may consider that $N = n\pi d^2 v$ collisions occur per unit time for a given particle. We here admit that the collided particles are fixed. We deduce that $1/N$ is the time interval between to collisions. v/N is therefore the distance swept between two collisions or the mean free path. Thus, we get

$$\ell = \frac{1}{\pi d^2 n}$$

for the mean free path ℓ. A more rigorous approach where one takes into account the motion of the collided spheres gives (see Sect. 11.4.3):

$$\ell = \frac{1}{\sqrt{2}\pi d^2 n} \tag{11.2}$$

Our simplified approach gives a very good order of magnitude.[2]

---

[2] The origin of the $\sqrt{2}$-factor may be understood with a simple argument. We first observe that the number of collisions is controlled by the relative velocity of the particles, i.e. $N = n\pi d^2 v_{rel}$. But $\mathbf{v}_{rel} = \mathbf{v} - \mathbf{v}_0$ where $\mathbf{v}_0$ is the velocity of the test particle. If we identify v_{rel} with the rms velocity $\sqrt{\langle v_{rel}^2 \rangle}$, we get

11.2.3   The Velocity of Particles

The characteristic velocity of the particles is that of thermal agitation v_{agit}: if we write the velocity of a gas particle as the sum of an average velocity $\langle \mathbf{v} \rangle$ and a random velocity of zero average $\mathbf{u}$, the microscopic transport is controlled by this random component. But various choices are possible for v_{agit}. Indeed, we may choose

$$v_{\text{agit}} = \bar{u} = \langle \|\mathbf{u}\| \rangle \qquad \text{or} \qquad v_{\text{agit}} = u_{\text{rms}} = \sqrt{\langle \|\mathbf{u}\|^2 \rangle}$$

that is to say the average of the norm of the velocity or the root-mean-square velocity. Other choices are also possible. As we shall see later, the velocity statistical distribution is close to the equilibrium Maxwell–Boltzmann one. Thus, we have

$$\bar{u} = \langle \|\mathbf{u}\| \rangle = \left(\frac{m}{2\pi kT} \right)^{3/2} \int_0^{+\infty} u e^{-\frac{mu^2}{2kT}} 4\pi u^2 du = \sqrt{\frac{8kT}{\pi m}} \qquad (11.3)$$

and

$$u_{\text{rms}} = \sqrt{\langle \|\mathbf{u}\|^2 \rangle} = \left(\frac{m}{2\pi kT} \right)^{3/2} \int_0^{+\infty} u^2 e^{-\frac{mu^2}{2kT}} 4\pi u^2 du = \sqrt{\frac{3kT}{m}} \qquad (11.4)$$

We choose $v = \bar{u}$, but the other choice is not very different because $u_{\text{rms}}/\bar{u} \simeq 1.085$.

11.2.4   Energy Transport

When we faced the problem of giving an expression to the (surface density of) heat flux $\mathbf{F}$, we assumed that the fluid was close to the thermodynamic equilibrium. We thus expanded this quantity in powers of the temperature gradient taken as the quantity measuring the distance to equilibrium.[3] We thus deduced Fourier's law:

$$\mathbf{F} = -\chi \nabla T$$

where χ is the thermal conductivity.

$$\langle v_{\text{rel}}^2 \rangle = \langle \mathbf{v}^2 \rangle + \langle \mathbf{v}_0^2 \rangle - 2 \langle \mathbf{v} \cdot \mathbf{v}_0 \rangle$$

The randomness and uncorrelation of velocities imply that $\langle \mathbf{v} \cdot \mathbf{v}_0 \rangle = 0$. Since the test particle is not different from other particles $\langle \mathbf{v}_0^2 \rangle = \langle \mathbf{v}^2 \rangle$. Thus $v_{\text{rel}} = \sqrt{2}v$.

[3] Actually, this expansion can be generalized to all the physical quantities that measure the distance to thermodynamic equilibrium like shear, gradient of concentration, current etc. We touch here Onsager's approach who worked out the general theory of slight deviations from thermodynamic equilibrium. Our purpose here is not that general and we shall consider only situations where fluxes depend only on a single quantity, which is correct in the simple cases that we are considering.

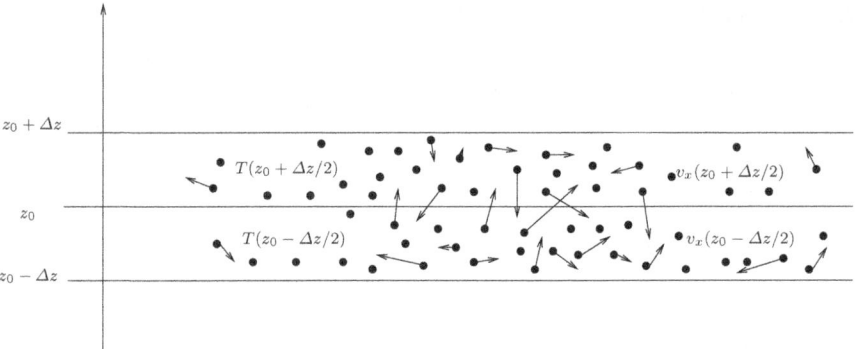

Fig. 11.3 The particles of a gas in two layers of thickness Δz. The *upper layer* has a mean temperature $T(z_0 + \Delta z/2)$ and a mean velocity $v_x(z_0 + \Delta z/2)$, resp. $T(z_0 - \Delta z/2)$ and $v_x(z_0 - \Delta z/2)$ for the lower layer

In order to have more information on this coefficient, we consider the case where the imposed temperature gradient is uniform along the z-axis, namely $\mathbf{F} = -\chi \partial_z T \mathbf{e}_z$. Dividing the gas into layers of thickness Δz (see Fig. 11.3), we first observe that in a steady state the mass of each layer is conserved. The particle flux across any plane $z = z_0$ is zero.

Let n be the number of particles per unit volume and v their typical velocity (for instance their mean velocity $\langle \|\mathbf{v}\| \rangle$). The number of particles crossing the plane $z = z_0$ in the upward direction through the surface element dS and during the lapse of time dt is

$$\beta n v \, dt \, dS$$

where β is a dimensionless constant of order unity. Each particle carries some momentum and kinetic energy. As a first step we consider their kinetic energy, which is a scalar quantity.

Setting to zero their mean velocities, the two layers at $z_0 - \Delta z/2$ and $z_0 + \Delta z/2$ differ by their temperatures $T(z_0 - \Delta z/2)$ and $T(z_0 + \Delta z/2)$ respectively. If we remember that temperature is just a measure of the mean kinetic energy of the gas particles, each particle of mass m moving from one layer to the other carries the kinetic energy $\frac{1}{2}mv^2$. On average, the flux of kinetic energy reads

$$F = \beta' \frac{n}{2} \left[\langle mv^2 v_z \rangle (z_0 - \Delta z/2) - \langle mv^2 v_z \rangle (z_0 + \Delta z/2) \right]$$

where β' is another dimensionless constant of order unity. Note that the flux is oriented positively with increasing z. Now, we wish to use temperature rather than kinetic energy, so we use the relation (see appendix):

$$\frac{3}{2}kT = \frac{1}{2}m\langle v^2\rangle$$

where k is the Boltzmann constant. Orders of magnitude say that $v_z \sim v$, hence $\langle mv^2 v_z \rangle \sim v \langle mv^2 \rangle$. This allows us to introduce the temperature in the expression of F and, with the help of an expansion to first order, to obtain:

$$F = -\beta'' n v \frac{3}{2} k_B \left(\frac{dT}{dz}\right)_{z_0} \Delta z$$

where β'' is again a dimensionless constant of order unity. This result leads to the following expression of χ:

$$\chi = \frac{3}{2}k\beta'' n v \Delta z \tag{11.5}$$

Now, the relation between density, mass of particles and number density is:

$$\rho = nm \tag{11.6}$$

Besides, the heat capacity of such an ideal gas is

$$c_v = \frac{3}{2}\frac{k}{m}.$$

In addition, the thickness of the layer should be something like the mean free path ℓ of the particles. Thus, we may rewrite (11.5) as

$$\chi = \rho c_v \beta_\chi v \ell \tag{11.7}$$

where β_χ is still a dimensionless constant of order unity. We could have written this expression directly, just using dimensional arguments from a set of dimensional quantities describing the microscopical model we are using. However, our analysis has the advantage to guide us towards this expression and to make us confident that it contains the relevant physics. At this point, two directions are possible. The first one is to use experimental results to derive β_χ. If this coefficient turns out to be of order unity, we have likely captured the relevant physics of the process leading to heat conduction. If this is not the case, an important physical process has been missed.

The second direction is to dig more into the theory in order to derive the mathematical expression of β_χ as a function of the microscopic characteristics of the model. This second way is challenging and will be detailed in the following sections. Before that we examine the other transport coefficients in a qualitative way.

11.2.5 Momentum Transport

We may now redo the same exercise for the momentum and derive an expression of the shear viscosity of the fluid. Just as for heat conductivity, we need considering a one-dimensional configuration. The easiest way is to consider a plane sheared gas flow like $\mathbf{v} = v(z)\mathbf{e}_x$.

We now need to estimate the force exerted by the layer at $z_0 - \Delta z/2$ on the layer at $z_0 + \Delta z/2$. This force comes from the momentum carried by the particles moving between the layers. Particles moving upward ($v_z > 0$) deposit momentum

$$m \langle v_x \rangle (z_0 - \Delta z/2)$$

in the layer at $z_0 + \Delta z/2$, while those moving downwards deposit $m \langle v_x \rangle (z_0 + \Delta z/2)$ in the layer at $z_0 - \Delta z/2$. The momentum flux (positive upwards) is also the tangential force exerted by the layer at $z_0 - \Delta z/2$ on the layer at $z_0 + \Delta z/2$ for a surface dS, in other words

$$df_x = \beta nvdS \left[m \langle v_x \rangle (z_0 - \Delta z/2) - m \langle v_x \rangle (z_0 + \Delta z/2) \right]$$

$$= -\beta nvm \frac{d \langle v_x \rangle}{dz} \Delta z dS = -\beta \rho v \Delta z \frac{d \langle v_x \rangle}{dz} dS \tag{11.8}$$

where β is $\mathcal{O}(1)$. Let us now come back to the definition of the stress. The force exerted on the surface element $d\mathbf{S}$ is:

$$d\mathbf{f} = [\sigma] d\mathbf{S}$$

where $[\sigma]$ is the stress tensor. Here we wish to know the force exerted by the layer at $z_0 - \Delta z/2$ on the upper layer. Hence, we have to consider the plane $z = z_0$ seen by the lower layer. $d\mathbf{S}$ is oriented towards the fluid that exert the stress, so $d\mathbf{S} = -dS\mathbf{e}_z$ and $df_x = -\sigma_{xz} dS$. Assuming the fluid is Newtonian, for a plane-parallel flow we have

$$\sigma_{xz} = \mu \frac{dv_x}{dz} \quad \Longrightarrow \quad df_x = -\mu \frac{dv_x}{dz} dS$$

where μ is the (dynamic) shear viscosity. Comparing this expression to (11.8), we find the expression of shear viscosity, namely:

$$\mu = \beta \rho v \Delta z$$

As for the heat conductivity, we replace $\beta \Delta z$ by $\beta_\nu \ell$ using the mean free path ℓ. The foregoing expression leads to that of the kinematic viscosity $\nu = \mu/\rho$,

$$\nu = \beta_\nu v \ell \tag{11.9}$$

As for the heat conductivity, we could have derived this expression by a straightforward dimensional analysis.

11.2.6 The Prandtl Number

The Prandtl number of a fluid is the ratio of its kinematic viscosity v to its heat diffusivity $\kappa = \chi/\rho/c_p$ (see Chap. 1, 1.46). From the expressions of χ and v that we derived previously, this dimensionless number reads

$$\mathcal{P} = \gamma \frac{\beta_v}{\beta_\kappa}$$

where $\gamma = c_p/c_v$. For a monatomic gas $\gamma = 5/3$ while experimental measurements show that $\mathcal{P} \simeq 2/3$ (see Table 11.1). The foregoing relation suggests that $\beta_v/\beta_\kappa \simeq 2/5$. Our naive model gives a Prandtl number of order unity, which is quite correct. We may have thought that $\beta_v \simeq \beta_\kappa$ implying that $\mathcal{P} \gtrsim 1$, which is contradicted by experiments. Observations thus show that $\beta_v < \beta_\kappa$, suggesting that kinetic energy is more efficiently transported than momentum. A posteriori, this is not so surprising since particles of high kinetic energy may indeed have a slightly larger mean free path than those of small mean free path. This remark shows that if we wish to go beyond simple orders of magnitude we need a more detailed statistical approach.

11.2.7 Comparing with Experimental Results

Let us end this section with a short discussion of experimental values compared to those of our simplistic model. For that, we focus on the shear viscosity. To be fair with the rigid elastic sphere model, we set $\beta_v = 0.491$ as given by the complete statistical approach (see Sect. 11.7.5). We get

$$\mu = \beta_v \rho \ell v = \frac{2\beta_v}{\pi^{3/2}d^2}\sqrt{mkT} \tag{11.10}$$

In this expression the diameter of the particles is a crucial quantity. However it is not well defined. If we think to an atom of helium, how can we define its radius? The only way is to study the potential of interaction between two such atoms and try to represent it by that of a hard sphere. This is not an easy matter. To circumvent this difficulty, we may go back to Thermodynamics and remember that a more precise model than that of ideal gases is the one of van der Waals. In the equation of state of this model, the volume occupied by the atoms is one parameter of the model. This is b in the equation of state

$$(P + \frac{a}{v^2})(V - b) = RT$$

Table 11.1 We give the co-volume b associated with the van der Waals equation of state for various gases and the van der Waals radius derived from (11.11)

Gas	b cm³/mole	r_w (nm)	ℓ (nm)	μ_{calc} (μPa s)	μ_{obs} (μPa s)	$\mathcal{P}_{obs}$
He	23.7	0.133	132	13.1	20.	0.680
Ne	17.1	0.119	164	36.5	32.	0.661
Ar	32.2	0.147	107	33.7	22.	0.672
H₂	26.6	0.138	122	8.6	9.	0.693
Air	36.4	0.153	99	26.5	18.	0.714

We give the mean free path from (11.2) and the dynamic viscosity from (11.10). The experimentally observed viscosity is also given as well as the experimental value of the Prandtl number. The experimental values of co-volumes and viscosities are from Gray (1975). All these values are computed at a temperature of 300 K and a pressure of 1 bar. The Prandtl numbers are measured at 0 °C (Chapman and Cowling 1970)

here written for a single mole. b is actually an "exclusion" volume, namely a volume that is not accessible by the atoms. It is also called the *co-volume* of the gas. If we assume that these atoms are spheres of radius r_w then

$$b = \frac{\mathcal{N}}{2}\frac{4}{3}\pi(2r_w)^3 \qquad (11.11)$$

where $\mathcal{N} = 6.022\ 10^{23}$ is the Avogadro number. The factor $\mathcal{N}/2$ comes from the fact that we are counting binary collisions only and therefore for the collisions of two particles, only one volume is excluded.[4] This radius is usually called the van der Waals radius of the atom (for monatomic gases of course).

We can now calculate the viscosity of a gas from the experimental values of the co-volume b. We did this exercise and reported the values in Table 11.1.

This table shows a first positive result: our simple model gives the right orders of magnitude of the viscosities and a very good value for neon and hydrogen. However, the matching is not that good with two other monatomic gases, helium and argon.

In fact the good results are an illusion. Our model predicts that the variations of viscosity with temperature are like $\sqrt{T}$ independently of the nature of the gas, while experiments show that helium viscosity varies like $T^{0.65}$ (see Fig. 11.8). We here reach the limits of our model and if we wish a better comparison between theory and experiments we need going deeper into the modeling, especially in its statistical sides.

---

[4] Another way of deriving this factor is to count the number of pairs for a set of N particles. There are $N(N - 1)/2$ pairs. Thus when $N \gg 1$, there are just $N/2$ pairs for each particles. Thus the exclusion volume is that indicated by (11.11).

11.3 Concepts and Questions for a Statistical Approach

The curiosity of the reader may remain unsatisfied by the preceding section and therefore we feel obliged to lead him or her in a more detailed discussion of microscopic transport. However, a full description would need a whole book like those of Vincenti and Kruger (1965) or Chapman and Cowling (1970). Our ambition is necessarily more modest: we wish to show the reader how the equations of fluid mechanics, which describe a continuous medium emerge from the dynamics of a set of gas atoms or molecules and how transport coefficients like viscosity or heat conductivity may be rigorously derived. Beside this aim, we'll discover also new questions that immediately come up from the microscopic side of the world.

11.3.1 The Distribution Function

11.3.1.1 Back on the Continuous Medium

When we derived the rheological laws of Newtonian fluids, we stressed the fact that their fluid particles are very close to thermodynamic equilibrium. In view of Fig. 11.1, it means that at scales L such that $L_m < L < L_M$, atoms in δV are very *close* to thermodynamic equilibrium. We underlined "close" because if the fluid particles are exactly at equilibrium, we know that no microscopic transport is possible: viscosity and heat diffusion disappear corresponding to the perfect fluid limit. The rest of the chapter aims at showing how a fluid flow generates this deviation from thermodynamic equilibrium and how viscosity and heat conductivity may be predicted.

To reach this goal, we need introducing a quantity that describes the statistical state of fluid particles for scales $L_m < L < L_M$. Hence, considering the δN particles that are in δV, we shall sort them according to their velocity $\mathbf{v}$, their spin $\boldsymbol{\ell}$, their excitation state, etc. We therefore assume that particles are numerous enough so that δN can be divided into subsets categorizing a class of particles. For instance, we may sort the gas particles according to their speed and count those whose speed belongs to $[v, v + dv]$. More generally, we introduce a new function, namely

$$f(\mathbf{x}, t; \mathbf{v}, \boldsymbol{\ell}, \ldots) \tag{11.12}$$

called a *distribution function*. This function informs us not only about the number density of atoms or molecules in space and time, which is the role of mass density $\rho(\mathbf{x}, t)$, but it also tells us about their distribution according to a given parameter specific to the microscopic constituents of the fluid. Consequently, if we sum over all the possible values of these parameters we should recover the number density:

$$n(\mathbf{x}, t) = \int f(\mathbf{x}, t; \mathbf{v}, \mathbf{l}, \ldots) d^3\mathbf{v} d^3\boldsymbol{\ell} \ldots \tag{11.13}$$

If all the particles have the same mass m, then

$$\rho(\mathbf{x}, t) = m\, n(\mathbf{x}, t) \tag{11.14}$$

In what follows we shall restrict us to a dependence of the distribution function with respect to $\mathbf{x}, t$ and $\mathbf{v}$ only, but we see that the concept is more general.

Finally, let us observe that the introduction of the distribution function f imposes us to work in a space with $4 + 3$ dimensions. To the usual space-time we have added three new dimensions: those of the velocity space, which is hidden from us in the macroscopic world.

11.3.1.2 The Limits of this New Description

As we may observe, the knowledge of the distribution function gives more details on the state of a gas. In fact, we have now the possibility to determine the behaviour of a gas beyond the local thermodynamic equilibrium. But its use requires that the gas particles are numerous enough so that the statistical distribution of some parameters makes sense and the first moments of the distribution are computable (for instance a stress comes from a second order moment as we shall see below).

To better appreciate this limit we consider a real situation like that of helium in the normal conditions (P $= 10^5$ Pa, T $= 300$ K). Then the mean free path is $\ell = 132$ nm and the number of helium atoms in a cube of volume ℓ^3 is

$$n = P\ell^3 / kT = 5.6\,10^4$$

which is large enough to make statistics, but we see that it won't be possible to consider much smaller scales. This is however sufficient for describing a gas within a shock wave whose thickness is of order of a few mean free paths. The continuous medium approximation represents the shock wave by a mere discontinuity.

11.3.2 An Equation Governing the Distribution Function

11.3.2.1 Back to Liouville Equation

In order to understand the origin of the equation that governs the distribution function f, it is useful to consider shortly an even more fundamental description of gas dynamics.

The ultimate description of a gas flow is of course that of the motion of each of its atoms or molecules. If this gas contains N such particles, then the knowledge of the 2N vectors $\{\mathbf{q}^i\}_{i=1,N}$ and $\{\mathbf{p}^i\}_{i=1,N}$, that are respectively their positions and momentum, allows us to predict (mathematically) the evolution of the system. Hence, considering a 6N-dimensional space, the famous phase space introduced by

Maxwell, the dynamical state of our N particles resumes to one point. Following a wording of statistical Physics, we shall call this point a *figurative point* underlining thus that it is just a mathematical representation of our system (Castaing 1970).

Now, just imagine that we prepare this system of N particles in $\mathcal{N}$ states or $\mathcal{N}$ initial conditions. Each state is represented by a figurative point which moves in the phase space according to the motion of the N particles. Hence, a "cloud of points" evolves in the phase space, each point representing a possible trajectory of the system.

The trajectory of the figurative points in the phase space is not random: it is governed by the laws of Mechanics that control the motion of each of its components. But the figurative points do not interact: they are just a representation of the dynamical state of the set of the N particles. Thus, we can fill the phase space with as many figurative points as we need. However, if we take $\mathcal{N}$ such points at $t = 0$, there are still $\mathcal{N}$ such points at any later time: their number is conserved. Thus, if we call ρ their numerical density we may say that this function

$$\rho(q_x^1, q_y^1, q_z^1, p_x^1, , p_y^1, p_z^1, \ldots, p_x^N, p_y^N, p_z^N, t)$$

obeys the law of local conservation

$$\frac{\partial \rho}{\partial t} + \nabla \cdot (\rho \mathbf{v}_\varphi) = 0 \qquad (11.15)$$

by analogy with mass conservation in a fluid flow. Here $(\mathbf{q}^n, \mathbf{p}^n)$ represents the position and momentum of the n-th particle. Of course the velocity $\mathbf{v}_\varphi$ is a velocity in the phase space and the divergence is taken with respect to the 6N coordinates of this space. Explicitly

$$\nabla \cdot (\rho \mathbf{v}_\varphi) = \sum_{i=1}^{N} \frac{\partial \rho \dot{q}_x^i}{\partial q_x^i} + \frac{\partial \rho \dot{q}_y^i}{\partial q_y^i} + \frac{\partial \rho \dot{q}_z^i}{\partial q_z^i} + \frac{\partial \rho \dot{p}_x^i}{\partial p_x^i} + \frac{\partial \rho \dot{p}_y^i}{\partial p_y^i} + \frac{\partial \rho \dot{p}_z^i}{\partial p_z^i}$$

where we noticed that the velocity in the phase space is simply

$$\mathbf{v}_\varphi = (\ldots, \dot{q}_x^i, \dot{q}_y^i, \dot{q}_z^i, \dot{p}_x^i, \dot{p}_y^i, \dot{p}_z^i, \ldots)$$

where the dot is for the time derivative. If we now assume that the system is hamiltonian, then the time derivatives of the positions and momenta can be explicitly expressed with Hamilton equations, namely

$$\dot{p}_k^i = \frac{\partial H}{\partial q_k^i}, \qquad \dot{q}_k^i = -\frac{\partial H}{\partial p_k^i}, \qquad \forall i \in [1, N], \ k \in [x, y, z]$$

where H is the hamiltonian of the system, which is the sum of the kinetic and potential energies expressed with the variables p and q. Using of this property we see that $\nabla \cdot \mathbf{v}_\varphi = 0$. In other words, the fluid of figurative points is incompressible. This is *Liouville's theorem*. With (11.15), we deduce this other expression of the theorem:

$$\frac{\partial \rho}{\partial t} + \mathbf{v}_\varphi \cdot \nabla \rho = \frac{D\rho}{Dt} = 0 \tag{11.16}$$

in other words the material derivative of ρ is vanishing.

Liouville's theorem has another interesting consequence: if we consider a volume of the phase space that is moving with the flow of figurative points, it implies that

$$\frac{d}{dt} \int_{V(t)} \rho dV = 0$$

since the number of points is conserved [one may also use the equality (1.10)]. Thus

$$\int_{V(t)} \rho dV = \int_{V(t_0)} \rho dV_0 \qquad \forall t$$

but this is true for any volume, thus

$$\rho_t dV = \rho_{t_0} dV_0$$

But at a point comoving with the flow $\rho_t = \rho_{t_0}$, which implies

$$dV = dV_0$$

or that the infinitesimal volume of phase space does not vary while following the evolution of the system. This result may also be obtained using the Lagrangian formulation of Fluid Mechanics. Indeed, the displacement of a figurative point from its initial position $\mathbf{q}$ (a 6N-dimensional vector) obeys

$$\mathbf{x} = \mathbf{q} + \boldsymbol{\xi}(\mathbf{q}, t)$$

Lagrangian kinematics says that $\mathbf{x}(\mathbf{q}, t)$ is a mapping of space from t_0 to t, the jacobian of which measures the contraction/dilation (see 1.85). If we apply this reasoning to the incompressible flow of figurative points, hence with a unit jacobian, we have

$$d^3\mathbf{q}_1 d^3\mathbf{p}_1 \cdots d^3\mathbf{q}_N d^3\mathbf{p}_N(t = 0) = d^3\mathbf{q}_1 d^3\mathbf{p}_1 \cdots d^3\mathbf{q}_N d^3\mathbf{p}_N(t) \tag{11.17}$$

which also expresses the constancy of the infinitesimal volume of the phase space along a trajectory.

11.3.2.2 The Boltzmann Equation for Non-Interacting Particles

Let us now consider a phase space for one particle. Such a space is six-dimensional with co-ordinates (x, y, z, p_x, p_y, p_z) for a point. Let us fill this space with N non-interacting particles. These particles may be considered as N real particles that do not "see" each other or as N representations of a unique particle, each with different initial conditions, or like N figurative points. From Liouville's theorem the density ρ of these points verifies Liouville's equation (11.16). But, up to the mass of particles, ρ is just identical to the distribution function f. Hence we may write that for non-interacting particles f obeys

$$\frac{\partial f}{\partial t} + \mathbf{v}_\varphi \cdot \nabla f = 0$$

with

$$\mathbf{v}_\varphi = (\dot{x}, \dot{y}, \dot{z}, \dot{p}_x, \dot{p}_y, \dot{p}_z)$$

and thus

$$\frac{\partial f}{\partial t} + \dot{x}\frac{\partial f}{\partial x} + \dot{y}\frac{\partial f}{\partial y} + \dot{z}\frac{\partial f}{\partial z} + \dot{p}_x\frac{\partial f}{\partial p_x} + \dot{p}_y\frac{\partial f}{\partial p_y} + \dot{p}_z\frac{\partial f}{\partial p_z} = 0$$

if the particles are in a force field $\mathcal{F}$, then using Newton's law $\dot{\mathbf{p}} = \mathcal{F}$ and the definition of time derivative $\dot{\mathbf{x}} = \mathbf{v}$, we finally get

$$\frac{\partial f}{\partial t} + \mathbf{v} \cdot \nabla_x f + \frac{\mathcal{F}}{m} \cdot \nabla_v f = 0 \qquad (11.18)$$

where ∇_x and ∇_v represent the gradients respectively taken with respect to space coordinates and velocity coordinates.

Equation (11.18) gives a first version of the equation that governs the distribution function f in the simple case of a gas made of non-interacting particles.

11.4 Boltzmann Equation ♠

Equation (11.18) is certainly too simple to describe the distribution function of a real gas where particles interact. This interaction is represented by a potential like the one shown in Fig. 11.4 where we see the one between two helium atoms.

A rigorous account of this interaction is very difficult but fortunately not necessary in the usual conditions of gas dynamics. Indeed, the distance between gas particles is large compared to the range of the interaction, which is usually of the order of a few radii of the atoms. Hence, the interaction between two particles can be described by a collision: a small part of the trajectory during which the interaction

Fig. 11.4 The potential of
interaction between to helium
atoms according to Aziz et al.
(1979)

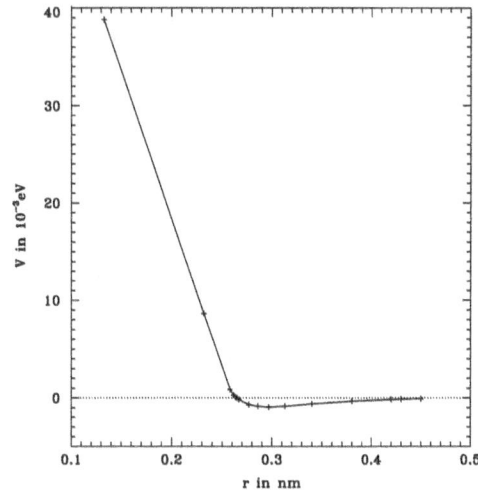

Fig. 11.5 Collisions and
interactions: the interaction
between two particles is
effective during a small
fraction of the trajectory: the
collision designate this small
fraction of the trajectory
when the motion is not
uniform

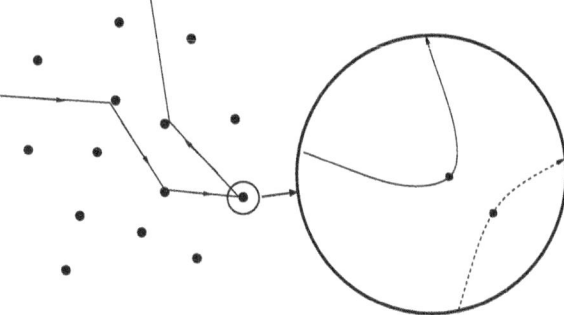

potential is non-vanishing as schematically represented by Fig.11.5. A collision is
therefore an approximate view of the interaction between particles: each particle
comes from infinity, interact and goes back to infinity. This is approximate because
infinity means the preceding or following collision. But the approximation remains
very good when the potential is of short range. To make this approximation more
concrete, let us consider helium in normal conditions (P = 101 325 Pa, T = 300 K).
The number of atoms per unit volume is $n = 2.4 \times 10^{25}$ m^{-3}, meaning a mean
distance of ~ 3.4 nm between atoms, which is $25r_w$ (r_w is the van der Waals radius).

Since in this case the interaction range is of the order of 0.5 nm (cf. Fig. 11.4),
we see that the use of collision is perfectly justified. However, in dense gases or
in liquids this concept is meaningless. For instance, in liquid water the number of
molecules per unit volume is $n = 3.3 \times 10^{28}$ (for a density of 1,000 kg/m^3 and mole
mass of 18 g) which is equivalent to an intermolecular distance of 0.3 nm while the
size of the water molecule is 0.3 nm. The notion of collision in this case makes no
sense because water molecules are always interacting with their neighbours.

The concept of collision is therefore useful in dilute gases. When the gas is sufficiently dilute, collisions are essentially *binary collisions*. We shall only consider this case. Since this means that the average distance between gas particles is large compared to the interaction range, it may be summarized by the inequality

$$n r_w^3 \ll 1 \qquad\qquad (11.19)$$

which we assume for the rest of the chapter.

Taking care of this constraint, the equation of the distribution function for non-interacting particles can be completed to take into account the collisions. We transform (11.18) into

$$\frac{Df}{Dt} = \mathcal{C}$$

where $\mathcal{C}$ is a *collision integral*. We shall now derive the expression of this integral in the case of binary collisions.

11.4.1   The Collision Integral

To derive and expression for $\mathcal{C}$ we have to count the collisions that add or remove gas particles from the small volume of velocity space $d^3\mathbf{v}$ around $\mathbf{v}$. For this, we set

$$\mathcal{C} d^3\mathbf{x} d^3\mathbf{v} = (C^+ - C^-) d^3\mathbf{x} d^3\mathbf{v}$$

Here, we introduced the number of collisions that replenish (C^+) or deplete (C^-), the volume of phase space $d^3\mathbf{x} d^3\mathbf{v}$ around $(\mathbf{x}, \mathbf{v})$.

Let us consider the collision of two gas particles of velocity $\mathbf{v}$ and $\mathbf{w}$ respectively. Following the wording of Vincenti and Kruger (1965), we shall call these particles of *class* $\mathbf{v}$ and $\mathbf{w}$ respectively, to emphasize that we are not dealing with a specific particle. Conservation laws during collision impose

$$\begin{cases} \mathbf{v} + \mathbf{w} = \mathbf{v}' + \mathbf{w}' \\ \frac{1}{2}v^2 + \frac{1}{2}w^2 = \frac{1}{2}v'^2 + \frac{1}{2}w'^2 \end{cases} \qquad\qquad (11.20)$$

We may first note that the 6 components of the pair of vectors $(\mathbf{v}', \mathbf{w}')$ cannot be computed from the four previous equations. Two other constraints are necessary to derive the post-collision parameters: conservation of angular momentum, which determines the plane of the trajectories, and the impact parameter b, which determines the relative positions of the trajectories (see Fig. 11.6). When these quantities are given then $\mathbf{v}'$ and $\mathbf{w}'$ can be expressed as functions of $\mathbf{v}$ and $\mathbf{w}$.

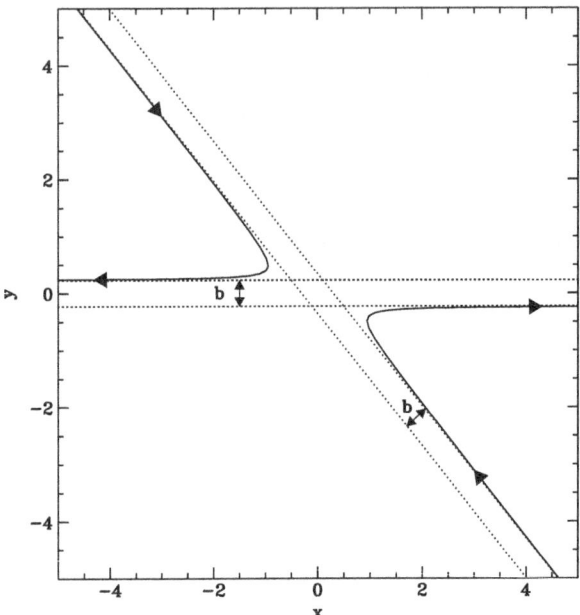

Fig. 11.6 The trajectory of two colliding particles with a potential of interaction in $1/r$. Note that even in this case of a long range potential, the part of the trajectory which is curved is small. b is the impact parameter

To derive an expression for C^+ and C^-, we need to evaluate the number of collisions. For that, we shall start with the depleting collisions and consider a particle of class $\mathbf{w}$. It is facing a beam of particles of class $\mathbf{v}$ of intensity

$$I = n_v \|\mathbf{v} - \mathbf{w}\| \ .$$

We recall that the intensity I is the number of particles crossing a unit surface per unit of time. Here n_v is the density of class-$\mathbf{v}$ particles. When the beam hits the particle of class $\mathbf{w}$, it is scattered and a fraction of the incident particles are deviated within the solid angle $d\Omega$ around the direction $\mathbf{n}$. Their number is

$$I\sigma(\mathbf{n})d\Omega$$

where $\sigma(\mathbf{n})$ is *the differential cross section* . A class-$\mathbf{w}$ particle therefore scatters

$$n_v\|\mathbf{v} - \mathbf{w}\|\sigma(\mathbf{n})d\Omega$$

class-$\mathbf{v}$ particles in the solid angle $d\Omega$. These particles thus leave class $\mathbf{v}$. Since there are $n_w d^3\mathbf{x}$ class-$\mathbf{w}$ particles in the volume $d^3\mathbf{x}$, the number of particles finally scattered per unit time is

$$n_v n_w \|\mathbf{v} - \mathbf{w}\| \sigma(\mathbf{n}) d\Omega d^3\mathbf{x}$$

Since $n_v = f(\mathbf{v})d^3\mathbf{v}$ and $n_w = f(\mathbf{w})d^3\mathbf{w}$, we find that the number of depleting collisions is

$$C^- d^3\mathbf{v} d^3\mathbf{x} = d^3\mathbf{v} d^3\mathbf{x} \int d^3\mathbf{w} \int_{4\pi} f(\mathbf{v}) f(\mathbf{w}) \|\mathbf{v} - \mathbf{w}\| \sigma(\mathbf{v}, \mathbf{w}|\mathbf{v}', \mathbf{w}') d\Omega \quad (11.21)$$

In this expression we replaced the scattering direction $\mathbf{n}$ by the quantities that governs its value, namely the velocities before and after the collision.

We need now to evaluated the C^+, which is the number of collisions that replenish the velocity range $[\mathbf{v}, \mathbf{v} + d^3\mathbf{v}]$. For that we need to consider the inverse collisions which start from $(\mathbf{v}', \mathbf{w}')$ and lead to $(\mathbf{v}, \mathbf{w})$ for all the possible $\mathbf{w}$. Following the foregoing reasoning, we consider a particle of class $\mathbf{w}'$ scattering particles of class $\mathbf{v}'$ with a beam of intensity

$$I' = n_{v'} \|\mathbf{v}' - \mathbf{w}'\| .$$

This beam impacts the $n_{w'} d^3\mathbf{x}$ particles of class $\mathbf{w}'$ located in $d^3\mathbf{x}$. The number of collisions which generate particles of class $\mathbf{v}$ and $\mathbf{w}$ is

$$n_{v'} \|\mathbf{v}' - \mathbf{w}'\| \sigma(\mathbf{v}', \mathbf{w}'|\mathbf{v}, \mathbf{w}) d\Omega n_{w'} d^3\mathbf{x}$$

where $\sigma(\mathbf{v}', \mathbf{w}'|\mathbf{v}, \mathbf{w})$ is the cross section of this type of collisions. Noting that $n_{v'} = f(\mathbf{v}')d^3\mathbf{v}'$ and $n_{w'} = f(\mathbf{w}')d^3\mathbf{w}'$, we derive the following expression of C^+:

$$C^+ d^3\mathbf{v} d^3\mathbf{x} = d^3\mathbf{v}' d^3\mathbf{x} \int d^3\mathbf{w}' \int_{4\pi} \|\mathbf{v}' - \mathbf{w}'\| \sigma(\mathbf{v}', \mathbf{w}'|\mathbf{v}, \mathbf{w}) f(\mathbf{v}') f(\mathbf{w}') d\Omega .$$

There we just integrated over $\mathbf{w}'$. We may have thought that the value of $\mathbf{v}'$ is not fixed either since the constraint is to produce a pair of particles of class $(\mathbf{v}, \mathbf{w})$. But from the relations between velocities in a collision (11.20), we see that if $\mathbf{v}$ is fixed and $\mathbf{w}$ free, then only $\mathbf{v}'$ or $\mathbf{w}'$ is free the other being fixed (up to collisions parameters like b that are taken into account in the cross section).

We shall now rewrite C^+ taking into account the invariants of binary elastic collisions. We first note that the symmetry $t/-t$ of these collisions implies

$$\sigma(\mathbf{v}', \mathbf{w}'|\mathbf{v}, \mathbf{w}) = \sigma(\mathbf{v}, \mathbf{w}|\mathbf{v}', \mathbf{w}')$$

in other words that the collision and the inverse collision have the same cross section. We can also show that (11.20) implies that the norm of the relative velocities is unchanged by a collision:

$$\|\mathbf{v} - \mathbf{w}\| = \|\mathbf{v}' - \mathbf{w}'\|$$

Lastly, from (11.17) applied to a phase space with two particles, the volume $d^3\mathbf{v}d^3\mathbf{w}$ is unchanged for a hamiltonian system. Hence

$$d^3\mathbf{v}d^3\mathbf{w} = d^3\mathbf{v}'d^3\mathbf{w}'$$

We are now in a position to give C^+ the following new expression:

$$C^+ = \int d^3\mathbf{w} \int_{4\pi} \|\mathbf{v} - \mathbf{w}\| \sigma(\mathbf{v}, \mathbf{w}|\mathbf{v}', \mathbf{w}') f(\mathbf{v}') f(\mathbf{w}') d\Omega$$

Gathering the parts of the collision integral we now write *Boltzmann equation*, which controls the evolution of the distribution function:

$$\frac{Df}{Dt} = \int d^3\mathbf{w} \int_{4\pi} \|\mathbf{v} - \mathbf{w}\| \left(f(\mathbf{v}') f(\mathbf{w}') - f(\mathbf{v}) f(\mathbf{w}) \right) \sigma(\mathbf{v}, \mathbf{w}|\mathbf{v}', \mathbf{w}') d\Omega$$

(11.22)

As may be observed this equation is a nonlinear integro-differential equation.

This equation is not universally valid and can be used when the following conditions are met:

1. The gas is dilute: binary collisions are dominant and the only ones taken into account. This is correct when the volume occupied by the particles is small compared to that occupied by the gas. In the framework of the van der Waals model, the co-volume b must be negligible. To fix ideas, let us consider air in normal conditions at 300 K and 10^5 Pa. The mean radius of a molecule of N_2 or O_2 is around 0.24 nm while the numerical density of particles is $n \sim 2.4 \times 10^{25}$ m^{-3}. The parameter measuring the dilution is therefore $n r_w^3 \sim 3 \times 10^{-4}$ which is very small indeed.
2. The effective interaction between particles is of short range so that the concept of collision is relevant.
3. There is no correlation between $\mathbf{v}$ and $\mathbf{w}$: the distribution of particles of class $\mathbf{v}$ and $\mathbf{w}$ are independent so that the number of collisions is proportional to $f(\mathbf{v}) f(\mathbf{w})$. In other words, there is no memory of collisions in the statistics.
4. The distribution function f does not vary on a scale shorter than the mean free path nor on a time less than the collision time.

11.4.2 Thermodynamic Equilibrium

Before pursuing our route to the equations of fluid dynamics, we shall briefly stop on the question of how Boltzmann equation describes thermodynamic equilibrium.

In this case the distribution function is independent of space and time variables. We may also consider the gas being free of any force field. In such conditions

$$\frac{\partial f}{\partial t} = 0, \ \nabla f = \mathbf{0}, \ \mathcal{F} = \mathbf{0} \quad \Longrightarrow \quad \frac{Df}{Dt} = 0 \quad \Longrightarrow \quad \mathcal{C} = 0$$

$$\int d^3\mathbf{w} \int_{4\pi} \|\mathbf{v} - \mathbf{w}\| \left(f(\mathbf{v}') f(\mathbf{w}') - f(\mathbf{v}) f(\mathbf{w}) \right) \sigma(\mathbf{v}, \mathbf{w}|\mathbf{v}', \mathbf{w}') d\Omega = 0$$

The result is that the collision integral is vanishing. One may then show, see the appendix at the end of the chapter, that the solution of this integral equation is unique and is the famous Maxwell–Boltzmann distribution:

$$f_0(\mathbf{v}) = n \left(\frac{m}{2\pi kT} \right)^{3/2} e^{-\frac{mv^2}{2kT}} \tag{11.23}$$

as we would have expected.

11.4.3   The Mean Free Path

Taking advantage of our knowledge of the collision integral we may have a closer look at the mean free path of atoms or molecules in a gas at equilibrium.

The concept of mean free path does really make sense if the concept of collision is relevant for the description of the interaction between the gas particles. In this case we may just focus on the model of rigid elastic spheres. There is no unique definition of the mean free path (cf. Chapman and Cowling 1970). Generally, the following definition is adopted. If n_c is the number of collisions per unit volume and unit time, then n_c/n is the number of collisions faced by a particle. n/n_c is the mean fraction of the time unit between two collisions for a gas particle. If $\bar{u}$ is the mean velocity of the particles then we may define the mean free path ℓ as

$$\ell = \frac{n\bar{u}}{n_c} \tag{11.24}$$

where $\bar{u}$ is given by (11.3). The expression of n_c is given by the summation of C^- (or C^+ since it is identical to C^- at equilibrium) over all the possible values of $\mathbf{v}$. Hence, we get

$$n_c = \int d^3\mathbf{v} \int d^3\mathbf{w} \int_{4\pi} f_0(\mathbf{v}) f_0(\mathbf{w}) \|\mathbf{v} - \mathbf{w}\| \sigma(\mathbf{v}, \mathbf{w}|\mathbf{v}', \mathbf{w}') d\Omega \tag{11.25}$$

where f_0 is the equilibrium distribution, namely that of Maxwell–Boltzmann. Integration over $d\Omega$ is straightforward because the integrant is independent of the angle between $\mathbf{v}$ and $\mathbf{w}$. The integral gives the cross section of rigid spheres that is

$$\int_{4\pi} \sigma(\mathbf{v}, \mathbf{w}|\mathbf{v}', \mathbf{w}')d\Omega = \pi d^2$$

where d is the diameter of the spheres (let us recall that the centres of the spheres cannot be closer than a diameter of the sphere, see Fig. 11.2).

Using Maxwell–Boltzmann distribution we get

$$n_c = \pi d^2 n^2 \left(\frac{m}{2\pi kT}\right)^3 \int d^3\mathbf{v}\, d^3\mathbf{w}\, \|\mathbf{v} - \mathbf{w}\|\, e^{-\frac{m}{2kT}(v^2 + w^2)}$$

However, because of the term $\|\mathbf{v} - \mathbf{w}\|$, we have to change variables in order to derive the expression of the integral. We set

$$\mathbf{g} = \mathbf{v} - \mathbf{w} \qquad \text{and} \qquad \mathbf{G} = \frac{\mathbf{v} + \mathbf{w}}{2}$$

where we introduced $\mathbf{G}$ the velocity of the centre of mass of the colliding particles. One may verify that

$$\mathbf{v} = \mathbf{G} + \frac{\mathbf{g}}{2} \qquad \text{and} \qquad \mathbf{w} = \mathbf{G} - \frac{\mathbf{g}}{2}$$

The jacobian of this change of variables is unity. Indeed, changing variables

$$dv_x dw_x = |\det[J]| dg_x dG_x$$

where the jacobian matrix is

$$[J] = \begin{pmatrix} \frac{\partial G_x}{\partial v_x} & \frac{\partial g_x}{\partial v_x} \\ \frac{\partial G_x}{\partial w_x} & \frac{\partial g_x}{\partial w_x} \end{pmatrix} = \begin{pmatrix} \frac{1}{2} & 1 \\ \frac{1}{2} & -1 \end{pmatrix}$$

from which we find $|\det[J]| = 1$. Consequently $d^3\mathbf{v}\, d^3\mathbf{w} = d^3\mathbf{g}\, d^3\mathbf{G}$ since this is the same for each component. In addition we have:

$$v^2 + w^2 = 2G^2 + \frac{g^2}{2}$$

Hence we get

$$n_c = \pi d^2 n^2 \left(\frac{m}{2\pi kT}\right)^3 \int_0^\infty 4\pi g^3 e^{-\frac{mg^2}{4kT}} dg \int_0^\infty 4\pi G^2 e^{-\frac{mG^2}{kT}} dG$$

Using the expressions of integrals of the products of a polynome and a Gaussian (cf. section "Gaussian Integrals" in appendix of Chap. 12), we obtain the number of collisions within a gas during a unit of time:

$$n_c = 4n^2 d^2 \sqrt{\frac{\pi kT}{m}} \qquad (11.26)$$

The expression of the mean free path follows from the use of the mean velocity (11.3). Hence,

$$\ell = \frac{1}{\sqrt{2}\pi d^2 n} \qquad (11.27)$$

11.5   Equations of Fluid Flow as Mean-Field Equations

Thanks to the foregoing section we have now the appropriate tools to investigate the expressions of microscopic transport like viscosity or heat diffusivity. However, the path is not short and a few steps are still in order before retrieving Navier–Stokes equation.

11.5.1   Mean Quantities

The distribution function that we introduced allows us to know the statistical distribution of velocities of gas particles inside a fluid element of volume $d^3\mathbf{x}$. Macroscopic quantities are then averages over the velocities. Thus

$$\rho(\mathbf{x}, t) = \int mf(\mathbf{x}, t, \mathbf{v}) d^3\mathbf{v}$$

where m is the mass of the gas particles. Indeed, the integral

$$\int f(\mathbf{x}, t, \mathbf{v}) d^3\mathbf{v} \equiv n(\mathbf{x}, t)$$

is just the numerical density n of particles.[5]

---

[5]In some textbooks like Vincenti and Kruger (1965), a normalized distribution function is used; it is such that $\int f(\mathbf{x}, t, \mathbf{v}) d^3\mathbf{v} = 1$ which is more practical in some cases.

The fluid velocity is the mean velocity of gas particles namely

$$\mathbf{V} = \langle \mathbf{v} \rangle = \frac{\int f \mathbf{v} d^3 \mathbf{v}}{\int f d^3 \mathbf{v}} \tag{11.28}$$

or

$$n(\mathbf{x}, t) \mathbf{V}(\mathbf{x}, t) = \int f(\mathbf{x}, t, \mathbf{v}) \mathbf{v} d^3 \mathbf{v}$$

equivalent to the expression of momentum per unit volume:

$$\rho \mathbf{V} = \int m \mathbf{v} f d^3 \mathbf{v}$$

We shall also need the expression of internal heat. In the frame of the dilute gas model, internal heat is just the kinetic energy of particles in the local frame moving with the fluid. Hence

$$n(\mathbf{x}, t) e(\mathbf{x}, t) = \int \frac{1}{2} (\mathbf{v} - \mathbf{V}(\mathbf{x}, t))^2 f(\mathbf{x}, t, \mathbf{v}) d^3 \mathbf{v} \tag{11.29}$$

11.5.2   Equation for a Quantity Conserved by Collisions

Quantities conserved during collisions are highly interesting. These are mass, momentum and kinetic energy also called *collisional invariants*. The first moments[6] of Boltzmann equation gives the evolution of their mean values, which is precisely what we are looking for. In addition, the conservation of mass, momentum and kinetic energy implies that the first moments (of order 0, 1 and 2) of the collision integral are zero, namely

$$\int \mathbf{v} \mathcal{C}(\mathbf{x}, t, \mathbf{v}) d^3 \mathbf{v} = \mathbf{0}$$

and obviously

$$\int m \mathcal{C}(\mathbf{x}, t, \mathbf{v}) d^3 \mathbf{v} = \int \frac{1}{2} v^2 \mathcal{C}(\mathbf{x}, t, \mathbf{v}) d^3 \mathbf{v} = 0$$

We shall not prove these mathematical equalities, which we consider as physically obvious (but see Vincenti and Kruger 1965, for a demonstration).

---

[6]The definition of the order n moment of a statistical distribution is given in Chap. 9 (9.3).

11.5.2.1 Mass Conservation

Mass conservation follows from the zero-order moment, with respect to velocity, of Boltzmann equation:

$$\int (\mathbf{v})^0 \frac{Df}{Dt} d^3\mathbf{v} = \int (\mathbf{v})^0 C d^3\mathbf{v}$$

where obviously $(\mathbf{v})^0 = 1$. From the properties of the collision integral, we therefore have

$$\int \frac{Df}{Dt} d^3\mathbf{v} = 0$$

Making explicit the operator D/Dt we obtain:

$$\frac{\partial}{\partial t} \int f d^3\mathbf{v} + \int \mathbf{v} \cdot \nabla_{\mathbf{x}} f d^3\mathbf{v} + \int \frac{\mathcal{F}}{m} \cdot \nabla_{\mathbf{v}} f d^3\mathbf{v} = 0$$

We note that $\mathbf{v} \cdot \nabla_{\mathbf{x}} f = \nabla_{\mathbf{x}} \cdot (f\mathbf{v})$ because $\mathbf{x}$ and $\mathbf{v}$ are independent variables. If we assume that the force field $\mathcal{F}$ felt by the particles is independent of $\mathbf{v}$ then

$$\int \mathcal{F} \cdot \nabla_{\mathbf{v}} f d^3\mathbf{v} = \int \nabla_{\mathbf{v}} \cdot (f\mathcal{F}) d^3\mathbf{v} = \int_{(S)} f\mathcal{F} \cdot d\mathbf{S}_\infty = 0$$

because the distribution function is such that $\lim_{v \to \infty} f(v) = 0$. Finally, we get:

$$\frac{\partial n}{\partial t} + \nabla \cdot (n\mathbf{V}) = 0 \tag{11.30}$$

If we multiply this equation by the mass of the gas particles, we retrieve the usual equation of mass conservation:

$$\frac{\partial \rho}{\partial t} + \nabla \cdot (\rho \mathbf{V}) = 0$$

11.5.3 Equation for Momentum

The next moment with respect to velocity of Boltzmann equation gives the equation of the mean momentum of particles and therefore leads to the equation of momentum for fluid particles. For the i-th component of the velocity we have

$$\int v_i \frac{Df}{Dt} d^3\mathbf{v} = 0$$

$$\frac{\partial}{\partial t} \int v_i f d^3\mathbf{v} + \int v_i \mathbf{v} \cdot \nabla_{\mathbf{x}} f d^3\mathbf{v} + \int v_i \frac{\mathcal{F}}{m} \cdot \nabla_{\mathbf{v}} f d^3\mathbf{v} = 0$$

Observing that v_i is an independent variable, just like t and $\mathbf{x}$, we find that the first term may be written as

$$\frac{\partial}{\partial t} \int v_i f d^3 \mathbf{v} = \frac{\partial}{\partial t} n \langle v_i \rangle = \partial_t (n V_i)$$

while the next term also reads

$$\int v_i v_j \partial_j f d^3 \mathbf{v} = \partial_j (n \langle v_i v_j \rangle)$$

As before we assume that the force field bathing the gas particles does not depend on their velocity. We can then rewrite the last term as:

$$\frac{1}{m} \int [\nabla_v \cdot (v_i f \mathcal{F}) - f \mathcal{F} \cdot \nabla_v v_i] d^3 \mathbf{v}$$

Here too the integral of the divergence vanishes because the distribution function decreases fast enough at infinity. The remaining term may be worked out as follows:

$$\frac{1}{m} \int \left[-f \mathcal{F}_j \frac{\partial}{\partial v_j} v_i \right] d^3 \mathbf{v} = \frac{1}{m} \int \left[-f \mathcal{F}_j \delta_{ji} \right] d^3 \mathbf{v} = -\frac{1}{m} n \langle \mathcal{F}_i \rangle$$

The mean field of the force applied to the particles appears. The equation of mean momentum finally reads:

$$\partial_t (n V_i) + \partial_i (n \langle v_i v_j \rangle) = n \langle \mathcal{F}_i \rangle / m \tag{11.31}$$

We observe that $n \langle \mathcal{F}_i \rangle$ is just the force per unit volume that we called $\mathbf{f}$ in Chap. 1 (1.22).

In order to retrieve the equation of momentum of Fluid Mechanics we have to introduce the fluid velocity, that is the mean velocity of the gas particles. We therefore set

$$\mathbf{v} = \langle \mathbf{v} \rangle + \mathbf{u} = \mathbf{V} + \mathbf{u}$$

where $\mathbf{u}$ is the velocity of the particles in the frame associated with the fluid particle. We deduce that

$$\langle v_i v_j \rangle = V_i V_j + \langle u_i u_j \rangle$$

which allows us to transform (11.31) into

$$\partial_t (\rho V_i) + \partial_i (\rho V_i V_j) = -\partial_j (\rho \langle u_i u_j \rangle) + f_i$$

or, using mass conservation,

$$\rho \frac{DV_i}{Dt} = \partial_j \sigma_{ij} + f_i \tag{11.32}$$

We give to this equation the same form as that of (1.25) which allows us to find the expression of the stress tensor:

$$\sigma_{ij} = -\rho \langle u_i u_j \rangle \tag{11.33}$$

The equation of momentum generates a new quantity: $\langle u_i u_j \rangle$, which is the one-point correlation of the particle velocity. We'll have to express this quantity as a function of other macroscopic quantities, namely derive the rheological law of the gas. However, unlike with the pure macroscopic approach of Chap. 1, we shall be able to express the microscopic transport (here viscosity) as a function of the parameters of the model. Before that we need to write down the equation associated with kinetic energy.

11.5.4   Kinetic Energy

Just like m and $\mathbf{v}$, the second moment of Boltzmann equation has no contribution from the collision term, thus

$$\int \frac{1}{2} v^2 \left[\frac{\partial f}{\partial t} + \mathbf{v} \cdot \nabla_x f + \frac{\mathcal{F}}{m} \cdot \nabla_v f \right] d^3\mathbf{v} = 0$$

which can be written as

$$\frac{\partial}{\partial t}\left(\frac{n}{2}\langle v^2 \rangle\right) + \nabla_x \cdot \left(\frac{n}{2}\langle v^2 \mathbf{v} \rangle\right) + \frac{1}{2}\int \left[\nabla_v \cdot \left(\frac{f v^2 \mathcal{F}}{m}\right) - \frac{f\mathcal{F}}{m}\cdot\nabla_v v^2\right] d^3\mathbf{v} = 0$$

and simplified into

$$\frac{\partial}{\partial t}\left(\frac{n}{2}\langle v^2 \rangle\right) + \nabla_x \cdot \left(\frac{n}{2}\langle v^2 \mathbf{v} \rangle\right) - \frac{n\mathcal{F}}{m}\cdot \mathbf{V} = 0$$

or else

$$\frac{\partial}{\partial t}\left(\frac{\rho}{2}\langle v^2 \rangle\right) + \nabla_x \cdot \left(\frac{\rho}{2}\langle v^2 \mathbf{v} \rangle\right) = \mathbf{V}\cdot\mathbf{f} \tag{11.34}$$

We may give a more familiar shape to this equation by introducing the vector

$$F_i = \frac{1}{2}\rho \langle u^2 u_i \rangle$$

and by remembering that specific internal energy reads (see also 11.29)

$$e = \frac{1}{2} \langle u^2 \rangle .$$

We first observe that

$$\langle v^2 v_i \rangle = V^2 V_i + 2 \langle u_i u_j \rangle V_j + \langle u^2 \rangle V_i + \langle u^2 u_i \rangle$$

and that

$$\frac{1}{2} \rho \langle v^2 v_i \rangle = \left(\frac{1}{2} V^2 + e \right) \rho V_i - \sigma_{ij} V_j + F_i$$

Using mass conservation we can transform (11.34) into

$$\rho \frac{D}{Dt} \left(\frac{1}{2} v^2 + e \right) = -\nabla \cdot \mathbf{F} + \partial_i (\sigma_{ij} V_j) + \mathbf{f} \cdot \mathbf{v}$$

which is identical to (1.23). Thus doing we derive the formal expression of the heat flux of microscopical origin, namely

$$\mathbf{F} = \frac{1}{2} \rho \langle u^2 \mathbf{u} \rangle \tag{11.35}$$

It arises as a triple correlation of the particle velocities or as the microscopic flux of kinetic energy.

11.6 Continuous Media, Perfect Fluids and Ideal Gases

At this stage, we may contemplate how the continuous medium arose: it is an average over the velocity space of some moments of the distribution function. Hence, although the continuous medium is unveiling now through mean quantities, the real mathematical step was accomplished when we admitted the existence of the distribution function. In fact we admitted that the local statistical properties of the particles making the fluid vary continuously and thus define a continuum that is supposed to reproduce all the properties of the fluid.

Before proceeding, we shall first examine the case of thermodynamic equilibrium. Thus, we first consider the case where the distribution function f is the maxwellian one f_0 as given by (11.23).

In this case the heat flux is exactly zero because $\langle u^2 u_i \rangle = 0$ as a consequence of (12.36). Besides, and for the same reason, all the off-diagonal components of $\langle u_i u_j \rangle$ are zero. We may also check that $\langle u_x^2 \rangle = \langle u_y^2 \rangle = \langle u_z^2 \rangle$, meaning that the three directions of space are equivalent, betraying the isotropy of the fluid. This property implies

$$\langle u_x^2 \rangle = \langle u_y^2 \rangle = \langle u_z^2 \rangle = \frac{1}{3} \mathrm{Tr} \langle u_i u_j \rangle = \frac{\langle u^2 \rangle}{3}$$

Finally, at equilibrium

$$\mathbf{F} = \mathbf{0} \quad \text{and} \quad \sigma_{ij} = -\rho \frac{\langle u^2 \rangle}{3} \delta_{ij}$$

Identifying the terms of this expression with the macroscopic form of the stress at equilibrium, namely $\sigma_{ij} = -p\delta_{ij}$, we deduce the expression of pressure

$$p = \rho \frac{\langle u^2 \rangle}{3} \tag{11.36}$$

If we use the Maxwell–Boltzmann distribution function, which characterizes thermodynamic equilibrium, from (11.4) we get

$$p = \rho \frac{kT}{m}$$

which is just the equation of state of ideal gases.

As may be expected, strict thermodynamic equilibrium implies no microscopic transport. To find viscosity or heat conductivity, we thus need to go further and consider non-equilibrium situations.

11.7 Gas Dynamics in a Newtonian Regime

11.7.1 Towards Navier–Stokes

While using a macroscopic description of matter (see Chap. 1), we learnt that transport coefficients measure the way the system is modified when external circumstances bring it out of equilibrium. To derive these coefficients we therefore only need to determine how the distribution function is modified when a small perturbation arises. We therefore write

$$f = f_0 + \delta f$$

where f_0 is the Maxwellian distribution. As an example, let us consider the heat flux. We have

$$F_i = \frac{1}{2} \rho \int u^2 u_i f d^3 \mathbf{v} = \frac{1}{2} \rho \int u^2 u_i \delta f d^3 \mathbf{v} \tag{11.37}$$

since the contribution of f_0 is zero. We therefore need to evaluate the deviation of f with respect to equilibrium.

11.7.2   The BGK54 Model and the Theory of Chapman–Enskog

In order to get acquainted with the way of deriving the equation of the continuous medium from that verified by the distribution function we need not deal with the full Boltzmann equation, which is very complicated as we already underlined. For a first pedagogical approach we propose to consider a simplified version of Boltzmann equation, namely the equation of Bhatnagar et al. (1954), where the collision integral is abruptly replaced by a damping term:

$$\frac{\partial f}{\partial t} + \mathbf{v} \cdot \nabla f + \frac{\mathcal{F}}{m} \cdot \nabla_v f = -\frac{f - f_0}{\tau} \qquad (11.38)$$

Here τ is the relaxation time, which is a model parameter. This modeling of the collision integral respects the vanishing moments of the integral due to collisional invariants (m, $\mathbf{v}$ and v^2). This model gives the same equations for the fluid but of course transport coefficients will be different.[7]

If we leave aside the force field $\mathcal{F}$, which is useless for the derivation of diffusion coefficients, and if we note that the distribution function varies on macroscopic scales L, we can write (11.38) like

$$\varepsilon \left(\frac{\partial f}{\partial \tilde{t}} + \tilde{v} \cdot \tilde{\nabla} f \right) = -f + f_0 \qquad (11.39)$$

with

$$\varepsilon = \frac{\tau V}{L}$$

where V is a typical fluid velocity. Tilded quantities are dimensionless. If the gas is weakly out of equilibrium, the relaxation time is that of the equilibrium configuration, namely of the order of a few collision time which is very short compared to the macroscopic advection time L/V so that $\varepsilon \ll 1$. One may expand the solution of (11.39) in powers of this small parameter, i.e.

$$f = f_0 + \varepsilon f_1 + \varepsilon^2 f_2 + \cdots$$

---

[7]The BGK model was proposed by Bhatnagar et al. (1954) to explore in a simplified way flows where the continuum approximation is no longer relevant, for instance in the case of very diluted gases where the Knudsen number (cf. 1.1) is no longer small compared to unity.

The expansion is usually referred to as the *Chapman–Enskog expansion* after the pioneering work of Sydney Chapman (1888–1970) and David Enskog (1884–1947) who derived independently (in 1916 and 1917 respectively) and for the first time the correct expressions of the transport coefficients from Boltzmann equation.

We observe that the first term f_1 describing the first order perturbation to equilibrium reads

$$- f_1 = \frac{\partial f_0}{\partial \tilde{t}} + \tilde{\mathbf{v}} \cdot \tilde{\nabla} f_0 = \frac{\partial f_0}{\partial \tilde{t}} + \tilde{v}_i \tilde{\partial}_i f_0 \tag{11.40}$$

If we go back to dimensional quantities, we note that to first order of perturbations we simply have

$$\delta f = -\tau \left(\frac{\partial f_0}{\partial t} + v_i \partial_i f_0 \right) \tag{11.41}$$

where f_0 is the local maxwellian distribution given by:

$$f_0(\mathbf{x}, \mathbf{v}, t) = n(\mathbf{x}, t) \left(\frac{m}{2\pi k T(\mathbf{x}, t)} \right)^{3/2} \exp \left[- \left(\frac{m(\mathbf{v} - \mathbf{V}(\mathbf{x}, t))^2}{2k T(\mathbf{x}, t)} \right) \right] \tag{11.42}$$

In this expression, we explicitly wrote the dependence of f_0 with respect to time and space coordinates. We observe that this dependence comes from that of the density, temperature and mean velocity of the fluid.

With (11.41) and (11.42), we can evaluate the correlations appearing in (11.37) and (11.33). For instance, the first term is

$$\frac{\partial f_0}{\partial t} = \frac{\partial f_0}{\partial n} \frac{\partial n}{\partial t} + \frac{\partial f_0}{\partial T} \frac{\partial T}{\partial t} + \frac{\partial f_0}{\partial V_i} \frac{\partial V_i}{\partial t}$$

with

$$\frac{\partial f_0}{\partial n} = \frac{f_0}{n}$$

$$\frac{\partial f_0}{\partial T} = \frac{f_0}{T} Q_{3/2}(u^2) \qquad \text{and} \qquad Q_{3/2}(u^2) = \frac{mu^2}{2kT} - \frac{3}{2}$$

$$\frac{\partial f_0}{\partial V_i} = f_0 \frac{mu_i}{kT}$$

Repeating this calculation with the other terms, we find

$$\frac{\partial f_0}{\partial t} + \mathbf{v} \cdot \nabla f_0 = f_0 \left[\frac{\partial \ln n}{\partial t} + v_k \partial_k \ln n + Q_{3/2}(u^2) \left(\frac{\partial T}{\partial t} + v_k \partial_k T \right) \right.$$
$$\left. + \frac{mu_i}{kT} \left(\frac{\partial V_i}{\partial t} + v_k \partial_k V_i \right) \right]$$

We again separate mean values and fluctuations for the velocity $v_i = V_i + u_i$, and get

$$\frac{\partial f_0}{\partial t} + \mathbf{v} \cdot \nabla f_0 = f_0 \left[\frac{D \ln \rho}{Dt} + Q_{3/2}(u^2) \frac{D \ln T}{Dt} + \frac{m u_i}{k \rho T} \partial_k \sigma_{ik} \right.$$
$$\left. + u_k \partial_k \ln \rho + Q_{3/2}(u^2) u_k \partial_{kT} + \frac{m u_i u_k}{kT} \partial_k V_i \right]$$

where we used the equation of mean momentum (11.32). Now, observing that $\sigma_{ik} = -p \delta_{ik}$ because viscous terms are of higher order in ε, and that $\ln p = \ln \rho + \ln T +$ Cst, we obtain

$$\frac{\partial f_0}{\partial t} + \mathbf{v} \cdot \nabla f_0 = f_0 \left[\frac{D \ln \rho}{Dt} + Q_{3/2}(u^2) \frac{D \ln T}{Dt} + Q_{5/2}(u^2) u_k \partial_k \ln T + \frac{m u_i u_k}{kT} \partial_k V_i \right]$$

where $Q_{5/2} = Q_{3/2} - 1$. Using the energy equation at zeroth order (without heat diffusion) and mass conservation, we may still simplify the foregoing expression into

$$\delta f = -\tau f_0 \left[Q_{5/2}(u^2) u_k \partial_k \ln T + \frac{m}{kT} \left(u_i u_k - \frac{u^2}{3} \delta_{ik} \right) \partial_k V_i \right] \qquad (11.43)$$

We have now the expression for the disturbance of the distribution function at hands. We can now focus on the calculation of the various moments that lead to viscosity and heat diffusion coefficients. Let us start with the heat flux.

11.7.3 Expression of the Heat Flux and of Thermal Conductivity

When evaluating $\langle u^2 u_i \rangle$, we observe that only terms with odd powers in (11.43) contribute to this average so that only the first term of (11.43) need to be considered. We get

$$F_i = -\frac{m\tau}{2T} \left[\int u^2 Q_{5/2}(u^2) u_i u_k f_0 d^3\mathbf{u} \right] \frac{\partial T}{\partial x_k}$$

The thermal conductivity tensor thus reads

$$\chi_{ik} = \frac{m\tau}{2T} \int u^2 Q_{5/2}(u^2) u_i u_k f_0 d^3\mathbf{u}$$

However, only diagonal terms are non-zero. In addition they are all the same. We can therefore write:

$$\chi_{ik} = \chi \delta_{ik} \quad \text{and} \quad \chi = \frac{1}{3}\text{Tr}[\chi] = \frac{\chi_{kk}}{3} = \frac{m\tau}{6T}\int u^4 Q_{5/2}(u^2) f_0 d^3\mathbf{u}$$

Using formulae on Gauss integrals (12.33 and following), we finally get the expression of thermal conductivity:

$$\chi = \frac{5}{2}\frac{n\tau k^2 T}{m} \tag{11.44}$$

or, dividing by ρc_p, the expression of thermal diffusivity:

$$\kappa = \frac{\tau kT}{m} \tag{11.45}$$

As expected, thermal conductivity is $\mathcal{O}(\varepsilon)$ in the Chapman–Enskog expansion.

11.7.4   Viscosity

To derive the expression of viscosity we start from σ_{ij}

$$\sigma_{ij} = -\rho \langle u_i u_j \rangle = -m \int u_i u_j (f_0 + \delta f) d^3\mathbf{u}$$

We leave aside the zero order term which, as shown before, gives the pressure, and focus on the next order term. Here, the second term of (11.43) is the only one that contributes and we can write:

$$\sigma_{ij}^{\text{visc}} = \frac{m^2 \tau}{kT}\int u_i u_j \left(u_k u_l - \frac{1}{3}u^2 \delta_{kl}\right) f_0 d^3\mathbf{u}\, \partial_l V_k$$

or else

$$\sigma_{ij}^{\text{visc}} = L_{ijkl}\partial_l V_k$$

where we introduced the fourth order tensor

$$L_{ijkl} = \frac{m^2 \tau}{kT}\int u_i u_j \left(u_k u_l - \frac{1}{3}u^2 \delta_{kl}\right) f_0 d^3\mathbf{u}$$

which already appeared in Chap. 1 when we were looking for rheological laws (see 1.36). Now we have its explicit expression! From the shape of this tensor we note that

$$L_{ijkk} = 0 \qquad\qquad (11.46)$$

by contraction on the last two indices. We easily check that the general form is indeed that given in Chap. 1 by (1.38), namely

$$L_{ijkl} = \mu(\delta_{ik}\delta_{jl} + \delta_{jk}\delta_{il}) + \lambda\delta_{ij}\delta_{kl}$$

because non-zero terms must have a pair of identical indices: F_{1212}, F_{1122} for example are non-zero while F_{1112} or F_{1231} are zero because at least one index cannot be paired.

Considering the component L_{ijij} with $i \neq j$, we find

$$\mu = L_{ijij} = \frac{m^2\tau}{kT} \int u_i^2 u_j^2 f_0 d^3\mathbf{u} = \frac{m^2\tau}{kT} n \left(\frac{kT}{m}\right)^2$$

from which we derive the dynamic and kinematic viscosities:

$$\mu = \tau n kT \qquad \text{and} \qquad \nu = \frac{\tau kT}{m}$$

We may also note that the second viscosity ζ is zero. Indeed, (11.46) implies that

$$L_{ijkk} = 2\mu\delta_{ik}\delta_{jk} + 3\lambda\delta_{ij} = 0$$

setting for instance $i = j = 1$ we deduce that

$$2\mu + 3\lambda = 0$$

since $\zeta = \lambda + 2\mu/3$, we conclude that $\zeta = 0$. This result is general for a gas that has only translational degrees of freedom. Second viscosity originates in the relaxation time needed to transfer energy between translational and rotational degrees of freedom in diatomic gases (see Chapman and Cowling 1970, for instance).

Lastly, we note that with this model the Prandtl number of the gas is unity:

$$\mathcal{P} = \frac{\nu}{\kappa} = \frac{\tau kT}{m} \frac{m}{\tau kT} = 1$$

11.7.4.1 Conclusions

The foregoing results may look rather deceptive: all our efforts have lead to a Prandtl number which is unity and to diffusion coefficient that are parametrized by an unknown quantity! However, as we mentioned it before the BGK model was not designed to give accurate values of diffusion coefficients but to study in a simpler manner than with the Boltzmann equation, the flows that cannot be investigated with the Navier–Stokes equation.

But the foregoing calculations are still interesting as they show us the way of deriving the viscosity or the thermal conductivity of gases. Let us consider Boltzmann equation and use the previous Chapman–Enskog expansion. We write it in the following way

$$f = f_0(1 + \delta f + \cdots)$$

This new form of the disturbance δf of the distribution function is a solution of

$$\int d^3 \mathbf{w} \int_{4\pi} \left[\delta f(\mathbf{v}') + \delta f(\mathbf{w}') - \delta f(\mathbf{v}) - \delta f(\mathbf{w}) \right]$$

$$\times f_0(\mathbf{v}) f_0(\mathbf{w}) \|\mathbf{v} - \mathbf{w}\| \sigma(\mathbf{v}, \mathbf{w}|\mathbf{v}', \mathbf{w}') d\Omega = \frac{D f_0}{Dt}$$

to first order. This equation replaces (11.41): δf is now the solution of an integral equation where the right-hand side is already known and given by (11.43) (up to the sign).

Solving the integral equation would need another few pages of analysis. The usual way is to expand the solution on the basis of the so-called Sonine polynomials. The interested reader may consult the textbooks given in reference at the end of the chapter. Here we shall be satisfied with the result given by Vincenti and Kruger (1965):

$$\mu_1 = \frac{5\sqrt{\pi m k T}}{8 \left(\frac{m}{4kT}\right)^4 \int_0^\infty g^7 \sigma_\mu(g) e^{-mg^2/4kT} dg} \quad \text{and} \quad \chi_1 = \frac{15}{4} \frac{k}{m} \mu_1 \qquad (11.47)$$

The index 1 appearing in the expression of the diffusion coefficients indicates that these expressions have been derived taking into account the first term of the polynomial expansion. The second term only gives a very small contribution. We set

$$\sigma_\mu = \int_{4\pi} \sigma(\Omega) \sin^2(2\theta) \, d\Omega$$

σ is the differential cross section of binary interactions. For rigid elastic spheres $\sigma(\Omega) = d^2 \cos\theta$ and $\sigma_\mu = 2\pi d^2/3$ (see exercises).

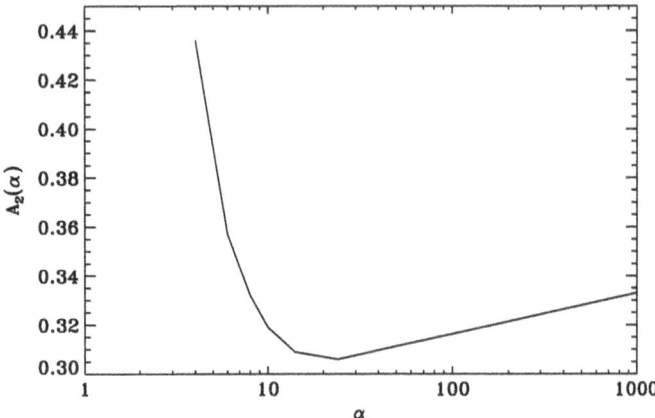

Fig. 11.7 Function $A_2(\alpha)$ sampled over the interval $[4, +\infty[$ after Chapman and Cowling (1970)

For a monatomic gas $c_p = 5k/2m$ so that the Prandtl number is:

$$\mathcal{P} = \frac{\mu c_p}{\chi} = \frac{2}{3}$$

which is now very close to the observed values. If we specify the interaction potential between gas particles, we may give a more specific expression of the dynamic viscosity at this level of approximation, namely:

$$\mu_1 = \frac{5}{16d^2}\sqrt{\frac{mkT}{\pi}} \qquad (11.48)$$

for elastic rigid spheres of diameter d, and

$$\mu_1 = \frac{5\sqrt{mkT/\pi}\,(2kT/\alpha a)^{2/\alpha}}{8A_2(\alpha)\Gamma(4-2/\alpha)} \qquad (11.49)$$

for particles interacting through a power law potential:

$$V(r) = \frac{a}{r^\alpha}$$

$\Gamma(x)$ is the gamma function and A_2 is a tabulated function that is plotted in Fig. 11.7.

11.7.5 Comparison with Experiments

To fully appreciate the foregoing results, it is useful to go back to experiments.

First we may be pleased that theory gives a value of the Prandtl number that is close to the observed values (see Table 11.1). The theoretical approach also says that kinetic energy is slightly more efficiently transported than momentum. As shown by Table 11.1 this effect is rather independent of the nature of the gas and related to statistics.

Expression (11.48) now gives a precise formula (the neglected terms in the expansion contribute only to a few percents) for the viscosity of a gas represented by elastic rigid spheres. Comparing this expression to the approximate one (11.10), we have

$$\mu_1 = \frac{5}{16d^2}\sqrt{\frac{mkT}{\pi}} = \frac{5\pi}{32}\left(\frac{2}{\pi^{3/2}d^2}\sqrt{mkT}\right)$$

We deduce that the coefficient β_v of the qualitative model reads:

$$\beta_v = \frac{5\pi}{32} \simeq 0.491$$

Besides, the elastic rigid sphere model predicts a $\sqrt{T}$ law with respect to temperature. In Fig. 11.8, we plotted the measured variations of the dynamic viscosity of helium between 15 K and $\sim$1,000 K. We note that these variations are rather well represented by a power law $\mu \propto T^s$ with $s \simeq 0.647$. Viscosity increases faster than $\sqrt{T}$ when temperature increases. Using the rigid sphere model, we may interpret this result by saying that it shows a decrease of the effective radii of helium atoms with increasing temperature. This is not surprising: at higher temperatures collisions are harder and the minimum distance between atoms is lower, implying a

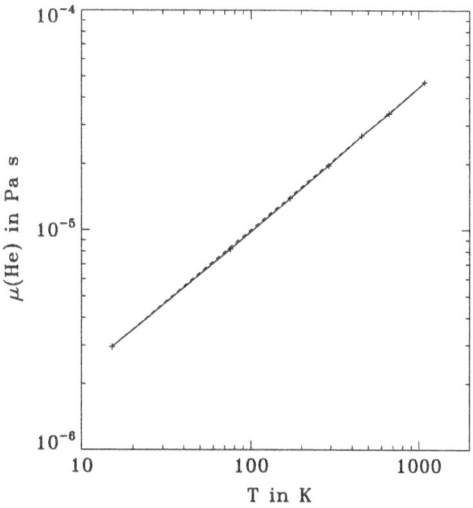

Fig. 11.8 Variations of helium viscosity with temperature (*solid line* and "+"). the *dashed line* shows the power law $T^{0.647}$. The discrepancy between this power law and experimental data is less than 3 % (data are from Chapman and Cowling 1970)

smaller effective radius for a rigid sphere. The dependence of viscosity with $1/d^2$ leads to a faster growth (compared to $\sqrt{T}$) with T.

Let us now use a more realistic potential of interaction such as the following one:

$$V(r) = \frac{a}{r^\alpha}$$

where a is positive (repulsive potential). Equation (11.49) shows that this type of potential implies a power law dependence of viscosity with an exponent related to that of the potential by

$$s = \frac{1}{2} + \frac{2}{\alpha}$$

If we use the experimental value of s for helium, we get $\alpha \simeq 13.6$. Back to (11.49), we can also deduce the value of a of the potential.

Even if the foregoing results do not predict a value of viscosity without adjustable parameters, they are still very interesting: we see that a power law potential of interaction may represent to a few percents the variations of viscosity with temperature and as a result gives us a first approximation of the effective potential of interaction between two colliding helium atoms. Deriving such a potential from first principles, namely from quantum mechanics, is difficult. We also see that through the temperature variations of the diffusion coefficients we may glimpse at some parameters of the microscopic world.

11.8   Conclusions

In this chapter we have seen how the diffusion coefficients such as viscosity or thermal conductivity emerge from the microscopic characteristics of a gas and especially from the interaction between its atoms or molecules.

We discovered how the combination of experimental results on temperature variations of viscosity with those of the theory can be used to derive some information on the potential of interaction between gas particles. Thus, looking for the statistical foundations of Fluid Mechanics, we immediately face the complexity of the microscopic world: as soon as we look at the temperature variations of viscosity, the rigid elastic spheres model appears too simple.

However, the foregoing approach of fluid dynamics using distribution functions is not restricted to fluids in a state slightly off the thermodynamic equilibrium. As the reader may have guessed when we introduced the BGK model, the knowledge of the distribution function permits the exploration of flows where fluid particles are far away from the equilibrium. One such regime is that of Knudsen where the mean free path is not small compared to the proper scales of the flow.[8] The dynamics of rarefied gases is the source of numerous problems: as we noticed, Boltzmann equation has its own limitation, boundary conditions that are met by such flows are not obvious, etc. These questions still motivate many investigations, both on the experimental/observational and theoretical sides.

11.9 Exercises

1. If I is the intensity of a particle beam (number of particles crossing a unit surface per unit time), $I\sigma(\Omega)d\Omega$ is the number of particles scattered in the solid angle $d\Omega$ by a target of differential cross section $\sigma(\Omega)$. If the target is a ball of diameter d hit by balls of the same diameter, show that the cross section of a ball is πd^2.
2. Give the cross section associated with momentum transfer $\sigma_\mu = \int \sigma(\Omega) \sin^2(2\theta)d\Omega$ where θ is defined in Fig. 11.2.
3. *Eucken relation.* Assuming that viscosity and thermal conductivity are of the following form:

$$\mu = \frac{1}{2}\rho v \ell \quad \text{and} \quad \chi = \frac{5}{4}\rho c_v v \ell$$

a) Show that the relation between μ, χ and c_v is

$$\chi = \frac{5}{2}c_v \mu \tag{11.50}$$

b) Express the Prandtl number as a function of the adiabatic index $\gamma = c_p/c_v$.
c) We now separate the heat capacity due to translational degrees of freedom from those due to other degrees of freedom, namely

$$c_v = c_v^{\text{trans}} + c_v^{\text{int}}$$

---

[8] A classical example of such flows is the one surrounding a spacecraft entering the atmosphere. At a height of 120 km the mean free path of air molecules is of the order of a metre. The flow around a space shuttle (wing span of 24 m) is no longer computable with solutions of Navier–Stokes equation. Other examples of such flows includes those of epitaxy by molecular jets that is used to make thin metallic layers in micro-electronic components.

We assume that the thermal conductivity can be written in the same way and that the part associated with translation verifies (11.50). Show that the Prandtl number is now given by Eucken's relation:

$$\mathcal{P} = \frac{4\gamma}{9\gamma - 5} \tag{11.51}$$

Compare this value for a diatomic gas with that measured for air $\mathcal{P}_{air} = 0.714$.

Appendix: Maxwell–Boltzmann Distribution

When a gas relaxes towards equilibrium, its distribution function relaxes towards the well-known Maxwell–Boltzmann distribution. One may derive its expression from thermodynamic considerations, but it is interesting to show that it is also the unique solution of the Boltzmann equation for a collisional gas at thermodynamic equilibrium.

The most direct demonstration of this result leans on Boltzmann H-theorem which says that the quantity

$$H(t) = \int f(\mathbf{r}, t, \mathbf{v}) \ln f(\mathbf{r}, t, \mathbf{v}) d^3\mathbf{r} d^3\mathbf{v} \tag{11.52}$$

cannot increase for an isolated system.

Boltzmann H-Theorem for an Isolate Gas in a Box

In order to demonstrate Boltzmann H-theorem in the case of a gas enclosed in a box, it is handy to consider the quantity

$$h(\mathbf{r}, t) = \int f(\mathbf{r}, t, \mathbf{v}) \ln f(\mathbf{r}, t, \mathbf{v}) d^3\mathbf{v}$$

whose temporal derivative is

$$\frac{\partial h}{\partial t} = \int (1 + \ln f) \frac{\partial f}{\partial t} d^3\mathbf{v} \tag{11.53}$$

$\frac{\partial f}{\partial t}$ is given by Boltzmann equation (11.22). We thus have to compute three integrals:

$$A = \int (1 + \ln f) \mathbf{v} \cdot \nabla_x f d^3 \mathbf{v} \tag{11.54}$$

$$B = \int (1 + \ln f) \frac{\mathcal{F}}{m} \cdot \nabla_v f d^3 \mathbf{v} \tag{11.55}$$

$$C = \int (1 + \ln f) d^3 \mathbf{v} \int d^3 \mathbf{w} \int_{4\pi} \|\mathbf{v} - \mathbf{w}\| \left(f(\mathbf{v}') f(\mathbf{w}') - f(\mathbf{v}) f(\mathbf{w}) \right) \sigma(\mathbf{v}, \mathbf{w}|\mathbf{v}', \mathbf{w}') d\Omega \tag{11.56}$$

The first integral may be rewritten as

$$A = \int \left[\partial_j \left((1 + \ln f) f v_j \right) - \partial_j (f v_j) \right] d^3 \mathbf{v} = \partial_j \int f \ln f v_j d^3 \mathbf{v} = \nabla \cdot \mathbf{Q}$$

where we introduce the current

$$\mathbf{Q} = \int \mathbf{v} f \ln f d^3 \mathbf{v}$$

If we assume that the force $\mathcal{F}$ does not depend on the velocity the second integral B is zero because it may be transformed into a surface integral at infinite velocity where the distribution function is vanishing (see the example of the derivation of mass conservation 11.30). Hence

$$B = 0$$

The remaining integral C is associated with collisions. Let us rewrite it as follows:

$$C = \int d^3 \mathbf{v} d^3 \mathbf{w} (1 + \ln f(\mathbf{v})) \int_{4\pi} \|\mathbf{v} - \mathbf{w}\| \left(f(\mathbf{v}') f(\mathbf{w}') - f(\mathbf{v}) f(\mathbf{w}) \right) \sigma(\mathbf{v}, \mathbf{w}|\mathbf{v}', \mathbf{w}') d\Omega$$

We note that $\mathbf{v}$ and $\mathbf{w}$ are integration variables that can be exchanged. Hence, C may also be written as

$$C = \int d^3 \mathbf{v} d^3 \mathbf{w} (1 + \ln f(\mathbf{w})) \int_{4\pi} \|\mathbf{v} - \mathbf{w}\| \left(f(\mathbf{v}') f(\mathbf{w}') - f(\mathbf{v}) f(\mathbf{w}) \right) \sigma(\mathbf{v}, \mathbf{w}|\mathbf{v}', \mathbf{w}') d\Omega$$

Adding the two expressions of C, we get a third one where the roles of $\mathbf{v}$ and $\mathbf{w}$ are perfectly symmetric:

$$C = \frac{1}{2} \int d^3 \mathbf{v} d^3 \mathbf{w} \left(2 + \ln [f(\mathbf{v}) f(\mathbf{w})] \right)$$

$$\times \int_{4\pi} \|\mathbf{v} - \mathbf{w}\| \left(f(\mathbf{v}') f(\mathbf{w}') - f(\mathbf{v}) f(\mathbf{w}) \right) \sigma(\mathbf{v}, \mathbf{w}|\mathbf{v}', \mathbf{w}') d\Omega \tag{11.57}$$

Lastly, as we did when deriving the expression of C^+ for the collision integral, we rewrite the foregoing expression inverting the direction of collisions. The foregoing integral C must remain unchanged because the integration is over all the possible values of $\mathbf{v}$ and $\mathbf{w}$ and thus of all possible values of $\mathbf{v}'$ and $\mathbf{w}'$. Thus doing we rewrite (11.57) as:

$$
C = \frac{1}{2} \int d^3\mathbf{v}' d^3\mathbf{w}' \left(2 + \ln\left[f(\mathbf{v}')f(\mathbf{w}')\right]\right)
$$
$$
\times \int_{4\pi} \|\mathbf{v}' - \mathbf{w}'\| \left(f(\mathbf{v})f(\mathbf{w}) - f(\mathbf{v}')f(\mathbf{w}')\right) \sigma(\mathbf{v}', \mathbf{w}'|\mathbf{v}, \mathbf{w}) d\Omega
$$

Now, we may use the collisional invariants

$$
\sigma(\mathbf{v}', \mathbf{w}'|\mathbf{v}, \mathbf{w}) = \sigma(\mathbf{v}, \mathbf{w}|\mathbf{v}', \mathbf{w}'), \qquad \|\mathbf{v} - \mathbf{w}\| = \|\mathbf{v}' - \mathbf{w}'\|, \qquad d^3\mathbf{v} d^3\mathbf{w} = d^3\mathbf{v}' d^3\mathbf{w}'
$$

which give to C the new following form:

$$
C = \frac{1}{2} \int d^3\mathbf{v} d^3\mathbf{w} \left(2 + \ln\left[f(\mathbf{v}')f(\mathbf{w}')\right]\right)
$$
$$
\times \int_{4\pi} \|\mathbf{v} - \mathbf{w}\| \left(f(\mathbf{v})f(\mathbf{w}) - f(\mathbf{v}')f(\mathbf{w}')\right) \sigma(\mathbf{v}, \mathbf{w}|\mathbf{v}', \mathbf{w}') d\Omega
$$

This new form can be added to (11.57) and leads to a fully symmetric expression of C:

$$
C = \frac{1}{4} \int d^3\mathbf{v} d^3\mathbf{w} \int_{4\pi} d\Omega \left(\ln\left[f(\mathbf{v})f(\mathbf{w})\right] - \ln\left[f(\mathbf{v}')f(\mathbf{w}')\right]\right)
$$
$$
\times \left(f(\mathbf{v}')f(\mathbf{w}') - f(\mathbf{v})f(\mathbf{w})\right) \|\mathbf{v} - \mathbf{w}\| \sigma(\mathbf{v}, \mathbf{w}|\mathbf{v}', \mathbf{w}') \qquad (11.58)
$$

This last expression shows that C is always negative. Indeed, all the terms in the integrant are positive except

$$
\left(f(\mathbf{v}')f(\mathbf{w}') - f(\mathbf{v})f(\mathbf{w})\right) \left(\ln\left[f(\mathbf{v})f(\mathbf{w})\right] - \ln\left[f(\mathbf{v}')f(\mathbf{w}')\right]\right)
$$

which may be rewritten

$$
f(\mathbf{v}')f(\mathbf{w}')(1 - X)\ln X \qquad \text{with} \qquad X = f(\mathbf{v})f(\mathbf{w})/[f(\mathbf{v}')f(\mathbf{w}')]
$$

However, for all $X > 0$, $(1 - X)\ln X$ is less than zero. Hence, we establish that

$$
C \leq 0 \qquad (11.59)
$$

Back to (11.53), we rewrite this equation as:

$$\frac{\partial h}{\partial t} = -\nabla \cdot \mathbf{Q} + C$$

and integrate it over the volume occupied by the gas. With the boundary conditions on the velocity, i.e. $\mathbf{v} \cdot \mathbf{n} = 0$ on the walls, the first term disappears and we deduce

$$\frac{dH}{dt} = \int_{(V)} C d^3\mathbf{r} \qquad \Longleftrightarrow \qquad \frac{dH}{dt} \leq 0$$

which is Boltzmann H-theorem.

We may remark that we can define Boltzmann entropy from H with the following formula

$$S(t) = -kH(t) + \text{Cst}$$

From the H-theorem, entropy is an increasing function of time. We here demonstrate the second principle of thermodynamics for this kind of systems (an isolated assembly of colliding particles).

The Maxwell–Boltzmann Distribution

The H-theorem indicates that at thermodynamic equilibrium

$$\frac{dH(t)}{dt} = 0$$

or $\int C d^3\mathbf{r} = 0$. From the previous expressions this integral is zero if and only if

$$\left(f(\mathbf{v}') f(\mathbf{w}') - f(\mathbf{v}) f(\mathbf{w}) \right)$$
$$\left(\ln\left[f(\mathbf{v}) f(\mathbf{w}) \right] - \ln\left[f(\mathbf{v}') f(\mathbf{w}') \right] \right) = 0 \qquad \Longleftrightarrow \qquad X = 1$$

which means that

$$f(\mathbf{v}) f(\mathbf{w}) = f(\mathbf{v}') f(\mathbf{w}')$$

or that

$$\ln f(\mathbf{v}) + \ln f(\mathbf{w}) = \ln f(\mathbf{v}') + \ln f(\mathbf{w}') \qquad\qquad (11.60)$$

This equality shows that when the gas is at equilibrium, $\ln f(\mathbf{v})$ verifies a conservation law in collisions just as momentum or kinetic energy. Since it cannot be other collisional invariants independent of energy and momentum (that would constrain in an impossible manner the collisions), $\ln f(\mathbf{v})$ is necessarily a linear combination of kinetic energy and momentum.[9] Thus we write

$$\ln f(\mathbf{v}) = Av^2 + B_1 v_x + B_2 v_y + B_3 v_z + C$$

where A, B_1, B_2, B_3, C are constants.

Without loss of generality we can rewrite this expression as follows:

$$\ln f(\mathbf{v}) = -A(\mathbf{v} - \mathbf{v}_0)^2 + \ln K$$

or

$$f(\mathbf{v}) = K e^{-A(\mathbf{v} - \mathbf{v}_0)^2}$$

where $A > 0$ since the distribution function must be integrable.

We are now getting closer to the Maxwell–Boltzmann distribution and the loop will be closed when the constants are identified. For that, we note that numerical density of gas particles is:

$$n = \int f(\mathbf{v}) d^3\mathbf{v} = K \left(\frac{\pi}{A}\right)^{3/2}$$

while $\mathbf{v}_0$ can be identified to the mean velocity. Indeed,

$$\langle \mathbf{v} \rangle = \frac{1}{n} \int \mathbf{v} f(\mathbf{v}) d^3\mathbf{v} = \frac{K}{n} \int (\mathbf{v} - \mathbf{v}_0) e^{-A(\mathbf{v} - \mathbf{v}_0)^2} d^3\mathbf{v} + \frac{\mathbf{v}_0}{n} \int K e^{-A(\mathbf{v} - \mathbf{v}_0)^2} d^3\mathbf{v} = \mathbf{v}_0$$

Let us now assume that this mean velocity is zero, then the mean kinetic energy is given by

$$\langle \mathbf{v}^2 \rangle = \frac{1}{n} \int v^2 f(\mathbf{v}) d^3\mathbf{v} = \frac{K}{n} \int v^2 e^{-Av^2} d^3\mathbf{v} = \frac{3K\pi^{3/2}}{2nA^{5/2}}$$

With the foregoing expression of n we deduce:

$$\langle \mathbf{v}^2 \rangle = \frac{3}{2A}$$

---

[9]The reader who prefers a mathematical demonstration of this result may find it in the book of Kennard (1938).

But the expression (11.36) of pressure shows that

$$p = \frac{\rho}{2A}$$

Since we should find here the equation of state of the ideal gas, we identify the constant A to

$$A = \frac{m}{2kT}$$

We find again the relation between absolute temperature and the mean kinetic energy of gas particles:

$$\frac{1}{2}m\langle \mathbf{v}^2 \rangle = \frac{3}{2}kT$$

Further Reading

There is an abundant literature of the dynamics of rarefied gases since the first important results go back to the beginning of the twentieth century. The curious reader may consult the monograph of Chapman and Cowling (1970) which remains a reference in the field. The textbook of Vincenti and Kruger (1965) gives a pedagogical account of the subject. A rapid presentation of the subject may be found in the theoretical Physics course of Landau and Lifchitz (Lifchitz and Pitaevskii 1981). More recently, two books of Carlo Cercignani (Cercignani 1988; Cercignani et al. 1994) propose a more mathematical discussion of Boltzmann equation and of its applications. For French readers the book of Noëlle Pottier (Pottier 2007) gives a presentation of gas kinetics in the wider context of nonequilibrium phenomena. Lastly, the reader may read the introduction of the paper of Venkattraman and Alexeenko (2012) to have a glimpse at nowadays investigations in the field of transport coefficients in rarefied gases.

References

Aziz, R. A., Nain, V. P. S., Carley, J. S., Taylor, W. L. & McConville, G. T. (1979). An accurate intermolecular potential for helium. *The Journal of Chemical Physics, 70*, 4330–4342.
Bhatnagar, P. L., Gross, E. P. & Krook, M. (1954). A model for collision processes in gases. I. Small amplitude processes in charged and neutral one-component systems. *Physical Review, 94*, 511–525.
Castaing, R. (1970). *Cours de Thermodynamique Statistique*. Paris: Masson.
Cercignani, C. (1988). *The Boltzmann equation and its applications*. New York: Springer.
Cercignani, C., Illner, R. & Pulvirenti, M. (1994). *The mathematical theory of dilute gases*. New York: Springer.

Chapman, S. & Cowling, T. (1970). *The mathematical theory of non-uniform gases*, 2nd edn. Cambridge: Cambridge University Press.

Gray, E. (ed.) (1975). *American institute of physics handbook*. New York: American Institute of Physics.

Guyon, E., Hulin, J.-P. & Petit, L. (2001). a *Hydrodynamique physique*, CNRS edn. Les Ulis: EDP Sciences

Kennard, E. (1938). *Kinetic theory of gases*. New York: McGraw-Hill.

Lifchitz, E. & Pitaevskii, L. (1981). *Physical kinetics*. Oxford: Butterworth-Heinemann.

Pottier, N. (2007). *Physique statistique hors d'équilibre*. Les Ulis: EDP Sciences.

Venkattraman, A. & Alexeenko, A. (2012). Binary scattering model for Lennard-Jones potential: Transport coefficients and collision integrals for non-equilibrium gas flow simulations. *Physics of Fluids, 24*, 027101.

Vincenti, W. & Kruger, C. (1965). *Gas dynamics*. Huntington: Krieger.

Chapter 12
Complements of Mathematics

12.1 A Short Introduction to Tensors

The idea of tensors arose when physicists started dealing with forces inside elastic solids (tensions lead to tensors). Mathematically speaking, tensors are multilinear forms. For instance, the scalar product between two vectors is a bilinear form:

$$\delta : (\mathbf{a}, \mathbf{b}) \longrightarrow \mathbf{a} \cdot \mathbf{b} = \sum_{i=1}^{3} a_i b_i$$

This is an application which associates a scalar with two vectors (hence bi-) and it is linear with respect to each vector. The foregoing expression could be rewritten using the components of the unit tensor, namely δ_{ij}, like:

$$\delta : (\mathbf{a}, \mathbf{b}) \longrightarrow \delta(\mathbf{a}, \mathbf{b}) = \sum_{i=1}^{3} \sum_{j=1}^{3} a_i \delta_{ij} b_j$$

This is the same as above since $\delta_{ij} = 0$ when $i \neq j$ and $\delta_{ii} = 1$. Very often the $\sum$ symbol is dropped, and Einstein notations are used: repeated indices meaning summation. Using these notation $\delta(\mathbf{a}, \mathbf{b}) = a_i \delta_{ij} b_j$. However, $\delta_{ij} b_j = b_i$ thus $\delta(\mathbf{a}, \mathbf{b}) = a_i b_i$.

To simplify the presentation of these notions, we stay in the metric space $\mathbb{R}^3$ with an orthonormal basis. This means that the notion of variance, namely contravariant and covariant tensors, is not useful at all. So the place of indices (up or down) is not meaningful here. We just use the Einstein convention of implicit summation on repeated indices. Thus,

$$a_i b_{ij} \qquad \text{means} \qquad \sum_{i=1}^{3} a_i b_{ij}$$

© Springer International Publishing Switzerland 2015
M. Rieutord, *Fluid Dynamics*, Graduate Texts in Physics,
DOI 10.1007/978-3-319-09351-2_12

12.1.1 Definitions

A multilinear form f may be simply defined by

$$f(\mathbf{X}^1, \mathbf{X}^2, \ldots, \mathbf{X}^n) = \sum_{i_1, i_2, \ldots, i_n} f_{i_1, i_2, \ldots, i_n} X_{i_1}^1 X_{i_2}^2 \ldots X_{i_n}^n$$

where the summation runs from 1 to 3 in a three-dimensional space. The set

$$\{ f_{i_1, i_2, \ldots, i_n} \,/\, i_1 = 1, 2, 3, i_2 = 1, 2, 3, \ldots i_n = 1, 2, 3 \}$$

is the set of the 3^n components of the nth-order tensor $[f]$. Very often, one calls, abusively, $f_{i_1, i_2, \ldots, i_n}$ a tensor, while this is only one of its components.

With the foregoing definition, we observe that *a zeroth order tensor is a scalar and a first order tensor is a vector.*

12.1.1.1 Notations

The cartesian coordinates are often denoted (x_1, x_2, x_3). The vectors are denoted $\mathbf{v}$ and their components v_i. Tensors of order greater or equal to 2 are denoted $[a]$ and their components $a_{ij\ldots}$.

The gradient of a scalar function f is ∇f and its components are:

$$(\nabla f)_i \equiv \frac{\partial f}{\partial x_i} \equiv \partial_i f$$

12.1.1.2 Contraction

The contraction of two indices is the summation over these two indices. For instance, the contraction of index i and j of the second order tensor a_{ij} gives a_{ii} which is the *trace* of the tensor: $a_{ii} = \text{Tr}[a]$. We may note that $\nabla \cdot \mathbf{v} = \partial_i v_i$ is also the trace of the velocity gradient tensor.

12.1.1.3 The Tensorial Product of Two Vectors

From vectors one can construct tensors of higher order. This operation is called the *tensorial product* and usually denoted $\otimes$. For example, from the velocity vector $\mathbf{v}$, we can build the Reynolds stress tensor $\rho[\mathbf{v} \otimes \mathbf{v}]$ whose components are $\rho v_i v_j$.

12.1.2 ϵ_{ijk}

To ease operations using the cross product, one uses the set of numbers ϵ_{ijk} which are defined as follows:

$$\epsilon_{ijk} = 1 \qquad \text{if } ijk \text{ is an even permutation of 123}$$
$$\epsilon_{ijk} = -1 \qquad \text{if } ijk \text{ is an odd permutation of 123}$$
$$\epsilon_{ijk} = 0 \qquad \text{if two indices are identical}$$

This is not a tensor but a pseudo-tensor often called the *completely antisymmetric pseudo-tensor*. It may indeed also be defined from a determinant:

$$\epsilon : \quad \mathbb{R}^3 \quad \longrightarrow \mathbb{R}$$

$$(\mathbf{a}, \mathbf{b}, \mathbf{c}) \longmapsto \epsilon_{ijk} a_i b_j c_k = Det(\mathbf{a}, \mathbf{b}, \mathbf{c})$$

ϵ is a pseudo-tensor because the sign of the determinant depends on the orientation of the basis. This is also the reason why the determinant is often called a pseudo-scalar, and the cross product a pseudo-vector.

In the following we give some useful relations for the manipulation of ϵ_{ijk}:

$$\epsilon_{ijk}\epsilon_{lmn} = \delta_{il}\delta_{jm}\delta_{kn} + \delta_{kl}\delta_{im}\delta_{jn} + \delta_{jl}\delta_{km}\delta_{in} - \delta_{kl}\delta_{jm}\delta_{in} - \delta_{jl}\delta_{im}\delta_{kn} - \delta_{il}\delta_{km}\delta_{jn}$$

$$= \begin{vmatrix} \delta_{il} & \delta_{jl} & \delta_{kl} \\ \delta_{im} & \delta_{jm} & \delta_{km} \\ \delta_{in} & \delta_{jn} & \delta_{kn} \end{vmatrix} \tag{12.1}$$

$$\epsilon_{ijk}\epsilon_{klm} = \delta_{il}\delta_{jm} - \delta_{im}\delta_{jl} \tag{12.2}$$

$$\epsilon_{ikl}\epsilon_{jkl} = 2\delta_{ij} \tag{12.3}$$

$$\epsilon_{ijk}\epsilon_{ijk} = 6 \tag{12.4}$$

Here are also the relation between some expressions using vectors and the same expressions using tensorial notation:

$$\mathbf{a} \cdot \mathbf{b} = a_i b_i$$

$$(\mathbf{a} \times \mathbf{b})_i = \epsilon_{ijk} a_j b_k$$

$$det(\mathbf{a}, \mathbf{b}, \mathbf{c}) = \epsilon_{ijk} a_j b_k c_i$$

If [A] is a symmetric tensor, then $A_{ij} = A_{ji}$ and

$$\epsilon_{ijk} A_{jk} = 0 \tag{12.5}$$

because terms cancel two by two. A more detailed and more formal presentation may be found in textbooks of Hladik (1993), Bass (1978) or Lebedev et al. (2010).

12.2   The Divergence Theorem

12.2.1   *Statement and Demonstration*

If $T_{i_1 i_2 \cdots i_n}$ is a tensorial field of order n, smooth enough so that derivatives are defined, then

$$\int_{(S)} T_{i_1 i_2 \cdots i_k \cdots i_n} dS_{i_k} = \int_{(V)} \partial_{i_k} T_{i_1 i_2 \cdots i_k \cdots i_n} dV \tag{12.6}$$

where the first integration is taken over the surface S surrounding the volume V. This relation is true in a space of dimension $n \geq 1$, but we shall give its demonstration only in the two-dimensional space (a plane). Its generalization to a higher dimension is straightforward (Fig. 12.1).

We first write the volume integral under the form:

$$\int_{(V)} (\partial_x A_x + \partial_y A_y) dx dy = \int_{y_{min}}^{y_{max}} (A_x(x_+) - A_x(x_-)) dy$$
$$+ \int_{x_{min}}^{x_{max}} (A_y(y_+) - A_y(y_-)) dx$$

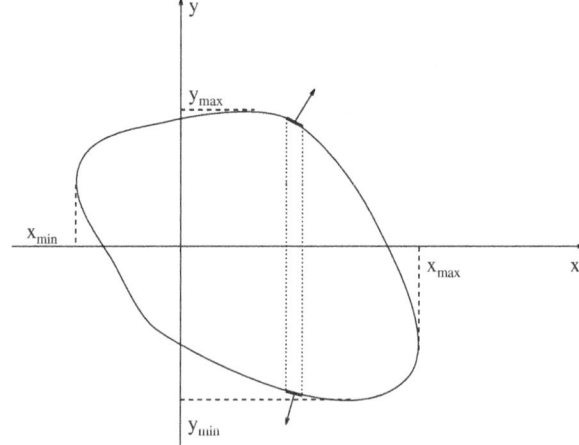

Fig. 12.1 The various quantities needed in the demonstration of the divergence theorem

where $x_{\pm}(y)$ and $y_{\pm}(x)$ are the curves defining the boundary of the "volume". Let us concentrate on one of the terms, for instance $A_y(y_+)dx$. This term also reads

$$A_y(y_+)\mathbf{e}_y \cdot \mathbf{n}\,dl$$

where $\mathbf{n}$ is the outward normal unit vector and dl is the "surface" element. Thus,

$$A_y(y_+)dx = (A_y(y_+)\mathbf{e}_y) \cdot d\mathbf{S}$$

but y_+ defines the upper part of the surface while $y_-(x)$ defines the lower part. Together, when x runs from x_{min} to x_{max}, y_- and y_+ describe the whole surface. We may therefore write

$$\int_{x_{min}}^{x_{max}} (A_y(y_+) - A_y(y_-))dx = \int_{(S)} (A_y \mathbf{e}_y) \cdot d\mathbf{S}$$

where we noted that $-A_y(y_-)dx = -(A_y(y_-)\mathbf{e}_y) \cdot (-\mathbf{n})dl = (A_y \mathbf{e}_y) \cdot d\mathbf{S}$.

Then, we follow the same reasoning for the other half of the integral finally get $\int_{(S)} A_i dS_i$. QED

12.2.2   Corollary

Similarly we may show the following relation:

$$\int_{(S)} \epsilon_{ji_km} T_{i_1 i_2 \cdots i_k \cdots i_n} dS_m = \int_{(V)} \epsilon_{ji_km} \partial_m T_{i_1 i_2 \cdots i_k \cdots i_n} dV \qquad (12.7)$$

12.2.3   A Few Consequences

The general result (12.6) has some interesting specific cases. In the usual three-dimensional space, we find again some classical formulae if we set the tensor $[T]$ to a scalar, a vector or a second order tensor, namely:

$$\int_{(S)} f d\mathbf{S} = \int_{(V)} \nabla f \, dV \qquad (12.8)$$

$$\int_{(S)} \mathbf{A} \cdot d\mathbf{S} = \int_{(V)} \nabla \cdot \mathbf{A} \, dV$$

$$\int_{(S)} [a] d\mathbf{S} = \int_{(V)} \mathbf{Div}[a] \, dV$$

The corollary (12.7) allows us to write:

$$\int_{(S)} \mathbf{A} \wedge d\mathbf{S} = -\int_{(V)} \nabla \times \mathbf{A} \, dV \tag{12.9}$$

In a two-dimensional space, volumes are surfaces and surfaces are contours, hence:

$$\oint_{(C)} f \, dl \, \mathbf{n} = \int_{(S)} \nabla f \, dS$$

$$\oint_{(C)} \mathbf{A} \cdot \mathbf{n} \, dl = \int_{(S)} \nabla \cdot \mathbf{A} \, dS$$

$$\oint_{(C)} \mathbf{A} \times \mathbf{n} \, dl = -\int_{(S)} \nabla \times \mathbf{A} \, dS$$

where $\mathbf{n}$ is the unit outward normal vector to the contour C.

12.3 Radius of Curvature

12.3.1 For a Plane Curve

The radius of curvature of a plane curve defined parametrically by the functions $x(t)$ and $y(t)$ is the radius of the osculating circle at the given point. It may be computed as follows.

Let M be a point of the curve and $M + dM$ another point infinitely close to M and on the same curve. If these two points also belong to the same osculating circle, then their distance is $Rd\theta$ where $d\theta$ is the angle between the vectors that are tangent to the curve in M and $M + dM$. Thus we have:

$$Rd\theta = \sqrt{\dot{x}^2 + \dot{y}^2} dt \tag{12.10}$$

where the dot means the derivative with respect to t. In addition

$$d\theta \simeq \sin d\theta = -\frac{(\mathbf{T} \times (\mathbf{T} + d\mathbf{T})) \cdot \mathbf{e}_z}{\|\mathbf{T}\| \, \|\mathbf{T} + d\mathbf{T}\|}$$

$$\Longleftrightarrow \quad d\theta \simeq -\frac{(\mathbf{T} \times d\mathbf{T}) \cdot \mathbf{e}_z}{\|\mathbf{T}\|^2}$$

where $\mathbf{T}$ is the tangent vector. Noting that

$$\mathbf{T}\begin{vmatrix} \dot{x} \\ \dot{y} \end{vmatrix}, \quad d\mathbf{T}\begin{vmatrix} \ddot{x}dt \\ \ddot{y}dt \end{vmatrix} \quad \text{and} \quad d\theta = -\frac{\dot{x}\ddot{y} - \dot{y}\ddot{x}}{\dot{x}^2 + \dot{y}^2}dt$$

(12.10) leads to the result:

$$\frac{1}{R} = \frac{\ddot{x}\dot{y} - \ddot{y}\dot{x}}{(\dot{x}^2 + \dot{y}^2)^{3/2}} \tag{12.11}$$

For a curve defined by the equation $y = y(x)$, then

$$\frac{1}{R} = -\frac{\ddot{y}}{(1 + \dot{y}^2)^{3/2}} \tag{12.12}$$

12.3.2   For a Curve in Space

Now, if the curve is defined in a three-dimensional space by its parametric equation $[x(s), y(s), z(s)]$, then we define the tangent vector as:

$$\mathbf{e}_s = \frac{1}{\sqrt{\dot{x}^2 + \dot{y}^2 + \dot{z}^2}}\begin{vmatrix} \dot{x}(s) \\ \dot{y}(s) \\ \dot{z}(s) \end{vmatrix}$$

The curvature radius R is defined as:

$$\frac{\partial \mathbf{e}_s}{\partial s} = \frac{\mathbf{n}}{R} \tag{12.13}$$

where $\mathbf{n}$ is a unit normal vector to the curve. The radius R is a positive number.

Finally, one then introduce the *binormal vector* $\mathbf{e}_b$, which defines, with $\mathbf{e}_s$ and $\mathbf{n}$, the Frenet frame. It is defined as

$$\frac{\partial \mathbf{n}}{\partial s} = \frac{\mathbf{e}_b}{T}$$

The length T is called the *torsion*.

12.4 The Boundary Layer Theory Viewed from Differential Equation Theory

We introduced in Chap. 4 the idea of boundary layers, basing our approach on physical considerations to find out the various possible balances of forces. We stressed the physical existence of this layer when we presented its destabilization that induces rapid variation of the drag coefficient C_x. However, the notion of boundary layer has deep roots in the theory of differential equations. Here, we shall briefly sketch out these ideas, but we strongly encourage the reader to have a look to the book of Bender and Orszag where many aspects of this technique are detailed.

One of the various ways of solving differential equations is to use the theory of singular perturbations. To illustrate this theory we readily take an example, namely the differential equation:

$$\varepsilon\frac{d^2 y}{dx^2} + (1 + \varepsilon)\frac{dy}{dx} + y = 0 \tag{12.14}$$

where ε is a small parameter. If we consider the terms factoring ε as a perturbation of an original equation where $\varepsilon = 0$, then this perturbation is said to be singular because the order of the differential equation is not the same when $\varepsilon = 0$ and $\varepsilon \neq 0$. This is said to be singular because the change of order leads to the appearance of a singularity when ε tends to zero.

Let us now study in more detail the example given by (12.14), which we complete by boundary conditions so as to fully determine the solution. We choose

$$y(0) = 0 \quad\text{and}\quad y(1) = 1 \tag{12.15}$$

(12.14) is a second order differential equation with constant coefficients. Its solutions are (in the general case) a linear combination of exponentials. Thus, we find:

$$y(x) = \frac{e^{-x} - e^{-x/\varepsilon}}{e^{-1} - e^{-1/\varepsilon}} \tag{12.16}$$

This solution is already quite interesting as it reveals the singularity which arises when ε goes to zero. The solution is discontinuous at the origin. In addition, when $\varepsilon \ll 1$, the solution varies very rapidly near this point (see Fig. 12.2). We see the presence of a boundary layer!

We shall now retrieve the solution (12.16) through a perturbative approach similar to the one we used for the boundary layer analysis. We need four steps:

1. First step: we determine the *outer solution* (outside the boundary layer). We set $\varepsilon = 0$ and we solve. However, we face a first difficulty: We have two boundary

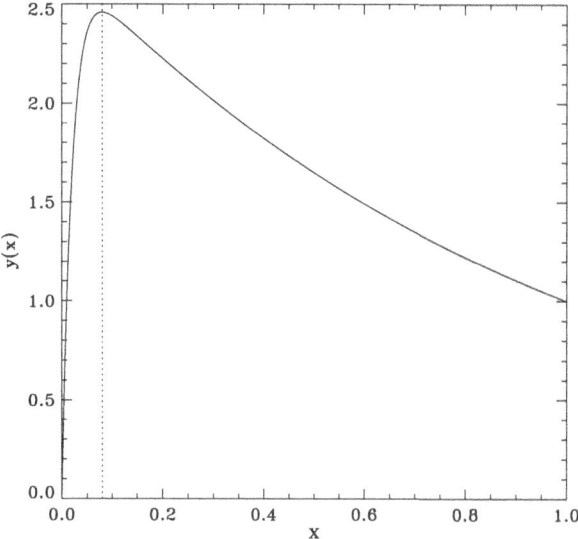

Fig. 12.2 A plot of the function $y(x) = (e^{-x} - e^{-x/\varepsilon})/(e^{-1} - e^{-1/\varepsilon})$ for $\varepsilon = 0.02$. The *vertical dotted line* shows the separation between the inner and outer regions

conditions for a first order differential equation. We have to dismiss one of the boundary conditions. In this case, the choice is quite simple since one of the boundary conditions is in the boundary layer. Thus we take the other one, namely $y(1) = 1$. Of course, this choice is not always so simple, and some trial and error is sometimes required.

We thus solve:

$$y' + y = 0 \qquad \text{with} \qquad y(1) = 1$$

The solution is $y_e(x) = e^{1-x}$. Let us point out that this is the solution we would have obtained if we had let $\varepsilon \to 0$ in the solution (12.16) for $x \neq 0$.

2. The second step is to determine the solution in the boundary layer. For this, we use the stretched coordinate $\zeta = x/\varepsilon$ (the boundary layer coordinate) and we rewrite the differential equation using ζ:

$$y'' + (1 + \varepsilon)y' + \varepsilon y = 0$$

Again, we let $\varepsilon \to 0$ and find:

$$y'' + y' = 0$$

This new differential equation is of second order and needs two boundary conditions. One of them is obvious: at $x = 0$, $y = 0$. We do not specify the second one at the moment and leave the constant undetermined. Thus, the boundary layer solution is

$$y_{bl} = C(1 - e^{-\zeta}) \tag{12.17}$$

where C is an arbitrary constant. We note that this solution can be derived from the exact solution (12.16) by making the change of variable $x \to \zeta$ and letting $\varepsilon \to 0$.

3. The third step is the asymptotic matching between the two previous solutions. We need to find an interval where both solutions are valid. In the present case, this is where:

$$\varepsilon \ll x \ll 1 \tag{12.18}$$

In this interval, the two functions y_e and y_{bl} must tend to the same limit. Hence

$$\lim_{\zeta \to \infty} y_{bl} = \lim_{x \to 0} y_e = e$$

This fixes the constant $C = e$.

4. The last step is to gather all the foregoing "pieces" and construct the function that solves the problem in the whole interval. This function is identical to y_e when $x \gg \varepsilon$ and equals y_{bl} when $x \leq \varepsilon$. It is obtained by the combination:

$$y_{unif} = y_e + y_{bl} - e = e^{1-x} - e^{1-\zeta} \tag{12.19}$$

The solution $y_{unif}(x)$ is said to be the *uniform approximation at order n* if, in the given interval,

$$y_{unif}(x) - y(x) = \mathcal{O}(\varepsilon^{n+1}) \qquad \text{when} \quad \varepsilon \to 0$$

We may observe that in our example $|y_{unif}(x) - y(x)|$ is $\mathcal{O}(e^{-1/\varepsilon})$ which is less than any power of ε. Our approximation is therefore valid at "infinite" order, however it always differs from the exact solution by an undetermined $\mathcal{O}(e^{-1/\varepsilon})$ quantity. This last point underlines the fact that *the boundary layer theory is a theory of singular perturbations and not of regular perturbations.*

12.5 The Sturm–Liouville Problem

The Sturm–Liouville problem is usually enounced in the following way:

Let three functions $p(x), q(x)$ and $w(x)$, continuous and defined on the interval $[a, b] \subset \mathbb{R}$, such that $p(x) > 0, q(x) \geq 0$ and $w(x) > 0$ in the whole interval. The Sturm–Liouville problem is the following boundary value problem:

$$(py')' - qy + \lambda wy = 0 \tag{12.20}$$

where the solution $y(x)$ verifies the boundary conditions:

$$y(a) = 0 \quad \text{or} \quad y'(a) = 0, \qquad y(b) = 0 \quad \text{or} \quad y'(b) = 0 \qquad (12.21)$$

It may be shown that the solutions of this problem form a denumerable suite of eigenfunctions y_n associated with real eigenvalues λ_n.

Two eigenfunctions associated with two distinct eigenvalues are orthogonal with respect to the scalar product:

$$\langle f | g \rangle = \int_a^b f(x)g(x)w(x)dx$$

Indeed, let y_m and y_n be two solutions associated respectively with λ_m and λ_n, then

$$(py'_m)' - qy_m + \lambda_m w y_m = 0$$

$$(py'_n)' - qy_n + \lambda_n w y_n = 0$$

Multiplying the first equation by y_n and the second by y_m, integrating and subtracting the two equations, one finds:

$$(\lambda_m - \lambda_n)\langle y_m | y_n \rangle = \left[y_m(wy'_n) - y_m(wy'_n) \right]_a^b = 0$$

where we used the boundary conditions at a and b. Since $\lambda_m \neq \lambda_n$, $\langle y_m | y_n \rangle = 0$.

This property of the Sturm–Liouville problem is at the origin of *spectral methods* for solving numerically differential equations. The unknown solutions are expanded on the basis formed by a set of functions solution of a Sturm–Liouville problem.

When one deals with a problem of stability, the sign of the real part of the eigenvalues is important. In a Sturm–Liouville problem, it can be determined quite easily. Multiplying (12.20) by y and integrating over $[a, b]$, one gets:

$$\lambda = \frac{\int_a^b (py'^2 + qy^2)dx}{\int_a^b y^2 w dx}$$

showing that all the eigenvalues are positive. A slightly more general result can be obtained if we note that $w(x) < 0$ in the whole interval, then all the eigenvalues are negative. A more complicated case is when w changes sign in the interval. It can be shown then that the eigenvalue spectrum spans from $-\infty$ to $+\infty$.

A complete study of the Sturm–Liouville problem may be found in the book of Ince (1956).

12.6   Second Order Partial Differential Equations

12.6.1   The Different Types

Second order partial differential equations are classified in four categories: they are either of *hyperbolic, parabolic, elliptic or mixed* type. They are categorized according to a property of the coefficients of the second order derivatives. The most general form of this kind of equation is:

$$A(x,y)\frac{\partial^2 f}{\partial x^2} + B(x,y)\frac{\partial^2 f}{\partial x \partial y} + C(x,y)\frac{\partial^2 f}{\partial y^2} + \cdots = 0$$

where the dots are for the first and zeroth order terms. The function $D(x,y) = B^2 - 4AC$ determines the type of the equation. If in the whole domain where f is defined,

- $D(x,y) > 0$, the equation is hyperbolic
- $D(x,y) = 0$ the equation is parabolic
- $D(x,y) < 0$ the equation is elliptic

If $D(x,y)$ changes sign in the domain, the equation is said to be of mixed type.

To be more familiar with the basic properties of these types of equations, we shall focus on three very classical examples: the wave equation, the heat equation and the Laplace equation. All these equations are met in Fluid Mechanics.

As a first step, we need to introduce the notion of characteristics.

12.6.2   An Introduction to Characteristics

Let $f(x,y)$ be a function defined on a plane (or in a domain of a plane) and verifying

$$A(x,y)\frac{\partial f}{\partial x} + B(x,y)\frac{\partial f}{\partial y} = 0 \qquad (12.22)$$

The characteristic curves are the curves where f is constant, namely where:

$$df = \frac{\partial f}{\partial x}dx + \frac{\partial f}{\partial y}dy = 0 \qquad (12.23)$$

Using (12.22) and (12.23), we derive the equation of characteristics:

$$\frac{dy}{dx} = \frac{B}{A}$$

The solution of this equation together with the boundary conditions permits, in some cases, a complete determination of f. Let us take an example of this favourable case. We assume that $f(x,0) = \cos x$, $A(x,y) = y$ and $B = 1$. The characteristics are such that:

$$y\frac{dy}{dx} = 1 \implies y = \pm\sqrt{2(x+K)}, \quad x \geq -K, \quad K \in \mathbb{R}$$

This is a family of parabola with Ox as their axis. Any point of the plane belongs to a unique characteristics and the value of f there just depends on the constant K defining the characteristics:

$$f(x,y) = f\left(\frac{1}{2}y^2 - x\right)$$

The boundary condition $f(x,0) = \cos x$ determines the function and we obtain the searched solution:

$$f(x,y) = \cos\left(\frac{1}{2}y^2 - x\right)$$

We may observe that the boundary condition on the x-axis, $f(x,0) = \cos x$, has been propagated in the whole plane. If the y-coordinate is replaced by the time, then we have a true propagation. This property is verified by all the equations having characteristics.

12.6.3   A Hyperbolic Equation: The Wave Equation

The equation of a wave propagating at the velocity c is given by:

$$\frac{\partial^2 f}{\partial x^2} - \frac{1}{c^2}\frac{\partial^2 f}{\partial t^2} = 0$$

The general solution of this equation is well-known:

$$f(x,t) = \Phi(x - ct) + \Psi(x + ct)$$

where Φ and Ψ are two arbitrary functions to be determined by the initial conditions. This solution is easily obtained after the change of variable: $u = x - ct$ and $v = x + ct$.

The solution is fully determined if we take into account the initial conditions. For instance, we may demand:

$$f(x,0) = \cos x \qquad \text{and} \qquad \left(\frac{\partial f}{\partial t}\right)_0 = 0$$

which leads to

$$f(x,t) = \frac{1}{2}\left[\cos(x - ct) + \cos(x + ct)\right]$$

Initial conditions are necessary to determined completely the solution. With these conditions the problem is well posed. To emphasize the importance of these conditions, let us try the exercise where, instead of imposing the value of the function and its time-derivative at an initial time, we impose the value of the function at two different times. For instance:

$$f(x,0) = I(x) \qquad \text{and} \qquad f(x,T) = F(x)$$

where $I(x)$ and $F(x)$ are given data. Such a problem is mathematically ill-posed. These conditions, which look like boundary conditions are not sufficient to fully determine the solution. It is easy to show that $\Psi(x)$ verifies:

$$\Psi(x) = \Psi(x + 2cT) + I(x) - F(x + cT)$$

It means that Ψ is undetermined in the interval $[0, 2cT]$. This kind of problem is faced when one tries to solve Poincaré equation (8.25) in general fluid domains.

12.6.4   A Parabolic Equation: The Diffusion Equation

The diffusion equation has the following general form:

$$\frac{\partial C}{\partial t} = \kappa \Delta C$$

where C is the concentration of a chemical element, for instance, and κ is its diffusion coefficient in the fluid.

To make things as simple as possible, we reduce this problem to one space dimension. $C(x,t)$ is therefore determined by

$$\frac{\partial C}{\partial t} = \kappa \frac{\partial^2 C}{\partial x^2} \tag{12.24}$$

Since this equation is of first order in time we just need one initial condition, namely $C(x,0) = C_0(x)$.

The general method to solve this kind of equation is to use the Laplace transform:

$$\tilde{C}(x, p) = \int_0^\infty C(x,t)e^{-pt}\,dt$$

The diffusion equation (12.24) changes into:

$$\frac{\partial^2 \tilde{C}}{\partial x^2} - \frac{p}{\kappa}\tilde{C} = -\frac{C_0(x)}{\kappa}$$

We shall solve this equation in the very simple case where $C_0(x) = C_0 \cos kx$. The general solution is then

$$\tilde{C}(x, p) = \frac{C_0 \cos kx}{p + \kappa k^2} + Ae^{-\sqrt{\frac{p}{\kappa}}x} + Be^{\sqrt{\frac{p}{\kappa}}x}$$

If the solution is finite at $\pm\infty$ then $A = B = 0$. We thus obtain the solution $C(x,t)$ by the Mellin–Fourier formula:

$$C(x,t) = \int_{c-i\infty}^{c+i\infty} \frac{T_0 \cos kx}{p + \kappa k^2}e^{pt}\,dp$$

which leads to:

$$C(x,t) = C_0 \cos kx\, e^{-\kappa k^2 t}$$

This solution shows us that the initial state described by $C_0(x)$, decreases exponentially on a time scale $1/(\kappa k^2)$ or $\lambda^2/4\pi^2\kappa$ if λ is the wavelength associated with k. This is the same result as the one we found when studying the diffusion of vorticity with (4.5).

Let us now consider the case where the initial conditions do not contain any specific length scale (contrary to the foregoing example). In this case, one looks for a self-similar solution. If $C(x,t)$ is a solution of (12.24), we look for a condition such that $C_1(x,t) = C(Lx, Tt)$ is also a solution. Reporting C_1 into (12.24) and using the fact that C is a solution leads to:

$$\frac{\partial C_1}{\partial t} = \frac{T\kappa}{L^2}\frac{\partial^2 C_1}{\partial x^2}$$

Since L and T are arbitrary, we may choose $L = \sqrt{T}$. C_1 is also a solution for any T. Choosing $T = 1/t$, we find $C_1(x/\sqrt{t}, 1) \equiv c(x/\sqrt{t})$. The solution only depends on a single variable $\eta = x/\sqrt{t}$ called the self-similarity variable.[1] $c(\eta)$ is a solution of the differential equation:

$$c''(\eta) = -\eta c'(\eta)/2\kappa$$

whose general solution is

$$c(\eta) = A \operatorname{erf}\left(\frac{\eta}{2\sqrt{\kappa}}\right) + B$$

where A et B are constants and erf is the error function.[2] Let us take the example where $c(x,t)$ is such that:

$$x > 0 \quad c(x,0) = 0 \qquad \text{and} \qquad c(0,t) = c_0$$

We find that

$$c(x,t) = c_0\left[1 - \operatorname{erf}\left(\frac{x}{2\sqrt{\kappa t}}\right)\right] \tag{12.25}$$

This example corresponds to the one-dimensional diffusion of a contaminant from a source with a unit concentration (see Fig. 12.3). This solution can of course be determined by the standard method using the Laplace transform. We may note that the iso-concentration lines, where $c(x,t)$ =Cst, move towards $+\infty$ *proportionally to the square root of time*. This law is typical of diffusion phenomena.

12.6.5 An Elliptic Equation: The Laplace Equation

12.6.5.1 Some General Properties

Laplace equation is the equation of potentials, which we find in electrostatics, magnetostatics, irrotational fluids, classic gravitation, etc. We met it several times either in perfect fluids (irrotational flows) or in very viscous ones (creeping flows). Thus, it is useful to know some of its basic properties.

---

[1]This kind of solution is used to solve the Prandtl equation describing boundary layer flows (see Sect. 4.3.6).

[2]erf is the error function defined by

$$\operatorname{erf}(x) = \frac{1}{\sqrt{\pi}} \int_0^x e^{-u^2}\, du$$

that is to say as the integral of a Gaussian (erf(∞)=1).

This equation reads:

$$\Delta f = 0 \tag{12.26}$$

It is *a linear equation* and its solutions are the *harmonic functions*. This equation contains no scale: if $f(\mathbf{r})$ is a solution then $f(\mathbf{r}/L)$ is also a solution. Scales only appear through the boundary conditions (the size of the domain) (Fig. 12.3).

If f and g are two harmonic functions, then

$$\nabla \cdot (f\nabla g) = \nabla f \cdot \nabla g$$

Integrating this equation over a volume V, bounded by a surface S characterized by the outward normal vector $\mathbf{n}$, we have

$$\int_{(S)} f\nabla g \cdot d\mathbf{S} = \int_{(V)} \nabla f \cdot \nabla g \, dV \tag{12.27}$$

Let us now assume that f or $\mathbf{n} \cdot \nabla f$ are vanishing on the bounding surface and let us use (12.27), namely

$$\int_{(S)} f\nabla f \cdot d\mathbf{S} = \int_{(V)} (\nabla f)^2 dV \tag{12.28}$$

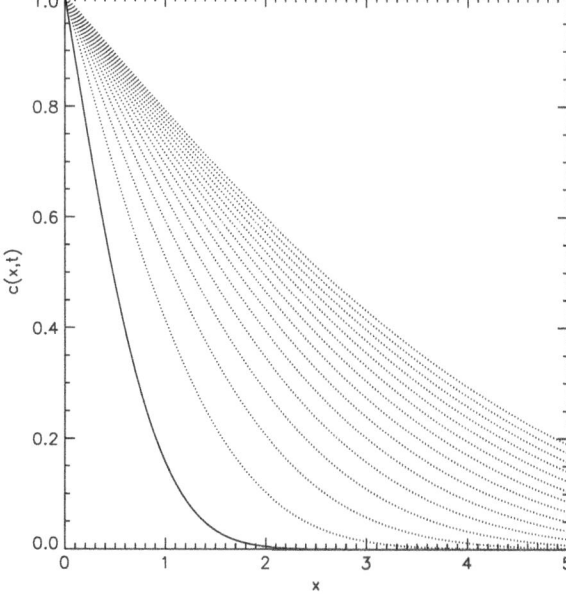

Fig. 12.3 Time evolution of the concentration according to (12.25). The various curves represent the solution at increasing time, the time difference being the same between one curve to the next. Note how they get closer with time because of the $\sqrt{t}$-law

From the boundary conditions met by f, the foregoing equation implies:

$$\nabla f = 0$$

everywhere in V, so that f is a constant in the volume V.

This result leads to the unicity of the solutions of Laplace equation. Indeed, let f_1 and f_2 be two solutions of this equation verifying the same boundary conditions, then $g = f_1 - f_2$ is also a solution of Laplace equation and meets the boundary condition $g = 0$ or $\mathbf{n} \cdot \nabla g = 0$. g is therefore a constant and if on a given point of the surface $g = 0$, this constant is zero. Thus, f_1 et f_2 are different at most by a constant. If f is a potential of a vector field, like the electric field $\mathbf{E} = \nabla f$, we see that the value of the constant has no importance and that the field is unique.

Laplace equation can be solved in several geometries. Here, we shall detail only two cases: the spherical geometry and its two-dimensional counterpart, the plane polar geometry.

12.6.5.2   The Method for Solving Laplace Equation

One classical way to solve Laplace equation is to try to separate the variables: If x and y are the coordinates in a plane, one looks for solutions of the form:

$$V(x, y) = f(x)g(y)$$

We thus form a *Laplace product*. Using this type of solutions leads to

$$\Delta V = 0 \quad \Longrightarrow \quad \frac{1}{f}\frac{\partial^2 f}{\partial x^2} + \frac{1}{g}\frac{\partial^2 g}{\partial y^2} = 0$$

The two parts of the equation depends on two different variables. Their sum can be zero if and only if they are both constants. Let A be this constant, hence:

$$\frac{d^2 f}{dx^2} = -Af \qquad \text{and} \qquad \frac{d^2 g}{dy^2} = Ag$$

If $A > 0$ then

$$V(x, y) = B \cos(\sqrt{A}x + \phi)\,\mathrm{sh}\,(\sqrt{A}y + \chi)$$

The four constants A, B, ϕ, χ are arbitrary. Thus we have at our disposal an infinite number of solutions that can be linearly combined. The boundary conditions are then used to fix the values of the constants and determine the unique final solution.

To illustrate this method, we study two examples that are met in Fluids Dynamics.

12.6.5.3 Solutions in Polar Coordinates

In these coordinates the Laplacian reads:

$$\Delta V = \frac{1}{r}\frac{\partial}{\partial r}\left(r\frac{\partial V}{\partial r}\right) + \frac{1}{r^2}\frac{\partial^2 V}{\partial \theta^2}$$

We thus write Laplace products as $V = F(r)G(\theta)$ and find:

$$\frac{r}{F}\frac{\partial}{\partial r}\left(r\frac{\partial F}{\partial r}\right) = C \qquad \text{and} \qquad \frac{1}{G}\frac{d^2 G}{d\theta^2} = -C$$

We may readily restrict the possible values of the constant C by demanding the periodicity of the solutions with respect to θ, namely $G(\theta) = G(\theta + 2\pi)$. We therefore set $C = n^2$, n being an integer. Hence,

$$G(\theta) = e^{in\theta} \qquad \text{and} \qquad F(r) = r^n$$

Thus, in the present case, we may write the solution of Laplace equation, in the following form:

$$V(r, \theta) = \sum_{n=-\infty}^{n=+\infty} A_n r^n e^{in\theta} \tag{12.29}$$

Here the constant A_n are determined by the boundary conditions.

We may note that $r^n e^{in\theta}$ is just z^n if $z = x + iy$. In other words, the foregoing solution is also the expansion in Laurent series of a function of a complex variable with an essential singularity at the origin. The solution (12.29) may also be expressed as

$$V(r, \theta) \equiv V(z) = \sum_{n=-\infty}^{n=+\infty} A_n z^n$$

In fact $V(z)$ is analytic, i.e. dV/dz exists, except at the origin and any analytic function is a solution of Laplace equation.

We demanded that the solution be periodic in θ but if $V(r, \theta)$ is a potential this condition can be relaxed and replaced by the periodicity of the field ∇V. The solution can include the solutions which follow when $C = 0$. In this case, $F(r) = P \ln r + Q$ and $G(\theta) = R\theta + S$ from which we derive the following linear combination $V = a\theta \ln r + b\theta + c \ln r + d$. If ∇V is periodic then we need to set $a = 0$. With complex number, we may rearrange this expression under the form: $A \ln z + B \ln z^*$ where z^* is the complex conjugate of z.

12.6.5.4 Solutions in Spherical Geometry

In spherical geometry the solution of Laplace equation are obtained in the same way but now the Laplace products are $F(r)G(\theta)H(\varphi)$. The developments are a bit more complicated, but lead to the construction of the *spherical harmonics* which replace the $e^{in\theta}$ in the previous solution. The spherical harmonics $Y_\ell^m(\theta, \varphi)$ are also the eigenfunctions of the horizontal (reduced to θ, ϕ) Laplacian operator. If we assume that the potential can be expanded on the basis of spherical harmonics (which indeed form a complete basis for all functions defined on the sphere), we can write:

$$V(r, \theta, \varphi) = \sum_{\ell=0}^{\ell=\infty} \sum_{m=-\ell}^{m=+\ell} V_m^\ell(r) Y_\ell^m(\theta, \varphi)$$

then, from (12.45) we get:

$$\Delta V = \frac{1}{r^2} \frac{\partial^2}{\partial r^2}(rV) + \frac{\Delta_{\theta\varphi} V}{r^2}$$

and

$$\frac{1}{r^2} \frac{d^2}{dr^2}(rV^\ell) - \frac{\ell(\ell+1)}{r^2} V^\ell = 0 \tag{12.30}$$

since

$$\Delta_{\theta\varphi} Y_\ell^m = -\ell(\ell+1) Y_\ell^m \tag{12.31}$$

The solutions of (12.30) are powers of r, namely r^n with $n = \ell$ or $n = -\ell - 1$. Finally

$$V(r, \theta, \varphi) = \sum_{\ell=0}^{\ell=\infty} \sum_{m=-\ell}^{m=+\ell} (A_{\ell m} r^\ell + B_{\ell m}/r^{\ell+1}) Y_\ell^m$$

These kind of solutions are used when the problem is close to sphericity.

12.7 Exercises

1. From (12.1) show the relations (12.2), (12.3) and (12.4).
2. Show the equivalence between the symmetry of a second order tensor and the relation (12.5).

3. Show that the harmonic oscillator equation

$$\frac{d^2 y}{dx^2} + \lambda y = 0$$

with boundary conditions $y(0) = 0$ and $y(1) = 0$, is a Sturm–Liouville problem. Determine the eigenvalue spectrum.

Appendix: Formulae

Gaussian Integrals

We set

$$I_n(\alpha) = \int_0^\infty x^n e^{-\alpha x^2} dx \qquad (12.32)$$

where $\alpha > 0$ et $n \neq 0$.
   We thus find

$$I_0(\alpha) = \frac{\sqrt{\pi}}{2\alpha^{1/2}}, \qquad I_1(\alpha) = \frac{1}{2\alpha} \qquad (12.33)$$

With the following equality

$$I_{n+2}(\alpha) = -\frac{dI_n}{d\alpha} \qquad (12.34)$$

we get:

$$I_2(\alpha) = \frac{\sqrt{\pi}}{4\alpha^{3/2}}, \qquad I_3(\alpha) = \frac{1}{2\alpha^2}$$

$$I_4(\alpha) = \frac{3\sqrt{\pi}}{8\alpha^{5/2}}, \qquad I_5(\alpha) = \frac{1}{\alpha^3}$$

$$I_6(\alpha) = \frac{15\sqrt{\pi}}{16\alpha^{7/2}}, \qquad I_7(\alpha) = \frac{3}{\alpha^4}$$

or else

$$I_{2n>0} = \int_0^\infty u^{2n} e^{-\alpha u^2} du = \frac{1 \times 3 \times 5 \times \cdots \times (2n-1)}{2^{n+1}} \sqrt{\frac{\pi}{\alpha^{2n+1}}} \qquad (12.35)$$

The symmetry of the gaussian function with respect to the origin implies that

$$\int_{-\infty}^{+\infty} x^{2p+1} e^{-\alpha x^2} dx = 0, \quad \forall p \in \mathbb{N} \tag{12.36}$$

Some Formula of Vectorial Analysis

The following formulae may be demonstrated using tensorial notations:

$$\Delta \mathbf{a} = \nabla \nabla \cdot \mathbf{a} - \nabla \times \nabla \times \mathbf{a} \tag{12.37}$$

$$\nabla \cdot (f\mathbf{a}) = \mathbf{a} \cdot \nabla f + f \nabla \cdot \mathbf{a} \tag{12.38}$$

$$\nabla \times (f\mathbf{a}) = \nabla f \times \mathbf{a} + f \nabla \times \mathbf{a} \tag{12.39}$$

$$\nabla \cdot (\mathbf{a} \times \mathbf{b}) = \mathbf{b} \cdot \nabla \times \mathbf{a} - \mathbf{a} \cdot \nabla \times \mathbf{b} \tag{12.40}$$

$$\nabla \times (\mathbf{a} \times \mathbf{b}) = \mathbf{a} \nabla \cdot \mathbf{b} - \mathbf{b} \nabla \cdot \mathbf{a} + (\mathbf{b} \cdot \nabla)\mathbf{a} - (\mathbf{a} \cdot \nabla)\mathbf{b} \tag{12.41}$$

$$\nabla(\mathbf{a} \cdot \mathbf{b}) = \mathbf{a} \times \nabla \times \mathbf{b} + \mathbf{b} \times \nabla \times \mathbf{a} + (\mathbf{a} \cdot \nabla)\mathbf{b} + (\mathbf{b} \cdot \nabla)\mathbf{a} \tag{12.42}$$

$$\nabla(\mathbf{v} \cdot \mathbf{v}) = 2\mathbf{v} \times \nabla \times \mathbf{v} + 2(\mathbf{v} \cdot \nabla)\mathbf{v} \tag{12.43}$$

The Operators in Various Coordinate Systems

Cylindrical Coordinates (s, φ, z)

$$\nabla f = \begin{vmatrix} \dfrac{\partial f}{\partial s} \\[2mm] \dfrac{1}{s}\dfrac{\partial f}{\partial \varphi} \\[2mm] \dfrac{\partial f}{\partial z} \end{vmatrix} \quad \nabla \times \mathbf{v} = \begin{vmatrix} \dfrac{1}{s}\dfrac{\partial v_z}{\partial \varphi} - \dfrac{\partial v_\varphi}{\partial z} \\[2mm] \dfrac{\partial v_s}{\partial z} - \dfrac{\partial v_z}{\partial s} \\[2mm] \dfrac{1}{s}\dfrac{\partial s v_\varphi}{\partial s} - \dfrac{1}{s}\dfrac{\partial v_s}{\partial \varphi} \end{vmatrix} \quad \Delta \mathbf{v} = \begin{vmatrix} \Delta v_s - \dfrac{v_s}{s^2} - \dfrac{2}{s^2}\dfrac{\partial v_\varphi}{\partial \varphi} \\[2mm] \Delta v_\varphi - \dfrac{v_\varphi}{s^2} + \dfrac{2}{s^2}\dfrac{\partial v_s}{\partial \varphi} \\[2mm] \Delta v_z \end{vmatrix}$$

$$\Delta f = \frac{1}{s}\frac{\partial}{\partial s}\left(s\frac{\partial f}{\partial s}\right) + \frac{1}{s^2}\frac{\partial^2 f}{\partial \varphi^2} + \frac{\partial^2 f}{\partial z^2}, \qquad \nabla \cdot \mathbf{v} = \frac{1}{s}\frac{\partial s v_s}{\partial s} + \frac{1}{s}\frac{\partial v_\varphi}{\partial \varphi} + \frac{\partial v_z}{\partial z}$$

$$(\mathbf{a} \cdot \nabla)\mathbf{b} = \begin{vmatrix} (\mathbf{a} \cdot \nabla)b_s - \dfrac{a_\varphi b_\varphi}{s} \\[2ex] (\mathbf{a} \cdot \nabla)b_\varphi + \dfrac{a_\varphi b_s}{s} \\[2ex] (\mathbf{a} \cdot \nabla)b_z \end{vmatrix} \qquad (\mathbf{v} \cdot \nabla)\mathbf{v} = \begin{vmatrix} (\mathbf{v} \cdot \nabla)v_s - \dfrac{v_\varphi^2}{s} \\[2ex] (\mathbf{v} \cdot \nabla)v_\varphi + \dfrac{v_s v_\varphi}{s} \\[2ex] (\mathbf{v} \cdot \nabla)v_z \end{vmatrix}$$

Remark $\qquad \dfrac{1}{s}\dfrac{\partial}{\partial s}\left(s\dfrac{\partial f}{\partial s}\right) - \dfrac{f}{s^2} = \dfrac{\partial}{\partial s}\left(\dfrac{1}{s}\dfrac{\partial sf}{\partial s}\right)$ $\qquad$ (12.44)

Spherical Coordinates (r, θ, φ)

$$\nabla f = \begin{vmatrix} \dfrac{\partial f}{\partial r} \\[2ex] \dfrac{1}{r}\dfrac{\partial f}{\partial \theta} \\[2ex] \dfrac{1}{r\sin\theta}\dfrac{\partial f}{\partial \varphi} \end{vmatrix} \qquad \nabla \times \mathbf{v} = \begin{vmatrix} \dfrac{1}{r\sin\theta}\dfrac{\partial}{\partial \theta}(\sin\theta v_\varphi) - \dfrac{1}{r\sin\theta}\dfrac{\partial v_\theta}{\partial \varphi} \\[2ex] \dfrac{1}{r\sin\theta}\dfrac{\partial v_r}{\partial \varphi} - \dfrac{1}{r}\dfrac{\partial r v_\varphi}{\partial r} \\[2ex] \dfrac{1}{r}\dfrac{\partial r v_\theta}{\partial r} - \dfrac{1}{r}\dfrac{\partial v_r}{\partial \theta} \end{vmatrix}$$

$$\Delta \mathbf{v} \begin{vmatrix} \Delta v_r - \dfrac{2}{r^2}\left(\dfrac{1}{\sin\theta}\dfrac{\partial(\sin\theta v_\theta)}{\partial \theta} + \dfrac{1}{\sin\theta}\dfrac{\partial v_\varphi}{\partial \varphi} + v_r\right) \\[3ex] \Delta v_\theta + \dfrac{1}{r^2}\left(2\dfrac{\partial v_r}{\partial \theta} - \dfrac{v_\theta}{\sin^2\theta} - \dfrac{2\cos\theta}{\sin^2\theta}\dfrac{\partial v_\varphi}{\partial \varphi}\right) \\[3ex] \Delta v_\varphi + \dfrac{1}{r^2}\left(\dfrac{2}{\sin\theta}\dfrac{\partial v_r}{\partial \varphi} + \dfrac{2\cos\theta}{\sin^2\theta}\dfrac{\partial v_\theta}{\partial \varphi} - \dfrac{v_\varphi}{\sin^2\theta}\right) \end{vmatrix}$$

where the Laplacian of a scalar field reads

$$\Delta f = \dfrac{1}{r}\dfrac{\partial^2 rf}{\partial r^2} + \dfrac{1}{r^2\sin\theta}\dfrac{\partial}{\partial \theta}\left(\sin\theta\dfrac{\partial f}{\partial \theta}\right) + \dfrac{1}{r^2\sin^2\theta}\dfrac{\partial^2 f}{\partial \varphi^2} \qquad (12.45)$$

and

$$\nabla \cdot \mathbf{v} = \dfrac{1}{r^2}\dfrac{\partial}{\partial r}(r^2 v_r) + \dfrac{1}{r\sin\theta}\dfrac{\partial \sin\theta v_\theta}{\partial \theta} + \dfrac{1}{r\sin\theta}\dfrac{\partial v_\varphi}{\partial \varphi}$$

$$(\mathbf{a} \cdot \nabla)\mathbf{b} = \left| \begin{array}{l} (\mathbf{a} \cdot \nabla)b_r - \dfrac{a_\theta b_\theta + a_\varphi b_\varphi}{r} \\[3mm] (\mathbf{a} \cdot \nabla)b_\theta + \dfrac{a_\theta b_r - a_\varphi b_\varphi \cot \theta}{r} \\[3mm] (\mathbf{a} \cdot \nabla)b_\varphi + \dfrac{a_\varphi b_r + a_\varphi b_\theta \cot \theta}{r} \end{array} \right.$$

The Stress-Tensor of a Newtonian Fluid

Cylindrical Coordinates

$$\sigma_{ss} = -P + 2\mu \frac{\partial v_s}{\partial s} + (\zeta - 2/3\mu)\nabla \cdot \mathbf{v}$$

$$\sigma_{\varphi\varphi} = -P + 2\mu \left(\frac{1}{s} \frac{\partial v_\varphi}{\partial \varphi} + \frac{v_s}{s} \right) + (\zeta - 2/3\mu)\nabla \cdot \mathbf{v}$$

$$\sigma_{zz} = -P + 2\mu \frac{\partial v_z}{\partial z} + (\zeta - 2/3\mu)\nabla \cdot \mathbf{v}$$

$$\sigma_{s\varphi} = \mu \left(\frac{1}{s} \frac{\partial v_s}{\partial \varphi} + \frac{\partial v_\varphi}{\partial s} - \frac{v_\varphi}{s} \right)$$

$$\sigma_{sz} = \mu \left(\frac{\partial v_z}{\partial s} + \frac{\partial v_s}{\partial z} \right)$$

$$\sigma_{\varphi z} = \mu \left(\frac{\partial v_\varphi}{\partial z} + \frac{1}{s} \frac{\partial v_z}{\partial \varphi} \right)$$

Spherical Coordinates (r, θ, φ)

$$\sigma_{rr} = -P + 2\mu \frac{\partial v_r}{\partial r} + (\zeta - 2/3\mu)\nabla \cdot \mathbf{v}$$

$$\sigma_{\theta\theta} = -P + 2\mu \left(\frac{1}{r} \frac{\partial v_\theta}{\partial \theta} + \frac{v_r}{r} \right) + (\zeta - 2/3\mu)\nabla \cdot \mathbf{v}$$

$$\sigma_{\varphi\varphi} = -P + 2\mu \left(\frac{1}{r \sin \theta} \frac{\partial v_\phi}{\partial \varphi} + \frac{v_r}{r} + \frac{v_\theta \cot \theta}{r} \right) + (\zeta - 2/3\mu)\nabla \cdot \mathbf{v}$$

$$\sigma_{r\theta} = \mu \left(\frac{1}{r} \frac{\partial v_r}{\partial \theta} + r \frac{\partial}{\partial r} \left(\frac{v_\theta}{r} \right) \right)$$

$$\sigma_{r\varphi} = \mu \left(\frac{1}{r \sin \theta} \frac{\partial v_r}{\partial \varphi} + r \frac{\partial}{\partial r} \left(\frac{v_\varphi}{r} \right) \right)$$

$$\sigma_{\theta\varphi} = \mu \left(\frac{1}{r \sin \theta} \frac{\partial v_\theta}{\partial \varphi} + \sin \theta \frac{\partial}{\partial \theta} \left(\frac{v_\varphi}{r \sin \theta} \right) \right)$$

Further Reading

An introduction to tensors may be found in Lebedev et al. (2010). The theory of singular perturbations, which is very useful to the study of boundary layers, may be found in various books, for instance in Bender and Orszag (1978) or O'Malley (1991). As far as partial differential equations are concerned, there are also a wealth of textbooks, from the classical Courant and Hilbert (1953) or, more recently, Dautray and Lions (1984–1985) or Zwillinger (1992). This last book also deals with all types of ordinary differential equations.

References

Bass, J. (1978). *Cours de Mathématiques*. Paris: Masson.
Bender, C. & Orszag, S. (1978). *Advanced Mathematical Methods for scientists and engineers*. New York: McGraw-Hill.
Courant, R. & Hilbert, D. (1953). *Methods of mathematical physics*. London: Interscience.
Dautray, R. & Lions, J.-L. 1984–1985 *Analyse mathématique et calcul numérique*. Paris: Masson.
Hladik, J. (1993). *Le calcul tensoriel en physique*. Paris: Masson.
Ince, E. L. (1956). *Ordinary differential equations*. New York: Dover.
Lebedev, L., Cloud, M. & Eremeyev, V. (2010). *Tensor analysis with applications in Mechanics*. Singapore: World Scientific.
O'Malley, R. (1991). *Singular Perturbation Methods for Ordinary Differential Equations*. Berlin: Springer.
Zwillinger, D. (1992). *Handbook of differential equations*. New York: Academic Press.

Chapter 13
The Solutions of Exercises

Chapter 1

1. The solution is $2\mathbf{\Omega}$; the velocity field of a solid rotation has a uniform vorticity.
2. One finds that all the components of $[s]$ are zero. The solid body rotation does not introduce any deformation. If $\mathbf{v} = \lambda\mathbf{r}$ then $s_{xx} = s_{yy} = s_{zz} = \lambda$ and $s_{xy} = s_{xz} = s_{yz} = 0$. A solid body rotation can be the velocity field of an incompressible fluid because $Tr([s]) = 0$, but the second velocity field cannot because it is not divergence-free.
3. We just need using the definition of the divergence. For the reciprocal, we observe that

$$\nabla \cdot \mathbf{v} = 0 \iff \frac{\partial v_x}{\partial x} + \frac{\partial v_y}{\partial y} = 0 \iff v_x = -\int \frac{\partial v_y}{\partial y} dx = -\frac{\partial}{\partial y}\int v_y dx$$

showing that the stream function exists and we just need to choose

$$\psi = -\int v_y dx$$

In polar coordinates, we easily find that

$$v_r = \frac{1}{r}\frac{\partial \psi}{\partial \theta}, \qquad v_\theta = -\frac{\partial \psi}{\partial r}$$

while for an axisymmetric flow in cylindrical coordinates

$$v_r = \frac{1}{r}\frac{\partial \psi}{\partial z}, \qquad v_z = -\frac{1}{r}\frac{\partial \psi}{\partial r}$$

© Springer International Publishing Switzerland 2015
M. Rieutord, *Fluid Dynamics*, Graduate Texts in Physics,
DOI 10.1007/978-3-319-09351-2_13

4. The mass of a volume moving with the fluid is constant. The integral of $\frac{\partial \rho}{\partial t} + \nabla \cdot (\rho \mathbf{v})$ is therefore always zero what ever the volume, hence we find the continuity equation again.

5. We have six equations

$$\partial_x v_x = \partial_y v_y = \partial_z v_z = 0$$

$$\partial_x v_y + \partial_y v_x = \partial_y v_z + \partial_z v_y = \partial_z v_x + \partial_x v_z = 0 \qquad (13.1)$$

The first three show that

$$v_x \equiv v_x(y, z), \quad v_y \equiv v_y(x, z), \quad v_z \equiv v_z(x, y)$$

The last three lead to

$$\partial_y^2 v_x = \partial_z^2 v_x = 0$$

with similar expressions for v_y and v_z. We derive that

$$\begin{cases} v_x = a_1 yz + b_2 y + c_1 z + d_1 \\ v_y = a_2 xz + b_3 z + c_2 x + d_2 \\ v_z = a_3 xy + b_1 x + c_3 y + d_3 \end{cases} \qquad (13.2)$$

We use again relations (13.1) to show that $a_1 = a_2 = a_3 = 0$ and $c_1 = -b_1$, $c_2 = -b_2, c_3 = -b3$. It implies that

$$v_i = d_i - \epsilon_{ijk} b_j x_k \qquad \text{or} \qquad \mathbf{v} = \mathbf{d} - \mathbf{b} \times \mathbf{r}$$

which shows that the velocity field is composed of a translation and a solid body rotation.

6. We start from the equation of kinetic energy (1.28) to which we add $\frac{1}{2} v^2$ times the continuity equation. We find

$$\frac{\partial \frac{1}{2} \rho v^2}{\partial t} + \nabla \cdot (\frac{1}{2} \rho v^2 \mathbf{v}) = v_i \partial_j \sigma_{ij}$$

We integrate this equation over the volume V. The integration of the second term in the left-hand side gives zero since $\mathbf{v} \cdot d\mathbf{S} = 0$, while the integral of the first term is the time derivative of the kinetic energy. The third integral needs to be transformed in the following way:

$$\int_V v_i \partial_j \sigma_{ij} dV = \int_V \partial_j (v_i \sigma_{ij}) dV - \int_V \sigma_{ij} \partial_j v_i dV$$

Here, the first integral may be transformed into an integral over the bounding surface, and therefore vanishes because of boundary conditions. The second integral is just the viscous dissipation within the volume. Thus

$$\frac{dE_c}{dt} = -\int_V \mathcal{D} dV$$

If the fluid is incompressible, using (1.48) gives the expected result.

7. We noticed that a biaxial extension can be derived from a uniaxial extension by just changing the sign of the shear.
   So we consider a biaxial extension characterized by a "shear time scale" equal to $-T$. The associated velocity field is $v_x = -x/T, v_y = -y/T, v_z = 2z/T$. Using the definition of the biaxial viscosity, we find

$$\sigma_{xx} - \sigma_{zz} = -\mu_{EB}(-T)/T . \tag{13.3}$$

However, in order to use the definition of the uniaxial viscosity (1.79), we have to exchange the axis $x \to y, y \to z, z \to x$. The velocity field is now $v_x = 2x/T, v_y = -y/T, v_z = -z/T$. From (1.79) we get $\sigma_{xx} - \sigma_{yy} = 2\mu_E(T/2)/T$. Taking into account the exchange of the axis in (13.3) we also get $\sigma_{xx} - \sigma_{yy} = \mu_{EB}(-T)/T$. The result follows.

Chapter 2

1. *About buoyancy*

   (a) Let M_g be the mass of the ice and V_{im} the volume below the water level. We also introduce the initial volume of water V_{e_i}, and the final volume of water V_{e_f}, when the ice is melted. Since the container is the same initially and finally, comparing the level of water is equivalent to comparing the volumes $V_{e_i} + V_{im}$ and V_{e_f}. From Archimedes theorem, the equilibrium of the ice cube means that

$$M_g \mathbf{g} - \rho_e V_{im} \mathbf{g} = \mathbf{0} \quad \Longrightarrow \quad M_g = \rho_e V_{im}$$

where ρ_e is the density of water. When the ice melts, it transforms into water so that

$$M_g = \rho_e (V_{e_f} - V_{e_i}) = \rho_e V_{im}$$

We therefore derive that $V_{e_i} + V_{im} = V_{e_f}$, showing that the level of water remains identical when the ice melts.

(b) The reasoning is the same as before, but introducing a piece of metal of mass M_m, we now have:

$$(M_g + M_m)\mathbf{g} - \rho_e V_{im}\mathbf{g} = 0$$

After the ice melting $V_F = V_{e_f} + M_m/\rho_m$, where V_F is the final volume of water plus that of the metal, which has to be compared to the initial volume $V_I = V_{e_i} + V_{im}$.

$$V_I = V_{e_i} + (M_g + M_m)/\rho_e$$

from the first equation. But $M_g = \rho_e(V_{e_f} - V_{e_i})$ is still true because the melting ice gives water. Thus

$$V_I = V_{e_f} + M_m/\rho_e > V_{e_f} + M_m/\rho_m = V_F$$

because the density of metal ρ_m is larger than the density of water ρ_e. Hence, the water level decreases.

(c) When the ice is melted, the cork is floating. Let V'_{im} be its volume under the water in the final state. Before the ice melts, the level of water in the glass is given by the volume $V_I = V_{e_i} + (M_g + M_l)/\rho_e$, following the same reasoning as above. M_l is the mass of the cork. Since the cork floats, its equilibrium implies that $V'_{im} = M_l/\rho_e$. It is then easy to check that

$$V_I = V_{e_i} + (M_g + M_l)/\rho_e = V_F = V_{e_f} + V'_{im}$$

since $M_g = \rho_e(V_{e_f} - V_{e_i})$. The level thus remains the same.

(d) This is because the air density decreases with altitude while that of water remains approximately constant.

(e) The balloon moves towards the front of the car. Indeed, when the car starts, the effective gravity has a small component towards the back of the car. The buoyancy force therefore has a small component in the opposite direction, namely towards the front. Do the experiment!

2.(a) The denser goes under the lighter (oil).

(b) $P(z)$ curve is made of two line segments of slope $-\rho_e g$ in the water and $-\rho_h g$ in the oil.

(c) The ball being denser than oil but less dense than water will stay at the interface of the two liquids. Although part of the ball is in the water and the other part in the oil, Archimedes theorem can be used because the pressure field is continuous everywhere over the ball. Thus

$$\rho_{wood} V_{ball} = \rho_{oil} V_{oil} + \rho_{water} V_{water}$$

Of course $V_{ball} = V_{oil} + V_{water}$. It follows that

$$\frac{V_{water}}{V_{ball}} = \frac{\rho_{wood} - \rho_{oil}}{\rho_{water} - \rho_{oil}}$$

Numerically $V_{water}/V_{ball} = 3/4$.

3. Let H_h be the height of the oil free surface, H_{he} be the height of the oil–water interface and H_e the height of the free surface of water in the other branch of the tube. We have $H_h - H_{he} = 2\,\text{cm}$ and $P(H_h) = P(H_e) = P_{atm}$. From Pascal Theorem (2.5),

$$P(H_h) + \rho_{oil}(H_h - H_{he})g = P(H_e) + \rho_{water}(H_e - H_{he})g$$

which allows us to express H_e as a function of H_{he}, namely $H_e = 2\,cm\rho_{oil}/\rho_{water} + H_{he}$. Now we need another equation. It is given by the fact that the liquids are incompressible, therefore if the level moves of x cm in one arm of the tube it moves of the same quantity in the other arm. Thus $H_e = H_0 + x$ and $H_{he} = H_0 - x$. This leads to $x = \rho_{oil}/\rho_{water} = 0.6\,\text{cm}$ and all the other heights, using $H_0 = 10\,\text{cm}$.

4.(a) The wood sphere being in equilibrium, the resultant of forces is vanishing, thus

$$M_b\mathbf{g} - \int_{(S)} P_{water}d\mathbf{S} + \mathbf{R} = 0$$

where M_b is the mass of the ball and $\mathbf{R}$ is the reaction of the reservoir floor. We therefore need to correctly evaluate the resultant of pressure forces. We note that we cannot use the Archimedes theorem because pressure is not continuous over the whole sphere surface. The integral needs to be computed explicitly. We may note that the contribution of the air pressure is vanishing since it is the same everywhere. We may note in addition that the resultant has just one component along $\mathbf{e}_z$ so that the evaluation of the integral may be done as follows:

$$\mathbf{e}_z \cdot \int_{(S)} Pd\mathbf{S} = -\int_0^{\theta_0} P(\theta)2\pi R \sin\theta \cos\theta R d\theta$$

where θ_0 is such that $\sin\theta_0 = r/R$ and $\cos\theta_0 = -\sqrt{1 - r^2/R^2}$; note the $\cos\theta$ factor due to the projection on $\mathbf{e}_z$. The expression of $P(\theta)$ comes from Pascal's formula : $P(\theta) = g\rho_{water}[H + R(\cos\theta_0 - \cos\theta)]$. It turns out that

$$\mathbf{e}_z \cdot \int_{(S)} P_{water}d\mathbf{S} = g\rho_{water}2\pi R^2 \left(\frac{1}{2}(H + R\cos\theta_0)\sin^2\theta_0 + \frac{R}{3}(\cos^3\theta_0 - 1)\right)$$

The force exerted by the sphere on the floor of the basin is therefore

$$\mathbf{R} = M_b \mathbf{g} \left\{ 1 + \frac{\rho_{\text{water}}}{\rho} \left[\frac{3}{4} \left(\frac{H}{R} + \cos \theta_0 \right) \sin^2 \theta_0 + \frac{1}{2} (\cos^3 \theta_0 - 1) \right] \right\}$$

(b) Numerically: $|\mathbf{R}| = 170$ N
(c) The sphere rises up before emerging at the surface. Indeed, this is determined by the vanishing of the resulting force, namely when

$$H \leq H_c = \frac{2R}{3} \frac{1 - \cos^3 \theta_0 - 2\rho/\rho_{\text{water}}}{\sin^2 \theta_0} - R \cos \theta_0$$

But with $H_c \geq R(1 - \cos \theta)$.
5. (a) The balloon takes off if its buoyancy is larger than its weight. We thus find that $(M_b + M_H)/V_b < \rho_0$.
   (b) The cruising altitude of the balloon is reached when its buoyancy compensates its weight or, in other words, when its mean density equals that of the air around, namely $\rho_b = (M_b + M_H)/V_b$. We thus find $z_{\text{flight}} = z_0(1 - \rho_b/\rho_0)^{(\gamma-1)/\gamma}$
6. (a) The hydrogen pressure is identical to that of the outside air.
   (b) Air and hydrogen being assumed ideal gases, the equality of pressures and temperatures implies that $V_H = m_a n_H / \rho_a$ where n_H is the mole number of hydrogen, m_a is the mass of an air mole and ρ_a is the air density. Since ρ_a decreases with altitude, η increases with altitude.
   (c) We just write that the buoyancy is larger than the weight so that $\rho_a V_H \geq M_b + M_H$. Using the expression of the hydrogen volume and that $M_H = n_H m_H$, we get the condition $n_H \geq M_b/(m_a - m_H)$.
   (d) The foregoing condition does not depend explicitly on the altitude so if the amount of hydrogen does not vary, the balloon rises. It will not rise indefinitely because, as we saw before, the volume of hydrogen increases with the altitude. When it exceeds $V_b = m_a n_H / \rho_a$, the balloon loses hydrogen. The altitude of the balloon stabilizes at a value such that $V_b = m_a n_H / \rho_a$ and $n_H = M_b/(m_a - m_H)$. One then finds that the cruising altitude is $z = z_0 \left(1 - [\rho_b/\rho_0/(1 - m_H/m_a)]^{(\gamma-1)/\gamma} \right)$, which is quite close to the value of the preceding problem.
7. We just need to compute the surface gravity, to which we withdraw the centrifugal acceleration. We find $g = GM/R^2 = 24.8$ m/s^2 and $\Omega^2 R = 2.25$ m/s^2. The effective gravity is therefore $g_e = 22.5$ m/s^2. The molecular mass of the gas is $M = 0.85 \times 2 + 0,15 \times 4 = 2.3$ g/mole; $c_p = 3.35\, R_*$. Finally, the gradient of temperature is $-g/c_p = -1.9$ K/km.
8. (a) The axial symmetry of the system implies that the resultant of pressure forces has just one component along $\mathbf{e}_z$, namely $F_z = \int_{(S)} P d\mathbf{S} \cdot \mathbf{e}_z$. We note that $dS_z = 2\pi \tan^2 \alpha\, (H - z) dz$. Finally, $F_z = \pi \rho g h^2 (H - h/3) \tan^2 \alpha$.
   (b) If $h = H$ then $F_z = 2M_{lg}$. We derive that if the mass of the funnel is smaller than twice the liquid mass, the liquid can move the funnel up and flow away.

9. *A polytropic model of the Sun*:

(a) We note that

$$\frac{dM}{dr} = 4\pi r^2 \rho$$

The equation of state together with the equation of mechanical equilibrium give

$$(1 + 1/n)K\rho^{1/n-1}\frac{d\rho}{dr} = -\frac{GM(r)}{r^2}$$

After multiplication by r^2 and taking the derivative, we introduce θ and get:

$$\frac{1}{r^2}\frac{d}{dr}r^2\frac{d\theta}{dr} + \frac{4\pi G\rho_c^{1-1/n}}{(n+1)K}\theta^n = 0$$

Making the change of variable $r = r_0\xi$ we obtain Emden equation.

(b) We just need to insert the expression of ρ as a function of θ in the equation of state.

(c) First insert the expression of P and ρ in the equation of mechanical equilibrium, then use this equation at the stellar surface. Central pressure is eliminated thanks to the expression of r_0.

(d) Use the expression of r_0.

(e) We find that $r_0 = 8.0\ 10^7$ m. $\rho_c = 1.61\ 10^5$ kg/m^3, $p_c = 3.2\ 10^{16}$ Pa.

(f) Helium ions, protons and electrons all contribute to the pressure. But electrons do not contribute to the molecular mass of the gas. Let x_p be the fractional number of protons, x_{He} that of helium ions and x_e that of electrons. The molecular mass of the gas is

$$\mathcal{M} = x_p M_p + x_{He} M_{He}$$

because the mass of electrons is negligible. Electrical neutrality imposes:

$$x_e = x_p + 2x_{He}$$

but we also have

$$x_p + x_e + x_{He} = 1$$

so that we can deduce

$$2x_p + 3x_{He} = 1$$

If Y is the mass fraction of helium and X that of protons then $Y = 1 - X$ and

$$Y = \frac{x_{He} M_{He}}{\mathcal{M}} \quad \text{and} \quad X = \frac{x_p M_p}{\mathcal{M}}$$

Using this relation and the previous one we get

$$\mathcal{M} = \frac{4}{8 - 5Y} \quad \text{g/mole}$$

If $Y = 0.28$, we obtain $\mathcal{M} = 0.61$ g/mole. This value is smaller than that of protons, in spite of the presence of helium ions. It shows the importance of electrons in a fully ionized plasma. We deduce

$$T_c = 1.43 \times 10^7 \text{ K}$$

The values of the thermodynamic conditions at the Sun's centre are therefore in the right order of magnitude compared to more elaborate models. Polytropic models can thus be used to study some properties of stars without having to deal with the complexity of a realistic equation of state or other specificities (like thermal conductivity).

Chapter 3

1. If the flow is irrotational $v = \nabla \Phi$, hence v is perpendicular to surfaces $\Phi = $ Cst. Since streamlines are parallel to v they are also perpendicular to the equipotentials of Φ.
2. We first project the equation of momentum along Oz, since v has no component along e_z, we find $\partial_z p = -\rho g \implies P_A = P_{atm} + \rho g h_A$ and $P_B = P_{atm} + \rho g h_B$. Focussing on the streamline going through A and B, Bernoulli's theorem shows that $V_A^2 + P_A/\rho = V_B^2 + P_B/\rho$, hence $h_A - h_B = (V_B^2 - V_A^2)/2g$. We have $V_A S_A = V_B S_B$.
3. (a) One may check that $\nabla \times v = 0$ or that $\Phi = \Omega a^2 \theta$ is a solution for a velocity potential.
   (b) In the outer domain the flow is irrotational and $p/\rho + \frac{1}{2}v^2 = $ Cst $\implies p = p_\infty - \rho \Omega^2 a^4/(2r^2)$. In the inner domain, we need Euler equation to derive the pressure field, which turns out to be $p = \frac{1}{2}\rho \Omega^2 r^2 + P_\infty - \rho \Omega^2 a^2$.
   (c) This quantity is constant in the outer domain, but depends on r in the inner domain. There it is constant, only *along* the streamlines. The constant changes from one streamline to the other. On the contrary, in the irrotational region, the constant is the same for all the streamlines.
   (d) The vortex central depression is given by $P_\infty - p(0) = \rho v_{max}^2$ because the velocity is maximum at $r = a$. If v=50 m/s then $P_\infty - p(0) \simeq 3250$ Pa.

(e) If the vortex is in the water, the shape of the air–water interface is given by $p(\text{surf})=p_{atm}$. The flow being purely horizontal, the vertical variations are given by the hydrostatic balance namely $\partial p/\partial z = -\rho g$. Therefore, the water pressure is given by the previous expressions to which $-\rho g z$ is added. Using $p(\text{surf})=p_{atm}$, we get the equation of the surface $z_{\text{surf}} = \Omega^2(r^2 - 2a^2)/(2g)$ for the inner domain: this is an axisymmetric paraboloid. For the outer region $z_{\text{surf}} = -\Omega^2 a^4/(2gr^2)$, which is the equation of an axisymmetric hyperboloid.

4.(a) i. The flow is irrotational because at $t = 0$ $\mathbf{v} = \mathbf{0}$, which is irrotational, and because driving forces derive from a potential.

ii. The flow is steady and irrotational. $\frac{1}{2}v^2+p/\rho+gz$ is constant everywhere in the fluid. Using this expression and its constancy at the reservoir's surface and in the outflow, we find

$$\frac{dh}{dt} = -\frac{s}{S}\sqrt{2gh} \implies t_{\text{purge}} = \frac{S}{s}\sqrt{\frac{2h_0}{g}}$$

where we noticed that the fluid pressure is the same in the two places, and that the reservoir surface velocity is $-dh/dt$.

iii. We need to show that the term $\partial\Phi/\partial t$ is very small compared to the others. The velocity in the reservoir is identical to that of the free surface dh/dt. Since $v_z = \partial\Phi/\partial z$, we determine Φ and $\partial\Phi/\partial t = (s/S)^2gz \ll gz$. The acceleration term is, as the kinetic energy, very small compared with the potential energy and pressure terms.

(b) i. The equation for the potential (3.22) written between A and B gives

$$\partial_t(\Phi_B - \Phi_A) + (v_B^2 - v_A^2)/2 + (p_B - p_A)/\rho + g(z_B - z_A) = 0.$$

We should note that $\Phi_B - \Phi_A = \oint_A^B \mathbf{v}\cdot d\mathbf{l} \sim v_B l$ Because the fluid velocity is much larger in the tube than in the reservoir, so the part of the path AB which is in the tube dominates the integral. With the assumptions of the text, we find that

$$l\frac{dv}{dt} + v^2/2 = v_\infty^2/2$$

This differential equation is easily solved if we note that $1/(x^2 - 1)$ is the derivative of $argth(x)$. Finally, $v = v_\infty$ th $(v_\infty t/(2l))$.

ii. The transient lasts $2l/v_\infty$. We should note that this transient corresponds to the starting motion of the water in the tube, because of its inertia, which grows with the tube length. Numerically we find 0.58 s.

5.(a) Imposing a jolt to the tube is similar to imposing the fluid an inertial force $\mathbf{f} = \rho\mathbf{a}(t)$, where $\mathbf{a}(t)$ is the tube's acceleration. The tube is assumed to be solid, $\mathbf{a}$ is independent of $\mathbf{x}$ and the force is a potential force. We may use

Lagrange theorem. We may observe that the velocity is almost uniform since the fluid is incompressible and the tube radius small compared to its length.

(b) The foregoing remark implies that $\Phi_A - \Phi_B = \oint_B^A \mathbf{v} \cdot d\mathbf{l} = VL$.

(c) We write (3.22) at the two free surfaces of the fluid (at A and B). We find that $\partial_t(\Phi_A - \Phi_B) + g(z_A - z_B) = 0$. Let δh be the level variation at A, then $V = d\delta h/dt$ and $z_A - z_B = 2\delta h$. We find that the surface of the fluid oscillates at a frequency

$$f = \frac{1}{2\pi}\sqrt{\frac{2g}{L}}$$

6.(a) Here too we may verify that $\nabla \times \mathbf{v} = \mathbf{0}$, but it is more efficient to note that the function $\Phi = \int v(r,t)dr$ is a potential for this velocity field.

(b) In the water $\nabla \cdot \mathbf{v} = 0$, consequently $v = C/r^2$. Noting that at $r = R(t)$, $v(r,t) = \dot{R}$, then $v(r,t) = \dot{R}R^2/r^2$.

(c) For an isentropic ideal gas $PV^\gamma = $ Cst, thus $PR^{3\gamma} = $ Cst.

(d) The velocity potential is $\Phi(r,t) = -\dot{R}R^2/r$. Equation (3.22) used at $r = R$ gives

$$\ddot{R}R + 3\dot{R}^2/2 = (p - p_0)/\rho = p_0/\rho\left((R_0/R)^{3\gamma} - 1\right) .$$

(e) If the radius of the bubble slightly oscillates around its equilibrium value then, setting $R(t) = R_0(1 + \varepsilon(t))$ with $\varepsilon \ll 1$, we find after linearization, that ε follows the harmonic oscillator equation with a frequency:

$$f = \frac{1}{2\pi R_0}\sqrt{\frac{3\gamma p_0}{\rho}}$$

Numerically, f(1 mm) = 3262 Hz and f(5 mm) = 652 Hz. These frequencies corresponds to sound waves that are readily audible and which give birth to the songs of springs.

7. For a barotropic fluid the equation of vorticity reads

$$\frac{D\omega}{Dt} = (\omega \cdot \nabla)\mathbf{v} - \omega\nabla \cdot \mathbf{v}$$

We divide this equation by ρ and subtract the continuity equation times ω/ρ^2. You should have noted that the continuity equation divided by ρ^2 may also be written:

$$\frac{D(1/\rho)}{Dt} + \frac{\nabla \cdot \mathbf{v}}{\rho} = 0$$

Chapter 4

1. The momentum flux is the same at the inlet and the outlet of the pipe. The resulting force therefore comes from the pressure field at entrance $\pi R^2 p_1 \mathbf{e}_z$ and outlet $(-\pi R^2 p_2 \mathbf{e}_z)$ of the pipe; the sum of it gives $-\pi R^2 G_p L \mathbf{e}_z$.

2. From (4.27) and (4.28), we get

$$F_t = -\frac{\alpha}{3} \frac{3r - 3 - (r+1)\ln r}{2r - 2 - (r+1)\ln r} F_p$$

where $r = h_1/h_2 = 1 + \varepsilon$. With a first order expansion in ε, we find

$$F_t \simeq \frac{\alpha F_p}{3\varepsilon} = \frac{h_2}{3\ell} F_p$$

The needed force is therefore $3.3\ 10^{-4}$ times the weight, so a force similar to that necessary to lift up a weight of $330\,g$!

3.(a) On the free surface of the fluid the pressure is constant. The z-component of Navier–Stokes equation shows that the pressure does not vary along Ox. The x-component of the Navier–Stokes equation thus gives

$$\nu \frac{\partial^2 V}{\partial z^2} + g \sin \alpha = 0$$

The boundary conditions are $v = 0$ at $z = 0$ and $\partial v/\partial z = 0$ at $z = h$, namely no-slip at the bottom and stress-free at the surface. The solution is easy to derive:

$$V(z) = \frac{g \sin \alpha}{2\nu}(h - z)z$$

(b) The volume flux through a cross section is

$$Q = \int_0^h V(z)dzS/h = \frac{g \sin \alpha}{12\nu} h^2 S$$

4.(a) The flow is axisymmetric (no dependance with respect to θ) and invariant with respect to translation along Oz. We project the Navier–Stokes equation along $\mathbf{e}_\theta$ and verify that the nonlinear terms disappear. The form of the vector Laplacian in cylindrical coordinates leads to the equation $\Delta v(r) - v(r)/r^2 = 0$. Using (12.44), we get

$$\frac{\partial}{\partial r}\left(\frac{1}{r} \frac{\partial r v(r)}{\partial r}\right) = 0$$

which is easily integrated, giving $v(r) = Ar + B/r$. The constants A and B are calculated using the boundary conditions. We find

$$A = \frac{\Omega_1 R_1^2 - \Omega_2 R_2^2}{R_1^2 - R_2^2} \quad \text{and} \quad B = \frac{(\Omega_1 - \Omega_2) R_1^2 R_2^2}{R_2^2 - R_1^2}$$

(b) The torque is defined by $\mathbf{C} = \int_{(S)} \mathbf{r} \times [\sigma] d\mathbf{S}$, thus

$$\mathbf{C} = -2\pi R_2 L \mu \left(\frac{\partial}{\partial r} \frac{v}{r} \right)_{r=R_2} \mathbf{e}_z = 4\pi L \mu B R_2^{-2} \mathbf{e}_z$$

$$\mathbf{C} = 4\pi \mu L \frac{(\Omega_1 - \Omega_2) R_1^2}{R_2^2 - R_1^2}$$

where L is the length of the cylinder.

(c) This is possible if one of the cylinders rotates at a given rate and if we measure the torque exerted on the other cylinder (this is the way viscosities are measured with the Couette viscometer).

5.(a) The pressure gradient outside the boundary layer is just $-UU'$.

(b) Mass conservation yields

$$g'(\eta) = \frac{U(x)b'(x)}{V(x)} \eta f'(\eta) - \frac{U'(x)b(x)}{V(x)} f(\eta) \tag{13.4}$$

while momentum conservation implies

$$f''(\eta) = b(x)V(x)gf' + b^2(x)U'(x)(f^2 - 1) - b(x)b'(x)U(x)\eta ff' \tag{13.5}$$

If self-similar solutions exist, then each coefficient depending on x is a constant. This leads to the requested solutions.

(c) Adding the first and second solutions, we observe that $Ub^2 = (c_1 + c_2)x + a$, where a is a new constant. Eliminating b^2 from the second relation, we obtain a differential equation for U whose solution is:

$$U(x) = Ax^{c_2/(c_1+c_2)} \quad \text{and} \quad b(x) = \left[\frac{c_1 + c_2}{A} \right]^{1/2} x^{c_1/2(c_1+c_2)}$$

(d) The three constants c_1, c_2, c_3 give the length and velocity scales. One of them is therefore arbitrary and we may require that $c_1/2 + c_2 = 1$ without loss of generality. With this relation (13.4) gives:

$$c_3 g = c_1 \eta f/2 - F$$

Inserting this in (13.5), we obtain Falkner–Skan equation:

$$f''' + FF'' - c_2(f^2 - 1) = 0$$

Then, it turns out that $m = c_2/(2 - c_2)$, so that $U(x) = Ax^m$ and $b(x) \propto x^{(1-m)/2}$

(e) From the form of $b(x)$, the thickness of the layer is constant if $c_1 = 0$ or if $c_2 = 1$ and $m = 1$. Blasius equation is found again when the velocity outside the boundary layer is uniform, namely when $m = 0$. Falkner–Skan equation then reads $F''' + FF'' = 0$. It differs from Blasius equation by a factor 2 because $b(x)$ also misses a factor 2.

Chapter 5

1. The flow associated with the eigenmode of the air column is purely radial, thus depending only on the distance to the centre of the "sphere". The pressure disturbance δp verifies the wave equation (5.17). Denoting the mode frequency ω, we have $(\Delta + \omega^2/c_s^2)\delta p = 0$ but $\delta p \equiv \delta p(r)$, so that

$$\frac{1}{r}\frac{d^2}{dr^2}(r\delta p) + \frac{\omega^2}{c_s^2}\delta p = 0$$

whose solution is $\delta p \propto \frac{\sin kr}{r}$ with $k = \omega/c_s$. At the end of the instrument, at $r = L$, $\delta p = 0$ so that the fundamental mode is such that $kL = \pi$. Let ν be its frequency ($\nu = \omega/2\pi$), then $L = c_s/2\nu$. The bassoon length should therefore be $L = 347/2 * 58.27 = 298$ cm, which is quite close to the real length. One may observe that the dispersion relation of the modes is exactly that of the flute.

2. The mode frequency is proportional to the sound velocity, which is itself proportional to the square root of the temperature. When the temperature increases from 10 to 30 °C (50 to 86 F), the frequency is increased by a factor $\sqrt{303/283} = 1.035$, which a quarter of tone (half a tone between two notes is $2^{1/12} = 1.06$). Thus the temperature variation induces a quarter of tone variation for all the notes of the instrument.

3. The dispersion relation $\omega = \sqrt{gk}$ implies that $\lambda = gT^2/2\pi$, where λ is the wavelength and T the wave period. Numerically, we find $\lambda = 351$ m and $v_\phi = 23$ m/s. In 57 h, these waves cross 4,800 km.

4. These are shallow water waves. Their velocity is $\sqrt{gh} = 221$ m/s. They need 6 h to cross the Atlantic ocean, thus after two crossing they are back, in phase with the tidal potential. There is a resonance. This is the reason why the tides on the Atlantic shores have an important amplitude.

5. We need to use expressions (5.37) and (5.42) with a finite depth. It follows that

$$\omega^2 = \frac{(\rho_e - \rho_a)gk + \gamma k^3}{\rho_e + \rho_a \, \text{th}\,(kH)}\, \text{th}\,(kH)$$

In the shallow water conditions $kH \ll 1$, so that

$$\omega^2 = (1 - \rho_a/\rho_e)gHk^2 + \frac{\gamma H}{\rho_e}k^4$$

6. From the equation of state of an ideal gas:

$$\frac{T_2}{T_1} = \frac{P_2}{P_1}\frac{\rho_1}{\rho_2} = \left(1 + \frac{2\gamma}{\gamma+1}(m-1)\right)\left(\frac{(\gamma-1)m+2}{(\gamma+1)m}\right)$$

where we set $m = M_1^2$. $T_2/T_1 > 1$ is equivalent to

$$(\gamma + 1 + 2\gamma(m-1))\left(\gamma - 1 + \frac{2}{m}\right) > (\gamma+1)^2$$

This inequality may also be written:

$$(\gamma - 1)(\gamma m + 1)(1 - 1/m) > 0$$

which is true when $m > 1$.

7.(a) Starting from the second jump condition (5.68), we eliminate v_2, thanks to the first condition (5.67). We thus derive the expression of v_1 then that of v_2.

 (b) Using the definition of Fr_2, (5.67) and (5.69) we can derive the requested expression. Then

$$\text{Fr}_2 \gtreqless 1 \iff 2\text{Fr}_1^{2/3} \gtreqless \sqrt{1 + 8\text{Fr}_1^2} - 1$$

Raising this expression to the cubic power, after simplification, we find

$$\sqrt{1 + 8\text{Fr}_1^2} \gtreqless 1 + 2\text{Fr}_1^2 \iff 1 \gtreqless \text{Fr}_1^2$$

which is the requested result.

8. We have

$$\frac{d}{d\tau}\int_{-\infty}^{+\infty}\zeta dx = \int_{-\infty}^{+\infty}\frac{\partial\zeta}{\partial\tau}dx = \int_{-\infty}^{+\infty}\left(-\frac{3}{2}\zeta\frac{\partial\zeta}{\partial x} - \frac{1}{6}\frac{\partial^3\zeta}{\partial x^3}\right)dx$$

Since the function and its second derivative are vanishing in $\pm\infty$, we note that the first integral is independent of time. It reflects mass conservation: the integral indeed measures the area below the curve $\zeta(x)$. This area is, in two-dimension, the volume and therefore also the mass since $\rho = $ Cst.

Proceeding in the same way for the second integral, it turns out that

$$\frac{d}{d\tau} \int_{-\infty}^{+\infty} \zeta^2 dx = -\left[\frac{\partial \zeta^3}{\partial x}\right]_{-\infty}^{+\infty} - \frac{1}{3}\left[\zeta \frac{\partial^2 \zeta}{\partial x^2}\right]_{-\infty}^{+\infty} + \frac{1}{6}\left[\left(\frac{\partial \zeta}{\partial x}\right)^2\right]_{-\infty}^{+\infty} = 0$$

The conserved quantity is here related to the mechanical energy of the system.

Chapter 6

1. The instability can develop only if Jeans length is smaller that the diameter of the sphere. Since the density is constant, $\rho = 3M/4\pi R^3$. Writing $\lambda_J < 2R$ we get

$$R < R_c = \frac{3GM}{\pi^2 c_s^2}$$

where G is the gravitation constant and c_s is the sound velocity. If we observe that GM/R is nothing but the order of magnitude of the escape velocity squared at the surface of the sphere, then the foregoing inequality just means that the molecular velocity (which is similar to the sound velocity), need to be smaller than the escape velocity. The system is gravitationally bound: one may check that the internal energy of the gas is indeed less than the absolute value of the gravitational energy of the sphere. Numerically, we find that $R_c = 1.2$ light-year. An interstellar cloud with diameter of one light-year is stable, while a bigger one with diameter of 10 light-years is unstable (according to this model of course).

2. (a) The specific angular momentum $s^2\Omega(s)$ needs to be a growing function of s (see Sect. 6.2.1).

   (b) We write Euler's equation in cylindrical coordinates. The disturbances of the flow $s\Omega(s)$ verify:

$$\begin{cases} (\lambda + im\Omega)u_s - 2\Omega u_\theta = -\frac{\partial p}{\partial s} \\ (\lambda + im\Omega)su_\theta + u_s \frac{\partial}{\partial s}(s^2\Omega) = -imp \\ \frac{\partial}{\partial s}(su_s) + imu_\theta = 0 \end{cases} \qquad (13.6)$$

   where the third equation is just $\nabla \cdot \mathbf{u} = 0$. We observe that in the first and third domain $\frac{\partial}{\partial s}(s^2\Omega) = 0$, while in the second $\frac{\partial}{\partial s}(s^2\Omega) = 2As$.

   (c) The radial component of the velocity and the pressure need to be continuous.

   (d) Using the three previous (13.6), we eliminate the pressure and u_θ. Then, we observe that in the equation of u the original flow intervene through $\frac{\partial}{\partial s}\frac{1}{s}\frac{\partial s^2\Omega}{\partial s}$, which is zero in the three domains.

(e) The solutions are :

$$\text{domain I} \quad u_{\mathrm{I}}(s) = A_1 s^{m-1}$$
$$\text{domain II} \quad u_{\mathrm{II}}(s) = A_2 s^{m-1} + B_2/s^{m+1}$$
$$\text{domain III} \quad u_{\mathrm{III}}(s) = B_3/s^{m+1}$$

(f) Using the last two equations of (13.6), we get the pressure perturbation:

$$m^2 p = im2asu - (\lambda + im\Omega)s\frac{\partial}{\partial s}(su)$$

With the continuity of pressure and u, we derive the interface conditions:

$$-(\lambda + im\Omega)_\eta \eta^2 u_{\mathrm{I}}'(\eta) = im2A\eta u_{\mathrm{II}}(\eta) - (\lambda + im\Omega)_\eta \eta^2 u_{\mathrm{II}}'(\eta)$$

$$-(\lambda + im\Omega)_1 u_{\mathrm{III}}'(1) = im2Au_{\mathrm{II}}(1) - (\lambda + im\Omega)_1 u_{\mathrm{II}}'(1)$$

$$u_{\mathrm{I}}(\eta) = u_{\mathrm{II}}(\eta) \quad \text{and} \quad u_{\mathrm{II}}(1) = u_{\mathrm{III}}(1)$$

(g) These four conditions lead to the requested relation. After elimination of A_1 and B_3, we verify that A_2 and B_2 are solutions of

$$\begin{cases} (\lambda + im\Omega_\eta)(A_2 - B_2\eta^{-2m}) = (2iA + \lambda)(A_2 + B_2\eta^{-2m}) \\ (\lambda + im\Omega_1)(A_2 - B_2) = (2iA - \lambda - im\Omega_1)(A_2 + B_2) \end{cases} \tag{13.7}$$

(h) If $m = 1$ we find that $(\lambda + i\Omega_0/2)^2 = -\Omega_0^2/4$ namely that $\lambda = 0$ or $\lambda = -i\Omega_0$. In the two cases the perturbation is neutral, hence the stability. If $m = 2$, we find again that $\lambda = -i\Omega_0$.

3. *Fjørtoft theorem* : Taking the real part of (6.21) we get:

$$\int_a^b \left[(|D\psi|^2 + k^2|\psi|^2) + \frac{k|\psi|^2 U''(-\lambda_I + kU)}{|\lambda + ikU|^2} \right] dz = 0$$

where λ_I is the imaginary part of λ. We consider the case where the instability develops so that $\lambda_R \neq 0$. In this case, because of Rayleigh condition (6.22), we know that

$$\int_a^b \frac{U''|\psi|^2}{|\lambda + ikU|^2} dz = 0$$

The λ_I-term therefore disappears. Hence, we can replace λ_I by any constant we wish. We choose to replace λ_I by kA, which leads to (6.82). We conclude that

$$\int_a^b \frac{k^2|\psi|^2 U''(U - A)}{|\lambda + ikU|^2} dz = -\int_a^b (|D\psi|^2 + k^2|\psi|^2) dz < 0$$

which is possible in $[a, b]$ only if there exists some interval where the quantity

$$U''(U - A)$$

is negative. The constant A is not fixed so we may choose the value of the velocity at the inflexion point. Since U'' and $U - A$ change sign at this point, we see that all the profiles where U'' and $U - A$ are of the same sign can be considered as stable. This stability criterion can be expressed in simpler way: If there exists a constant A such that $U''(U - A) > 0$ in the whole interval $[a, b]$ then $\psi = 0$ and the solution is stable.

Chapter 7

1. Writing $dh = Tds + dP/\rho = c_p dT$, we get a relation on the gradients. Noting that $\nabla P = \rho \mathbf{g}$ we get:

$$\nabla T - \frac{\mathbf{g}}{c_p} = T\nabla s/c_p$$

Since $(\nabla T)_{ad} = \frac{\mathbf{g}}{c_p}$, we derive the requested relation.

Using (7.8) we derive $\nabla T_{pot} = \frac{T_{pot}}{c_p} \nabla s$ and thus (7.9).

2. In the standard model of the stratosphere the temperature gradient is positive or zero. This is certainly larger than the adiabatic, which is negative. The stratosphere is therefore a stable layer as its name reflects.

3. Let A_v be amplitude of the dimensionless velocity $\mathbf{u}$. From the chosen scales we get:

$$\mathrm{Re} = \frac{Vd}{\nu} = \frac{\kappa}{\nu} A_v = \frac{A_v}{\mathcal{P}}$$

From the definition of the stream function, we have

$$u^2 = |D\psi|^2 + k^2|\psi|^2 = 3\pi^2/2|\psi|^2 = A_v^2 .$$

The amplitude for ψ is given by (7.71) and thus

$$\mathrm{Re} = \frac{6\pi}{\mathcal{P}} \sqrt{\varepsilon} = 6\pi \sqrt{\frac{\mathrm{Ra} - \mathrm{Ra}_c}{\mathrm{Ra}_c \mathcal{P}^2}}$$

4. We linearize the Lorenz equations around zero. The perturbations $\delta X, \delta Y, \delta Z$
verify

$$
\begin{cases}
\dfrac{d\delta X}{dt} = \mathcal{P}(\delta Y - \delta X) \\[2mm]
\dfrac{d\delta Y}{dt} = r\delta X - \delta Y \\[2mm]
\dfrac{d\delta Z}{dt} = -b\delta Z
\end{cases}
\tag{13.8}
$$

We look for a solution proportional to $e^{\lambda t}$, and get λ by setting the determinant
of the system to zero. Hence,

$$
\lambda^2 + (\mathcal{P}+1)\lambda + \mathcal{P}(1-r) = 0 \implies 2\lambda = -(\mathcal{P}+1) \pm \sqrt{(\mathcal{P}+1)^2 + 4\mathcal{P}(r-1)}
$$

When $r > 1$ one of the root has a positive real part showing that the system is
unstable.

Chapter 8

1. The demonstration is carried out in three steps:

- We break the velocity field into two parts: the relative velocity $\mathbf{v}_r$ and the
  background velocity $\boldsymbol{\Omega} \times \mathbf{r}$

$$
\mathbf{v} = \mathbf{v}_r + \boldsymbol{\Omega} \times \mathbf{r}
$$

The acceleration reads

$$
\frac{D\mathbf{v}}{Dt} = \frac{\partial \mathbf{v}_r}{\partial t} + (\mathbf{v}_r \cdot \boldsymbol{\nabla})\mathbf{v}_r + (\boldsymbol{\Omega} \times \mathbf{r}) \cdot \boldsymbol{\nabla}\mathbf{v}_r + \boldsymbol{\Omega} \times \mathbf{v}_r + \boldsymbol{\Omega} \times (\boldsymbol{\Omega} \times \mathbf{r})
$$

where we used

$$
(\mathbf{v}_r \cdot \boldsymbol{\nabla})\boldsymbol{\Omega} \times \mathbf{r} = \boldsymbol{\Omega} \times \mathbf{v}_r \qquad \text{and} \qquad [(\boldsymbol{\Omega} \times \mathbf{r}) \cdot \boldsymbol{\nabla}]\,\boldsymbol{\Omega} \times \mathbf{r} = \boldsymbol{\Omega} \times (\boldsymbol{\Omega} \times \mathbf{r})
$$

- In the second step, we observe that $\mathbf{v}_r$ should be expressed with the vectors
  attached to the rotating frame, which are time-dependent. In other words, we
  are not interested in $\mathbf{v}_r$ but in $\mathbf{v}'$ which is such that $\mathbf{v}_r = [\Omega]\mathbf{v}'$. $[\Omega]$ represents
  the rotation which transform the new coordinates into the old ones. We now
  assume that $\boldsymbol{\Omega} = \Omega\mathbf{e}_z$. We have

$$\mathbf{r}=[\Omega]\mathbf{r}', \quad \nabla=[\Omega]\nabla', \quad \mathbf{v}_r \cdot \nabla=\mathbf{v}' \cdot \nabla' \quad \text{and} \quad [\Omega]=\begin{bmatrix} \cos \Omega t & -\sin \Omega t \\ \sin \Omega t & \cos \Omega t \end{bmatrix}$$

It is easily verified that:

$$\frac{\partial \mathbf{v}_r}{\partial t} = [\Omega]\frac{\partial \mathbf{v}'}{\partial t} + [\dot{\Omega}]\mathbf{v}' = [\Omega]\left(\frac{\partial \mathbf{v}'}{\partial t} + \mathbf{\Omega} \times \mathbf{v}'\right)$$

Applying the rotation to all the vectors of the equation, we find the equation verified by $\mathbf{v}'$, namely

$$\frac{D\mathbf{v}'}{Dt} + 2\mathbf{\Omega} \times \mathbf{v}' + (\mathbf{\Omega} \times \mathbf{r}) \cdot \nabla \mathbf{v}' + \mathbf{\Omega} \times (\mathbf{\Omega} \times \mathbf{r}') = \mathbf{f}'$$

- In the last step, we observe that the new coordinates depend on the old ones and on time. Hence

$$\frac{\partial \mathbf{v}'}{\partial t}(\mathbf{r}, t) = \frac{\partial \mathbf{v}'}{\partial t}(\mathbf{r}', t) + \frac{\partial \mathbf{r}'(\mathbf{r}, t)}{\partial t} \cdot \nabla' \mathbf{v}'$$

but

$$\mathbf{r}' = [\Omega]^{-1}\mathbf{r} = [-\Omega]\mathbf{r} \quad \Longrightarrow \quad \frac{\partial \mathbf{r}'(\mathbf{r}, t)}{\partial t} = -\mathbf{\Omega} \times \mathbf{r}'$$

which removes the term $(\mathbf{\Omega} \times \mathbf{r}) \cdot \nabla \mathbf{v}'$ and yields the expected result.

2. When viscosity is taken into account the following term

$$E \int_{(V)} \mathbf{u}^* \cdot \Delta \mathbf{u} \, dV$$

is important. Introducing $[\sigma^v]$ as the viscous stress tensor, the foregoing integral may also be written

$$\int_{(V)} u_i^* \partial_j \sigma_{ij}^v dV = \int_{(S)} u_i^* \sigma_{ij}^v dS_j - \int_{(V)} \partial_j u_i^* \sigma_{ij}^v dV$$

Because of boundary conditions, the surface integral is always zero (either for no-slip or stress-free conditions). The volume integral is real. Observing that $\sigma_{ij}^v = E(\partial_j u_i + \partial_i u_j)$, we get

$$\int_{(V)} \partial_j u_i^* \sigma_{ij}^v dV = \frac{E}{2} \int_{(V)} |\partial_j u_i + \partial_i u_j|^2 dV$$

This integral is truly real, therefore the bounds of the eigenspectrum are not modified.

3. The easiest way is to start from the expressions of v_r and v_z given by (8.29) and (8.30), which we combine to eliminate the pressure field. We get

$$\frac{1-\omega^2}{i\omega}\frac{\partial v_s}{\partial z} = i\omega\frac{\partial v_z}{\partial s}$$

but $v_s = \frac{1}{s}\frac{\partial \psi}{\partial z}$, $v_z = -\frac{1}{s}\frac{\partial \psi}{\partial s}$.

Making the substitution, we obtain the equation for ψ :

$$\frac{\partial^2 \psi}{\partial s^2} - \frac{1}{s}\frac{\partial \psi}{\partial s} - \left(\frac{1-\omega^2}{\omega^2}\right)\frac{\partial^2 \psi}{\partial z^2} = 0$$

which is similar to the Poincaré equation. The only change is the sign of $\frac{1}{s}\frac{\partial \psi}{\partial s}$. The equation is however still of hyperbolic type with the same characteristics.

Chapter 9

1.(a) From the definitions

$$\langle \hat{v}_i^*(\mathbf{k})\hat{v}_j(\mathbf{k}')\rangle = \left\langle \int v_i(\mathbf{x})e^{i\mathbf{k}\cdot\mathbf{x}}\frac{d^3\mathbf{x}}{(2\pi)^3}\int v_j(\mathbf{x})e^{-i\mathbf{k}'\cdot\mathbf{x}}\frac{d^3\mathbf{x}}{(2\pi)^3}\right\rangle$$

where we noted that v_i is real. Here we transform this expression into a double integral:

$$= \int\int \langle v_i(\mathbf{x})v_j(\mathbf{x}')\rangle e^{i\mathbf{k}\cdot\mathbf{x}-i\mathbf{k}'\cdot\mathbf{x}'}\frac{d^3\mathbf{x}}{(2\pi)^3}\frac{d^3\mathbf{x}'}{(2\pi)^3}$$

setting $\mathbf{x}' = \mathbf{x} + \mathbf{r}$ and splitting the integrals over $\mathbf{x}$ and $\mathbf{r}$, we get

$$= \underbrace{\int Q_{ij}(\mathbf{r})e^{-i\mathbf{k}'\cdot\mathbf{r}}\frac{d^3\mathbf{r}}{(2\pi)^3}}_{\phi_{ij}(\mathbf{k}')}\underbrace{\int e^{-i(\mathbf{k}'-\mathbf{k})\cdot\mathbf{x}}\frac{d^3\mathbf{x}}{(2\pi)^3}}_{\delta(\mathbf{k}'-\mathbf{k})}$$

hence the result.

(b) We have

$$Z_{ij} = \int \langle \omega_i'(\mathbf{x})\omega_j'(\mathbf{x}')\rangle e^{-i\mathbf{k}\cdot\mathbf{r}}\frac{d^3\mathbf{r}}{(2\pi)^3}$$

or

$$= \epsilon_{ilm}\epsilon_{jnp} \int \left\langle \partial_l v'_m(\mathbf{x})\partial_n v'_p(\mathbf{x}')\right\rangle e^{-i\mathbf{k}\cdot\mathbf{r}} \frac{d^3\mathbf{r}}{(2\pi)^3}$$

But

$$\partial_l v'_m(\mathbf{x}) = \int ik_l \hat{v}_m e^{i\mathbf{k}\cdot\mathbf{x}} d^3\mathbf{k}, \quad \partial_n v'_p(\mathbf{x}) = \partial_n v'^*_p = -\int ik'_l \hat{v}_m e^{-i\mathbf{k}'\cdot\mathbf{x}} d^3\mathbf{k}'$$

Using the result of the previous exercise, we have

$$\left\langle \partial_l v'_m(\mathbf{x})\partial_n v'_p(\mathbf{x}')\right\rangle = \int k_l k_n \phi_{mp}(\mathbf{k}) e^{i\mathbf{k}\cdot\mathbf{r}} d^3\mathbf{k}$$

Using this expression in that of Z_{ij} we get the requested expression. The final formula (9.28) is derived after an expansion of $\epsilon_{ilm}\epsilon_{jnp}$ (given in the complements of mathematics) and after the use of the incompressibility relation (9.23).

(c) We note that

$$\phi_{ii}(k) = \frac{1}{(2\pi)^3} \int Q_{ii}(r) e^{-ikr\cos\theta} r^2 dr \sin\theta d\theta d\varphi$$

Using $\int_0^\pi e^{-ikr\cos\theta} \sin\theta d\theta = 2\sin kr/kr$ and (9.33) we derive the requested result.

2. We start from the definition (9.46) and modify it into

$$\ell_Z = \Omega(0)^{-1} \int_0^\infty \Omega(r) dr$$

where $\Omega(r) = Q_{ii}(r) = -6\Delta R(r)$, Δ being the Laplacian. It is sufficient to use (9.35) and the expression of the Laplacian in spherical coordinates (see 12.45), to get the desired expression. To show the similarity between ℓ_D and ℓ_Z, we may consider the Kolmogorov spectrum where $k_D \gg k_0$. Its energy density is negligible beyond k_D. Then it is easy to show that $\ell_Z \sim \ell_D$.

3. (a) A short integration by part yields the result.
   (b) A Taylor expansion of the functions at the origin yields $(\sin kr - kr\cos kr)/(kr)^3 \sim 1/3$. The result follows.
   (c) From $E(k) = C_K \langle\varepsilon\rangle^{2/3} k^{-5/3}$, if we set $x = kr$ then

$$S_2 = C_K \langle\varepsilon\rangle^{2/3} r^{2/3} \int_0^\infty q(x) dx$$

One should check that the integral converges. C_2 and C_K are proportional since $C_2 = C_K \int_0^\infty q(x) x^{-5/3} dx$.

(d) If $S_2 \propto r^{2/3+m}$, then $E(k) \propto k^{-5/3-m}$. In the case of the law (9.76), $m = \frac{14}{9} - 2\left(\frac{2}{3}\right)^{2/3} = 0.02927$ if $\beta = 2/3$.

4. From the Definition (9.49) and after splitting over the various domains

$$\ell_T^2 = \frac{\displaystyle\int_0^{k_0} i(k)dk + \int_{k_0}^{k_D} k^{-5/3} + \int_{k_D}^{\infty} d(k)dk}{\displaystyle\int_0^{k_0} k^2 i(k)dk + \int_{k_0}^{k_D} k^{1/3}dk + \int_{k_D}^{\infty} k^2 d(k)dk}$$

Taking an upper bound of the numerator and a lower bound of the denominator, and then doing the opposite, we can construct a lower and upper bound for ℓ_T^2, namely:

$$\frac{2/3 k_0^{-2/3}}{k_0^{4/3} + 3/4 k_D^{4/3} + 5 k_D^{4/3}} \le \ell_T^2 \le \frac{k_0^{-2/3} + 2/3(k_0^{-2/3} - k_D^{-2/3}) + k_D^{-2/3}}{3/4(k_D^{4/3} - k_0^{4/3})}$$

The result follows since $k_D \gg k_0$.

5. We note that

$$\langle x^p \rangle = \langle e^{p \ln x} \rangle = \langle e^{p(\ln x - \langle \ln x \rangle)} \rangle e^{p \langle \ln x \rangle}$$

setting $y = (\ln x - \langle \ln x \rangle)$, we have

$$\langle e^{py} \rangle = \sum_{n=0}^{+\infty} \frac{p^n \langle y^n \rangle}{n!}$$

however, y is a random variable with a normal distribution, thus $\langle y^n \rangle = 0$ if n is odd. It follows that

$$\langle e^{py} \rangle = \sum_{m=0}^{+\infty} \frac{p^{2m} \langle y^{2m} \rangle}{(2m)!}.$$

For a normal law of probability even order moments just depend on the variance of the random variable, namely $\langle y^{2m} \rangle = (2m - 1)!!\sigma^{2m}$ (which can be demonstrated by induction) where

$$(2m - 1)!!/(2m)! = 2^{-m}/m!.$$

It follows that

$$\langle e^{py} \rangle = \sum_{m=0}^{+\infty} \frac{1}{m!}\left(\frac{p^2\sigma^2}{2}\right)^m = e^{p^2\sigma^2/2}$$

hence the result.

Chapter 10

1. The electric current generated by free charges of density ρ_e comes from the matter motion and reads $\rho_e \mathbf{v}$. From Gauss equation $\nabla \cdot \mathbf{E} = \rho_e/\varepsilon_0$ and the induction equation $\partial_t \mathbf{B} = -\nabla \times \mathbf{E}$, we find the following order of magnitude relations:

$$\rho_e \mathbf{v} \sim \varepsilon_0 EV/L \sim V^2/c^2 (B/\mu_0 L) \sim V^2/c^2 \mathbf{j}$$

Thus $j \gg \rho_e V$ and $|\mathbf{j} \times \mathbf{B}| \gg \rho_e VB \sim \rho_e E$. The Laplace force is always larger than the Coulomb force coming from the local electric field.

2. The mean magnetic field verifies:

$$\frac{\partial \mathbf{B}}{\partial t} = \nabla \times (\alpha \mathbf{B}) + \eta_{turb} \Delta \mathbf{B}$$

Taking $\mathbf{B} = \mathbf{B}_0 \exp i(\omega t + \mathbf{k} \cdot \mathbf{r})$, we get

$$i\omega \mathbf{B} = \alpha i \mathbf{k} \times \mathbf{B} - \eta_{turb} k^2 \mathbf{B}$$

Taking the cross product of this expression with $\mathbf{k}$ and reporting the new expression in the first, we derive the dispersion relation:

$$(i\omega + \eta_{turb} k^2)^2 = \alpha^2 k^2$$

which shows that all the waves whose wavenumber is smaller than $|\alpha|/\eta_{turb}$ are amplified.

Complements of Mathematics

1. To derive one expression from the preceding one we just need to contract two indices and note that $\delta_{ii} = 3$.

2. If [A] is symmetric then $A_{ij} = A_{ji}$, thus

$$\epsilon_{ijk} A_{jk} = \epsilon_{ijk} A_{kj} = -\epsilon_{ijk} A_{kj} = -\epsilon_{ijk} A_{jk}$$

The reciprocal is also true: if $\epsilon_{ijk} A_{jk} = 0$ then

$$\epsilon_{lmi} \epsilon_{ijk} A_{jk} = 0 \iff (\delta_{lj}\delta_{mk} - \delta_{lk}\delta_{mj}) A_{jk} = 0 \iff A_{lm} - A_{ml} = 0$$

thus [A] is symmetric.

3. This is indeed a Sturm–Liouville problem. Identifying the terms, we find $p(x) = 1$, $q(x) = 0$ and $w(x) = 1$. The general solution is:

$$y(x) = Ae^{ax} + Be^{-ax}$$

with $a = \sqrt{-\lambda}$. The boundary conditions $y(0) = y(1) = 0$ imply that $A + B = 0$ and $\operatorname{sh} a = 0$. Thus $a = in\pi$ with $n \in \mathbb{N}$. The eigenvalue spectrum is therefore the set $\{n^2\pi^2 \,|\, n \in \mathbb{N}\}$. It is discrete and has a lower bound as expected.

Index

© Springer International Publishing Switzerland 2015
M. Rieutord, *Fluid Dynamics*, Graduate Texts in Physics,
DOI 10.1007/978-3-319-09351-2

The manufacturer's authorised representative in the EU is Springer
Nature Customer Service Centre GmbH, Europaplatz 3, 69115 Heidelberg,
Germany. If you have any concerns regarding our products, please
contact ProductSafety@springernature.com

Printed and bound by CPI Group (UK) Ltd, Croydon, CR0 4YY
27/04/2026
02097573-0009